Modeling and Simulation of Systems Using MATLAB® and Simulink®

Modeling and Simulation of Systems Using MATLAB® and Simulink®

Devendra K. Chaturvedi

CRC Press is an imprint of the
Taylor & Francis Group, an **informa** business

MATLAB® and Simulink® are trademarks of the MathWorks, Inc. and are used with permission. The MathWorks does not warrant the accuracy of the text or exercises in this book. This book's use or discussion of MATLAB® and Simulink® software or related products does not constitute endorsement or sponsorship by the MathWorks of a particular pedagogical approach or particular use of the MATLAB® and Simulink® software.

CRC Press
Taylor & Francis Group
6000 Broken Sound Parkway NW, Suite 300
Boca Raton, FL 33487-2742

© 2010 by Taylor and Francis Group, LLC
CRC Press is an imprint of Taylor & Francis Group, an Informa business

No claim to original U.S. Government works

Printed in the United States of America on acid-free paper
10 9 8 7 6 5 4 3 2 1

International Standard Book Number: 978-1-4398-0672-2 (Hardback)

This book contains information obtained from authentic and highly regarded sources. Reasonable efforts have been made to publish reliable data and information, but the author and publisher cannot assume responsibility for the validity of all materials or the consequences of their use. The authors and publishers have attempted to trace the copyright holders of all material reproduced in this publication and apologize to copyright holders if permission to publish in this form has not been obtained. If any copyright material has not been acknowledged please write and let us know so we may rectify in any future reprint.

Except as permitted under U.S. Copyright Law, no part of this book may be reprinted, reproduced, transmitted, or utilized in any form by any electronic, mechanical, or other means, now known or hereafter invented, including photocopying, microfilming, and recording, or in any information storage or retrieval system, without written permission from the publishers.

For permission to photocopy or use material electronically from this work, please access www.copyright.com (http://www.copyright.com/) or contact the Copyright Clearance Center, Inc. (CCC), 222 Rosewood Drive, Danvers, MA 01923, 978-750-8400. CCC is a not-for-profit organization that provides licenses and registration for a variety of users. For organizations that have been granted a photocopy license by the CCC, a separate system of payment has been arranged.

Trademark Notice: Product or corporate names may be trademarks or registered trademarks, and are used only for identification and explanation without intent to infringe.

Visit the Taylor & Francis Web site at
http://www.taylorandfrancis.com

and the CRC Press Web site at
http://www.crcpress.com

Dedicated to

The cherished memories of my guru and guide

Most Revered Dr. Makund Behari Lal Sahab

DSc (Lucknow), DSc (Edinburgh)

(1907–2002)

August Founder of Dayalbagh Educational Institute

Contents

Preface xvii
Acknowledgments xxi
Author xxiii

1. Introduction to Systems 1
    1.1   System 1
            1.1.1   System Boundary 3
            1.1.2   System Components and Their Interactions 3
            1.1.3   Environment 4
    1.2   Classification of Systems 5
            1.2.1   According to the Time Frame 5
            1.2.2   According to the Complexity of the System 6
            1.2.3   According to the Interactions 6
            1.2.4   According to the Nature and Type of Components 7
            1.2.5   According to the Uncertainties Involved 7
                    1.2.5.1 Static vs. Dynamic Systems 8
                    1.2.5.2 Linear vs. Nonlinear Systems 8
    1.3   Linear Systems 9
            1.3.1   Superposition Theorem 9
            1.3.2   Homogeneity 10
            1.3.3   Mathematical Viewpoint of a Linear System 10
                    1.3.3.1 Linear Differential Equation 10
                    1.3.3.2 Nonlinear Differential Equations 11
    1.4   Time-Varying vs. Time-Invariant Systems 12
    1.5   Lumped vs. Distributed Parameter Systems 13
    1.6   Continuous-Time and Discrete-Time Systems 13
    1.7   Deterministic vs. Stochastic Systems 15
            1.7.1   Complexity of Systems 15
    1.8   Hard and Soft Systems 16
    1.9   Analysis of Systems 18
    1.10 Synthesis of Systems 18
    1.11 Introduction to System Philosophy 18
            1.11.1 Method of Science 20
                    1.11.1.1 Reductionism 20
                    1.11.1.2 Repeatability 20
                    1.11.1.3 Refutation 20
            1.11.2 Problems of Science and Emergence of System 20
    1.12 System Thinking 21
    1.13 Large and Complex Applied System Engineering:
           A Generic Modeling 24
    1.14 Review Questions 29
    1.15 Bibliographical Notes 30

2. Systems Modeling 31
2.1 Introduction 31
2.2 Need of System Modeling... 33
2.3 Modeling Methods for Complex Systems 34
2.4 Classification of Models 35
    2.4.1 Physical vs. Abstract Model 35
    2.4.2 Mathematical vs. Descriptive Model 36
    2.4.3 Static vs. Dynamic Model 37
    2.4.4 Steady State vs. Transient Model 37
    2.4.5 Open vs. Feedback Model.. 37
    2.4.6 Deterministic vs. Stochastic Models 37
    2.4.7 Continuous vs. Discrete Models 37
2.5 Characteristics of Models... 38
2.6 Modeling 38
    2.6.1 Fundamental Axiom (Modeling Hypothesis) 40
    2.6.2 Component Postulate (First Postulate) 40
    2.6.3 Model Evaluation 41
    2.6.4 Generic Description of Two-Terminal
            Components 41
            2.6.4.1 Dissipater Type Components....................................... 41
            2.6.4.2 Delay Type Elements 42
            2.6.4.3 Accumulator Type... 42
            2.6.4.4 Sources or Drivers 43
2.7 Mathematical Modeling of Physical Systems 43
    2.7.1 Modeling of Mechanical Systems... 46
            2.7.1.1 Translational Mechanical Systems.............................. 47
            2.7.1.2 Rotational Mechanical Systems 64
    2.7.2 Modeling of Electrical Systems... 78
    2.7.3 Modeling of Electromechanical Systems 84
    2.7.4 Modeling of Fluid Systems.. 87
            2.7.4.1 Hydraulic Systems 87
    2.7.5 Modeling of Thermal Systems 92
2.8 Review Questions 99
2.9 Bibliographical Notes 102

3. Formulation of State Space Model of Systems 103
3.1 Physical Systems Theory.. 103
3.2 System Components and Interconnections.. 103
3.3 Computation of Parameters of a Component 105
3.4 Single Port and Multiport Systems 109
    3.4.1 Linear Perfect Couplers.. 110
    3.4.2 Summary of Two-Terminal and Multiterminal
            Components 113
    3.4.3 Multiterminal Components... 113
3.5 Techniques of System Analysis.. 114
    3.5.1 Lagrangian Technique 115
    3.5.2 Free Body Diagram Method... 115
    3.5.3 Linear Graph Theoretic Approach 115
3.6 Basics of Linear Graph Theoretic Approach 116

3.7	Formulation of System Model for Conceptual System		119
	3.7.1	Fundamental Axioms	121
	3.7.2	Component Postulate	121
	3.7.3	System Postulate	121
		3.7.3.1 Cutset Postulate	122
		3.7.3.2 Circuit Postulate	124
3.8	Formulation of System Model for Physical Systems		127
3.9	Topological Restrictions		131
	3.9.1	Perfect Coupler	131
	3.9.2	Gyrator	131
	3.9.3	Short Circuit Element ("A" Type)	132
	3.9.4	Open Circuit Element ("B" Type)	132
	3.9.5	Dissipater Type Elements	132
	3.9.6	Delay Type Elements	132
	3.9.7	Accumulator Type Elements	132
	3.9.8	Across Drivers	133
	3.9.9	Through Drivers	133
3.10	Development of State Model of Degenerative System		144
	3.10.1	Development of State Model for Degenerate System	146
	3.10.2	Symbolic Formulation of State Model for Nondegenerative Systems	151
	3.10.3	State Model of System with Multiterminal Components	157
	3.10.4	State Model for Systems with Time Varying and Nonlinear Components	162
3.11	Solution of State Equations		166
3.12	Controllability		180
3.13	Observability		181
3.14	Sensitivity		182
3.15	Liapunov Stability		184
3.16	Performance Characteristics of Linear Time Invariant Systems		186
3.17	Formulation of State Space Model Using Computer Program (SYSMO)		187
	3.17.1	Preparation of the Input Data	187
	3.17.2	Algorithm for the Formulation of State Equations	187
3.18	Review Questions		208
3.19	Bibliographical Notes		217

4. Model Order Reduction 219

4.1	Introduction		219
4.2	Difference between Model Simplification and Model Order Reduction		220
4.3	Need for Model Order Reduction		221
4.4	Principle of Model Order Reduction		221
4.5	Methods of Model Order Reduction		223
	4.5.1	Time Domain Simplification Techniques	223
		4.5.1.1 Dominant Eigenvalue Approach	223
		4.5.1.2 Aggregation Method	228
		4.5.1.3 Subspace Projection Method	232
		4.5.1.4 Optimal Order Reduction	233
		4.5.1.5 Hankel Matrix Approach	233
		4.5.1.6 Hankel–Norm Model Order Reduction	234

	4.5.2	Model Order Reduction in Frequency Domain 234
		4.5.2.1 Pade Approximation Method 234
		4.5.2.2 Continued Fraction Expansion 235
		4.5.2.3 Moment-Matching Method 235
		4.5.2.4 Balanced Realization-Based Reduction Method 236
		4.5.2.5 Balanced Truncation 238
		4.5.2.6 Frequency-Weighted Balanced Model Reduction 244
		4.5.2.7 Time Moment Matching 249
		4.5.2.8 Continued Fraction Expansion 252
		4.5.2.9 Model Order Reduction Based on the Routh Stability Criterion 259
		4.5.2.10 Differentiation Method for Model Order Reduction 263
4.6	Applications of Reduced-Order Models 273
4.7	Review Questions 273
4.8	Bibliographical Notes 275

5. Analogous of Linear Systems 277
- 5.1 Introduction 277
  - 5.1.1 D'Alembert's Principle 277
- 5.2 Force–Voltage (f–v) Analogy 278
  - 5.2.1 Rule for Drawing f–v Analogous Electrical Circuits 278
- 5.3 Force–Current (f–i) Analogy 279
  - 5.3.1 Rule for Drawing f–i Analogous Electrical Circuits 279
- 5.4 Review Questions 298

6. Interpretive Structural Modeling 301
- 6.1 Introduction 301
- 6.2 Graph Theory 301
  - 6.2.1 Net 305
  - 6.2.2 Loop 305
  - 6.2.3 Cycle 305
  - 6.2.4 Parallel Lines 306
  - 6.2.5 Properties of Relations 306
- 6.3 Interpretive Structural Modeling 307
- 6.4 Review Questions 323
- 6.5 Bibliographical Notes 325

7. System Dynamics Techniques 327
- 7.1 Introduction 327
- 7.2 System Dynamics of Managerial and Socioeconomic System 327
  - 7.2.1 Counterintuitive Nature of System Dynamics 327
  - 7.2.2 Nonlinearity 328
  - 7.2.3 Dynamics 328
  - 7.2.4 Causality 328
  - 7.2.5 Endogenous Behavior 328
- 7.3 Traditional Management 328
  - 7.3.1 Strength of the Human Mind 328
  - 7.3.2 Limitation of the Human Mind 328

7.4	Sources of Information		329
	7.4.1	Mental Database	329
	7.4.2	Written/Spoken Database	330
	7.4.3	Numerical Database	330
7.5	Strength of System Dynamics		331
7.6	Experimental Approach to System Analysis		332
7.7	System Dynamics Technique		332
7.8	Structure of a System Dynamic Model		333
7.9	Basic Structure of System Dynamics Models		334
	7.9.1	Level Variables	334
	7.9.2	Flow-Rate Variables	334
	7.9.3	Decision Function	335
7.10	Different Types of Equations Used in System Dynamics Techniques		342
	7.10.1	Level Equation	342
	7.10.2	Rate Equation (Decision Functions)	343
	7.10.3	Auxiliary Equations	343
7.11	Symbol Used in Flow Diagrams		344
	7.11.1	Levels	344
	7.11.2	Source and Sinks	345
	7.11.3	Information Takeoff	345
	7.11.4	Auxiliary Variables	345
	7.11.5	Parameters (Constants)	345
7.12	Dynamo Equations		345
7.13	Modeling and Simulation of Parachute Deceleration Device		376
	7.13.1	Parachute Inflation	377
	7.13.2	Canopy Stress Distribution	378
	7.13.3	Modeling and Simulation of Parachute Trajectory	378
7.14	Modeling of Heat Generated in a Parachute during Deployment		382
	7.14.1	Dynamo Equations	384
7.15	Modeling of Stanchion System of Aircraft Arrester Barrier System		385
	7.15.1	Modeling and Simulation of Forces Acting on Stanchion System Using System Dynamic Technique	387
	7.15.2	Dynamic Model	389
	7.15.3	Results	390
7.16	Review Questions		395
7.17	Bibliographical Notes		399

8. Simulation ... 401

8.1	Introduction		401
8.2	Advantages of Simulation		402
8.3	When to Use Simulations		403
8.4	Simulation Provides		403
8.5	How Simulations Improve Analysis and Decision Making?		404
8.6	Application of Simulation		404
8.7	Numerical Methods for Simulation		405
	8.7.1	The Rectangle Rule	406
	8.7.2	The Trapezoid and Tangent Formulae	406
	8.7.3	Simpson's Rule	407
	8.7.4	One-Step Euler's Method	410

		8.7.5	Runge–Kutta Methods of Integration	410
			8.7.5.1 Physical Interpretation	411
		8.7.6	Runge–Kutta Fourth-Order Method	411
		8.7.7	Adams–Bashforth Predictor Method	412
		8.7.8	Adams–Moulton Corrector Method	413
	8.8	The Characteristics of Numerical Methods		413
	8.9	Comparison of Different Numerical Methods		413
	8.10	Errors during Simulation with Numerical Methods		414
		8.10.1	Truncation Error	414
		8.10.2	Round Off Error	415
		8.10.3	Step Size vs. Error	418
		8.10.4	Discretization Error	418
	8.11	Review Questions		430

9. Nonlinear and Chaotic System 433

	9.1	Introduction		433
	9.2	Linear vs. Nonlinear System		434
	9.3	Types of Nonlinearities		434
	9.4	Nonlinearities in Flight Control of Aircraft		435
		9.4.1	Basic Control Surfaces Used in Aircraft Maneuvers	435
		9.4.2	Principle of Flight Controls	435
		9.4.3	Components Used in Pitch Control	437
		9.4.4	Modeling of Various Components of Pitch Control System	438
		9.4.5	Simulink Model of Pitch Control in Flight	440
			9.4.5.1 Simulink Model of Pitch Control in Flight Using Nonlinearities	440
		9.4.6	Study of Effects of Different Nonlinearities on Behavior of the Pitch Control Model	440
			9.4.6.1 Effects of Dead-Zone Nonlinearities	440
			9.4.6.2 Effects of Saturation Nonlinearities	441
			9.4.6.3 Effects of Backlash Nonlinearities	442
			9.4.6.4 Cumulative Effects of Backlash, Saturation, Dead-Zone Nonlinearities	443
		9.4.7	Designing a PID Controller for Pitch Control in Flight	445
			9.4.7.1 Designing a PID Controller for Pitch Control in Flight with the Help of Root Locus Method (Feedback Compensation)	445
			9.4.7.2 Designing a PID Controller (Connected in Cascade with the System) for Pitch Control in Flight	454
			9.4.7.3 Design of P, I, D, PD, PI, PID, and Fuzzy Controllers	456
		9.4.8	Design of Fuzzy Controller	461
			9.4.8.1 Basic Structure of a Fuzzy Controller	462
			9.4.8.2 The Components of a Fuzzy System	462
		9.4.9	Tuning Fuzzy Controller	469
	9.5	Conclusions		473
	9.6	Introduction to Chaotic System		478
		9.6.1	General Meaning	478
		9.6.2	Scientific Meaning	478
		9.6.3	Definition	479

9.7	Historical Prospective	481
9.8	First-Order Continuous-Time System	484
9.9	Bifurcations	487
	9.9.1 Saddle Node Bifurcation	488
	9.9.2 Transcritical Bifurcation	488
	9.9.3 Pitchfork Bifurcation	490
	9.9.3.1 Supercritical Pitchfork Bifurcation	490
	9.9.4 Catastrophes	492
	9.9.4.1 Globally Attracting Point for Stability	492
9.10	Second-Order System	493
9.11	Third-Order System	496
	9.11.1 Lorenz Equation: A Chaotic Water Wheel	498
9.12	Review Questions	501
9.13	Bibliographical Notes	501

10. Modeling with Artificial Neural Network 503
10.1	Introduction	503
	10.1.1 Biological Neuron	503
	10.1.2 Artificial Neuron	504
10.2	Artificial Neural Networks	505
	10.2.1 Training Phase	505
	10.2.1.1 Selection of Neuron Characteristics	505
	10.2.1.2 Selection of Topology	505
	10.2.1.3 Error Minimization Process	506
	10.2.1.4 Selection of Training Pattern and Preprocessing	506
	10.2.1.5 Stopping Criteria of Training	506
	10.2.2 Testing Phase	506
	10.2.2.1 ANN Model	506
	10.2.2.2 Building ANN Model	508
	10.2.2.3 Backpropagation	509
	10.2.2.4 Training Algorithm	509
	10.2.2.5 Applications of Neural Network Modeling	510
10.3	Review Questions	526

11. Modeling Using Fuzzy Systems 527
11.1	Introduction	527
11.2	Fuzzy Sets	528
11.3	Features of Fuzzy Sets	531
11.4	Operations on Fuzzy Sets	532
	11.4.1 Fuzzy Intersection	532
	11.4.2 Fuzzy Union	532
	11.4.3 Fuzzy Complement	532
	11.4.4 Fuzzy Concentration	533
	11.4.5 Fuzzy Dilation	534
	11.4.6 Fuzzy Intensification	535
	11.4.7 Bounded Sum	536
	11.4.8 Strong α-Cut	537
	11.4.9 Linguistic Hedges	538
11.5	Characteristics of Fuzzy Sets	540

		11.5.1	Normal Fuzzy Set	540
		11.5.2	Convex Fuzzy Set	540
		11.5.3	Fuzzy Singleton	540
		11.5.4	Cardinality	540
	11.6	Properties of Fuzzy Sets		541
	11.7	Fuzzy Cartesian Product		541
	11.8	Fuzzy Relation		542
	11.9	Approximate Reasoning		545
	11.10	Defuzzification Methods		554
	11.11	Introduction to Fuzzy Rule-Based Systems		556
	11.12	Applications of Fuzzy Systems to System Modeling		558
		11.12.1	Single Input Single Output Systems	559
		11.12.2	Multiple Input Single Output Systems	564
		11.12.3	Multiple Input Multiple Output Systems	566
	11.13	Takagi–Sugeno–Kang Fuzzy Models		567
	11.14	Adaptive Neuro-Fuzzy Inferencing Systems		568
	11.15	Steady State DC Machine Model		574
	11.16	Transient Model of a DC Machine		579
	11.17	Fuzzy System Applications for Operations Research		592
	11.18	Review Questions		602
	11.19	Bibliography and Historical Notes		603

12. Discrete-Event Modeling and Simulation 605
12.1	Introduction		605
12.2	Some Important Definitions		606
12.3	Queuing System		609
12.4	Discrete-Event System Simulation		611
12.5	Components of Discrete-Event System Simulation		611
12.6	Input Data Modeling		615
12.7	Family of Distributions for Input Data		615
12.8	Random Number Generation		616
	12.8.1	Uniform Distribution	616
	12.8.2	Gaussian Distribution of Random Number Generation	617
12.9	Chi-Square Test		619
12.10	Kolomogrov–Smirnov Test		619
12.11	Review Questions		619

Appendix A 621
A.1	What Is MATLAB®?		621
A.2	Learning MATLAB		621
A.3	The MATLAB System		621
	A.3.1	Development Environment	622
	A.3.2	The MATLAB Mathematical Function Library	622
	A.3.3	The MATLAB Language	622
	A.3.4	Handle Graphics	622
	A.3.5	The MATLAB Application Program Interface (API)	623
A.4	Starting and Quitting MATLAB		623
A.5	MATLAB Desktop		623

A.6	Desktop Tools	623
	A.6.1 Command Window	623
	A.6.2 Command History	624
	A.6.2.1 Running External Programs	624
	A.6.2.2 Launch Pad	625
	A.6.2.3 Help Browser	625
	A.6.2.4 Current Directory Browser	626
	A.6.2.5 Workspace Browser	626
	A.6.2.6 Array Editor	626
	A.6.2.7 Editor/Debugger	627
	A.6.2.8 Other Development Environment Features	627
A.7	Entering Matrices	627
A.8	Subscripts	630
A.9	The Colon Operator	630
A.10	The Magic Function	631
A.11	Expressions	631
	A.11.1 Variables	632
	A.11.2 Numbers	632
	A.11.3 Operators	632
	A.11.4 Functions	632
	A.11.4.1 Generating Matrices	633
A.12	The Load Command	633
A.13	The Format Command	635
A.14	Suppressing Output	635
A.15	Entering Long Command Lines	635
A.16	Basic Plotting	636
	A.16.1 Creating a Plot	636
	A.16.2 Multiple Data Sets in One Graph	636
	A.16.3 Plotting Lines and Markers	636
	A.16.4 Adding Plots to an Existing Graph	637
	A.16.5 Multiple Plots in One Figure	637
	A.16.6 Setting Grid Lines	638
	A.16.7 Axis Labels and Titles	638
	A.16.8 Saving a Figure	638
	A.16.9 Mesh and Surface Plots	638
A.17	Images	640
A.18	Handle Graphics	640
	A.18.1 Setting Properties from Plotting Commands	640
	A.18.2 Different Types of Graphs	640
	A.18.2.1 Bar and Area Graphs	641
A.19	Animations	644
A.20	Creating Movies	644
A.21	Flow Control	645
	A.21.1 If	645
	A.21.2 Switch and Case	646
	A.21.2.1 For	646
	A.21.2.2 While	647
	A.21.2.3 Continue	647
	A.21.2.4 Break	647

- A.22 Other Data Structures ... 648
  - A.22.1 Multidimensional Arrays ... 648
  - A.22.2 Cell Arrays ... 650
  - A.22.3 Characters and Text ... 650
- A.23 Scripts and Functions ... 654
  - A.23.1 Scripts ... 654
  - A.23.2 Functions ... 655
    - A.23.2.1 Global Variables ... 655
    - A.23.2.2 Passing String Arguments to Functions ... 656
    - A.23.2.3 Constructing String Arguments in Code ... 656
    - A.23.2.4 A Cautionary Note ... 656
    - A.23.2.5 The Eval Function ... 657
    - A.23.2.6 Vectorization ... 657
    - A.23.2.7 Preallocation ... 658
    - A.23.2.8 Function Handles ... 658
    - A.23.2.9 Function Functions ... 658

Appendix B: Simulink ... 661
- B.1 Introduction ... 661
- B.2 Features of Simulink ... 661
- B.3 Simulation Parameters and Solvers ... 661
- B.4 Construction of Block Diagram ... 663
- B.5 Review Questions ... 667

Appendix C: Glossary ... 671
- C.1 Modeling and Simulation ... 671
- C.2 Artificial Neural Network ... 676
- C.3 Fuzzy Systems ... 678
- C.4 Genetic Algorithms ... 680
- Bibliography ... 681

Index ... 693

Preface

Systems engineering has great potential for solving problems related to physical, conceptual, and esoteric systems. The power of systems engineering lies in the three R's of science, namely, reductionism, repeatability, and refutation. Reductionism recognizes the fact that any system can be decomposed into a set of components that follow the fundamental laws of physics. The diversity of the real world can be reduced into laboratory experiments, which can be validated by their repeatability, and one can make intellectual progress by the refutation of hypotheses. Due to advancements in systems engineering for handling complex systems, modeling and simulation have, of late, become popular.

Modeling and simulation are very important tools of systems engineering that have now become a central activity in all disciplines of engineering and science. Not only do they help us gain a better understanding of the functioning of the real world, they are also important for the design of new systems as they enable us to predict the system behavior before the system is actually built. Modeling and simulation also allow us to analyze systems accurately under varying operating conditions.

This book aims to provide a comprehensive, state-of-the-art coverage of all important aspects of modeling and simulation of physical as well as conceptual systems. It strives to motivate beginners in this area through the modeling of real-life examples and their simulation to gain better insights into real-world systems. The extensive references and related literature at the end of every chapter can also be referred to for further studies in the area of modeling and simulation.

This book aims to

- Provide a basic understanding of systems and their modeling and simulation
- Explain the step-by-step procedure for modeling using top-down, bottom-up, and middle-out approaches
- Develop models for complex systems and reduce their order so as to use them effectively for online applications
- Present the simulation code in MATLAB®/Simulink® for gaining quick and useful insights into real-world systems
- Apply soft computing techniques for modeling nonlinear, ill-defined, and complex systems

The book will serve as a primary text for a variety of courses. It can be used as a first course in modeling and simulation at the junior, the senior, or the graduate levels in engineering, manufacturing, business, or computer science (Chapters 1 through 3, 5, and 8), providing a broad idea about modeling and simulation. At the end of such a course, the student would be prepared to carry out complete and effective simulation studies, and to take advanced modeling and simulation courses.

This book will also serve as a second course for more advanced studies in modeling and simulation for graduate students (Chapters 6 through 11) in any of the above disciplines. After completing this course, the student should be able to comprehend and conduct simulation research.

Finally, this book will serve as an introduction to simulation as part of a general course in operations research or management science (Chapters 1 through 3 and 6 through 8, Chapter 12).

Organization of the Book

Early chapters deal with the introduction of systems and includes concepts and their underlying philosophy; step-by-step procedures for the modeling of different types of systems using appropriate modeling techniques such as the graph-theoretic approach, interpretive structural modeling, and system dynamics modeling, are also discussed.

Focus then moves to the state of the art of simulation and how simulation evolved from the pre-computer days into the modern science of today. In this part, MATLAB/Simulink programs are developed for system simulation.

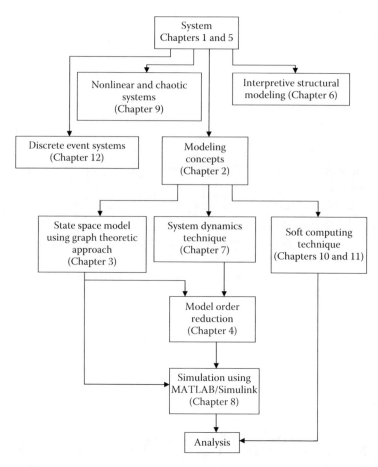

FIGURE P1
Schematic outline of the book.

Preface xix

We then take a fresh look at modern soft computing techniques (such as artificial neural networks [ANN], fuzzy systems, and genetic algorithms, or their combinations) for the modeling and simulation of complex and nonlinear systems.

Finally, chapters address the discrete systems modeling. The schematic outline of this book is shown in Figure P1.

Software Background

The key to the successful application of modeling and simulation techniques depends on the effective use of their software. For this, it is necessary that the student be familiar with their fundamentals. In this book, MATLAB/Simulink programming software are used. Appendix B provides some background information on MATLAB/Simulink. The MATLAB programs included in this book are easier to understand than the programs written in other programming languages such as C/C++.

MATLAB® is a registered trademark of The MathWorks, Inc. For product information, please contact:

The MathWorks, Inc.
3 Apple Hill Drive
Natick, MA 01760-2098 USA
Tel: 508 647 7000
Fax: 508-647-7001
E-mail: info@mathworks.com
Web: www.mathworks.com

Acknowledgments

I express my profound and grateful veneration to Prof. P. S. Satsangi, chairman, advisory committee on education in the Dayalbagh Educational Institute, Agra, Uttar Pradesh, India. Prof. Satsangi taught me modeling and simulation methodologies, and software engineering at the postgraduate (MTech) level and guided me during my MTech and PhD in the area of modeling and simulation. He is the fountain source of my inspiration and intuition in this area. I thank him with due reverence. I am also thankful to Prof. P. K. Kalra, Indian Institute of Technology, Kanpur, India, for the encouragement, support, and visionary ideas he provided during my research. I am very deeply indebted to Prof. O. P. Malik, University of Calgary, Alberta, Canada, for providing all the facilities in his laboratory to pursue my experimental work on modeling and control during my BOYSCAST (Better Opportunities for Young Scientists in Chosen Areas of Science and Technology) fellowship, offered by the Department of Science and Technology, Government of India, during 2001–2002, and, later, in the summer of 2005 and 2007.

I wish to express my deep gratitude to Prof. Helmut Röck and Prof. Anand Srivastava, Universitate Zu Kiel, Kiel, Germany, and to Prof. Chaudhary, Imperial College London, United Kingdom, for providing all the facilities and help during my visit to these institutions in the summer of 2007.

I would also like to thank Prof. Westwick and Garwin Hancock, University of Calgary, Alberta, Canada; Prof. P.H. Roe, system design engineering, and Prof. Shesha Jayaram, High Voltage Lab, University of Waterloo, Ontario, Canada; and S. Krishnananda at the Dayalbagh Educational Institute, Agra, Uttar Pradesh, India, for their good wishes and help extended during the preparation of this book. Thanks are also due to my colleagues Anand Sinha, Gaurav Rana, Himanshu Vijay, and Ashish Chandiok for all their help.

I am thankful to Jessica Vakili and Amber Donley, project coordinators; Nora Konopka, publisher of engineering and environmental science books; and other team members from Taylor & Francis who directly or indirectly helped in realizing this book.

Finally, I am grateful to my wife, Dr. Lajwanti, and my daughters, Jyoti and Swati, who had to endure many inconveniences during the course of writing this book.

I also express my deepest gratitude to many, not mentioned here, for their support in countless ways when this book was written.

Devendra K. Chaturvedi
Dayalbagh Educational Institute (Deemed University)
Agra, Uttar Pradesh, India

Author

Devendra K. Chaturvedi was born in Madhya Pradesh, India, on August 3, 1967. He graduated in electrical engineering from the Government Engineering College, Ujjain, Madhya Pradesh, India in 1988, and did his MTech in engineering systems and management in 1993. He then pursued his PhD in electrical power systems in 1998 from the Dayalbagh Educational Institute (Deemed University), Agra, India. Currently, he is working as a professor in the Department of Electrical Engineering, Dayalbagh Educational Institute (Deemed University). He has won several awards and recognition including the President's Gold Medal and the Director's Medal of the Dayalbagh Educational Institute (Deemed University) in 1993, the Tata Rao Medal in 1994, the Dr. P. S. Nigam U.P. State Power Sector Award in 2005 and 2007, the Musaddi Lal Memorial Award in 2007, and the institutional prize award in 2005 from the Institution of Engineers, India. He was awarded a BOYSCAST fellowship of the Department of Science and Technology, Government of India, in 2001.

He is a regular visiting fellow at the University of Calgary, Alberta, Canada. He has many national and international research collaborations in the area of modeling and simulations, soft computing, intelligent adaptive control systems, and optimization. He has also organized many short-term courses in the area of modeling and simulation of systems, and fuzzy systems and its applications. He serves as a consultant at the Aerial Delivery Research and Development Establishment, Agra, and at the Defense Research and Development Organisation (DRDO) lab, Government of India. He has organized many national seminars and workshops on theology; ethics, values, and social service; professional ethics; ethics, agriculture, and religion; the relationship between religion and the future of mankind; and the teachings of the *Bhagavad Gita* and the religions of saints.

He has authored a book, *Soft Computing and Its Applications to Electrical Engineering*, published by Springer, Germany (2008). He has also edited a book, *Theology, Science, and Technology: Ethics and Moral Values*, published by Vikas Publishing House, Delhi (2005).

His name is included in the *Marquis Who's Who in Engineering and Science* in Asia (2006–2007), the *Marquis Who's Who in Engineering and Science* in America (2006–2007), and the *Marquis Who's Who in World* (2006–2007). He is a fellow of the Institution of Engineers, India, and a member of many professional bodies such as the IEEE, the IEE, the Indian Society for Technical Education (ISTE), the Indian Society of Continuing Engineering Education (ISCEE), the Aeronautical Society of India, and the System Society of India.

1
Introduction to Systems

> Science teaches us to search for "truth," think with reason and logic; accept your mistakes; tolerate other's point of view and suggestions and make an effort to shift from diversity to unity.
>
> **Dr. M. B. Lal Sahab**

> A complex system that works is invariably found to have evolved from a simple system that works.
>
> **John Gaule**

> To manage a system effectively, you might focus on the interactions of the parts rather than their behavior taken separately.
>
> **Russell L. Ackoff**

1.1 System

Most systems that surround us are multidimensional, extremely complex, time varying, and nonlinear in nature as they are comprised of large varieties of actively or passively interacting subsystems. These systems consist of interacted subsystems, which have separate and conflicting objectives. The term *system* is derived from the Greek word *systema*, which means an organized relationship among functioning units or components. It is used to describe almost any orderly arrangement of ideas or construct.

According to the *Webster's International Dictionary*, "A system is an aggregation or assemblage of objects united by some form of regular interaction or interdependence; a group of diverse units so combined by nature or art as to form an integral; whole and to function, operate, or move in unison and often in obedience to some form of control...."

A system is defined to be a collection of entities, for example, people or machines that act and interact together toward the accomplishment of some logical end. In practice, what is meant by "the system" depends on the objectives of a particular study. The collection of entities that compose a system for one study might be only a subset of another larger system. For example, if one wants to study a banking system to determine the number of tellers needed to provide adequate service for customers who want just to encash or deposit, the system can be defined to be that portion of the bank comprising of the tellers and the customers. Additionally, if, the loan officer and the safety deposit counters are to be included, then the definition of the system must be more inclusive accordingly.

The state of a system is to be defined as an assemblage of variables necessary to describe a system at a particular instant of time with respect to the objectives of the study. In this case of study of a banking system, possible state variables are the number of busy tellers, the number of customers in the bank, and the line of arrival of each customer in the bank.

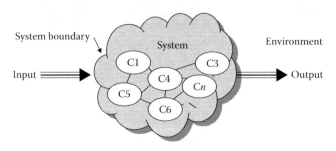

FIGURE 1.1
System as collection of interconnected components.

The fundamental feature in the system's concept is that all the aggregation of entities united, have a regular interaction, as a finite number of interfaces as shown in Figure 1.1. Considering a hierarchy among systems, a system can also be expressed as a collection of various subsystems and the subsystem is a further collection of interconnected components. The system behavior can be comprehended as combined interconnected components behavior. So, a large system can be regarded as a collection of different interconnected components.

More appropriately, a large-scale system may be viewed at the supremum of the hierarchy and components at the bottom most level (root level). The power of the system's concept is its sheer generality, which can be emphasized by general systems theory.

Some examples of the systems are

- Esoteric systems
- Medical/biological systems
- Socioeconomic systems
- Communication and information systems
- Planning systems
- Solar system
- Environmental systems
- Manufacturing systems
- Management systems
- Transportation systems
- Physical systems—electrical, mechanical, thermal, hydraulic systems, and combinations of them

Every system consists of subsystem or components at lower levels and supersystems at higher levels, as shown in Figure 1.2. One needs to be extremely careful to define such a hierarchically nested system because this will determine the kind of results one will obtain.

A system is characterized by the following attributes:

- System boundary
- System components and their interactions
- Environment

Introduction to Systems

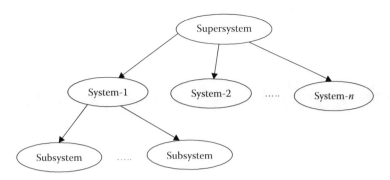

FIGURE 1.2
Hierarchically nested set of systems.

1.1.1 System Boundary

To study a given system, it is necessary to determine what comprises (falls inside and what falls outside) a system. For this a demarcation is required to differentiate entities from the environment. Such a partition is called a system boundary. The system boundaries are observer-dependent, time-dependent, and most importantly system-dependent. The different observers may draw different boundaries for the same system. Also, the same observer may draw the system boundaries differently for different times.

Finally, they may also be drawn differently with respect to the nature of the study, that is, steady state or transient. For example, in case of steady-state study of series R–L circuit, only R is to be included in the system boundary, but in transient study, both R and L must be considered in the system.

Some salient points about the system boundary are

- It is a partitioning line between the environment and the system.
- System is inside the boundary and environment is outside the system.
- A real or imaginary boundary separates the system from the rest of the universe, which is referred to as the environment or surroundings.
- System exchanges input–output from its environment.
- This boundary might be material boundary (like the skin of a human body) or immaterial boundary (like the membership to a certain social group).
- Considering a system boundary in systems analysis and evaluation is of immense importance as it helps in identifying the system and its components. The interaction between a system and its environment takes place mainly at the boundaries. It determines what can enter or leave a system (input and output).
- System boundary may be crisp (clearly defined) or fuzzy (ill defined). In crisp boundaries, it is quite clear that what is inside the boundary (i.e., part of system) and what is outside the boundary (i.e., part of environment). In fuzzy boundaries, it is not very clear whether a particular component belongs to the environment or the system.

1.1.2 System Components and Their Interactions

System component is a fundamental building block. It is quite easy to find the input–output relations for the system components with the help of some fundamental laws of physics, which is called the mathematical model for components. It may be written in the form of difference or differential equations. They are pretty simple and easily understandable.

Business system environment includes customers, suppliers, other industries, and government. Its inputs include materials, services, new employees, new equipments, facilities, etc. Output includes product, waste materials, money, etc.

- It is static or dynamically changing with time, input, or state of the system.
- Interaction may be constrained or nonconstrained type.
- The component interaction may be unidirectional or bidirectional.
- Interaction strength may be 0, 1, or between 0 and 1.
    a. If interaction strength is zero (0) then there is no interaction.
    b. If interaction strength is one (1) it means full interaction and if the interaction strength lies between zero and one, then the interaction is partial interaction.

1.1.3 Environment

A living organism is a system. Organisms are open systems: they cannot survive without continuously exchanging matter and energy with their environment. When we separate a living organism from its surrounding, it will die shortly due to lack of oxygen, water, and food. The peculiarity of open systems is that they interact with other systems outside of themselves. This interaction has two components: input, that is, what enters the system from outside the boundary, and output, that is, what leaves the system boundary to the environment. In order to speak about the inside and the outside of a system, we need to be able to distinguish between the system and its environment, which is in general separated by a boundary (for example, living systems, skin is the boundary). The output of a system is generally a direct or indirect result to a given input. For example, the food, drink, and oxygen we consume are generally separated by a boundary and discharged as urine, excrements, and carbon dioxide. The transformation of input into output by the system is usually called throughput.

A system is intended to "absorb" inputs and process them in some way in order to produce outputs. Outputs are defined by goals, objectives, or common purposes. In order to understand the relationship between inputs, outputs, and processes, you need to understand the environment in which all of this occurs. The environment represents everything that is important to understand the functioning of the system, but is not a part of the system. The environment is that part of the world that can be ignored in the analysis except for its interaction with the system. It includes competition, people, technology, capital, raw materials, data, regulation, and opportunities.

When we are concerned only with the input and corresponding output of a system, while undermining the internal intricacies of component-level dynamics of the system, such study may be called as black box study. For example, if we consider a city, we may safely measure the total fuel consumption (input) of the city and the level of emissions (output) out of such consumptions without actually bothering about trivial details like who/what consumed more and who/what emitted or polluted the most. Such point of view considers the system as a "black box," that is, something that takes input, and produces output, without looking at what happens inside the system during process. Contrary

Introduction to Systems

to the former, when we are equally concerned about the internal details of the system and its processes besides the input and output variable, such an approach of system is considered as white box. For example, when we model a city as a pollution production system, regardless of which chimney emitted a particular plume of smoke, it is sufficient to know the total amount of fuel that enters the city to estimate the total amount of carbon dioxide and other gases produced. The "black box" view of the city will be much simpler and easier to use for the calculation of overall pollution levels than the more detailed "white box" view, where we trace the movement of every fuel tank to every particular building in the city.

The system as a whole is more than the sum of its parts. For example, if person A alone is too short to reach an apple on a tree and person B is too short as well, once person B sits on the shoulders of person A, they are more than tall enough to reach the apple. In this example, the product of their synergy would be one apple. Another case would be two politicians. If each is able to gather 1 million votes on their own, but together they were able to appeal to 2.5 million voters, their synergy would have produced 500,000 more votes than had they each worked independently.

1.2 Classification of Systems

Systems can be classified on the basis of time frame, type of measurements taken, type of interactions, nature, type of components, etc.

1.2.1 According to the Time Frame

Systems can be categorized on the basis of time frame as

Discrete
Continuous
Hybrid

A *discrete* system is one in which the state variables change instantaneously at separated points in time, for example, queuing systems (bank, telephone network, traffic lights, machine breakdowns), card games, and cricket match. In a bank system, state variables are the number of customers in the bank, whose value changes only when a customer arrives or when a customer finishes being served and departs.

A *continuous* system is one in which the state variables change continuously with respect to time, for example, solar system, spread of pollutants, charging a battery. An airplane moving through the air is an example of a continuous system, since state variables such as position and velocity can change continuously with respect to time.

Few systems in practice are wholly discrete or wholly continuous, but since one type of change predominates for most systems, it will usually be possible to classify a system as being either discrete or continuous.

A *hybrid* system is a combination of continuous and discrete dynamic system behavior. A hybrid system has the benefit of encompassing a larger class of systems within its structure, allowing more flexibility in modeling continuous and discrete dynamic phenomena, for example, traffic along a road with traffic lights.

1.2.2 According to the Complexity of the System

Systems can be classified on the basis of complexity, as shown in Figure 1.3.

Physical systems
Conceptual systems
Esoteric systems

Physical systems can be defined as systems whose variables can be measured with physical devices that are quantitative such as electrical systems, mechanical systems, computer systems, hydraulic systems, thermal systems, or a combination of these systems. Physical system is a collection of components, in which each component has its own behavior, used for some purpose. These systems are relatively less complex. Some of the physical systems are shown in Figure 1.4a and b.

Conceptual systems are those systems in which all the measurements are conceptual or imaginary and in qualitative form as in psychological systems, social systems, health care systems, and economic systems. Figure 1.4c shows the transportation system. Conceptual systems are those systems in which the quantity of interest cannot be measured directly with physical devices. These are complex systems.

Esoteric systems are the systems in which the measurements are not possible with physical measuring devices. The complexity of these systems is of highest order.

1.2.3 According to the Interactions

Interactions may be unidirectional or bidirectional, crisp or fuzzy, static or dynamic, etc. Classification of systems also depends upon the degree of interconnection of events from

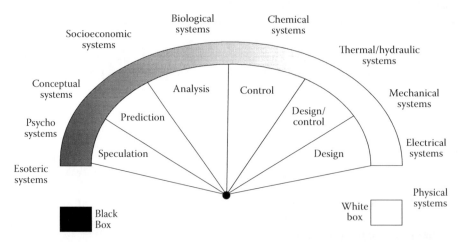

FIGURE 1.3
Classification of system based on complexity.

Introduction to Systems

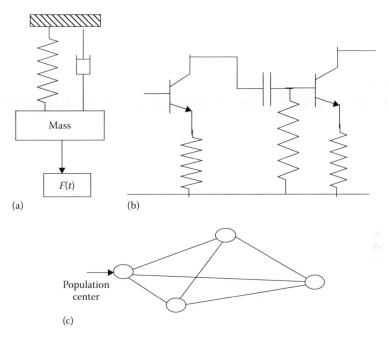

FIGURE 1.4
Different types of systems. (a) Mechanical system. (b) Electronic circuit. (c) Transportation system.

none to total. Systems will be divided into three classes according to the degree of interconnection of events.

1. *Independent*—If the events have no effect upon one another, then the system is classified as independent.
2. *Cascaded*—If the effects of the events are unilateral (that is, part A affects part B, B affects C, C affects D, and not vice versa), the system is classified as cascaded.
3. *Coupled*—If the events mutually affect each other, the system is classified as coupled.

1.2.4 According to the Nature and Type of Components

1. Static or dynamic components
2. Linear or nonlinear components
3. Time-invariant or time-variant components
4. Deterministic or stochastic components
5. Lumped parametric component or distributed parametric component
6. Continuous-time and discrete-time systems

1.2.5 According to the Uncertainties Involved

Deterministic—No uncertainty in any variables, for example, model of pendulum.

Stochastic—Some variables are random, for example, airplane in flight with random wind gusts, mineral-processing plant with random grade ore, and phone network with random arrival times and call lengths.

Fuzzy systems—The variables in such type of systems are fuzzy in nature. The fuzzy variables are quantified with linguistic terms.

1.2.5.1 Static vs. Dynamic Systems

Normally, the system output depends upon the past inputs and system states. However, there are certain systems whose output does not depend on the past inputs called static or memoryless systems. On the other hand, if the system output depends on the past inputs and earlier system states which essentially implied that the system has some memory elements, it is called a dynamic system. For example, if an electrical system contains inductor or capacitor elements, which have some finite memory, due to which the system response at any time instant is determined by their present and past inputs.

1.2.5.2 Linear vs. Nonlinear Systems

The study of linear systems is important for two reasons:

1. Majority of engineering situations are linear at least within specified range.
2. Exact solutions of behavior of linear systems can usually be found by standard techniques.

Except, a handful special types, there are no standard methods for analyzing nonlinear systems. Solving nonlinear problems practically involves graphical or experimental approaches. Approximations are often necessary, and each situation usually requires special handling. The present state of art is such that there is neither a standard technique which can be used to solve nonlinear problems exactly, nor is there any assurance that a good solution can be obtained at all for a given nonlinear system.

The *Ohm's law* governs the relation between the voltage across and the current through a resistor. It is a linear relationship because voltage across a resistor is linearly proportional to the current through it.

$$V \propto I$$

But even for this simple situation, the linear relationship does not hold good for all conditions. For instance, as the current in a resistor increases exceedingly, the value of its resistance will increase due to increase in temperature of the resistor:

$$R_t = R(1 + \alpha T)$$

The amount of change in resistance is being dependent upon the magnitude of the current, and it is no longer correct to say that the voltage across the resistor bears a linear relationship to current through it.

Similarly, the *Hooke's law* states that the stress is linearly proportional to the strain in a spring. But this linear relationship breaks down when the stress on the spring is too great. When the stress exceeds the elastic limit of the material of which the spring is made, stress

Introduction to Systems

and strain are no longer linearly related. The actual relationship is much more complicated than the Hooke's law situation, that is,

$$\text{Stress}(\sigma) \propto \text{Strain}(\varepsilon)$$

Therefore, we can say that restrictions always exist for linear physical situation, saturation, breakdown, or material changes with ultimate set in and destroy linearity. Under ordinary circumstances physical conditions in many engineering problems stay well within the restrictions and the linear relationship holds good.

Ohm's law and Hooke's law describe only special linear systems. There exist systems that are much more complicated and are not conveniently described by simple voltage–current or stress–strain relationships.

1.3 Linear Systems

An engineer's interest in a physical situation is very frequently the determination of the response of a system to a given excitation. Both the excitation and the response may be any physically measurable quantity, depending upon the particular problem, as shown in Figure 1.5. The linear system obeys superposition and homogeneity theorems.

1.3.1 Superposition Theorem

Suppose that an excitation function, $e_1(t)$, which varies with the time in a specified manner, produces a response function, $\omega_1(t)$, and a second excitation function, $e_2(t)$ produces a response function, $\omega_2(t)$.

Symbolically, the above-mentioned observation may be written as

$$e_1(t) \to \omega_1(t)$$

and

$$e_2(t) \to \omega_2(t)$$

For the linear system, if these excitations applied simultaneously

$$e_1(t) + e_2(t) \to \omega_1(t) + \omega_2(t)$$

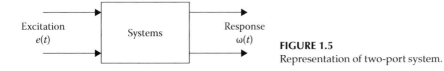

FIGURE 1.5
Representation of two-port system.

The above equation shows the superposition theorem, which can be described as a superposition of excitation functions results in a response which is the superposition of the individual response functions.

1.3.2 Homogeneity

If there are n identical excitations applied to the same part of the system, that is, if

$$e_1(t) = e_2(t) = e_3(t) = \cdots = e_n(t)$$

then for a linear system

$$\sum_{k=1}^{n} e_k(t) = ne_1(t) \rightarrow \sum_{k=1}^{n} \omega_k(t) = n\omega_1(t)$$

Hence, a characteristic of linear systems is that the magnitude is preserved. This characteristic is referred to as the property of homogeneity.

A system is said to be linear if and only if both the properties of superposition and homogeneity are satisfied.

1.3.3 Mathematical Viewpoint of a Linear System

Mathematically, we can define linear systems as those whose behavior is governed by linear equations (whether linear algebraic equations, linear difference equations, or linear differential equations).

1.3.3.1 Linear Differential Equation

Consider the following differential equation:

$$\frac{d^2\omega}{dt^2} + a_1 \frac{d\omega}{dt} + a_0 \omega = e(t)$$

where
   t is used as the independent variable
   e is excitation function
   ω is the response function

Coefficients a_1 and a_0 are system parameters determined entirely by the number, type, and the arrangement of elements in the system; they may or may not be functions of the independent variable t. Since, there are no partial derivatives in the above equation and the highest order of the derivative is 2, the equation above is an ordinary differential equation of second order. This equation is a linear differential equation of second order because neither the dependent variable ω nor any of its derivatives is raised to a product of two or more derivatives of dependent variable or a product of the dependent variable and one of its derivatives.

The validity of the principle of superposition here can be verified as follows. We assume that the excitations $e_1(t)$ and $e_2(t)$ give rise to responses $\omega_1(t)$ and $\omega_2(t)$, respectively. Hence

Introduction to Systems

$$\frac{d^2\omega_1}{dt^2} + a_1\frac{d\omega_1}{dt} + a_0\omega_1 = e_1(t)$$

$$\frac{d^2\omega_2}{dt^2} + a_1\frac{d\omega_2}{dt} + a_0\omega_2 = e_2(t)$$

Adding these equations, we have

$$\frac{d^2}{dt^2}(\omega_1+\omega_2) + a_1\frac{d}{dt}(\omega_1+\omega_2) + a_0(\omega_1+\omega_2) = (e_1(t)+e_2(t))$$

This shows that the principle of superposition applies for a linear system even when the coefficients a_1 and a_0 are function of the independent variable t.

1.3.3.2 Nonlinear Differential Equations

The differential equation becomes nonlinear if there is a product of the dependent variable and its derivative; power of the dependent variable; or power of a derivative of dependent variable. The existence of powers or other nonlinear functions of the independent variable does not make an equation nonlinear. Some nonlinear differential equations are

$$4\frac{d^2y}{dt^2} + y\frac{dy}{dt} + 5y = 7t^2 \qquad \text{(product of dependent variable and its derivative)}$$

$$\frac{du}{d\theta} + u + u^2 = \sin^3\theta \qquad \text{(second power of dependent variable)}$$

$$t\left(\frac{d^2y}{dt^2}\right)^2 + 4\frac{dy}{dt} + t^2y = e^{-t} \qquad \text{(power of a derivative of dependent variable)}$$

Notes:

1. Order of a differential equation is the highest order derivative present in the differential equation.
2. Degree of a differential equation is the power of highest order derivative of the equation.
3. An ordinary linear differential equation of an arbitrary order n may be written as

$$a_n(t)\frac{d^n\omega}{dt^n} + a_{n-1}(t)\frac{d^{n-1}\omega}{dt^{n-1}} + \cdots + a_1(t)\frac{d\omega}{dt} + a_0(t)\omega = e(t)$$

where the coefficients $a_n(t), a_{n-1}(t), \ldots, a_1(t), a_0(t)$ derived from system parameters and the right hand side of the equation, $e(t)$ are given as functions of the independent variable t. The response $\omega(t)$ is determined by the system and the excitation function, respectively. The equation is said to be homogeneous if $e(t) = 0$, and nonhomogeneous if $e(t) \neq 0$.

1.4 Time-Varying vs. Time-Invariant Systems

A system whose parameters change with time is called time-varying system. A familiar example of a time-varying system is the carbon microphone, in which the resistance is a function of mechanical pressure. Similarly, the mileage of a brand new car increases gradually as it is used progressively for sometime, and after attaining its maximum mileage, it steadily decreases.

If system parameters do not change with time then such systems are called time-invariant (or constant parameter) systems.

If the excitation function $e(t)$ applied to such a system is an alternating function of time with frequency f, then the steady-state response $\omega(t)$, after the initial transient has died out, appearing at any part of the system will also be alternating with the frequency f. In other words, time-invariant nonlinear systems create no new frequencies. For time-invariant systems if

$$e(t) \to \omega(t)$$

then

$$e(t - \tau) \to \omega(t - \tau)$$

where τ is an arbitrary time delay.

Hence, the output of time-invariant system depends upon the shape and magnitude of the input and not on the instant at which the input is applied. If the input is delayed by time τ, the output is the same as before but is delayed by τ, as shown in Figure 1.6.

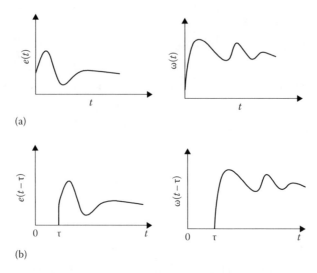

FIGURE 1.6
Time-invariant system responses. Application of input at time instant (a) t and (b) $(t - \tau)$.

1.5 Lumped vs. Distributed Parameter Systems

A lumped system is one in which the components are considered to be concentrated at a point. For example, the mass of a pendulum in simple harmonic motion is considered to be concentrated at a point in space. This is a lumped parametric system because the mass is a point mass. This assumption is justified at lower frequencies (higher wavelengths). Therefore, in lumped parameter models, the output can be assumed to be functions of time only. Hence, the system can be expressed with ordinary differential equations.

In contrast, distributed parametric systems such as the mass or stiffness of mechanical power transmission shaft cannot be assumed to concentrate at a point; thus the lumped parameter assumption breaks down. Therefore, the system output is a function of time and one or more spatial variables (space), which results in a mathematical model consisting of partial differential equations. Figure 1.7 shows lumped and distributed parameter electrical systems.

1.6 Continuous-Time and Discrete-Time Systems

Systems whose inputs and outputs are defined over a continuous range of time (i.e., continuous-time signals) are continuous-time systems. On the other hand, the systems whose inputs and outputs are signals defined only at discrete instants of time $t_0, t_1, t_2, \ldots, t_k$ are called discrete systems, as shown in Figure 1.8. The digital computer is a familiar example of this type of systems.

The discrete-time signals arise naturally in situations which are inherently discrete time such as population in a particular town and customers served at ATM counter. Sometimes, we want to process continuous-time signals with discrete-time systems. In such situations it is necessary to convert continuous-time signals to discrete signals using analog-to-digital converters (ADC) and process the discrete signals with discrete systems. The output of discrete-time system is again converted back into continuous-time signals using digital-to-analog converters (DAC), as shown in Figure 1.9.

The terms discrete-time and continuous-time signals qualify the nature of signal along the time axis (x-axis) and the terms analog and digital, on the other hand, qualify the nature of signal amplitude (y-axis), as shown in Figure 1.10.

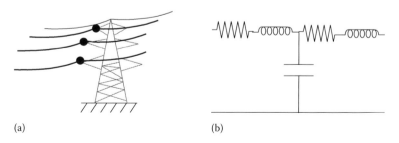

FIGURE 1.7
Lumped- and distributed-parameter-electrical systems. (a) Transmission line parameters are distributed and (b) electrical circuit parameters are lumped.

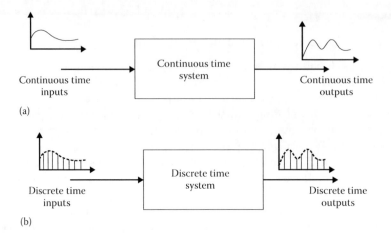

FIGURE 1.8
Analog and digital systems. (a) Continuous time system and (b) discrete time system.

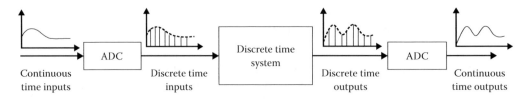

FIGURE 1.9
Processing continuous-time signals by discrete-time systems.

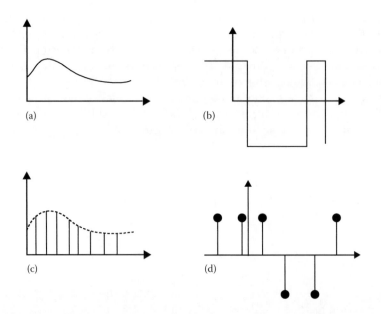

FIGURE 1.10
Different types of signals. (a) Analog and continuous signal, (b) digital and continuous signal, (c) analog and discrete signal, and (d) digital and discrete signal.

1.7 Deterministic vs. Stochastic Systems

A system that will always produce the same output for a given input is said to be deterministic. Determinism is the philosophical proposition that every event, including human cognition and behavior, decision, and action, is causally determined by an unbroken chain of prior occurrences.

A system that will produce different outputs for a given input is said to be stochastic. A stochastic process is one whose behavior is nondeterministic and is determined both by the process's predictable actions and by a random element. Classical examples of this are medicine: a doctor can administer the same treatment to multiple patients suffering from the same symptoms, however, the patients may not all react to the treatment the same way. This makes medicine a stochastic process. Additional examples are warfare, meteorology, and rhetoric, where success and failure are so difficult to predict that explicit allowances are made for uncertainty.

1.7.1 Complexity of Systems

Another basic issue is the complexity of a system. Complexity of a system depends on the following factors:

- Number of interconnected components
- Type/nature of component
- Number of interactions
- Strength of the interaction
- Type/nature of interactions
  a. Static or dynamic
  b. Unidirectional or bidirectional
  c. Constrained or nonconstraint interaction

Consider an example of a family. If there is only one person (component) in the family (system), then there is no interaction. Therefore, the number of interactions is zero. If there are two persons, say, husband and wife (two components) in the family, then they may talk each other. In this case, the number of interactions is two, that is, husband can say something to his wife and wife can reply to her husband. Similarly, if there are three persons in the family, say, husband, wife, and a child, then each one may interact with the rest two. The number of interactions in this case is 6 as shown in Figure 1.11. From the discussion it is clear that the number of interactions increases exponentially as the number of components is increased in the system and sometimes they become unmanageable, as shown in Figure 1.12 and Table 1.1.

The complexity of any system in modern times is quite large, whether it is an industrial organization, an engineering system (electrical, mechanical, thermal, hydraulic, etc.), or a social system. The efficient utilization of inputs to these systems calls for a thorough understanding of the basic structure of the system with a view to develop suitable control strategy for the application of engineering techniques employing computers to achieve optimum performance in industrial organizations, equipment design, socio-economic activities, as well as management systems. The systems' engineering based on system thinking and systems approach facilitates us to have a thorough understanding of the basic structure of the system.

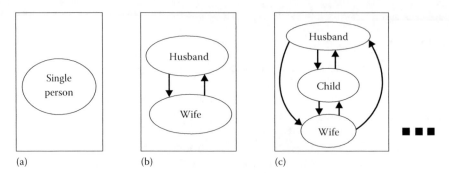

FIGURE 1.11
Number of components and number of interactions (a) 0, (b) 2, and (c) 6 in the family system.

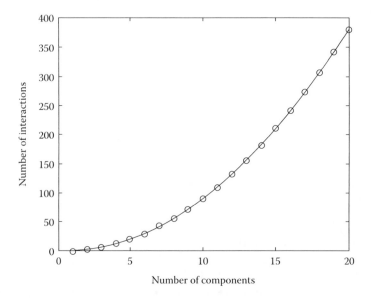

FIGURE 1.12
Relation between number of components and their interactions.

1.8 Hard and Soft Systems

Peter Checkland is a British management scientist and professor of Systems at Lancaster University. He developed soft systems methodology (SSM): a methodology-based systems thinking. Prof. Checkland became interested in applying systems ideas to messy management problems while working as a manager in industry. His ideas for SSM emerged from the failure of the application of, what he called, "hard" systems engineering in messy management problems. SSM developed

TABLE 1.1

Effect of Number of Interactions with Number of Components

S.No	Number of Components	Number of Interactions
1	1	0
2	2	2
3	3	6
4	4	12
5	5	20
n	n	$n(n-1)$

from this continuous cycle of intervention in ill-structured management problems and learning from the results (Checkland, 1999, 2006; Checkland and Winter, 2000; Winter and Checkland, 2003).

Soft system is a branch of systems thinking specifically designed for use and application in a variety of real-world contexts. David Brown stated that a key factor in its development was the recognition that purposeful human activity can be modeled systemically. Peter Checkland's work has influenced the development of "soft" operations research (OR), which joins optimization, mathematical programming, and simulation as part of the OR topography.

Just like Boland (1985) who has brought phenomenology in the field of information systems, to critically examine this field, raise consciousness, and clarify its path, so has Checkland (1981) done in the field of systems thinking. In so doing, sense-making and the social construction of reality have become central notions in their respective fields. System thinking as comprehended by great thinkers and scientists is not only considering reality independent of observer but also the description of the participation of interconnected entities, cybernetic process, and their emergent properties. In fact, the system thinking is about how we attribute the world and construct the unity of our ever-changing reality.

In systems science jargon hard and soft systems are used to differentiate between different nature of system's problems. The characteristics of soft system are

- Ill-defined problems
- Nonquantified components/systems
- No definite solution
- Qualitative input and outputs

Similarly hard systems may be characterized by the following things:

- The problems associated with such systems are well defined
- They have a single, optimum solution
- A scientific approach to problem solving will work well
- Technical factors will tend to predominate

In hard systems approaches (or structured systems analysis and design methodology, SSADM), rigid techniques and procedures are used to provide unambiguous solutions to well-defined data and processing problems. These focus on computer implementations.

Checkland draws attention to these two alternative paradigms to explain the nature and significance of systems thinking:

> Paradigm 1—The world is considered to be systemic and is studied systematically.
> Paradigm 2—The world is chaotic (i.e., it admits to many different interpretations) and we study it systemically.

Now the first paradigm reflects the notion of hard systems thinking and the second reflects the notion of soft systems thinking. Hard systems thinking can be characterized as having an objective or end to be achieved, and a system can be engineered to achieve the stated objective. Soft systems thinking can be characterized as having a desirable end, but the means to achieve it and the actual outcome are not easily quantified (Wilson, 2001).

1.9 Analysis of Systems

Analysis is the separation of the subject into its constituent parts or elements in an effort to either clarify or to enhance the understanding of the subject.

1.10 Synthesis of Systems

Synthesis is the study of composition or combination of parts or elements so that the system performs, behaves, or responds according to a given set of specifications.

1.11 Introduction to System Philosophy

Generally speaking, any entity, which may be subdivided into components or parts of a constituent, may be considered a system. We have been hearing about biological systems, physical systems, social systems, ecological systems, transportation systems, economic systems, linguistic systems, health care systems, etc. Systems concept enriched by Ludwig Von Bertalanffy, who was the first mastermind to advocate general system theory or systemology. When science failed to explain open systems, the concept of systems science came with rational explanation. In essence, he postulated, on the basis of formal analogies between physical, chemical, and biological systems, the existence of general systems law, which is equally applicable to physical systems as well as human and social systems. This new world view reaches farther than science or arts by itself and should ultimately include science, arts, ethics, politics, and spirituality (Satsangi, 2008).

Karl Ludwig von Bertalanffy (September 19, 1901, Vienna, Austria–June 12, 1972, New York, United States) was an Austrian-born biologist known as one of the founders of "general systems theory" or "general systemology." He advocated the concepts of system to explain the commonalities (Wilson, 1999; Weinberg, 2001; Skyttner, 2006; Vester, 2007) and is often regarded as the father of modern system thinking.

Bertalanffy occupies an important position in the intellectual history of the twentieth century. His contributions went beyond biology, and extended to cybernetics, education, history, philosophy, psychiatry, psychology, and sociology. Some of his admirers even believe that von Bertalanffy's general systems theory provides a conceptual framework for all these disciplines to integrate for holistic comprehension of what we see and perceive. The individual growth model published by von Bertalanffy in 1934 is widely used in biological models and exists in a number of permutations.

Bertalanffy's contribution to systems theory is widely acclaimed for his theory of open systems. The system theorist argued that traditional closed system models based on classical science and the second law of thermodynamics were untenable. Bertalanffy maintained that "the conventional formulation of physics are, in principle, inapplicable to the living organism being open system having steady state. We may well suspect that many characteristics of living systems which are paradoxical in view of the laws of physics are a consequence of this fact." (Bertalanffy, 1969a) However, while closed physical systems were questioned, questions equally remained over whether or not open physical systems could justifiably lead to a definitive science for the application of an open systems view to a general theory of systems.

In Bertalanffy's model, the theorist defined general principles of open systems and the limitations of conventional models. He ascribed applications to biology, information theory, and cybernetics. Concerning biology, examples from the open systems view suggested they "may suffice to indicate briefly the large fields of application" that could be the "outlines of a wider generalization"; (Bertalanffy, 1969b) from which, a hypothesis for cybernetics. Although potential applications exist in other areas, the theorist developed only the implications for biology and cybernetics. Bertalanffy also noted unsolved problems, which included continued questions over thermodynamics, thus the unsubstantiated claim that there are physical laws to support generalizations (particularly for information theory), and the need for further research into the problems and potential with the applications of the open system view from physics.

Pre-Greek era was gripped in the realness of religion and magic. The main contribution of Greek philosophers was to remove the world from the pangs of magic and religion and to create an explanation to the working of the real-world rational explanation which was the subject of a new kind of inquiry.

Aristotle, one of the greatest Greek philosophers, was born at Stagira in northern Greece. He included his investigations of an amazing range of subjects, from logic, philosophy, and ethics to physics, biology, psychology, politics, and rhetoric. Aristotle appears to have thought through his views as he wrote, returning to significant issues at different stages of his own development. The result is less a consistent system of thought than a complex record of Aristotle's thinking about many significant issues.

Aristotelian explanation for the world with its holistic notion is that "a whole is more than the sum of its parts." This holistic notion forms a methodological and conceptual foundation of systems philosophy. He advanced the teleological outlook that objects in this world fulfilled their inner nature or purpose.

Teleology is based on belief that things happen because of the purpose or design that will be fulfilled by them. The scientific revolution started in the seventeenth century due to the pioneering work of Copernicus and Kepler pertaining to solar system and that of Galileo and Newton pertaining to terrestrial and celestial dynamics is proved to be the most powerful activity man has discovered. Copernicus and Kepler's heliocentric model of the solar system demolished the Aristotelian solar system, which consisted of crystalline sphere upon which the planets moved in perfect circles. Galileo propounded that the earth is not the center of the world but the sun is the center and immovable. His work directly challenged Aristotelian view that motion needed a force to maintain it to produce acceleration. Newton developed the concepts and definitions to be used and stated the three laws of motion governing the classical dynamics. He discovered the motion of bodies in vacuum thereby proving the basis of celestial mechanics, discussed the modifications introduced in fluids, and ultimately demonstrated "the frame of the system of the world," applied the ideas to solar system to accurately predict the facts about the motion of planets and developed laws of gravitation. He advocated the importance of expressing nature's behaviors in the language of mathematics. Thus, the present day science may be seen as applications of Galileo Newtonian methodology to study of natural phenomena. Newton's physics provided a mechanical picture of the universe, which survived severe tests and Aristotle's teleological outlook and his holistic notion, seemed an unnecessary doctrine.

The Newtonian mechanics succeeded because its protagonists asked questions which in the range of experimental answer, by limiting their inquiries to physical, rather metaphysical problems, and in particular on those aspects of the physical world, which could be expressed in terms of mathematics. The core of scientific revolution was a belief in rational modes of thought applied to design and the subsequent analysis of experimentation is outstanding in that it works. Within well-defined limits, scientific evidence is trust worthy and that is why it has created the present world.

1.11.1 Method of Science

Science is a way of acquiring publicly testable knowledge of the world. It is characterized by the application of rational thinking to experience gained from observation and from deliberately designed experiments with an aim to develop the concise expression of the laws, which govern the regularities of the universe, expressing them mathematically if possible. This particular pattern of human activity can be summarized in three crucial characteristics as 3Rs, namely, reductionism, repeatability, and refutation.

1.11.1.1 Reductionism

There are three senses in which science is a reductionism:

1. The real world is so rich in variety, so many, that in order to make coherent investigations of it, it is necessary to select some items only to examine, out of all those, which could be looked at. To define an experiment is to define a reduction of the world, one made for a particular purpose.
2. Reductionism in explanation, that is, accepting the minimum explanation required by the facts.
3. Breakdown problems and analyze piecemeal.

1.11.1.2 Repeatability

Repeatability refers to a characteristic of a system for a given excitation and the system output remains same, independent of time and location. Repeatability of experiments makes the knowledge public or scientific and is its strength, for example, literary knowledge is a private knowledge. A critic, who wishes to convince us that Mr. X is a great novelist will explain why he thinks so. He will try to influence our opinion about a book of Mr. X. We may get convinced or not. Whether we agree or not depends upon our latter data. Such knowledge is private whereas scientific experiments may repeat by one and all and verified. Thus they are public knowledge.

1.11.1.3 Refutation

The method of science is the method of bold conjectures with ingenious and severe attempt to refute them. The important message is—Do not be satisfied with normal science; try to find ways of challenging the paradigm. This makes intellectual progress by the refutation of the hypothesis.

In brief, we may reduce the complexity of the variety of the real world in experiments, whose results are validated by repeatability and we may build knowledge by refutation of hypothesis.

1.11.2 Problems of Science and Emergence of System

Science is a great success in solving the problems of regular physical and materialistic sciences and well-structured and explicitly defined problems of laboratory have resulted in a comfortable life style. However, science has its own limitations. The reductionism which is a powerful tool also happens to be its weakness.

Scientists assume that the components of the whole are the same even when examined in isolation. This is true only for well-structured and regular physical system. However, if we move beyond them to study complex phenomena such as human society, the scientific methods fail. This is more so in biological systems. The 3Rs of science is not capable to cope up with complexities. In unstructured sciences, progress is slow and methodological problems are there. This requires that different and other ways of thinking are needed to be explored. It is in this connection that Von Bertalanffy tried to exploit Aristotelian holistic notion that "A whole is more than some of its parts" to advance system philosophy.

1.12 System Thinking

Bertalanffy argues that Aristotle's holistic notion forms a definition of the basic system, which is still valid. According to him, order or aggregation of a whole or a system, exceeding its parts when these are considered in isolation, is nothing metaphysical, not a superstition or a philosophical speculation. It is a fact of observation encountered whenever we look at a living organization, a social group, or even an atom. Articulates of Ludwig Von Bertalanffy are

1. The fact that the study of living systems is a study of commonalties shared by systems of differing physical structures
2. That physics cannot encompass organic phenomenon without fundamental modification and extension
3. That the commonalties' characteristics of biological systems are exemplary of other kinds of commonalties, which could form the basis for the general systems theory
4. That biological systems (developmental systems) are open systems, able to exchange matter, energy, and information with their environments, and consequently the second law of thermodynamics (that material systems should proceed from ordered to disordered states) is not even directly applicable to them
5. That apparently purposive telic behavior of developing systems is entirely plausible

Therefore, we see that there is a sense in which we can learn something about a particular system S, such as a developing organism, by studying some other purely reductionistic or physical sense, but nonetheless manifesting the same behavior is shared by systems of the utmost physical diversity. That is, physically disparate systems can nevertheless be models or analogs of each other. Thus the two distinct systems, S_1 and S_2, can behave similarly only to the extent that they comprise alternate realizations of a common mathematical or formal structure.

The crucial difference between magic and science resides in the manner in which models are generated and through which specific properties of the system are captured. The hypothesis of Kabala, that makes it mysticism rather than science, is that the name of a system are manifested in the corresponding numerical properties of the associated number and can be studied thereby. A Mathematical object with system captures some aspects of the reality of the system. We seek to learn about the system by studying the properties of the associated model.

It is perhaps ironic that few achievements of scientific world should be nonreductionist character typical of system theoretic arguments. Hamilton showed that two distinct and independent branches of physics, namely, optics and mechanics could be conceptually unified not through reduction of optics to mechanics or vice versa but rather through the fact that both obeyed the same formal law, that is, realized the same formal system. Hamilton stopped there and 100 years later, Schrödinger exploited the commonalties between the two and developed wave mechanics.

Even mathematicians often construct mathematical models of other mathematical systems. The relation between a topological space (a differential and abstract subject) and an associated group (an algebraic system much simpler than topology) is a modeling relation and it enabled us to learn about the homomorphism topological spaces by studying entirely different isomorphic groups of algebra system. This was further exploited to develop "Theory of Categories" which cuts across all other.

It was clearly demonstrated that where science fails to explain, system approach of exploiting commonalties applies. A.D. Hall and R.E. Fagan have thoughtfully defined systems thinking as below:

Systems thinking is the art and science of linking structure to performance, and performance to structure—often for purposes of changing relationships so as to improve performance.

But, how should we look at a complex world? One approach is to break the complex world into smaller, more manageable pieces. The argument goes that if we can understand the separate pieces, then we can put our separate understandings together to understand the whole. This is reductionism, or Cartesian thinking. It works for simple things. Cartesian thinking fails to address complex problems because, in the process of breaking up the overall system into parts, the connections and interactions between those parts get lost. Consider a comparison:

1. If you break a brick wall into parts, you end up with a pile of bricks, with which you can rebuild the brick wall—nothing lost, and perhaps even something gained in an improved wall, as shown in Figure 1.13.
2. Now if you break a human being into parts, you end up with a pile of organs, bones, muscles, sinews … but you can never reconstitute the human being.

Where does the difference lie?

The whole human being depends on the continued interaction between all its parts—in fact, the parts are all mutually dependent. So it turns out to be with complex systems. They are made up from many interacting, mutually dependent parts. Because of this it is often impractical to conduct experiments on them.

So, we need to be able to think about complex issues, partly because we cannot use Cartesian methods to "reduce" them, and partly because we cannot conduct controlled experiments upon them. The implications of that is simply this: System thinking must be rigorous if it is to be both credible and useful. But, is all systems thinking rigorous?

When learning new concepts and ways of thinking, a picture can be worth a thousand words. When we look at how people learn new things, the graphical aspect of systems thinking helps us visually see how systems work and how we might be able to work through them in better ways. The term "systems thinking" was first associated with Jay Forrester from MIT in the 1940s to refer to a different way of looking at problems and goals not as isolated events, but as parts of interrelated structures.

When we look at business and human endeavors as systems, we need to understand the complete picture, the interrelated variables, and the effects they have on each other.

Introduction to Systems

FIGURE 1.13
Example of a system and its component.

We cannot understand the whole of any system by studying its parts. For example, if you want to understand how a car works and you take it apart and study each of its parts (engine, tires, a drive shaft, a carburetor, transmission, etc.), you would have no clue as to how a car works. To understand an automobile, you must study the relationship of the parts and how they work together. The holistic approach for explaining the system working is important.

> System thinking is a conceptual framework, a body of knowledge and tools, developed over the past many years, to explicitly underline the effects of emergent properties and visualize this implicit and explicit effect.

System thinking can be used to understand highly complex systems; it also can help us understand day-to-day issues. As an example, many of us would like to lose weight.

What is the nature of the system in which we find ourselves? Let us identify the variables in this system: unhappiness with weight, amount of food consumed, and degree of hunger. What is the relationship between these variables?

From Figure 1.14, one may understand the nature of systems thinking. How could this help in an everyday problem solving like dieting by seeing the structure of the underlying system.

Another example: Any company which finds labor costs are too high and wants to reduce them. So, they simply lay off 10% of their work force. Costs might go down right away but the workload on the remaining work force increases manyfold. Those workers feel stressed and cannot get all the work done. As a result, the company hires outside people to help them and reduce the workload.

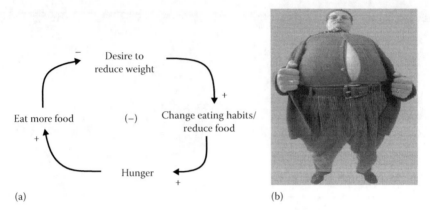

FIGURE 1.14
Relationship diagram for weight reduction.

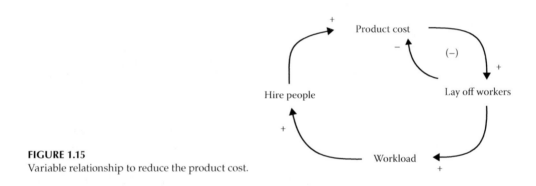

FIGURE 1.15
Variable relationship to reduce the product cost.

Step 1. The following variables could be identified.
  Product manufacturing cost
  Lay off workers
  Hire outside people
  Workload

Step 2. Find relationships between these variables shown in Figure 1.15.

Systems thinking is a wonderful tool for understanding the environment. It provides a visual tool for learning. It was not until I saw the picture that I began to understand some of the critical elements of thinking systemically. A systems picture can indeed be worth a thousand words!

1.13 Large and Complex Applied System Engineering: A Generic Modeling

In this century, we live in a large and complex world perceived to be a result of a variety of human activity systems from domains such as science, technology, economics, ecology, environment, management, psychology, sociology, anthropology, geography, philosophy, mathematics, arts, literature, ethics, spirituality, and so on.

The field of systems analysis and design is concerned with unstructured, multidisciplinary, or large-scale problems in industry and government that require skills and methods from more than one specific discipline. Identifying, modeling, and solving systems problems may require combinations of methods of mathematical, economic, behavioral, and engineering analysis. The focus may range from normative to empirical, from design to testing, and from analysis to evaluation. Specific real-world problems, for example, industrial, governmental, technical, information (computer) may require methods of acquiring information of the system and its components and environment, introducing changes (experimental interventions), and evaluating the effects (Thompson, 1991).

There seems to be a general agreement that information society (or postindustrial society) will be fundamentally different environment for organizations than was industrial society (Klir and Lowen, 1991). It is expected that the amount of available knowledge, the level of complexity, and the degree of turbulence will be significantly greater in the information society than they were in the industrial society. In addition, it is also expected that even the absolute growth rates of these three factors will be significantly greater than in the past. To keep organization compatible with the environment, a substantial portion of decision making in information society will be concerned with organizational innovations, that is, radical changes in produced goods or services, as well as in the technologies, processes, and structures of the organizations themselves. In general, demands on organizational innovations will be more frequent, more extensive, and will have to be implemented faster than in the past.

All these demands on organizations in information society indicate that organizations will be required to function as anticipatory systems and the associated problem of systems modeling and decision making is at the heart of systems science. Bahm (1981) has sketched five types of systems philosophy:

1. Atomism (the world is an aggregate of elements without wholes to be understood by analysis).
2. Holism (ultimate reality is a whole without parts, except as illusory manifestations, apprehended intuitively).
3. Emerqentism (parts exist together and their relations, connections, and organized interaction constitute whole that continue to depend upon them for their existence and nature, understood first analytically and then synthetically).
4. Structuralism (the universe is a whole within which all systems and their processes exist as depending parts, understanding can be aided by creative deduction).
5. Organicism (every existing system has both parts and whole, and is part of a larger whole, etc., understanding the nature of whole part polarities is a clue to understanding the nature of systems). These five types of systems philosophies have emphasized characteristics of existing systems. Philosophies of conceptual systems, of relations between conceptual and existing systems, and of methodologies may also correlate with them.

Many real-life problems are brought forth by present day technology and by societal and environmental processes, which are highly complex, "large" in dimension and stochastic or fuzzy in nature (Singh and Titlied, 1979; Jamshidi, 1983). One viewpoint as to how large is large, has been that a system is considered large in scale if it can be decoupled or partitioned into a number of interconnected subsystems for either computational or practical reasons. Another viewpoint considers large-scale systems to be simply those whose dimensions are so large that conventional techniques of modeling, analysis, control, and optimization fail to give reasonable solutions with reasonable computational efforts.

Many real problems are considered to be "large scale" by nature and not by choice. Two important attributes of large-scale systems are

1. They often represent complex real-life systems.
2. Their hierarchical multilevel and decentralized information structures depict systems dealing with societal, business, and management organizations, the economy, the environment, data networks, electric power, transportation, aerospace including space structures, water resources, energy, and so on.

What is going on in the dynamics of any enterprise is not merely the manipulation of material, energy (physical, human, solar) and information but a more fundamental aspect, namely, management of complexity which is measured by variety (the number of possible states of a system) (Beer, 1985; Satsangi, 1985). The basic axioms will assuredly hold, that the variety of the environment greatly exceeds that of the operation that serves or exploits it which will in turn greatly exceed the variety of the management that regulates or controls it. Hence, variety engineering (manipulation of varieties by design through attenuation and amplification) is invariably required to satisfy Ashby's law of requisite variety states that only variety can absorb variety. The essence of viability leads to the following principle of organization: managerial, operational, and environmental varieties, diffusing through an institutional system, tend to equate; they should be designed to do so with minimum damage to people and to cost. The hierarchical concept is so very central in systems science because it is strongly connected to other basic concepts such as complexity, synergetics, and autonomy (Auger, 1991).

Indeed, large-scale systems containing a large number of units and elements have a spontaneous tendency to subdivide into smaller subsystems, quasi-autonomous, themselves able to be divided into still smaller subsystems and so on. The subsystems behavior is mainly governed by intra subsystems interactions, the elements inside the same subsystems being often strongly interacting. This is the essence of the large-scale system modeling philosophy through sub-sub-system-to-sub-system-to-system modeling construct of the physical system theory (Satsangi and Ellis, 1971). In many examples and cases, the subsystems at each level are quasi-autonomous with stable internal processes responsible for realizing a given function. Autonomy and hierarchy are very connected and a hierarchical structure of the system implies a decomposition into many levels in which stable and autonomous units can be found. The subsystems are self-functioning with quasiautonomy and also can often be characterized in "upper level" by a few global variables. Thus, at each jump in upper level, these subsystems containing a large number of degrees of freedom (variety) are represented and described by a small number of global variables. As a consequence when integrating the levels, the complexity of the system is much reduced. There is a strong link between hierarchy and complexity. The complexity of large-scale system is reduced and in a certain way solved by hierarchical organization. Interpretive structural modeling, which transforms unclear, poorly articulated mental models of a system into visible well-defined, hierarchical models, assumes importance in this context (Warfield, 1974, 1990; Sage, 1977).

Moreover, internal dynamics of the subsystems are governed by intra-sub-systems interactions. When the whole system is rebuilt from its isolated sub-systems, inter-sub-systems interaction must be added. As a consequence, the behavior of the coupled subsystem is different from the sum of the behaviors of the isolated subsystems. The behavior of the whole system is different from the sum of the behaviors of the isolated parts, as in the case of Kron's tearing and reconstruction (Bowden, 1991) and physical system theory (Koenig et al., 1966). Two different forms of hierarchy theory have been frequently propounded (Salthe, 1991). One, the scalar hierarchy is used in systems science and ecology. It describes constraint relations on dynamics between systems of different scale. It also describes parts and wholes and processes taking place in such complex structures as described above. The other, the specification hierarchy is an older discourse dating back to Plato and as a theory of developments, to Aristotle. In the twentieth century it was relegated to peripheral importance in biology, but has continued to inform social science and psychology. It can be represented as a system of nested classes, the outermost class containing the most general phenomena, the inner most the most highly specified. These integrative levels can be considered as stages of development as well. Each new stage, transcends the one before it and integrates it into a new whole. Incredibly, the same statements can be made about a system from these two viewpoints even though their meanings are logically different. One often speaks of synergetic effects to represent cooperative effects between the parts. Synergetics and hierarchy seem to be strongly connected and inter-sub-system interactions are responsible for these synergetic couplings. A basic distinction is made between the lower levels of control and the higher level of management (Wilson, 1984; Satsangi, 1985). Thus

1. A process control system is a designed physical system (containing only inert physical elements) based on material flow and energy costs.
2. A management control system is a human activity system (containing autonomous human beings).

Metagame theory (Hipel and Fraser, 1980; Satsangi, 1985) is used to analyze and find stable solution to the problems that involve conflict between the strategies of the interested parties. This could be a war situation between two countries, management and labor problems, landlord and tenant disputes, national issues involving various political parties, and so on. Conflict is virtually inevitable in most situations where humans interact, either individually or else in group. Conflict resolution is required by all decision makers.

Numbers just do not tell the whole story anymore in systems such as economic, urban, biological, educational, and disaster management, which are basically human-based systems. The hard and fast is making room for the soft and fuzzy. Zadeh has argued that the applications of the techniques and tools of "hard" systems to the study of systems involving humans are often unsuccessful primarily because the methodologies used demand very high levels of precision in measuring the variables and parameters used in describing the system. Zadeh stressed that these levels of precision were often neither attainable nor, more importantly, required for effective analysis of these systems. To overcome the need of undue precision, he introduced the concepts of fuzzy set theory (Zadeh, 1965; Satsangi, 1985). Accidents like the one at Three Mile Island a few years ago show the weakness of the classical approach. There engineers had rigorously measured the risks, set up fault trees, calculated every possible combination of human factors. However human error was not sufficiently figured in. Just as the human eye can immediately read a message scrawled sloppily on a note pad making a dazzling set of associations and interfaces from context that the most of sophisticated computer scanner is helpless to emulate, so people who understand other people know best how human error at Three Mile Island, Chernobyl, Bhopal Gas Leak, and Uttarkashi earthquake can be avoided next time. While a statistician might ask the likelihood based on the occurrences in comparable situations, a fuzzy theorist might gather a variety of opinions and viewpoints and predict something the statistics cannot show. A key difference lies in the acknowledgment by fuzzy mathematics that society is not a series of random occurrences following the laws of chance. Rather, they maintain, it is the sometimes surprising outcome of improvising and purposeful, albeit a little fuzzy, people trying to beat the odds.

The development of qualitative models of physical systems is currently attracting much interest from the artificial intelligence research community. It consists of a set of eclectic techniques designed to generate a qualitative description of the behavior of physical system from a description of its structure and some initial "disturbance." The term "qualitative" has been used in many ways and generally to mean "nonnumerical," "symbolic models." Qualitative modeling is used here to mean reasoning about systems characterized by continuously changing variables of time by discrete abstractions of the values of such variables that allow identifying the important distinctions or landmarks, in the behavior to be computed. This requires quantization of the real number line into a finite set of distinctions for particular generic tasks. Abstract descriptions of state make it possible to have more concise representations of behavior. However, the generation of the behavior from such description tends not to produce a unique solution. This of course, is to be expected, as the information required to produce a unique description has been eliminated in the intentional abstraction. Therefore, qualitative models produce ambiguous description of behavior. However, such ambiguous behavior can still contain useful information for some tasks. For example, if it is required to predict whether the current state can lead to a critical or faulty condition, it may be sufficient to show that none of the possible behaviors leads to a critical situation. It is important to show, therefore, that the set of possible behaviors includes the actual behavior of the system. In this way, a task can be satisfied even with incomplete descriptions, whereas in traditional method, all of the information needs to be available and it needs to be precisely and uniquely characterized before any inference can be made (Leitch, 1989).

Intelligent knowledge-based expert systems, knowledge-based expert systems, or, simply, expert systems are a product of artificial intelligence (AI). Artificial intelligence that branch of computer science which deals with development of programs that exhibit intelligent human behavior to arrive at decisions in a complex environment. In essence, an expert system can be defined as a sophisticated computer program that is designed to replicate the expertise of humans in diverse fields. The expertise to solve a problem depends upon the available knowledge (Zadeh, 1983; Satsangi, 1985; Kalra and Batra, 1990; Kalra and Srivastav, 1990).

Analogical reasoning is paradigm for problem solving within the field of AI (Sage, 1990). A system which employs analogical reasoning operates by transferring knowledge from past problem-solving cases to new problems that are similar to the past cases. The past cases known to the system are referred to as analogs. Several approaches to analogical reasoning include associative, distributed, and connectionist, or neural network (Sage, 1990; Kumar and Satsangi, 1992; Chaturvedi, 2008). A connectionist system is a network of a very large number of simple processors, usually called units or "neurons," which are highly interconnected and operate in parallel. Each unit has a numeric activation value, which is communicated to other units along connections of varying strengths. As the network operates, the activation value of each unit continuously changes in response to the activity of the units to which it is connected. This process is called spreading activation and is the fundamental mechanism underlying the operation of all connectionist networks. All the knowledge held in such a network is stored in the numerical strengths of the connections between units. There are two major types of connectionist representations, localist and distributed. A future challenge for a logical reasoning system is to integrate in some manner the necessary parallel techniques including those discussed above, into a cooperative problem-solving system.

Software system engineering is the application of system engineering principles, activities, tasks, and procedures to the development of a software in a computer-based system. This application is the overall concept that integrates the managerial and technical activity that controls the cost, schedule, and technical achievement of the developing software system throughout its lifecycle (Thayer, 1990).

Applied systems engineering is rather a new dimension in science and engineering. According to this new dimension, systems are recognized, classified, and dealt with in terms of their structural properties while the nature of the entities is that these properties are de-emphasized. Such an alternative point of view transcends the artificial boundaries between the traditional disciplines of science and engineering and makes it possible to develop a genuine cross-disciplinary methodology more adequate for dealing with large and complex socio-technological problems inherent in the information society.

To conclude, in a rapidly changing world, the most efficient planning and management strategies are likely to be soft planning approaches, not attempting to determine the future completely, but to steer the whole system toward basic mode of desirable behavior and allowing the system itself to adjust in its minor aspects, according to its own organization and dynamics. Such planning approach should include a continuous monitoring of the most important variables with early detection of tendencies of the system to move toward undesirable behavior modes. It should also emphasize disperse, diffuse, and loose controls, rather than tight, concentrated ones, and explicitly favor development of the generalized intelligent capability of the system to react to new situations, thus increasing rather than reducing the future degrees of freedom. This requires not only technical capability, but more important an applied system way of thinking, which is more holistic, more interdisciplinary and capable of dealing with the behavior and characteristics of incompletely known complex systems (Satsangi, 1985, 2006).

1.14 Review Questions

1. Define system.
2. What are the salient aspects of systems approach first articulated by Ludwig Von Bertalanffy.
3. Distinguish between feedback control and feedforward control systems.
4. a. Give a classification of systems.
   b. What is the difference between a set and a system?
   c. What do you understand by "whole is more than sum of its parts?"
5. Explain 3Rs of science.
6. What do you mean by the complexity of systems?
7. Differentiate the linear and nonlinear systems.
8. Explain the natural systems and manmade systems.
9. Explain in brief about system thinking.
10. Differentiate between hard and soft systems.

1.15 Bibliographical Notes

Basics of systems and systems theory are explained by Forrester (1968), Padulo and Arbib (1974), and Roe (2009). Kenneth (1994) defines systems in terms of conceptual, concrete, and abstract systems either isolated, closed, or open. Walter (1967) defines social systems in sociology in terms of mechanical, organic, and process models. Klir (1969) well classified the systems and defines systems in terms of abstract, real, and conceptual physical systems, bounded and unbounded systems, discrete to continuous, pulse to hybrid systems, etc. Important distinctions have also been made between hard and soft systems in Sterman (2000). Hard systems are associated with areas such as systems engineering, operations research, and quantitative systems analysis. Soft systems are commonly associated with concepts involving methods such as action research and emphasizing participatory designs. Distinction between hard and soft systems is given in Zadeh (1969).

It is always intended to integrate in a formal mode nature and culture in a single cosmos of which the inner purity and unity are simultaneously ensured (Pouvreau and Drack, 2007).

2

Systems Modeling

Art is the lie that helps us to see the truth.

Picasso

The sciences do not try to explain, they hardly even try to interpret, they mainly make models. By a model is meant a mathematical construct which, with the addition of certain verbal interpretations, describes observed phenomena. The justification of such a mathematical construct is solely and precisely that it is expected to work.

John Von Neumann

Physical models are as different from the world as a geographical map is from the surface of earth.

L. Brillouin

2.1 Introduction

Why is modeling required? Because ... modeling may be quite useful:

1. To find the height of a tower, say the *Kutub Minar* of Delhi or the Leaning Tower at Pisa without actually climbing it
2. To measure the width of a river without actually crossing it
3. To gauge the mass of the Earth, not using any balance

4. To find the temperature at the surface or at the centre of the sun
5. To estimate the yield of wheat in India from the standing crop
6. To quantify the amount of blood inside a living human body
7. To predict the population of China for the year 2050
8. To determine the time required by a satellite to complete one orbit around the earth, say at the height of about 10,000 km above the ground
9. To assess the impact of 30% reduction in income tax over the national economy
10. To ascertain the optimally efficient gun whose performance depends on 10 parameters, each of which can take 10 different values, without actually manufacturing 10^{10} guns
11. To determine the mean time between failures (MTBF) or average life span of an electric bulb
12. To forecast the total amount of insurance claims a company has to pay next year

Similarly, for a given physical, biological, or social problem, we may first develop a mathematical model for it, and then solve the model, and interpret its solution with respect to the problem statement.

Man has been modeling and simulating ever since his brain developed power to image. Children start modeling from birth. We are all simulating—like a child with a doll, an architect with a model, and a business man with a business plan, etc.

What is modeling?

Modeling is a process of abstraction of a real system. A model portrays a conceptual framework to describe a system and can be viewed as an abstraction (essence) of an actual system or a physical replica of a system or a situation. It is a factual representation of reality. The word model is derived from Latin and its meaning is mould or pattern (physical model).

The abstracted model may be logical or mathematical. A mathematical model is a mathematical description of properties and interactions in the system. The development of a mathematical model depends on the system boundary, system components, and their interactions. It also depends upon the type of analysis that we want to perform, like steady state or transient analysis and the assumptions that we will consider while model development. If assumptions are more then the model will be simpler, but the accuracy of the response of the model would be less. If there are fewer assumptions, the model will be complex and the accuracy will be better. Hence, during model development, it is necessary to optimize two things:

1. Simplicity of the model
2. Accuracy of the model or faithfulness of model

We know that the accuracy of a model is complementary to its simplicity.

Often, when engineers analyze a system or are supposed to control a system, they use a mathematical model. In the analysis, an engineer can build a descriptive model of the system as a hypothesis of how the system would work, or try to estimate how an unforeseeable event could affect the system. Similarly, in control of a system the engineer can try out different control approaches in simulations.

A mathematical model is a mathematical description of properties and interactions in the system. Mathematical modeling is the use of mathematical language to describe the behavior of a system, be it biological, economic, electrical, mechanical, thermodynamic, or one of many other examples.

A mathematical model usually describes a system by means of variables. The values of the variables can be practically anything; real or integer numbers, Boolean values, strings, etc. The variables represent some properties of the system, for example, measured system outputs often in the form of signals, timing data, counters, and event occurrence (yes/no).

The actual model is the set of functions that describe the relations between the different variables. Mathematical modeling problems are often classified into white-box or black-box models, according to how much prior information is available for the system. A black-box model is a system of which there is no prior information available, and a white-box model is a system where all necessary information is available. Practically, all systems fall somewhere in between the lines of white-box and black-box models, so this concept only works as an intuitive guide for approach.

Usually it is preferable to use more a priori information as possible to make the model more accurate. Therefore white-box models are usually considered easier, because if you have used the information correctly, then the model will behave correctly. Often prior information comes in the form of knowing the type of functional relationship between different variables. For example, if we make a model of how a medicine works in a human system, we know that usually the amount of medicine in the blood is an exponentially decaying function. But we are still left with several unknown parameters: how rapidly does the medicine amount decay, and what is the initial amount of medicine in blood? This example is therefore not a completely white-box model. These parameters have to be estimated through some means before one can use the model.

In black-box models one tries to estimate both the functional relationship between variables and the numerical parameters in those functions. Using prior information we could end up, for example, with a set of functions that probably could describe the system adequately. If there is no a priori information, we would try to use functions as general as possible to cover all different models. An often used approach for black-box models are artificial neural networks (ANNs), which usually do not need anything except the input and output data sets. ANN models are good for complex systems, especially when input–output patterns known to us are in quantitative form. If the input and output information are not in quantitative form, but in qualitative or fuzzy form, then ANN cannot be used. For such situations fuzzy models are good.

2.2 Need of System Modeling

Models are used to mimic the behavior of systems under different operating conditions. This may also be done with the help of experimentation on the system. But, sometimes it is inappropriate or impossible to do experiments on real systems due to the following reasons.

1. *Too expensive*: Experimenting with a real system is an extremely costly affair. For example, the physical experimentation of a complex system like the satellite system is quite expensive and time consuming.
2. *Risky*: Risk involved in experimentation is another factor. In some systems there is a risk of damaging the system, or a risk of life. For example, training a person for operating the nuclear plant in a dangerous situation would be inappropriate and life threatening.

Modeling is an essential requirement in certain situations, such as the following:

1. *Abstract specifications of the essential features of a system*: When a system does not exist and a designer wants to design a new system like a missile or an airplane. The model will help in knowing, prior to the development of the system, how that system will work for different environmental conditions and inputs.
2. *Modeling forces us to think clearly before making a physical model*: One has to be clear about the structure and the essentials of the situation.
3. *To guide the thought process*: It helps in refining ideas or decisions before implementing it in the real world.
4. It is a tool that improves the understanding about a system, and allows us to demonstrate and interact with what we design, and not just describe it.
5. *To improve system performance*: Models will help in changing the system structure to improve its performance.
6. *To explore the multiple solutions economically*: It also allows us to find many alternate solutions for the improvement in system performance.
7. To create virtual environments for training purpose or entertainment purposes.

2.3 Modeling Methods for Complex Systems

It is possible to acquire an almost white-box model of a fighter jet, by modeling it with every mechanical part of such a jet embedded into the model. However, the computational cost of adding such a huge amount of detail would effectively inhibit the usage of such a model. Additionally, the uncertainty would increase due to an overly complex system, because each separate part induces some amount of variance into the model. It is therefore usually appropriate to make some approximations to reduce the model to a sensible size. Engineers often can accept some approximations in order to get a more robust and simple model. For example, Newton's classical mechanics is only an approximated model of the real world. Still, Newton's model is quite sufficient for most ordinary-life situations, that is, as long as particle speeds are well below the speed of light, and as long as we study macroparticles only.

Mathematical models of such systems are most accurate and precise, but can handle the system complexity only up to certain limit. Simple systems are easy to model mathematically. As the system's complexity increases, mathematical model development becomes quite cumbersome. At the same time, it is also difficult and time consuming to simulate complex system models. In such situations, ANN models are better in comparison to mathematical models. As it is evident from literature that for good ANN model development, it is necessary to have accurate and sufficient training data, and this is really difficult for real-life problems. Most of the real-life problems have qualitative information, which is either difficult or impossible to translate into quantitative form. Hence, fuzzy modeling is the only option for such circumstances. The modeling using fuzzy logic is quite useful for highly complex systems as shown

Systems Modeling

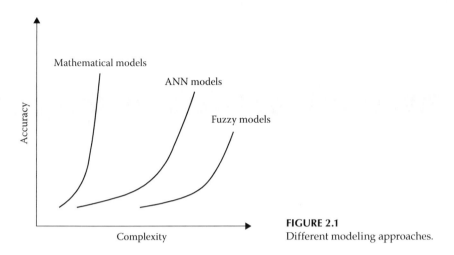

FIGURE 2.1
Different modeling approaches.

in Figure 2.1. Hence, according to the complexity of systems, the following modeling techniques may be used:

1. Less complex systems—Mathematical modeling techniques.
2. Medium complex systems—ANN modeling technique.
3. Highly complex systems—Fuzzy systems modeling technique.

2.4 Classification of Models

Models have been widely accepted as a means for studying complex phenomena for experimental investigations at a lower cost and in less time, than trying changes in actual systems. Knowledge can be obtained more quickly, and for conditions not observable in real life. Models tell us about our ignorance and give better insights into the system. System models may be classified as shown in Figure 2.2.

2.4.1 Physical vs. Abstract Model

To most people, the word "model" evokes images of clay cars in wind tunnels, cockpits disconnected from their airplanes to be used in pilot training, or miniature supertankers scurrying about in a swimming pool. These are examples of physical models (also called iconic models), and are not typical of the kinds of models that are of interest in operations research and system analysis. Physical models are most easily understood. They are usually physical replicas, often on a reduced scale. Dynamic physical models are used as in wind tunnels to show the aerodynamic characteristics of proposed aircraft designs. Occasionally, however, it has been found useful to build physical models to study engineering or management systems; examples include tabletop scale models of material-handling systems, and in at least one case a full-scale physical model of a fast food restaurant inside a warehouse, complete with full-scale, and, presumably hungry humans. But the vast majority of models built for such purposes are abstracted,

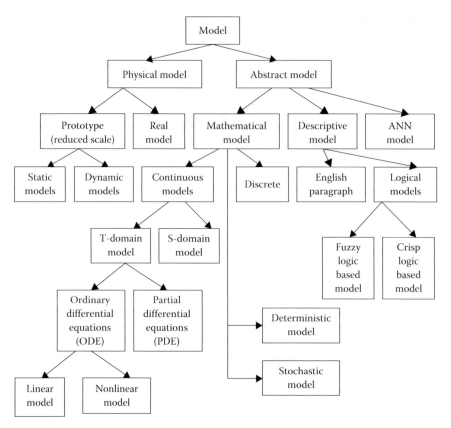

FIGURE 2.2
Pictorial representation of the classification of models.

representing a system in terms of logical or quantitative relationships that are then manipulated and changed to see how the model reacts, and thus, how the system would react—if the abstract model is a valid one. An abstract model is one in which symbols, rather than physical devices, constitute the model. The abstract model is more common but less recognized. The symbolism used can be a written language or a thought process.

2.4.2 Mathematical vs. Descriptive Model

A mathematical model is a special subdivision of abstract models. The mathematical model is written in the language of mathematical symbols. Perhaps the simplest example of an abstracted mathematical model is the familiar relation

$$\text{Distance} = \text{Acceleration} \times \text{Time}$$

$$d = a * t \qquad (2.1)$$

This might provide a valid model in one instance (e.g., a space probe to another planet after it has attained its flight velocity) but a very poor model for other purposes (e.g., rush-hour commuting on congested urban freeways).

2.4.3 Static vs. Dynamic Model

Static models are quite common for architectural works to visualize the floor plane. A static simulation model is a representation of a system at a particular time, or one that may be used to represent a system in which time simply plays no role. On the other hand, a dynamic simulation model represents a system as it evolves over time, such as a conveyor system in a factory. A dynamic model deals with time-varying interactions.

2.4.4 Steady State vs. Transient Model

A steady state pattern is one that is representative with time and in which the behavior in one time period is of the same nature as any other period.

Transient behavior describes those changes where the system response changes with time. A system that exhibits growth would show transient behavior, as it is a "one-time" phenomena, and cannot be repeated.

2.4.5 Open vs. Feedback Model

The distinction is not as clear as the word suggests. Different degrees of openness can exist. The closed model is one that internally generates the values of variables through time by the interaction of variables one on another. The closed model can exhibit interesting and informative behavior without receiving an input variable from an external source. Information feed back systems are essentially closed systems.

2.4.6 Deterministic vs. Stochastic Models

If a simulation model does not contain any probabilistic (i.e., random) components, it is called deterministic; a complicated (and analytically intractable) system of differential equations describing a chemical reaction might be such a model. In deterministic models, the output is "determined" once the set of input quantities and relationships in the model have been specified; even though it might take a lot of computer time to evaluate what it is. Many systems, however, must be modeled as having at least some random input components; and these give rise to stochastic simulation models. Most queuing and inventory systems are modeled stochastically. Stochastic simulation models produce an output that is by itself random, and must therefore be treated as only an estimate of the true characteristics of the model. This is one of the main disadvantages of simulation.

2.4.7 Continuous vs. Discrete Models

Loosely speaking, we define discrete and continuous simulation models analogously to the way discrete and continuous systems were defined. It should be mentioned that a discrete model is not always used to model a discrete system and vice versa. The decision whether to use a discrete or a continuous model for a particular system depends on the specific objectives of the study. For example, a model of traffic flow on a freeway would be discrete if the characteristics and movement of individual cars are important. Alternatively, if the cars can be treated "in the aggregate," the flow of traffic can be described by differential equations in a continuous model. The continuous and discrete functions are shown in Figure 2.3.

FIGURE 2.3
(A) Continuous and (B) discrete functions.

2.5 Characteristics of Models

Model	Descriptive Capability	Ambiguity	Manipulation Capability	Implementation Capability	Primary Function
English text	Good	Very ambiguous	None	Limited	Descriptive explanation and directions
Drawings and block diagrams	Good	Not ambiguous	None	Good	Design, assembly and construction
Logical flow charts and decision tables	Fair	Not ambiguous	None	Good	Computer programming
Curves, tables monographs	Fair	Not ambiguous	Good	None	Express simple relations between a few variables
Mathematical	Poor	Not ambiguous	Excellent	Good	Problem solution and optimization

2.6 Modeling

Modeling is the art/process of developing a system model. The purpose of modeling a system is to expose its internal working and to present it in a form useful to science and engineering studies. In other words, modeling means the process of organizing knowledge about a given system. Various inputs required for model development are shown in Figure 2.4a, and the methodology of modeling a system is shown in Figure 2.4b. For the same system we may develop different models depending upon the purpose and an analyst's viewpoint. Consider an aircraft shown in Figure 2.5, which could be modeled as

1. A particle
2. A system of rigid bodies
3. A system of deformable bodies

and the choice depends on the viewpoint of analysts as follows:

1. If the analyst is interested in the trajectory of flight to find the fuel consumption, then the particle model of aircraft is good, simple, and sufficient.

Systems Modeling

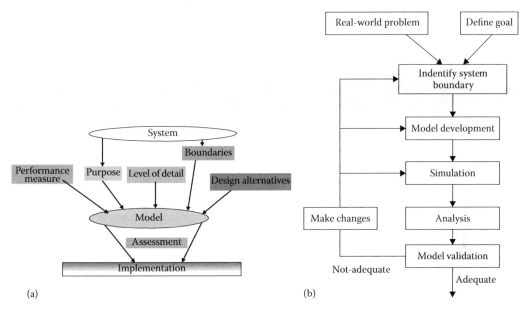

FIGURE 2.4
(a) Inputs for model development. (b) Modeling process.

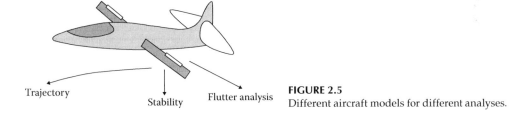

FIGURE 2.5
Different aircraft models for different analyses.

2. When the analyst is interested in flight stability, i.e., aircraft behavior for small disturbances, then the aircraft is considered as a rigid body system.
3. Finally, when performing flutter analysis, i.e., determining the so-called critical speed of flutter, the deformable body system is a good model.

The very first step in modeling is to identify a system (i.e., collection of components) and interaction among components. System boundary affects the system model.

Assumptions made during modeling also affect the system model. More assumptions increase the simplicity of the model by reducing complexity but at the same time these also reduce the accuracy of the system model.

So there is a trade off between the simplicity, accuracy and computation time. Accuracy tells about the faithfulness of the model. The degree of faithfulness implies that up to what extent the system is accurate.

2.6.1 Fundamental Axiom (Modeling Hypothesis)

Mathematical model of a component characterizes it's behavior as an independent entity of a system, and how it is interconnected with the other components to form a system.

It implies that the various components can be removed either literally or conceptually from the remaining components and can be studied in isolation to establish a model of their characteristics. This is a tool of science which makes the system theory universal.

Analysts can go as far as they wish in breaking down the system in search of building blocks that are sufficiently simple to model and which identify a structure upon which alteration can be made.

2.6.2 Component Postulate (First Postulate)

The pertinent performance (behavior) characteristics of each n terminal component in an identified system structure can be specified completely by a set of $(n - 1)$ equations in $(n - 1)$ pairs of oriented complementary variables (i.e., across variables $x_i(t)$ and through variables $y_i(t)$) identified by n arbitrarily chosen terminal graph as shown in Figure 2.6. The variables x_i and y_i may be vectors if necessary. The complementary pair of variables for different systems are shown in Table 2.1.

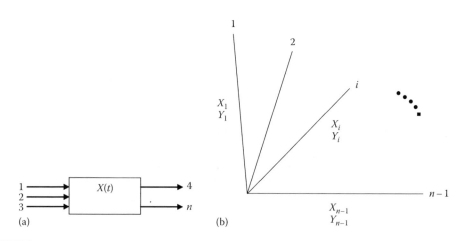

FIGURE 2.6
(a) n-terminal component and (b) terminal graph of n-terminal component.

TABLE 2.1

Complimentary Pair of Variables for Different Systems

	Physical Systems			Conceptual Systems		Esoteric Systems
	Mechanical System	Hydraulic System	Electrical System	Transportation System	Economic System	Spiritual System
Across Variables X_i	Velocity (linear, angular)	Pressure	Voltage	Traffic density	Unit price	Potential
Through Variables Y_i	Force, Torque	Flow-rate of liquid	Current	Traffic flow rate	Flow-rate of goods	Current

2.6.3 Model Evaluation

An important part of the modeling process is the evaluation of an acquired model. *How do we know if a mathematical model describes the system well?* This is not an easy question to answer. Usually the engineer has a set of measurements from the system which is used while creating the model. If the model was built well, the model will adequately show the relations between system variables for the measurements at hand. The question then becomes: *How do we know that the measured data is a representative set of possible values?* Does the model describe the properties of the system between the measured data well (interpolation)? Does the model describe events outside the measured data well (extrapolation)? A common approach is to split the measured data into two parts; training data and verification data. The training data is used to train the model, that is, to estimate the model parameters. The verification data is used to evaluate model performance. Assuming that the training data and verification data are not the same, we can assume that if the model describes the verification data well, then the model will describe the real system well.

However, this still leaves the extrapolation question open. How well does this model describe events outside the measured data? Consider again Newtonian classical mechanics model. Newton made his measurements without advanced equipment, so he could not measure properties of particles traveling at speeds close to the speed of light. Likewise, he did not measure the movements of molecules and other small particles, but macroparticles only. It is then not surprising that his model does not extrapolate well into these domains, even though his model is quite sufficient for ordinary-life physics.

2.6.4 Generic Description of Two-Terminal Components

The mathematical model of the components identified in a system structure serves as a building block in the analysis and design of physical systems. These mathematical models must be established from empirical tests on the components or calculated from constructional features of the components such as their geometric dimensions and material composition.

In passive element the direction change does not cause changes in the direction terminal equation, but in active elements terminal equations are changed with direction.

The fundamental two-terminal components may be classified as

A-type elements: dissipater or algebraic components
B-type elements: delay type components
C-type elements: accumulator or capacitive type components
D-type elements: drivers or sources as shown in Table 2.2

2.6.4.1 Dissipater Type Components

These are the components in which power loss takes place and the terminal equation may be written in algebraic form explicitly with across variables on one side and through variables on another side of the equation.

The terminal equation for A-type components may be written in impedance form as

$$\text{Across variable} = \text{Impedance} * \text{Through variable}$$

$$x(t) = ay(t) \tag{2.2}$$

TABLE 2.2

Two-Terminal Components for Different Systems

Systems \ Components	Dissipater Type $x(t) = ay(t)$	Delay Type $x(t) = b\dfrac{dy(t)}{dt}$	Accumulator Type $y(t) = c\dfrac{dx(t)}{dt}$	Source or Driver $x(t)$ or $y(t)$ Specified
Electrical	Resistor	Inductor	Capacitor	Voltage or current drivers
Mechanical (translatory motion)	Damper	Spring	Mass	Velocity or force
Mechanical (rotary motion)	Rotary damper	Torsional spring	Inertia	Angular velocity or torque
Hydraulic	Pipe resistance	Inertial delay component	Liquid tank	Pressure or flow rate

or in admittance form as

$$\text{Through variable} = \text{Admittance} * \text{Across variable}$$

$$y(t) = \frac{1}{a} x(t) \tag{2.3}$$

In the impedance form of a terminal equation the across variable is dependent on the through variable, and vice versa in the admittance form.

2.6.4.2 Delay Type Elements

The terminal equation for delay type components can be written as

$$y(t) = \frac{1}{b}\int_{t_0}^{t} x(t) \quad \text{or} \quad x(t) = b\frac{dy(t)}{dt} \tag{2.4}$$

Power and energy can be respectively shown for a delay type component as

$$P = x(t) \cdot y(t)$$

$$= b\frac{dy}{dt} \cdot y(t) \tag{2.5}$$

$$e = \int_{t_0}^{t} P(t)$$

$$= \frac{b}{2}\left[y^2(t) - y^2(t_0)\right]$$

The energy or average power for infinite time is zero for delay type components. Therefore these components are called nondissipative type elements.

2.6.4.3 Accumulator Type

The terminal equation for accumulator or storage type components may be written as

$$x(t) = \frac{1}{C}\int_{t_0}^{t} y(t) \quad \text{or} \quad y(t) = C\frac{dx(t)}{dt} \tag{2.6}$$

Similar to the delay type components the energy or average power over infinite time is zero for accumulator type components. Therefore, these components are also called nondissipative type elements.

2.6.4.4 Sources or Drivers

The drivers or sources are those, whose across variable or through variable is specified, and the different operating conditions have no effect on these variables. The terminal equations for the drivers may be written as

$x(t)$ = specified for across drivers
$y(t)$ = specified for through drivers

Ideal across driver: The magnitude is perfectly specified and it will not change whatever be the value of the through variable.

Ideal through driver: The magnitude of the through variable is perfectly specified and it will not change whatever be the value of the across variable—but, practically it is not possible to manufacture any source whose value will be unaffected by the operating conditions. The ideal sources are only used for theoretical applications.

2.7 Mathematical Modeling of Physical Systems

The task of mathematical modeling is an important step in the analysis and design of physical systems. In this chapter, we will develop mathematical models for the mechanical, electrical, hydraulic, and thermal systems. The mathematical models of systems are obtained by applying the fundamental physical laws governing the nature of the components making these systems. For example, Newton's laws are used in the mathematical modeling of mechanical systems. Similarly, Kirchhoff's laws are used in the modeling and analysis of electrical systems.

Our mathematical treatment will be limited to linear, time-invariant ordinary differential equations whose coefficients do not change in time. In real life many systems are nonlinear, but they can be linearized around certain operating ranges about their equilibrium conditions.

Real systems are usually quite complex and exact analysis is often impossible. We shall make approximations and reduce the system components to idealized versions whose behaviors are similar to the real components.

In this chapter we shall look only at the passive components. These components are of two types: those storing energy, e.g., the accumulator type or delay type system components, and those dissipating energy, e.g., the dissipater type components.

The mathematical model of a system is one which comprises of one or more differential equations describing the dynamic behavior of the system. The Laplace transformation is applied to the mathematical model and then the model is converted into an algebraic equation. The properties and behavior of the system can then be represented as a block diagram, with the transfer function of each component describing the relationship between its input and output behavior as shown in Figure 2.6a.

We can use state-space model for all types of systems that consist of state variables.

State

State refers to the past, present, and future condition of the system from a mathematical cell. State could be defined as a set of state variables and state equations to model the dynamic system. All the state equations are first-order differential equations.

State variables

It is defined as the minimal set of variables $[x_1(t), x_2(t),...,x_n(t)]$ such that the knowledge of these variables at any time $t=0$ and information on the input excitation subsequently applied are sufficient to determine the state of the system at time $t > t_0$. One is likely to confuse state variables with output variables. Output variables can be measured but state variables do not always satisfies this condition.

The variable $U(t)$ is controllable at all time instants $t > t_0$. The input $U(t)$ is taken from the input space. The output variable $Z(t)$ is observable at all time instants $t > t_0$, but there is no direct control on the output variables. $Z(t)$ is taken from the universe of output Z.

State equation

$$\dot{X}(t) = AX(t) + BU(t) \tag{2.7}$$

Output equation

$$Z(t) = CX(t) + DU(t) \tag{2.8}$$

Example 2.1

Consider a tank of volume V which is full of a solution of a material A at concentration C. A solution of the same material at concentration C_0 is flowing into the tank at flow rate F_0 and a solution is flowing out the top of the tank at flow rate F_1 as shown in Figure 2.7.
Determine the dynamic response to a step change in the inlet concentration C_0.

Solution

The following assumptions are to be considered and some data gathered for modeling the above mentioned hydraulic system.

1. Assumptions:
   Well mixed solution
   Density of solution is constant
   Level is constant in the tank

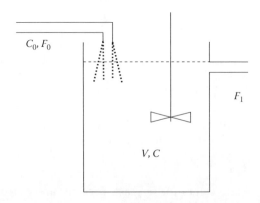

FIGURE 2.7
Hydraulic system.

Systems Modeling

2. Data:
$F_0 = 0.085 \, \text{m}^3/\text{min}, \, V = 2.1 \, \text{m}^3$
$C_{init} = 0.925 \, \text{kg/m}^3 \quad t <= 0$
$C_0 = 1.85 \, \text{kg/m}^3 \quad t > 0$

The system is initially at steady state i.e.,

$$C(t) = C_0(t) = C_{init} \quad \text{for } t <= 0$$

To formulate the model the following fundamental principles are used:

- Conservation laws
- Component terminal equation

Conservation laws are the laws for conservation of material, energy, momentum, and electrons, which have the general form

$$\text{Rate of accumulation} = (\text{rate})_{in} - (\text{rate})_{out}$$

Component terminal equations relate quantities of different kinds, e.g., Hooke's law for a spring: $F = kx$ relates the spring force F to the spring's displacement x, where k is the spring constant. Similarly, Ohm's law for a resistor $V = Ri$, where V is the voltage drop, i is the current, and R is the resistance.

Note these are only approximate relations. For a real spring or resistor the relation holds well for a certain range only; outside this range the relation is nonlinear.

Important variables/constants

Input variables: C_0, F_0
State Variables: C, F_1
Constants: V, ρ
Initial value C_{int}

Rate equation for the flow of fluid is

$$\text{Rate of change of mass} = \text{flow rate in} - \text{flow rate out}$$

$$\frac{d}{dt}(\rho V) = \rho F_0 - \rho F_1$$

$$\frac{dV}{dt} = 0 \quad \text{since } \rho \text{ is constant.}$$

$$\rho F_0 - \rho F_1 = 0$$

$$F_0 = F_1 = F$$

$$\frac{d}{dt}(CV) = C_0 F_0 - C F_1$$

$$V\frac{d}{dt}C = F(C_0 - C)$$

$$\frac{dC}{dt} = \frac{F}{V}C_0 - \frac{F}{V}C \qquad (2.9)$$

The general solution of the nonhomogeneous equation (Equation 2.9) is

$$C(t) = \overline{C}_0 + ke^{-\frac{F}{V}t}$$

Initially at $t = 0$

$$C(0) = \overline{C}_0 + k = C_{init} \rightarrow k = C_{init} - \overline{C}_0$$

Therefore, $C(t) = \overline{C}_0 + (C_{init} - \overline{C}_0)e^{-\frac{F}{V}t}$ when $t \geq 0$

MATLAB® program

```
% Computer program for simulating hydraulic system
% Initialization
clear all;
F=0.085;      % cubic m/min
V=2.1;        % cubic m
C0=1.85;      % kg/cubic m
C=0.925;      % kg/cubic m
t=0;
dt=0.01;
for n=1:10000
          X1(n,:)=[C t];
          dC=F*(C0-C)/V;
          C=C+dt*dC;
          t=t+dt;
end
plot(X1(:,2),X1(:,1))
xlabel('Time (sec.)')
ylabel('C')
```

The system response will be as shown in Figure 2.8.

2.7.1 Modeling of Mechanical Systems

Models of mechanical systems are important in control engineering because a mechanical system may be a vehicle, a robotic arm, a missile, or any other system which incorporates a mechanical component. Mechanical systems can be divided into two categories: translational systems and rotational systems. Some systems may be either purely translational or purely rotational, whereas others may be hybrid, incorporating both translational and rotational components.

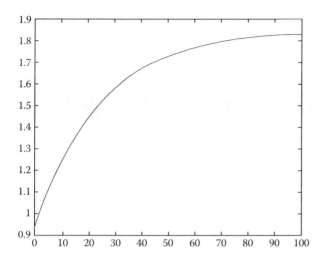

FIGURE 2.8
Dynamic response of tank system.

2.7.1.1 Translational Mechanical Systems

The basic building blocks of translational mechanical systems are mass, spring, and dashpot (Figure 2.9). The input to a translational mechanical system may be a force F and the output the displacement y.

Springs

Springs store energy as shown in Figure 2.9, and are used in most mechanical systems as delay element. As shown in Figure 2.10 some springs are hard, some are soft, and some are linear. A hard or a soft spring can be linearized for small deviations from its equilibrium condition. In the analysis in this section, a spring is assumed to be massless, or of negligible mass, i.e., the forces at both ends of the spring are assumed to be equal in magnitude but opposite in direction. The springs in different states such as compression, tension, and torsion are shown in Figure 2.9a. For a linear spring, the extension y is proportional to the applied force F as given in equation

$$F = ky \tag{2.10}$$

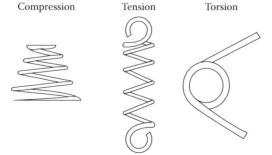

FIGURE 2.9
Springs under different types of forces.

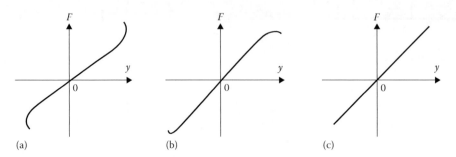

FIGURE 2.10
(a) Hard spring, (b) Soft spring, (c) Linear spring.

where k is known as the stiffness constant. The spring when stretched stores energy E, given by

$$E = \frac{1}{2}ky^2 \qquad (2.11)$$

This energy is released when the spring contracts back to its original length.

In some applications springs can be in parallel or in series. When n springs are in parallel, then the equivalent stiffness constant k_{eq} is equal to the sum of all the individual spring stiffness k_i:

$$k_{eq} = k_1 + k_2 + \cdots + k_n \qquad (2.12)$$

Similarly, when n springs are in series, then the reciprocal of the equivalent stiffness constant k_{eq} is equal to the sum of all the reciprocals of the individual spring stiffness k_i:

$$\frac{1}{k_{eq}} = \frac{1}{k_1} + \frac{1}{k_2} + \cdots + \frac{1}{k_n} \qquad (2.13)$$

Dashpot

A dashpot element is a form of damping and can be considered to be represented by a piston moving in a viscous medium in a cylinder. As the piston moves the liquid passes through the edges of the piston, damping the movement of the piston. The force F which moves the piston is proportional to the velocity of the piston movement and is given by

$$F = b\frac{dy}{dt} \qquad (2.14)$$

A dashpot does not store energy.

Mass

When a force is applied to a mass, the relationship between the force F and the acceleration of the mass is given by Newton's second law as $F = ma$. Since acceleration is the rate of change of velocity and the velocity is the rate of change of displacement,

$$F = m\frac{d^2y}{dt^2} \qquad (2.15)$$

Systems Modeling

The energy stored in a mass when it is moving is the kinetic energy which is dependent on the velocity of the mass and is given by

$$E = \frac{1}{2}mv^2 \qquad (2.16)$$

This energy is released when the mass stops.

Some examples of translational mechanical system models are given below.

Example 2.2

Figure 2.11 shows a simple mechanical translational system with a mass, a spring, and a dashpot. A force F is applied to the system. Derive a mathematical model for the system.

Solution

As shown in Figure 2.11, the net force on the mass is the applied force minus the forces exerted by the spring and the dashpot. Applying Newton's second law, we can write the system equation as

$$F - ky - b\frac{dy}{dt} = m\frac{d^2y}{dt^2} \qquad (2.17)$$

or

$$F = m\frac{d^2y}{dt^2} + b\frac{dy}{dt} + ky \qquad (2.18)$$

Taking the Laplace transform of Equation 2.18, the transfer function of the system may be written as

$$F(s) = ms^2Y(s) + bsY(s) + kY(s)$$

or

$$\frac{Y(s)}{F(s)} = \frac{1}{ms^2 + bs + k} \qquad (2.19)$$

The transfer function in Equation 2.19 is represented by the block diagram shown in Figure 2.12. The step response of this mechanical system may be determined by simulating the following MATLAB codes and the response obtained is shown in Figure 2.13.

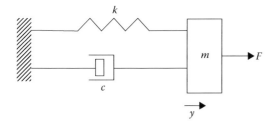

FIGURE 2.11
Mechanical system with mass, spring, and dashpot.

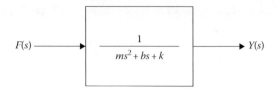

FIGURE 2.12
Block diagram of a simple mechanical system.

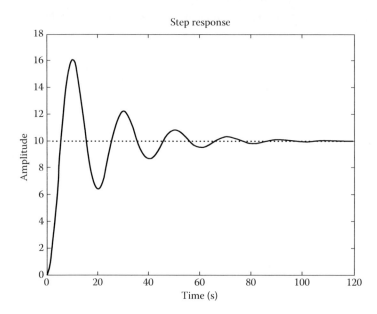

FIGURE 2.13
Results of step input to mechanical system.

MATLAB codes

```
% Simulation program for given mechanical system shown in Figure 2.11
% Force is increased in unit step
m=1.0;      % kg
c=0.1;      %
k=0.1;      %
Num = [1];
Den= [m c k];
Step (Num, Den)
```

Example 2.3

Determine the state model for the mechanical system shown in Figure 2.14.
The state variable vector for the given mechanical system may be displacements and velocities

$$[X_1 \ X_2 \ X_3 \ X_4] = [x_1 \ \overset{o}{x}_1 \ x_2 \ \overset{o}{x}_2] \tag{2.20}$$

Systems Modeling

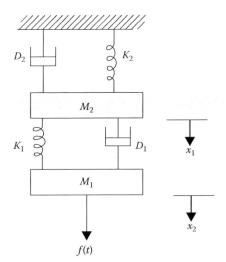

FIGURE 2.14
Mechanical system.

The force equation for mass M_1 and M_2 may be written as

$$\frac{d\dot{x}_1}{dt} = \frac{k_1}{M_1}x_2 + \frac{D_1}{M_1}\dot{x}_2 - \frac{D_1}{M_1}\dot{x}_1 - \frac{k_1}{M_1}(x_1 - x_2) \qquad (2.21)$$

$$k_2 x_2 + M_2 \frac{d^2 x_2}{dt^2} + D_2 \frac{dx_2}{dt} = k_1(x_1 - x_2) + D_1 \frac{d(x_1 - x_2)}{dt}$$

$$\frac{d\dot{x}_2}{dt} = -\frac{k_2}{M_2}x_2 + \frac{D_2}{M_2}\dot{x}_2 + \frac{D_1}{M_2}(\dot{x}_1 - \dot{x}_2) + \frac{k_1}{M_2}(x_1 - x_2) \qquad (2.22)$$

The state-space model for the mechanical system may be obtained after combining the above equation as

$$\frac{d}{dt}\begin{bmatrix} X_1 \\ X_2 \\ X_3 \\ X_4 \end{bmatrix} = \begin{bmatrix} 0 & 1 & 0 & 0 \\ -k_1/M_1 & -D_1/M_1 & k_1/M_1 & D_1/M_1 \\ 0 & 0 & 0 & 1 \\ k_1/M_2 & D_1/M_2 & -(k_1+k_2)/M_2 & -(D_1+D_2)/M_1 \end{bmatrix} \begin{bmatrix} X_1 \\ X_2 \\ X_3 \\ X_4 \end{bmatrix} + \begin{bmatrix} 0 \\ 1/M_1 \\ 0 \\ 0 \end{bmatrix} f(t) \qquad (2.23)$$

The simulation results of the state model mentioned above are shown in Figure 2.15.

MATLAB codes

```
% Simulation program for state space model
clear all
t=0;                    % Initial time
dt=0.01;                % step size
tsim=10.0;              % Simulation time
n=round((tsim-t)/dt);   % number of iterations
%system parameters
k1=5;
k2=7;
m1=2;
m2=3;
```

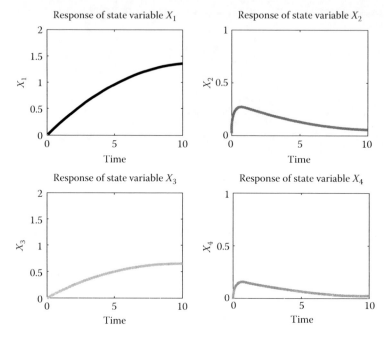

FIGURE 2.15
Simulation results.

```
d1=40;
d2=30;

A=[0 1 0 0; -k1/m1 -d1/m1 k1/m1 d1/m1; 0 0 0 1; k1/m2 d1/m2 -(k1+k2)/m2-
(d1+d2)/m2];
B=[0; 1/m1; 0; 0];
% C=[0 1];
% D=[0 0];
X=[0 0 0 0]';
u=5;
for i=1:n;
    dx=A*X+B*u;
    X=X+dx*dt;
    X1(i,:)=[t,X'];
    t=t+dt;
end
subplot(2,2,1)
plot(X1(:,1),X1(:,2),'b.')
axis([0 10 0 2])
xlabel('time')
ylabel('X1')
title('Response of state variable X1')

subplot(2,2,2)
plot(X1(:,1),X1(:,3),'r.')
axis([0 10 0 1])
xlabel('time')
ylabel('X2')
title('Response of state variable X2')
```

```
subplot(2,2,3)
plot(X1(:,1),X1(:,4),'c.')
axis([0 10 0 2])
xlabel('time')
ylabel('X3')
title('Response of state variable X3')

subplot(2,2,4)
plot(X1(:,1),X1(:,5),'g.')
axis([0 10 0 1])
xlabel('time')
ylabel('X4')
title('Response of state variable X4')
```

Example 2.4

Figure 2.16 shows a mechanical system with two masses and two springs. Obtain the mathematical model of the system.

Solution

The free body diagrams for mass M_1 and M_2 are shown in Figures 2.17 and 2.18.
Applying Newton's second law to get the force Equation 2.24 for the mass m_1,

$$-K_2(y_1 - y_2) - B\left(\frac{dy_1}{dt} - \frac{dy_2}{dt}\right) - k_1 y_1 = m_1 \frac{d^2 y_1}{dt^2} \tag{2.24}$$

Similarly, the force equation for the mass m_2 may be written as

$$F - K_2(y_1 - y_2) - B\left(\frac{dy_2}{dt} - \frac{dy_1}{dt}\right) = m_2 \frac{d^2 y_2}{dt^2} \tag{2.25}$$

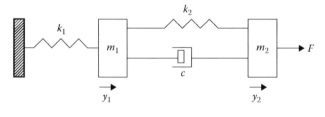

FIGURE 2.16
Example mechanical system.

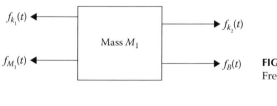

FIGURE 2.17
Free body diagram for mass M_1.

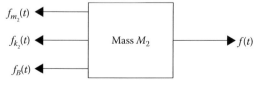

FIGURE 2.18
Free body diagram for mass M_2.

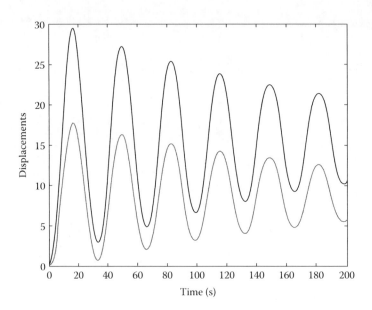

FIGURE 2.19
Displacement vs. time curve.

These equations can be rewritten as Equations 2.26 and 2.27.

$$m_1 \overset{oo}{y_1} + B(\overset{o}{y_1} - \overset{o}{y_2}) + (k_2 + k_1)y_1 - k_2 y_2 = 0 \tag{2.26}$$

$$m_2 \overset{oo}{y_2} + B(\overset{o}{y_2} - \overset{o}{y_1}) + k_2 y_2 - k_2 y_1 = F \tag{2.27}$$

Equations 2.26 and 2.27 can be written in matrix form as

$$\begin{bmatrix} m_1 & 0 \\ 0 & m_2 \end{bmatrix} \begin{bmatrix} \overset{oo}{y_1} \\ \overset{oo}{y_2} \end{bmatrix} + \begin{bmatrix} B & -B \\ -B & B \end{bmatrix} \begin{bmatrix} \overset{o}{y_1} \\ \overset{o}{y_2} \end{bmatrix} + \begin{bmatrix} k_1 + k_2 & -k_2 \\ -k_2 & k_2 \end{bmatrix} \begin{bmatrix} y_1 \\ y_2 \end{bmatrix} = \begin{bmatrix} 0 \\ F \end{bmatrix} \tag{2.28}$$

The state-space model may be obtained from Equation 2.28 as

$$\begin{bmatrix} \overset{oo}{y_1} \\ \overset{oo}{y_2} \end{bmatrix} = \begin{bmatrix} 0 \\ \dfrac{F}{m_2} \end{bmatrix} - \begin{bmatrix} \dfrac{B}{m_1} & \dfrac{-B}{m_1} \\ \dfrac{-B}{m_2} & \dfrac{B}{m_2} \end{bmatrix} \begin{bmatrix} \overset{o}{y_1} \\ \overset{o}{y_2} \end{bmatrix} - \begin{bmatrix} \dfrac{k_1 + k_2}{m_1} & \dfrac{-k_2}{m_1} \\ \dfrac{-k_2}{m_2} & \dfrac{k_2}{m_2} \end{bmatrix} \begin{bmatrix} y_1 \\ y_2 \end{bmatrix} \tag{2.29}$$

The MATLAB script for simulating the above developed model is given below and the simulation results are shown in Figures 2.19 and 2.20.

MATLAB codes

```
% Simulation program for the state model given in Equation 2.29
% Initialization
```

Systems Modeling

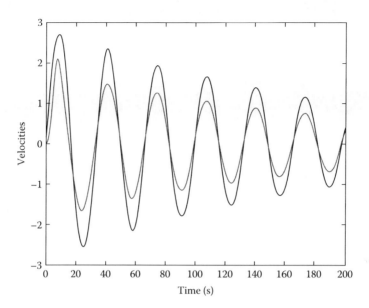

FIGURE 2.20
Velocities vs. time curves.

```
clear all;
F=1;
M1=1;          %kg
M2=1.5;        %kg
B=0.1;         %
K1=0.2;
K2=0.15;
Coef_1=[B/M1     -B/M1
    -B/M2     B/M2];
Coef_2=[(K1+K2)/M1    -K1/M1
    -K2/M2      K2/M2];
Y=[0.1; 0.1];
dY=[0; 0];
dt=0.1;          % step size
t=0;       % Initial time
tsim=200;        % Simulation time
n=round(tsim-t)/dt;
for i=1:n
        X1(i,:)=[Y' dY' t];
    ddY=[0;F/M2] - Coef_1*dY - Coef_2*Y;
    dY=dY+dt*ddY;
    Y=Y+dt*dY;
    t=t+dt
end
plot(X1(:,5),X1(:,1:2))
xlabel('Time (sec.)')
ylabel('Displacements')
figure(2)
plot(X1(:,5),X1(:,3:4))
xlabel('Time (sec.)')
ylabel('Velocities')
```

Example 2.5: Modeling of Train System

In this example, we will consider a toy train consisting of an engine and a car as shown in Figure 2.21. Assuming that the train travels only in one direction, we want to apply control to the train so that it has a smooth start-up and stop, and a constant-speed ride.

The mass of the engine and the car will be represented by M_1 and M_2, respectively. The two are held together by a spring, which has the stiffness coefficient of k. F represents the force applied by the engine, and the Greek letter μ represents the coefficient of rolling friction.

Free body diagram and Newton's law

Where $B_1 = \mu M_1 g$ and $B_2 = \mu M_2 g$

From Newton's law, we know that the sum of the forces acting on a mass equals mass times its acceleration as shown in Figure 2.22. In this case, the forces acting on M_1 are the spring, the friction, and the force applied by the engine. The forces acting on M_2 are the spring and the friction. In the vertical direction, the gravitational force is canceled by the normal force applied by the ground, so that there will be no acceleration in the vertical direction. The equations of motion in the horizontal direction are the following:

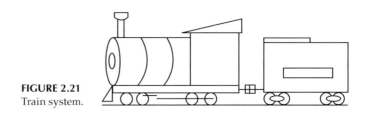

FIGURE 2.21 Train system.

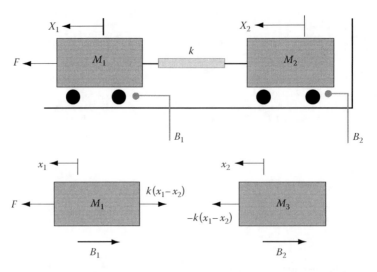

FIGURE 2.22 Free body diagram.

Systems Modeling

$$M_1 \ddot{X}_1 = F - k(X_1 - X_2) - \mu g M_1 \dot{X}_1$$

$$M_2 \ddot{X}_2 = k(X_1 - X_2) - \mu g M_2 \dot{X}_2$$

State-variable and output equations

This set of system equations can now be manipulated into state-variable form. Knowing state variables are X_1 and X_2 and the input is F, state-variable equations will look like the following:

$$\frac{d}{dt} X_1 = V_1$$

$$\frac{d}{dt} V_1 = \frac{-k}{M_1} X_1 - \mu g V_1 + \frac{k}{M_1} X_2 + \frac{1}{M_1} F$$

$$\frac{d}{dt} X_2 = V_2$$

$$\frac{d}{dt} V_2 = \frac{k}{M_2} X_1 - \mu g V_2 - \frac{k}{M_1} X_2$$

Let the output of the system be the velocity of the engine. Then the output equation will become

$$y = V_1$$

1. Transfer function

   To find the transfer function of the system, first, we take the Laplace transforms of the above state-variable and output equations.

   $$sX_1(s) = V_1(s)$$

   $$sV_1(s) = \frac{-k}{M_1} X_1(s) - \mu g V_1(s) + \frac{k}{M_1} X_2(s) + \frac{1}{M_1} F(s)$$

   $$sX_2(s) = V_2(s)$$

   $$sV_2(s) = \frac{k}{M_2} X_1(s) - \mu g V_2(s) - \frac{k}{M_1} X_2(s)$$

   $$Y(s) = V_1(s)$$

   Using these equations, we derive the transfer function $Y(s)/F(s)$ in terms of constants. When finding the transfer function, zero initial conditions must be assumed. The transfer function should look like the one shown below.

   $$\frac{Y(s)}{F(s)} = \frac{M_2 s^2 + M_2 \mu g s + 1}{M_1 M_2 s^3 + 2 M_1 M_2 \mu g s^2 + (M_1 k + M_1 M_2 (\mu g)^2 + M_2 k)s + (M_1 + M_2)k \mu g}$$

2. State space
   Another method to solve the problem is to use the state-space form. Four matrices A, B, C, and D characterize the system behavior, and will be used to solve the problem. The state-space form manipulated from the state-variable and output equations is shown below.

$$\frac{d}{dt}\begin{bmatrix} x_1 \\ v_1 \\ x_2 \\ v_2 \end{bmatrix} = \begin{bmatrix} 0 & 1 & 0 & 0 \\ \frac{-k}{M_1} & -\mu g & \frac{-k}{M_1} & 0 \\ 0 & 0 & 0 & 1 \\ \frac{k}{M_2} & 0 & \frac{-k}{M_2} & -\mu g \end{bmatrix} \begin{bmatrix} x_1 \\ v_1 \\ x_2 \\ v_2 \end{bmatrix} + \begin{bmatrix} 0 \\ \frac{1}{M_1} \\ 0 \\ 0 \end{bmatrix} F(t)$$

$$y = \begin{bmatrix} 0 & 1 & 0 & 0 \end{bmatrix} \begin{bmatrix} x_1 \\ v_1 \\ x_2 \\ v_2 \end{bmatrix} + [0]F(t)$$

The above state model simulated in MATLAB is given below and the results are shown in Figure 2.23.

```
% Matlab Codes
clear all
clc
k=35; m1=1800; m2=1000;
u=0.05; g=9.81; F=100;
A=[0 1 0 0; -k/m1 -u*g -k/m1 0; 0 0 0 1; k/m2 0 -k/m2 -u*g];
B=[0 1/m1 0 0]';
X=[0 0 0 0]';
t=0;
tsim=3;
```

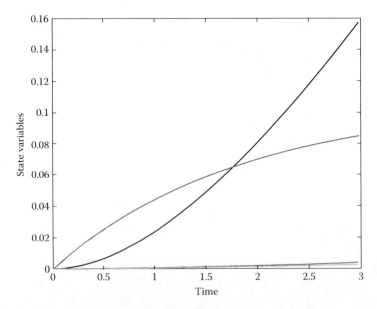

FIGURE 2.23
Simulation results of train system.

Systems Modeling

```
dt=0.02;
n=round((tsim-t)/dt);
for i=1:n
    X1(i,:)=[t X'];
    dvdx=A*X+B*F;
    X=X+dt*dvdx ;
    t=t+dt;
end
plot(X1(:,1),X1(:,2:5))
xlabel('x(m)');ylabel('state variables')
```

Example 2.6

Consider a mechanical coupler normally used for coupling of two railway coaches as shown in Figure 2.24. The equivalent system of railway coupling is shown in Figure 2.25, which consists of two masses, a spring, a dashpot, and forces applied to each mass. Derive an expression for the mathematical model of the system.

SOLUTION

Draw the free body diagram for mass m_1 as shown in Figure 2.26.

Applying Newton's second law to the mass m_1 write down the force

$$F_1 - f_k(t) - f_b(t) - f_m(t) = 0$$

$$F_1 - k(y_1 - y_2) - b\left(\frac{dy_1}{dt} - \frac{dy_2}{dt}\right) = m_1 \frac{d^2 y_1}{dt^2} \tag{2.30}$$

FIGURE 2.24
A mechanical coupler.

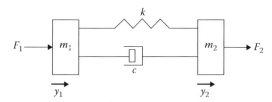

FIGURE 2.25
Equivalent diagram of mechanical coupler.

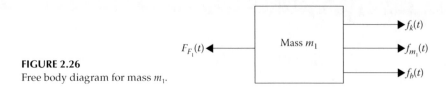

FIGURE 2.26
Free body diagram for mass m_1.

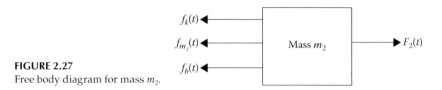

FIGURE 2.27
Free body diagram for mass m_2.

and draw the free body diagram for m_2 as shown in Figure 2.27, and its equation to the mass m_2,

$$F_1 - f_k(t) - f_b(t) - f_m(t) = 0$$

$$F_2 - k(y_1 - y_2) - b\left(\frac{dy_2}{dt} - \frac{dy_1}{dt}\right) = m_2 \frac{d^2 y_2}{dt^2} \qquad (2.31)$$

we can then write Equations 2.30 and 2.31 as

$$m_1 \ddot{y}_1 + b(\dot{y}_1 - \dot{y}_2) + k y_1 - k y_2 = F_1 \qquad (2.32)$$

$$m_2 \ddot{y}_2 + b(\dot{y}_2 - \dot{y}_1) + k_2 y_2 - k_2 y_1 = F_2 \qquad (2.33)$$

Equations 2.32 and 2.33 can be written in matrix form to get the state-space model for the given system:

$$\begin{bmatrix} m_1 & 0 \\ 0 & m_2 \end{bmatrix} \begin{bmatrix} \ddot{y}_1 \\ \ddot{y}_2 \end{bmatrix} + \begin{bmatrix} b & -b \\ -b & b \end{bmatrix} \begin{bmatrix} \dot{y}_1 \\ \dot{y}_2 \end{bmatrix} + \begin{bmatrix} k & -k \\ -k & k \end{bmatrix} \begin{bmatrix} y_1 \\ y_2 \end{bmatrix} = \begin{bmatrix} F_1 \\ F_2 \end{bmatrix}$$

or

$$\begin{bmatrix} \ddot{y}_1 \\ \ddot{y}_2 \end{bmatrix} = \begin{bmatrix} \frac{F_1}{m_1} \\ \frac{F_2}{m_2} \end{bmatrix} - \begin{bmatrix} \frac{b}{m_1} & \frac{-b}{m_1} \\ \frac{-b}{m_2} & \frac{b}{m_2} \end{bmatrix} \begin{bmatrix} \dot{y}_1 \\ \dot{y}_2 \end{bmatrix} - \begin{bmatrix} \frac{k}{m_1} & \frac{-k}{m_1} \\ \frac{-k}{m_2} & \frac{k}{m_2} \end{bmatrix} \begin{bmatrix} y_1 \\ y_2 \end{bmatrix} \qquad (2.34)$$

The MATLAB script for simulating the above system is given below and the results are shown in Figure 2.28.

Systems Modeling

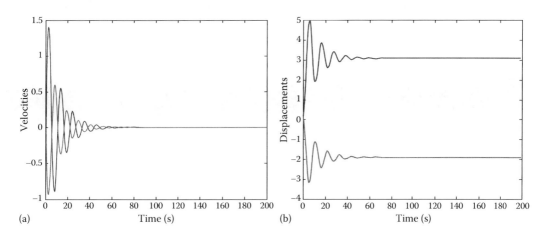

FIGURE 2.28
Simulation results of mechanical coupling. (a) Velocity variation with time. (b) Displacement variation with time.

MATLAB codes

```
% Simulation program for the state model given in Equation 2.34
% Initialization
clear all;
F1=1;
F2=-1;
M1=1;              %kg
M2=1.5;            %kg
B=0.1;       %
K=0.2;
Coef_1=[B/M1    -B/M1
    -B/M2   B/M2];
Coef_2=[K/M1    -K/M1
    -K/M2    K/M2];
Y=[0.1; 0.1];
dY=[0; 0];
dt=0.1;            % step size
t=0;          % Initial time
tsim=200;          % Simulation time
n=round(tsim-t)/dt;
for i=1:n
      X1(i,:)=[Y' dY' t];
    ddY=[F1/M1;F2/M2] - Coef_1*dY - Coef_2*Y;
    dY=dY+dt*ddY;
    Y=Y+dt*dY;
    t=t+dt
end
figure(1)
plot(X1(:,5),X1(:,1:2))
xlabel('Time (sec.)')
ylabel('Displacements')
figure(2)
plot(X1(:,5),X1(:,3:4))
xlabel('Time (sec.)')
ylabel('Velocities')
```

Example 2.7

Figure 2.29 shows a mechanical system with three masses, two springs, and a dashpot. A force is applied to mass m_3 and a displacement is applied to spring k_1. Drive an expression for the mathematical model of the system.

Solution

Applying Newton's second law to the mass m_1,

$$k_1(y - y_1) - k_2(y_1 - y_2) = m_1 \frac{d^2 y_1}{dt^2} \tag{2.35}$$

to the mass m_2,

$$-b\left(\frac{dy_2}{dt} - \frac{dy_3}{dt}\right) - k_2(y_2 - y_1) - k_3(y_2 - y_3) = m_2 \frac{d^2 y_2}{dt^2} \tag{2.36}$$

and to the mass m_3,

$$F - b\left(\frac{dy_3}{dt} - \frac{dy_2}{dt}\right) - k_3(y_3 - y_2) = m_3 \frac{d^2 y_3}{dt^2} \tag{2.37}$$

we can write Equations 2.35 through 2.37 as

$$m_1 \ddot{y}_1 + (k_1 + k_2) y_1 - k_2 y_2 = k_1 y \tag{2.38}$$

$$m_2 \ddot{y}_2 + b \dot{y}_2 - b \dot{y}_3 - k_2 y_1 + (k_2 + k_3) y_2 - k_3 y_3 = 0 \tag{2.39}$$

$$m_3 \ddot{y}_3 + b \dot{y}_3 - b \dot{y}_2 + k_3 y_3 - k_3 y_2 = F \tag{2.40}$$

The above equations can be written in matrix form as

$$\begin{bmatrix} m_1 & 0 & 0 \\ 0 & m_2 & 0 \\ 0 & 0 & m_3 \end{bmatrix} \begin{bmatrix} \ddot{y}_1 \\ \ddot{y}_2 \\ \ddot{y}_3 \end{bmatrix} + \begin{bmatrix} 0 & 0 & 0 \\ 0 & b & -b \\ 0 & -b & b \end{bmatrix} \begin{bmatrix} \dot{y}_1 \\ \dot{y}_2 \\ \dot{y}_3 \end{bmatrix} + \begin{bmatrix} k_1 + k_2 & -k_2 & 0 \\ -k_2 & k_2 + k_3 & -k_3 \\ 0 & -k_3 & k_3 \end{bmatrix} \begin{bmatrix} y_1 \\ y_2 \\ y_3 \end{bmatrix} = \begin{bmatrix} k_1 y \\ 0 \\ F \end{bmatrix} \tag{2.41}$$

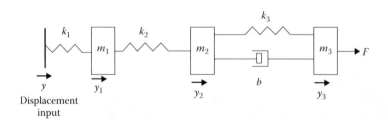

FIGURE 2.29
Example mechanical system.

Systems Modeling

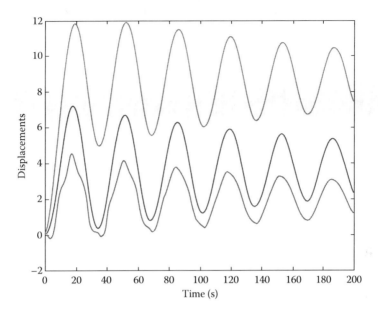

FIGURE 2.30
The simulation results of MATLAB program.

or

$$\begin{bmatrix} \overset{\circ\circ}{y_1} \\ \overset{\circ\circ}{y_2} \\ \overset{\circ\circ}{y_3} \end{bmatrix} = \begin{bmatrix} \dfrac{k_1 y}{m_1} \\ 0 \\ \dfrac{F}{m_3} \end{bmatrix} - \begin{bmatrix} 0 & 0 & 0 \\ 0 & \dfrac{b}{m_2} & \dfrac{-b}{m_2} \\ 0 & \dfrac{-b}{m_3} & \dfrac{b}{m_3} \end{bmatrix} \begin{bmatrix} \overset{\circ}{y_1} \\ \overset{\circ}{y_2} \\ \overset{\circ}{y_3} \end{bmatrix} - \begin{bmatrix} \dfrac{(k_1+k_2)}{m_1} & \dfrac{-k_2}{m_1} & 0 \\ \dfrac{-k_2}{m_2} & \dfrac{(k_2+k_3)}{m_2} & -\dfrac{k_3}{m_2} \\ 0 & \dfrac{-k_3}{m_3} & \dfrac{k_3}{m_3} \end{bmatrix} \begin{bmatrix} y_1 \\ y_2 \\ y_3 \end{bmatrix} \quad (2.42)$$

The MATLAB codes for simulating the above mentioned mechanical system are given below and the simulation results are shown in Figure 2.30.

MATLAB codes

```
clear all;
F=0.25;
y=-0.5;
M1=0.5;              %kg
M2=1.0;              %kg
M3=0.5;
B=0.5;      %
K1=0.1;
K2=0.15;
K3=0.05;
Coef_1=[0 0 0;
0 B/M2  -B/M2
  0 -B/M3   B/M3];
Coef_2=[(K1+K2)/M1  -K2/M1 0
    -K2/M2    (K2+K3)/M2 -K3/M2
0 -K3/M3 K3/M3];
```

```
Y=[0.1; 0.1; 0.05];
dY=[0; 0;0];
dt=0.1;            % step size
t=0;         % Initial time
tsim=200;        % Simulation time
n=round(tsim-t)/dt;
for i=1:n
      X1(i,:)=[Y' dY' t];
  ddY=[K1*y/M1;0 ;F/M3] - Coef_1*dY - Coef_2*Y;
  dY=dY+dt*ddY;
  Y=Y+dt*dY;
  t=t+dt
end
figure(1)
plot(X1(:,7),X1(:,1:3))
xlabel('Time (sec.)')
ylabel('Displacements')
```

2.7.1.2 Rotational Mechanical Systems

The basic building blocks of rotational mechanical systems are the moment of inertia, the torsion spring (or rotational spring), and the rotary damper (Figure 2.31). The input to a rotational mechanical system may be the torque T and the output the rotational displacement, or angle.

Torsional spring

A rotational spring is similar to a translational spring, but here the spring is twisted. The relationship between the applied torque T and the angle θ rotated by the spring is given by

$$T = k \cdot \theta \qquad (2.43)$$

where θ is known as the rotational stiffness constant. In our modeling we are assuming that the mass of the spring is negligible and the spring is linear.

The energy stored in a torsional spring when twisted by an angle θ is given by

$$E = \frac{1}{2}k\theta^2 \qquad (2.44)$$

A rotary damper element creates damping as it rotates. For example, when a disk rotates in a fluid we get a rotary damping effect. The relationship between the applied torque T and the angular velocity of the rotary damper is given by

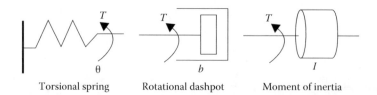

Torsional spring   Rotational dashpot   Moment of inertia

FIGURE 2.31
Rotational mechanical system components.

Systems Modeling

$$T = b\omega = b\frac{d\theta}{dt} \tag{2.45}$$

In our modeling the mass of the rotary damper will be neglected, or will be assumed to be negligible. A rotary damper does not store energy.

Moment of inertia refers to a rotating body with a mass. When a torque is applied to a body with a moment of inertia we get an angular acceleration, and this acceleration rotates the body.

The relationship between the applied torque T, angular acceleration a, and the moment of inertia I, is given by

$$T = Ia = I\frac{d\omega}{dt} \tag{2.46}$$

or

$$T = I\frac{d^2\theta}{dt^2} \tag{2.47}$$

The energy stored in a mass rotating with an angular velocity can be written as

$$E = \frac{1}{2}I\omega^2 \tag{2.48}$$

Some examples of rotational system models are given below.

Example 2.8

A disk of moment of inertia I is rotated (refer Figure 2.32) with an applied torque of T. The disk is fixed at one end through an elastic shaft. Assuming that the shaft can be modeled with a rotational dashpot and a rotational spring, derive an equation for the mathematical model of this system.

Solution

The damper torque and the spring torque oppose the applied torque. If θ is the angular displacement from the equilibrium, we can write the torque balancing equation for the given system:

$$T(t) - T_b(t) - T_k(t) - T_I(t) = 0$$

$$T - b\frac{d\theta}{dt} - k\theta = I\frac{d^2\theta}{dt^2} \tag{2.49}$$

or

$$I\frac{d^2\theta}{dt^2} + b\frac{d\theta}{dt} + k\theta = T \tag{2.50}$$

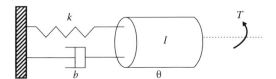

FIGURE 2.32
Rotational mechanical system.

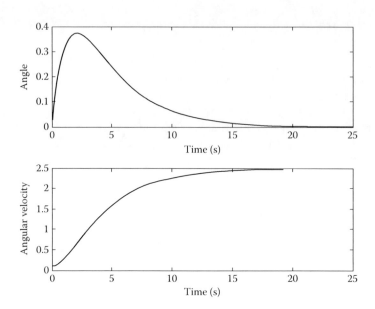

FIGURE 2.33
Simulation results of Example 2.5.

Equation 2.50 may be written in the following form:

$$I\ddot{\theta} + b\dot{\theta} + k\theta = T \qquad (2.51)$$

or

$$\ddot{\theta} = \frac{T - b\dot{\theta} - k\theta}{I}$$

This second-order differential equation may be decomposed into two first-order differential equations and may be written in state-space form as

$$\frac{d}{dt}\begin{bmatrix}\dot{\theta}\\\theta\end{bmatrix} = \begin{bmatrix}\frac{-b}{I} & \frac{-k}{I}\\1 & 0\end{bmatrix}\begin{bmatrix}\dot{\theta}\\\theta\end{bmatrix} + \begin{bmatrix}\frac{1}{I}\\0\end{bmatrix}T \qquad (2.52)$$

$$\dot{X} = AX + BU$$

This state model may be simulated using MATLAB codes given below and the results are given in Figure 2.33.

MATLAB codes

```
% Simulation program for the state model given in Equation 2.52
% Initialization
clear all;
T=0.25;
```

Systems Modeling

```
I=0.5;
b=0.5;
k=0.1;
A=[-b/I  -k/I
    1     0];
B=[1/I
   0];
X=[0; 0.1;];
dt=0.1;      % step size
t=0;         % Initial time
tsim=200;              % Simulation time
n=round(tsim-t)/dt;
for i=1:n
       X1(i,:)=[X' t];
    dX=A*X+B*T;
    X=X+dt*dX;
    t=t+dt;
end
Subplot(2,1,1)     % Divides the graphics window into sub windows
plot(X1(:,3),X1(:,1))
xlabel('Time (sec.)')
ylabel('Angle')
Subplot(2,1,2)     % Divides the graphics window into sub windows
plot(X1(:,3),X1(:,2))
xlabel('Time (sec.)')
ylabel('Angular velocity')
```

Example 2.9

Figure 2.34 shows a rotational mechanical system with two moments of inertia and a torque applied to each one. Derive a mathematical model for the system.

Solution

For the system shown in Figure 2.34 we can write the following equations for disk 1,

$$T_1 - T_k(t) - T_b(t) - T_{I1}(t) = 0$$

$$T_1 - k(\theta_1 - \theta_2) - b\left(\frac{d\theta_1}{dt} - \frac{d\theta_2}{dt}\right) = I_1 \frac{d^2\theta_1}{dt^2} \quad (2.53)$$

and for disk 2,

$$T_2 - T_k(t) - T_b(t) - T_{I2}(t) = 0$$

$$T_2 - k(\theta_2 - \theta_1) - b\left(\frac{d\theta_2}{dt} - \frac{d\theta_1}{dt}\right) = I_2 \frac{d^2\theta_2}{dt^2} \quad (2.54)$$

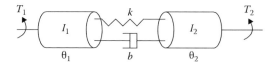

FIGURE 2.34
Rotational mechanical system.

Equations 2.53 and 2.54 can be written as

$$I_1 \ddot{\theta}_1 + b\dot{\theta}_1 - b\dot{\theta}_2 + k\theta_1 - k\theta_2 = T_1 \tag{2.55}$$

and

$$I_2 \ddot{\theta}_2 - b\dot{\theta}_1 + b\dot{\theta}_2 - k\theta_1 + k\theta_2 = T_2 \tag{2.56}$$

Equations 2.55 and 2.56 may be written in the matrix form to get state-space model as in Equation 2.57.

$$\begin{bmatrix} I_1 & 0 \\ 0 & I_2 \end{bmatrix} \begin{bmatrix} \ddot{\theta}_1 \\ \ddot{\theta}_2 \end{bmatrix} + \begin{bmatrix} b & -b \\ -b & b \end{bmatrix} \begin{bmatrix} \dot{\theta}_1 \\ \dot{\theta}_2 \end{bmatrix} + \begin{bmatrix} k & -k \\ -k & k \end{bmatrix} \begin{bmatrix} \theta_1 \\ \theta_2 \end{bmatrix} = \begin{bmatrix} T_1 \\ T_2 \end{bmatrix}$$

or

$$\begin{bmatrix} \ddot{\theta}_1 \\ \ddot{\theta}_2 \end{bmatrix} = \begin{bmatrix} \dfrac{T_1}{I_1} \\ \dfrac{T_2}{I_2} \end{bmatrix} - \begin{bmatrix} \dfrac{b}{I_1} & \dfrac{-b}{I_1} \\ \dfrac{-b}{I_2} & \dfrac{b}{I_2} \end{bmatrix} \begin{bmatrix} \dot{\theta}_1 \\ \dot{\theta}_2 \end{bmatrix} - \begin{bmatrix} \dfrac{k}{I_1} & \dfrac{-k}{I_1} \\ \dfrac{-k}{I_2} & \dfrac{k}{I_2} \end{bmatrix} \begin{bmatrix} \theta_1 \\ \theta_2 \end{bmatrix} \tag{2.57}$$

2.7.1.2.1 Rotational Mechanical Systems with Gear Train

Gear-train systems are very important in many mechanical engineering systems. Figure 2.35 shows a simple gear train, consisting of two gears, each connected to two masses with moments of inertia I_1 and I_2. Suppose that gear 1 has n_1 teeth and radius r_1, and that gear 2 has n_2 teeth and radius r_2. Assume that the gears have no backlash, they are rigid bodies, and the moment of inertia of the gears is negligible.

The rotational displacement of the two gears depends on their radii and is given by

$$r_1 \theta_1 = r_2 \theta_2 \tag{2.58}$$

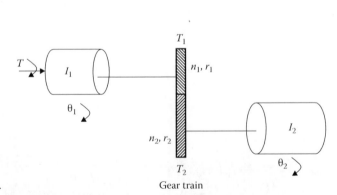

FIGURE 2.35
A two-gear train system.

Systems Modeling

or

$$\theta_2 = \frac{r_1}{r_2} \theta_1 \qquad (2.59)$$

where θ_1 and θ_2 are the rotational displacements of gear 1 and gear 2, respectively.

The ratio of the teeth numbers is equal to the ratio of the radii and is given by

$$\frac{r_1}{r_2} = \frac{n_1}{n_2} = n \qquad (2.60)$$

where n is the gear teeth ratio.

Let a torque T be applied to the system, then torque equation may be written as

$$I_1 \frac{d^2\theta_1}{dt^2} = T - T_1 \qquad (2.61)$$

and

$$I_2 \frac{d^2\theta_2}{dt^2} = T_2 \qquad (2.62)$$

Equating the power transmitted by the gear train,

$$T_1 \frac{d\theta_1}{dt} = T_2 \frac{d\theta_2}{dt}$$

or

$$\frac{T_1}{T_2} = \frac{d\theta_1/dt}{d\theta_2/dt} = n \qquad (2.63)$$

Substituting Equation 2.63 into Equation 2.61, we obtain

$$I_1 \frac{d^2\theta_1}{dt^2} = T - nT_2 \qquad (2.64)$$

or

$$I_1 \frac{d^2\theta_1}{dt^2} = T - n\left(I_2 \frac{d^2\theta_2}{dt^2}\right) \qquad (2.65)$$

then, since $\theta_2 = n\theta_1$, we obtain $(I_1 + n^2 I_2)\dfrac{d^2\theta_1}{dt^2} = T \qquad (2.66)$

It is clear from Equation 2.66 that the moment of inertia of the load I_2 is reflected to the other side of the gear train as $n^2 I_2$.

Example 2.10

Figure 2.36 shows a rotational mechanical system coupled with a gear train. Derive an expression for the model of the system.

Solution

Assuming that a torque T is applied to the system, then the system equation may be written as

$$I_1 \frac{d^2\theta_1}{dt^2} + b_1 \frac{d\theta_1}{dt} + k_1\theta_1 = T - T_1 \tag{2.67}$$

and

$$c = T_2 \tag{2.68}$$

Equating the power transmitted by the gear train,

$$\frac{T_1}{T_2} = \frac{d\theta_2/dt}{d\theta_1/dt} = n \tag{2.69}$$

Substituting Equation 2.69 into 2.67, we obtain

$$I_1 \frac{d^2\theta_1}{dt^2} + b_1 \frac{d\theta_1}{dt} + k_1\theta_1 = T - nT_2 \tag{2.70}$$

or

$$I_1 \frac{d^2\theta_1}{dt^2} + b_1 \frac{d\theta_1}{dt} + k_1\theta_1 = T - n\left(I_2 \frac{d^2\theta_2}{dt^2} + b_2 \frac{d\theta_2}{dt} + k_2\theta_2\right) \tag{2.71}$$

Since $\theta_2 = n \cdot \theta_1$, this gives

$$(I_1 + n^2 I_2)\frac{d^2\theta_1}{dt^2} + (b_1 + n^2 b_2)\frac{d\theta_1}{dt} + (k_1 + n^2 k_2)\theta_1 = T \tag{2.72}$$

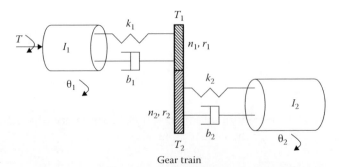

FIGURE 2.36
Mechanical system with gear train.

Example 2.11

A safety bumper is placed at the end of a racetrack to stop out-of-control cars as shown in Figure 2.37. The bumper is designed in such a way that the force applied is a function of the velocity v and the displacement x of the front edge of the bumper according to the equation
$F = Kv^3(x + 1)^3$ where $K = 35$ s-kg/m^5 is a constant.

A car with a mass m of 1800 kg hits the bumper at a speed of 60 km/h. Determine and plot the velocity of the car as a function of its position for $0 \leq x \leq 3$ m.

SOLUTION

The declaration of the car once it hits the bumper can be calculated from Newton's second law of motion.

$$Ma = -Kv^3(x + 1)^3$$

Which can be solved for the acceleration a as a function of v and x:

$$a = \frac{-Kv^3(x + 1)^3}{m}$$

The velocity as a function of x can be calculated by substituting the acceleration in the equation

$$v\,dv = a\,dx$$

which gives

$$\frac{dv}{dx} = \frac{-Kv^2(x + 1)^3}{m}$$

The ordinary differential equation may be solved for the interval $0 \leq x \leq 3$ with the initial condition: $v = 60$ km/h at $x = 0$.

A numerical solution of the differential equation with MATLAB is given in the following program:

```
global k m
k=35; m=1800; v0=60;
xspan = [0:0.2:3];
v0mps =v0*1000/3600;
[x v] =ode45('bumper',xspan,v0mps)
plot(x,v)
xlabel('x(m)');ylabel('velocity(m/s)')
```

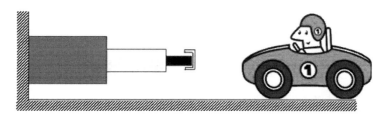

FIGURE 2.37
Safety bumper at the end of the racing track.

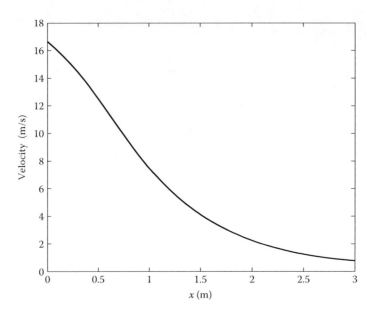

FIGURE 2.38
Velocity as a function of distance.

The function file contains the differential equation of bumper as below:

```
function dvdx=bumper(x,v)
global k m
dvdx=-(k*v^2*(x+1)^3)/m;
```

When the above MATLAB code is executed the following results have been obtained and the results as graphically shown in Figure 2.38:

```
X             V
     0    16.6667
0.2000    15.3330
0.4000    13.5478
0.6000    11.4947
0.8000     9.4189
1.0000     7.5234
1.2000     5.9165
1.4000     4.6209
1.6000     3.6066
1.8000     2.8256
2.0000     2.2281
2.2000     1.7706
2.4000     1.4191
2.6000     1.1475
2.8000     0.9358
3.0000     0.7696
```

Example 2.12: 3-D Projectile Trajectory

A projectile is fired with a velocity of 300 m/s at an angle of 65° relative to the ground. The projectile is aimed directly north. Because of the strong wind blowing to the west, the projectile also

Systems Modeling

moves in this direction at a constant speed of 35 m/s. Determine and plot the trajectory of the projectile until it hits the ground. Also simulate the projectile when there is no wind.

SOLUTION

The coordinate system is set up such that the x and y axes point to the east and north directions, respectively. Then, the motion of the projectile can be analyzed by considering the vertical direction z and the two horizontal components x and y. Since the projectile is fired directly north, the initial velocity v_0 can be resolved into a horizontal y component and a vertical z component:

$$v_{0y} = v_0 \cos(\theta) \quad \text{and} \quad v_{0z} = v_0 \sin(\theta)$$

In addition, due to the wind the projectile has a constant velocity in the negative x direction, $v_x = -30$ m/s.

The initial position of the projectile (x_0, y_0, z_0) is at point (3000, 0, 0). In the vertical direction the velocity and position of the projectile are given by

$$v_z = v_{0z} - gt \quad \text{and} \quad z = z_0 + v_{0z}t - \frac{1}{2}gt^2$$

The time it takes for the projectile to reach the highest point ($v_z = 0$) is $t_{hmax} = \dfrac{v_{0z}}{g}$.

The total flying time is twice this time, $t_{tot} = 2t_{hmax}$, in the horizontal direction the velocity is constant (both in the x and y direction), and the position of the projectile is given by

$$x = x_0 + v_x t \quad \text{and} \quad y = y_0 + v_{0y}t$$

The following MATLAB program written in a script file solves the problem by following the equation above.

```
v0=250; g=9.81; theta=65;
x0=3000;vx=-30;
v0z=v0*sin(theta*pi/180);
v0y=v0*cos(theta*pi/180);
t=2*v0z/g;
tplot=linspace(0,t,100);
z=v0z*tplot-0.5*g*tplot.^2;
y=v0y*tplot;
x=x0+vx*tplot;
xnowind(1:length(y))=x0
plot3(x,y,z,'k-',xnowind,y,z,'k--')
grid on
axis([0 6000 0 6000 0 2500])
xlabel('x(m)'); ylabel('y(m)'); zlabel('z(m)')
```

The results of the above program are shown in Figure 2.39.

Example 2.13: Determining the Size of a Capacitor

An electrical capacitor has an unknown capacitance. In order to determine its capacitance it is connected to the circuit shown in Figure at the right hand side. The switch is first connected to B and the capacitor is charged. Then, the switch is switched to A and the capacitor discharges through the resistor. As the capacitor is discharging, the voltage across the

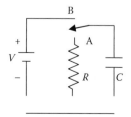

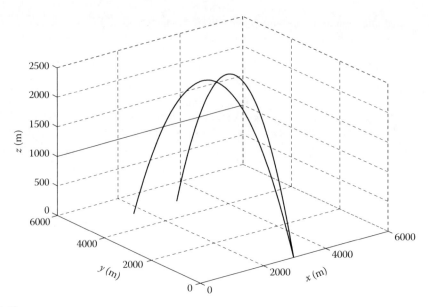

FIGURE 2.39
Results of projectile system.

capacitor is measured for 10 s in intervals of 1 s. The recorded measurements are given in the table below. Plot the voltage as a function of time and determine the capacitance of the capacitor by fitting an exponential curve to the data point.

t(s)	1	2	3	4	5	6	7	8	9	10
V(V)	9.4	7.31	5.51	3.55	2.81	2.04	1.26	0.97	0.74	0.58

Solution

When a capacitor discharges through a resistor, the voltage of the capacitor as a function of time is given by

$$V = V_0 e^{(-t)/(RC)}$$

where
 V_0 is the initial voltage
 R the resistance of the resistor
 C the capacitance of the capacitor

The exponential function can be written as a linear equation for $\ln(V)$ and t in the form

$$\ln(V) = \frac{-1}{RC} t + \ln(V_0)$$

This equation which has the form $y = mx + b$ can be fitted to the data points by using the *polyfit* (x, y, 1) function with t as the independent variable x and $\ln(V)$ as the dependent variable y. The coefficients m and b determined by the polyfit function are then used to determine C and V_0 by

$$C = \frac{-t}{Rm} \quad \text{and} \quad V_0 = e^b$$

Systems Modeling

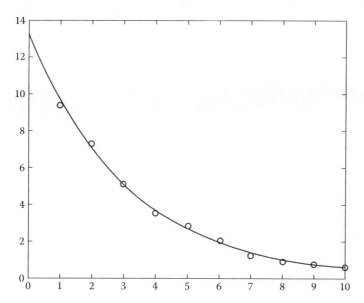

FIGURE 2.40
Simulation results of capacitor selection problem.

The following program written in a MATLAB script file determines the best fit exponential function to the data points, determines C and V_0, and plots the points and the fitted function.

```
R = 2000;
t = 1:10;
v = [9.4 7.31 5.15 3.55 2.81 2.04 1.26 0.97 0.74 0.58];
p = polyfit(t,log(v),1);
C = -1/(R*p(1))
V0 = exp(p(2))
tplot = 0:0.1:10;
vplot = V0*exp(-tplot./(R*C));
plot(t,v,'o',tplot,vplot)
```

The program creates also the plot shown in Figure 2.40.

Example 2.14: Flight of Model Rocket

The flight of a model rocket can be developed as follows. During the first 0.15 s the rocket is propelled up by the rocket engine with a force of 16 N. The rocket then flies up while slowing down under the force of gravity. After it reaches its peak, the rocket starts to fall back. When its down velocity reaches 20 m/s a parachute opens (assumed to open instantly) and the rocket continues to move down at a constant speed of 20 m/s until it hits the ground. Write a program that calculates and plots the speed and altitude of the rocket as a function of time during the flight.

SOLUTION
The rocket is assumed to be a particle that moves along a straight line in the vertical plane. For motion with constant acceleration along a straight line, the velocity and position as a function of time are given by

$$v(t) = v_0 + at \quad \text{and} \quad s(t) = s_0 + v_0 t + \frac{1}{2}at$$

where v and s are the initial velocity and position, respectively. In the computer program the flight of the rocket is divided into three segments. Each segment is calculated in a while loop. In every pass the time increases by an increment.

Segment 1: The first 0.15 s when the rocket engine is on. During this period, the rocket moves up with a constant acceleration. The acceleration is determined by drawing a free body and a mass acceleration diagrams. From Newton's second law, the sum of the forces in the vertical direction is equal to the mass times the acceleration (equilibrium equation):

$$\sum F = F_E - mg = ma$$

Solving the equation for the acceleration gives

$$a = \frac{F_E - mg}{m}$$

The velocity and the height as a function of time are

$$v(t) = 0 + at \quad \text{and} \quad h(t) = 0 + 0 + \frac{1}{2}at^2$$

where the initial velocity and the initial position are both zero. In the computer program this segment starts when $t = 0$, and the looping continues as long as $t < 0.15$ s. The time, velocity, and height at the end of this segment are t_1, v_1, and h_1.

Segment 2: The motion from when the engine stops until the parachute opens. In this segment the rocket moves with a constant deceleration g. The speed and height of the rocket as a function of time are given by

$$v(t) = v_1 - g(t - t_1) \quad \text{and} \quad h(t) = h_1 + v_1(t - t_1) - \frac{1}{2}g(t - t_1)^2$$

In this segment the looping continues until the velocity of the rocket is −20 m/s (negative since the rocket moves down). The time and height at the end of this segment are t_2 and h_2.

Segment 3: The motion from when the parachute opens until the rocket hits the ground. In this segment the rocket moves with constant velocity (zero acceleration). The height as a function of time is given by $h(t) = h_2 - v_{chute}(t - t_2)$, where v_{chute} is the constant velocity after the parachute opens. In this segment the looping continues as long as the height is greater than zero.
A program in a script file that carries out the calculation is shown below.

```
m=0.05; g=9.81; tEngine=0.15; Force=16; vChute=-20; Dt=0.01;
clear t v h
n=1;
t(n)=0; v(n)=0; h(n)=0;
% Segment 1
a1=(Force-m*g)/m;
while t(n)<tEngine &n< 50000
    n=n+1;
    t(n)=t(n-1)+Dt;
    v(n)=a1*t(n);
    h(n)=0.5*a1*t(n)^2;
end
```

Systems Modeling

```
v1=v(n); h1=h(n); t1=t(n);
% Segment 2
while v(n)>=vChute &n<50000
    n=n+1;
    t(n)=t(n-1)+Dt;
    v(n)=v1-g*(t(n)-t1);
    h(n)=h1+v1*(t(n)-t1)-0.5*g*(t(n)-t1)^2;
end
v2=v(n); h2=h(n); t2=t(n);
% Segment 3
while h(n)>0 & n<50000
    n=n+1;
    t(n)=t(n-1)+Dt;
    v(n)=vChute;
    h(n)=h2+vChute*(t(n)-t2);
end
subplot(1,2,1)
plot(t,h,t2,h2, 'o')

subplot(1,2,2)
plot(t,v,t2,v2, 'o')
```

The accuracy of the result depends on the magnitude of the time increment Dt. An increment of 0.01 s appears to give good results. The conditional expression in the while commands also includes a condition for *n* (if *n* is larger than 50,000 the loop stops). This is done as a precaution to avoid an infinite loop in case there is an error in the statements inside the loop. The plots generated by the program are shown in Figure 2.41.

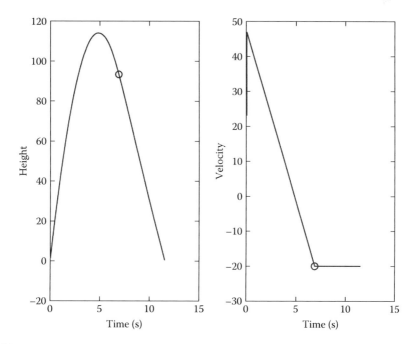

FIGURE 2.41
Simulation results of parachute system.

2.7.2 Modeling of Electrical Systems

The basic building blocks of electrical systems are the resistor, the inductor, and the capacitor (Figure 2.42). The input to an electrical system may be the voltage V and current i.

The relationship between the voltage across a resistor and the current through it is given by Ohm's law as mentioned in Equation 2.73.

$$v_R = R * i \tag{2.73}$$

where R is the resistance.

For an inductor, the potential difference across the inductor depends on the rate of change of current through the inductor, and is given by

$$v_L = L \frac{di}{dt} \tag{2.74}$$

where L is the inductance.

Equation 2.74 can also be written as

$$i = \frac{1}{L} \int v_L dt \tag{2.75}$$

The energy stored in an inductor is given by

$$E = \frac{1}{2} L i^2 \tag{2.76}$$

The potential difference across a capacitor depends on the charge the plates hold, and is given by

$$v_c = \frac{q}{C} \tag{2.77}$$

The relationship between the current through the capacitor and the voltage across it is given by

$$i_c = C \frac{dv_c}{dt} \tag{2.78}$$

or

$$v_c = \frac{1}{C} \int i_c dt \tag{2.79}$$

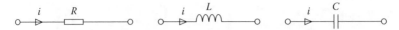

FIGURE 2.42
Electrical system components.

Systems Modeling

The energy stored in a capacitor depends on the capacitance and the voltage across the capacitor and is given by

$$E = \frac{1}{2}Cv_c^2 \tag{2.80}$$

Electrical circuits are modeled using Kirchhoff's laws. There are two laws: Kirchhoff's current law and Kirchhoff's voltage law. To apply these laws effectively, a sign convention should be employed.

Kirchhoff's current law

The sum of the currents at a node in a given network is zero, i.e., the total current flowing into any junction in a circuit is equal to the total current leaving the junction. Figure 2.43 shows the sign convention that can be employed when using Kirchhoff's current law. We can write
$i_1 + i_2 + i_3 = 0$ for the circuit in Figure 2.43a,
$-(i_1 + i_2 + i_3) = 0$ for the circuit in Figure 2.43b, and
$i_1 + i_2 - i_3 = 0$ for the circuit in Figure 2.43c.

Kirchhoff's voltage law

The sum of voltages around any loop in a circuit is zero, i.e., in a circuit containing a voltage source, the algebraic sum of the potential drops across each circuit element is equal to the algebraic sum of the applied emfs.

It is important to observe the sign convention when applying Kirchhoff's voltage law. An example circuit is given in Figure 2.44. The voltage equation for the given circuit is $v_R + v_L + v_C = V$.

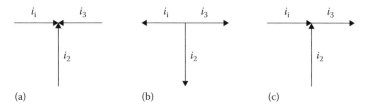

FIGURE 2.43
Applying Kirchhoff's current law.

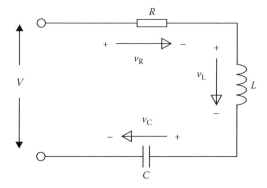

FIGURE 2.44
Applying Kirchhoff's voltage law.

Example 2.15

Determine the state model for the electrical system shown in Figure 2.45.

$$i(t) = i_L(t) = i_C(t) = i_R(t)$$

$$V_L(t) = V - V_R - V_C$$

$$\frac{di_L(t)}{dt} = \frac{1}{L}(V - R.i(t) - V_C) \quad (1)$$

$$\frac{dV_C(t)}{dt} = \frac{i_C(t)}{C} = \frac{i(t)}{C} \quad (2)$$

Number of state equations = number of dynamic elements = 2

Here we may choose $\begin{bmatrix} i_L(t) \\ V_C(t) \end{bmatrix}$ as the state variables. Therefore, the state and output equations respectively can be written as

$$\frac{d}{dt}\begin{bmatrix} i_L(t) \\ V_C(t) \end{bmatrix} = \begin{bmatrix} -R/L & -1/L \\ 1/C & 0 \end{bmatrix}\begin{bmatrix} i_L(t) \\ V_C(t) \end{bmatrix} + \begin{bmatrix} 1/L \\ 0 \end{bmatrix}e(t)$$

$$e_C(t) = \begin{bmatrix} 0 & 1 \end{bmatrix}\begin{bmatrix} i_L(t) \\ V_C(t) \end{bmatrix}$$

MATLAB codes

```
% simulation program for state space model
clear all
t=0;%simulation starting time
dt=0.01;%step size
tsim=10.0;%finish time
n=round((tsim-t)/dt);%no. of iterations
%system parameters
R=10;
L=15;
C=.05;
A=[-R/L -1/L; 1/C 0];
B=[1/L; 0];
C=[0 1];
D=[0 0];
X=[0 0]';
```

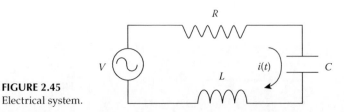

FIGURE 2.45
Electrical system.

Systems Modeling

```
for i=1:n;
  u=5*exp(-t);%exponential input
% u=4*sin(t);%sinusoidal input
% u=4;%fixed input

  dx=A*X+B*u;
  X=X+dx*dt;
  Y=C*X;
  Y1(i,:)=[t,Y];
  X1(i,:)=[t,X'];
  t=t+dt;
end
subplot(2,1,1)
plot(X1(:,1),X1(:,2:3))
axis([0 10 -10 10])
xlabel('time')
ylabel('state variables')
title('Response of state variables')

subplot(2,1,2)
plot(Y1(:,1),Y1(:,2))
axis([0 10 -10 10])
xlabel('time')
ylabel('output variable')
title('Response of output variable')
```

The simulation results of above mentioned MATLAB program are shown in Figure 2.46.

There is no unique way of selecting the state variable, i.e., we can select another set of state variables and therefore we will have different state and output equations. This can be illustrated by choosing $X_1 = i(t)$ and $X_2 = \dfrac{di}{dt}$ as the state variables in the previous example.

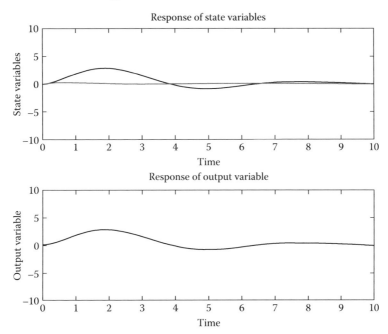

FIGURE 2.46
Results of state model simulation.

The Kirchhoff's voltage law can be used to write the loop equation:

$$e(t) = Ri(t) + L\frac{di(t)}{dt} + \frac{1}{C}\int i_C(t)dt \qquad (2.81)$$

$$\frac{de(t)}{dt} = R\frac{di(t)}{dt} + L\frac{d^2i(t)}{dt^2} + \frac{1}{C}i(t)$$

Example 2.16

Figure 2.47 shows an electrical circuit consisting of a capacitor, an inductor, and a resistor. The inductor and the capacitor are connected in parallel. A voltage V_a is applied to the circuit. Derive a mathematical model for the system.

SOLUTION

Applying Kirchhoff's current law, we can write

$$i_1 = i_2 + i_3 \qquad (2.82)$$

Now, the potential difference across the inductor and also across the capacitor is v_c. Similarly, the potential difference across the resistor is $V_a - v_c$. Thus,

$$i_1 = \frac{V_a - v_c}{R} \qquad (2.83)$$

$$i_2 = C\frac{dv_c}{dt} \qquad (2.84)$$

$$i_3 = \frac{1}{L}\int v_c dt \qquad (2.85)$$

Substituting Equations 2.83 through 2.85 in Equation 2.82 we get

$$\frac{V_a - v_c}{R} = C\frac{dv_c}{dt} + \frac{1}{L}\int v_c dt$$

or

$$\frac{R}{L}\int v_c dt + RC\frac{dv_c}{dt} + v_c = V_a$$

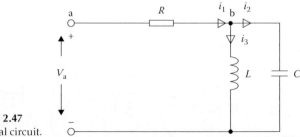

FIGURE 2.47
Electrical circuit.

Example 2.17

Figure 2.48 shows an electrical circuit consisting of two inductors, two resistors, and a capacitor. A voltage V_a is applied to the circuit. Derive an expression for the mathematical model for the circuit.

SOLUTION

The circuit consists of two nodes and two loops. We can apply Kirchhoff's current law to the nodes. For node 1,

$$i_1 + i_2 + i_3 = 0$$

or

$$\frac{V_a - v_1}{R_1} + \frac{1}{L_1}\int(0 - v_1)dt + \frac{1}{L_2}\int(v_2 - v_1)dt = 0 \qquad (2.86)$$

Differentiating Equation 2.86 with respect to time, we get

$$\frac{\dot{V}_a}{R_1} - \frac{\dot{v}_1}{R_1} - \frac{v_1}{L_1} + \frac{v_2}{L_2} - \frac{v_1}{L_2} = 0$$

$$\frac{\dot{V}_a}{R_1} = \frac{\dot{v}_1}{R_1} + \left(\frac{1}{L_1} + \frac{1}{L_2}\right)v_1 - \frac{v_2}{L_2} \qquad (2.87)$$

For node 2,

$$i_4 + i_5 + i_6 = 0$$

or

$$\frac{1}{L_2}\int(v_1 - v_2)dt + C\frac{d(0 - v_2)}{dt} + \frac{0 - v_2}{R_2} = 0 \qquad (2.88)$$

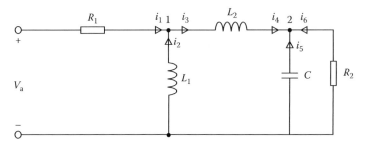

FIGURE 2.48
Electrical circuit.

Differentiating Equation 2.88 with respect to time, we obtain

$$\frac{(v_1 - v_2)}{L_2} - C\overset{oo}{v_2} - \frac{\overset{o}{v_2}}{R_2} = 0$$

which can be written as

$$C\overset{oo}{v_2} + \frac{\overset{o}{v_2}}{R_2} - \frac{v_1}{L_2} + \frac{v_2}{L_2} = 0 \qquad (2.89)$$

Equations 2.87 and 2.89 describe the state of the system. These two equations can be represented in matrix form as in equation

$$\begin{bmatrix} 0 & 0 \\ 0 & C \end{bmatrix} \begin{bmatrix} \overset{oo}{v_1} \\ \overset{oo}{v_2} \end{bmatrix} + \begin{bmatrix} \frac{1}{R_1} & 0 \\ 0 & \frac{1}{R_2} \end{bmatrix} \begin{bmatrix} \overset{o}{v_1} \\ \overset{o}{v_2} \end{bmatrix} + \begin{bmatrix} \frac{1}{L_1} + \frac{1}{L_2} & -\frac{1}{L_2} \\ -\frac{1}{L_2} & \frac{1}{L_2} \end{bmatrix} \begin{bmatrix} v_1 \\ v_2 \end{bmatrix} = \begin{bmatrix} \frac{\overset{o}{V_a}}{R_1} \\ 0 \end{bmatrix} \qquad (2.90)$$

2.7.3 Modeling of Electromechanical Systems

Electromechanical systems such as electric motors and electric pumps are used in most industrial and commercial applications. Figure 2.49 shows a simple dc motor circuit. The torque produced by the motor is proportional to the applied current and is given by

$$T = k_t i \qquad (2.91)$$

where
T is the torque produced
k_t is the torque constant
i is the motor current

Assuming there is no load connected to the motor, the motor torque can be expressed as given in Equation 2.92.

$$T = I\frac{d\omega}{dt} \quad \text{or} \quad I\frac{d\omega}{dt} = k_t i \qquad (2.92)$$

As the motor armature coil is rotating in a magnetic field there will be a back emf induced in the coil in such a way as to oppose the change producing it. This emf is proportional to the angular speed of the motor and is given by Equation 2.93.

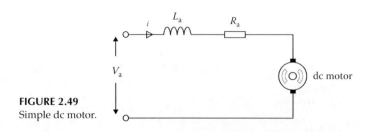

FIGURE 2.49
Simple dc motor.

Systems Modeling

$$v_b = k_e \cdot \omega \tag{2.93}$$

where
v_b is the back emf
k_e is the back emf constant
ω is the angular speed of the motor

Using Kirchhoff's voltage law, we can write the following equation for the motor circuit:

$$V_a - v_b = L_a \frac{di}{dt} + R_a i \tag{2.94}$$

where
V_a is the applied voltage
L_a and R_a are the inductance and the resistance of the armature circuit, respectively

From Equation 2.92,

$$i = \frac{I}{K_t} \frac{d\omega}{dt} \tag{2.95}$$

Substituting Equation 2.95 into Equation 2.94, we get

$$\frac{L_a I}{K_t} \frac{d^2\omega}{dt^2} + \frac{R_a I}{K_t} \frac{d\omega}{dt} + K_e \omega = V_a \tag{2.96}$$

Equation 2.96 is the model for a simple dc motor, describing the change of angular velocity with applied voltage. In many applications the motor inductance is small and can be neglected. The model then becomes

$$\frac{RI}{K_t} \frac{d\omega}{dt} + K_e \omega = V_a \tag{2.97}$$

Models of more complex dc motor circuits are given in the following examples.

Example 2.18

Figure 2.50 shows a dc motor circuit with a load connected to the motor shaft. Assume that the shaft is rigid, has negligible mass, and has no torsional spring effect or rotational damping associated with it. Derive an expression for the mathematical model for the system.

Solution

Since the shaft is assumed to be massless, the moments of inertia of the rotor and the load can be combined into I

$$I = I_M + I_L \tag{2.98}$$

where
I_M is the moment of inertia of the motor
I_L is the moment of inertia of the load

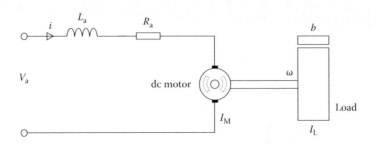

FIGURE 2.50
Direct current motor circuit.

Using Kirchhoff's voltage law, we can write the circuit equation for the motor:

$$V_a - v_b = L_a \frac{di}{dt} + R_a i \tag{2.99}$$

where
V_a is the applied voltage
L and R are the inductance and the resistance of the armature circuit, respectively

Substituting Equation 2.93, we obtain

$$V_a = L_a \frac{di}{dt} + R_a i + k_e \omega$$

or

$$V_a = L_a \frac{di}{dt} + R_a i + k_e \dot{\theta} \tag{2.100}$$

We can also write the torque equation as

$$T + T_L - b\omega = I \frac{d\omega}{dt} \tag{2.101}$$

Using Equation 2.91,

$$I \frac{d\omega}{dt} + b\omega - k_t i = T_L$$

or

$$I \ddot{\theta} + b \dot{\theta} - k_t i = T_L \tag{2.102}$$

Equations 2.100 and 2.102 describe the model of the circuit. These two equations can be represented in matrix form as

$$\begin{bmatrix} I & 0 \\ 0 & L \end{bmatrix} \begin{bmatrix} \ddot{\theta} \\ \dot{i} \end{bmatrix} + \begin{bmatrix} b & -k_t \\ k_e & R_a \end{bmatrix} \begin{bmatrix} \dot{\theta} \\ i \end{bmatrix} = \begin{bmatrix} T_L \\ V_a \end{bmatrix}$$

Systems Modeling

The above equation may be written in state-space form as Equation 2.103

$$\begin{bmatrix} \overset{\circ\circ}{\theta} \\ \overset{\circ}{i} \end{bmatrix} = \begin{bmatrix} T_L \\ V_a \end{bmatrix} - \begin{bmatrix} \dfrac{b}{I} & \dfrac{-k_t}{I} \\ \dfrac{k_e}{L} & \dfrac{R_a}{L} \end{bmatrix} \begin{bmatrix} \overset{\circ}{\theta} \\ i \end{bmatrix} \tag{2.103}$$

2.7.4 Modeling of Fluid Systems

Gases and liquids are collectively referred to as fluids. Fluid systems are used in many industrial as well as commercial applications. For example, liquid level control is a well-known application of liquid systems. Similarly, gas systems are used in robotics and in industrial movement control applications.

2.7.4.1 Hydraulic Systems

The basic elements of hydraulic systems are resistance, capacitance, and inertance (refer Figure 2.51). These elements are similar to their electrical equivalents of resistance, capacitance, and inductance. Similarly, electrical current is equivalent to volume flow rate, and the potential difference in electrical circuits is similar to pressure difference in hydraulic systems.

1. Hydraulic resistance
   Hydraulic resistance occurs whenever there is a pressure difference, such as liquid flowing from a pipe of one diameter to one of a different diameter. If the pressures at either side of a hydraulic resistance are P_1 and P_2, then the hydraulic resistance R is defined as

   $$P_1 - P_2 = R \dfrac{dg}{dt} \tag{2.104}$$

   where $\dfrac{dg}{dt} = q$ is the volumetric flow rate of the fluid.

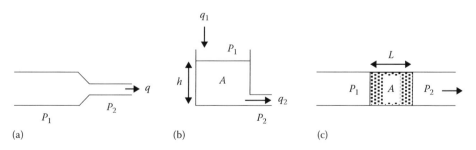

FIGURE 2.51
Hydraulic system elements. (a) Hydraulic resistance. (b) Hydraulic capacitance. (c) Hydraulic inertance.

2. **Hydraulic capacitance**

   Hydraulic capacitance is a measure of the energy storage in a hydraulic system. An example of hydraulic capacitance is a tank which stores energy in the form of potential energy. Consider the tank shown in Figure 2.51b. If q_1 and q_2 are the inflow and outflow, respectively, and V is the volume of the fluid inside the tank, we can write

   $$q_1 - q_2 = \frac{dV}{dt} = A\frac{dh}{dt} \tag{2.105}$$

   Now, the pressure difference is given by

   $$P_1 - P_2 = h\rho g = p \quad \text{or} \quad h = \frac{p}{\rho g} \tag{2.106}$$

   Substituting Equation 2.106 in Equation 2.105, we obtain

   $$q_1 - q_2 = \frac{A}{\rho g}\frac{dp}{dt} \tag{2.107}$$

   Writing Equation 2.107 as

   $$q_1 - q_2 = C\frac{dp}{dt} \tag{2.108}$$

   the hydraulic capacitance can be defined as

   $$C = \frac{A}{\rho g} \tag{2.109}$$

   Note that Equation 2.108 is similar to the expression for a capacitor and can be written as Equation 2.110.

   $$p = \frac{1}{C}\int (q_1 - q_2)dt \tag{2.110}$$

3. **Hydraulic inertance**

   Hydraulic inertance is similar to the inductance in electrical systems and is derived from the inertia force required to accelerate fluid in a pipe.

   Let $P_1 - P_2$ be the pressure drop that we want to accelerate in a cross-sectional area A, where m is the fluid mass and v is the fluid velocity. Applying Newton's second law, we can write

   $$m\frac{dv}{dt} = A(P_1 - P_2) \tag{2.111}$$

Systems Modeling

If the pipe length is L, then the mass is given by $m = L\rho A$

We can now write Equation 2.111 as $L\rho A \dfrac{dv}{dt} = A(P_1 - P_2)$

$$(P_1 - P_2) = L\rho \dfrac{dv}{dt} \tag{2.112}$$

but the rate of flow is given by $q = Av$, so Equation 2.112 can be written as

$$(P_1 - P_2) = \dfrac{L\rho}{A} \dfrac{dq}{dt} \tag{2.113}$$

The inertance I is then defined as $I = \dfrac{L\rho}{A}$, and thus the relationship between the pressure difference and the flow rate is similar to the relationship between the potential difference and the current flow in an inductor, i.e.,

$$(P_1 - P_2) = I \dfrac{dq}{dt} \tag{2.114}$$

Example 2.19

Figure 2.52 shows a liquid level system where liquid enters a tank at the rate of q_i and leaves at the rate of q_o through an orifice. Derive the mathematical model for the system, showing the relationship between the height h of the liquid and the input flow rate q_i.

Solution

From Equation 2.107, $q_i - q_o = \dfrac{A}{\rho g} \dfrac{dp}{dt}$

$$q_i = \dfrac{A}{\rho g} \dfrac{dp}{dt} + q_o \tag{2.115}$$

Recalling that $p = h\rho g$, Equation 2.115 becomes

$$q_i = A \dfrac{dh}{dt} + q_o \tag{2.116}$$

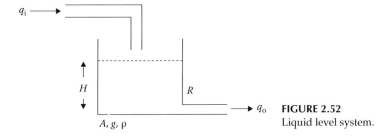

FIGURE 2.52
Liquid level system.

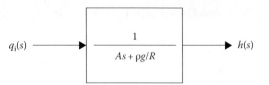

FIGURE 2.53
Block diagram of liquid level system.

Since $p_1 - p_2 = Rq_o$

$$q_o = \frac{p_1 - p_2}{R} = \frac{h\rho g}{R}$$

so that substituting in Equation 2.116 gives

$$q_i = A\frac{dh}{dt} + \frac{\rho g}{R}h \qquad (2.117)$$

Equation 2.117 shows the variation of the height of the water with the inflow rate. If we take the Laplace transform of both sides, we obtain

$$q_i(s) = Ash(s) + \frac{\rho g}{R}h(s) \qquad (2.118)$$

and the transfer function of the system can be written as

$$\frac{h(s)}{q_i(s)} = \frac{1}{As + \frac{\rho g}{R}} \qquad (2.119)$$

The block diagram is shown in Figure 2.53.

Example 2.20

Figure 2.54 shows a two-tank liquid level system where liquid enters the first tank at the rate of q_i and then flows to the second tank at the rate of q_1 through an orifice of radius R_1. Water then leaves the second tank at the rate of q_o through an orifice of radius R_2. Derive the mathematical model for the system.

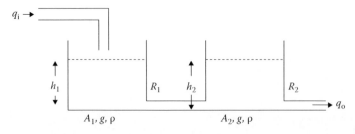

FIGURE 2.54
Two tank liquid level system.

Systems Modeling

SOLUTION

The solution is similar to the previous example of hydraulic system, but we have to consider both tanks.

For tank 1,

$$q_i - q_1 = \frac{A_1}{\rho g}\frac{dp}{dt}$$

or

$$q_i = \frac{A_1}{\rho g}\frac{dp}{dt} + q_1 \quad (2.120)$$

But

$$P = h\rho g,$$

thus Equation 2.120 becomes

$$q_i = A_1 \frac{dh_1}{dt} + q_1 \quad (2.121)$$

Since

$$p_1 - p_2 = R_1 q_1$$

or

$$q_1 = \frac{p_1 - p_2}{R_1} = \frac{h_1 \rho g - h_2 \rho g}{R_1}$$

we have

$$q_i = A_1 \frac{dh_1}{dt} + \frac{h_1 \rho g - h_2 \rho g}{R_1} \quad (2.122)$$

For tank 2,

$$q_1 - q_o = \frac{A_2}{\rho g}\frac{dp}{dt} \quad (2.123)$$

and with $P = h\rho g$
Equation 2.123 becomes

$$q_1 - q_o = A_2 \frac{dh_2}{dt} \quad (2.124)$$

But $q_i = \dfrac{p_1 - p_2}{R_1}$ and $q_o = \dfrac{p_2 - p_3}{R_2}$

so $q_i - q_o = \dfrac{p_1 - p_2}{R_1} - \dfrac{p_2 - p_3}{R_2} = \dfrac{h_1 \rho g - h_2 \rho g}{R_1} - \dfrac{h_2 \rho g}{R_2}$

Substituting in Equation 2.124, we obtain

$$A_2 \frac{dh_2}{dt} - \frac{\rho g h_1}{R_1} + \left(\frac{1}{R_1} + \frac{1}{R_2}\right) \rho g h_2 = 0 \qquad (2.125)$$

Equations 2.122 and 2.125 describe the behavior of the system. These two equations can be represented in matrix form as

$$\begin{bmatrix} A_1 & 0 \\ 0 & A_2 \end{bmatrix} \frac{d}{dt} \begin{bmatrix} h_1 \\ h_2 \end{bmatrix} + \begin{bmatrix} \dfrac{\rho g}{R_1} & \dfrac{-\rho g}{R_1} \\ \dfrac{-\rho g}{R_1} & \dfrac{\rho g}{R_1} + \dfrac{\rho g}{R_2} \end{bmatrix} \begin{bmatrix} h_1 \\ h_2 \end{bmatrix} = \begin{bmatrix} q_i \\ 0 \end{bmatrix}$$

or

$$\frac{d}{dt} \begin{bmatrix} h_1 \\ h_2 \end{bmatrix} = \begin{bmatrix} \dfrac{q_i}{A_1} \\ 0 \end{bmatrix} - \begin{bmatrix} \dfrac{\rho g}{A_1 R_1} & \dfrac{-\rho g}{A_1 R_1} \\ \dfrac{-\rho g}{A_2 R_1} & \dfrac{\rho g}{A_2 R_1} + \dfrac{\rho g}{A_2 R_2} \end{bmatrix} \begin{bmatrix} h_1 \\ h_2 \end{bmatrix} \qquad (2.126)$$

2.7.5 Modeling of Thermal Systems

Thermal systems are encountered in chemical processes like heating, cooling, and air conditioning systems, power plants, etc. Thermal systems have two basic components: thermal resistance and thermal capacitance. Thermal resistance is similar to the resistance in electrical circuits. Similarly, thermal capacitance is similar to the capacitance in electrical circuits. The across variable, which is measured across an element, is the temperature, and the through variable is the heat flow rate. In thermal systems there is no concept of inductance or inertance. Also, the product of the across variable and the through variable is not equal to power. The mathematical modeling of thermal systems is usually complex because of the complex distribution of the temperature. Simple approximate models can, however, be derived for the systems commonly used in practice.

Thermal resistance R is the resistance offered to the heat flow, and is defined as

$$R = \frac{T_2 - T_1}{q} \qquad (2.127)$$

where
T_1 and T_2 are the temperatures
q is the heat flow rate

Thermal capacitance is a measure of the energy storage in a thermal system. If q_1 is the heat flowing into a body and q_2 is the heat flowing out then the difference $q_2 - q_1$ is stored by the body, and we can write

Systems Modeling

$$q_2 - q_1 = mc\frac{dT}{dt} \qquad (2.128)$$

If we let the heat capacity be denoted by C, then

$$q_2 - q_1 = C\frac{dT}{dt} \qquad (2.129)$$

where
$C = mc$
m is the mass
c is the specific heat capacity of the body

Example 2.21

Figure 2.55 shows a room heated with an electric heater. The inside of the room is at temperature T_r and the walls are assumed to be at temperature T_w. If the outside temperature is T_o, develop a model of the system to show the relationship between the supplied heat q and the room temperature T_r.

Solution

The heat flow from inside the room to the walls is given by

$$q_{rw} = \frac{T_r - T_w}{R_r} \qquad (2.130)$$

where R_r is the thermal resistance of the room.
Similarly, the heat flow from the walls to the outside is given by

$$q_{wo} = \frac{T_w - T_o}{R_w} \qquad (2.131)$$

where R_w is the thermal resistance of the walls.
Heating equation may be written as

$$q - q_{rw} = C_1 \frac{dT_r}{dt} \qquad (2.132)$$

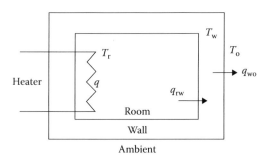

FIGURE 2.55
Simple thermal system.

where q is the heat flow rate from heater:

$$q - \frac{T_r - T_w}{R_r} = C_1 \frac{dT_r}{dt}$$

or

$$C_1 \overset{o}{T_r} + \frac{T_r - T_w}{R_r} = q \qquad (2.133)$$

Heat equation for

$$q_{rw} - q_{wo} = C_2 \frac{dT_w}{dt}$$

$$\frac{T_r - T_w}{R_r} - \frac{T_w - T_o}{R_w} = C_2 \frac{dT_w}{dt}$$

or

$$C_2 \overset{o}{T_w} - \frac{T_r}{R_r} + \left(\frac{1}{R_r} + \frac{1}{R_w}\right) T_w = \frac{T_o}{R_w} \qquad (2.134)$$

Equations 2.111 and 2.112 describe the behavior of the system, and they can be written in matrix form as

$$\begin{bmatrix} C_1 & 0 \\ 0 & C_2 \end{bmatrix} \begin{bmatrix} \overset{o}{T_r} \\ \overset{o}{T_w} \end{bmatrix} + \begin{bmatrix} \frac{1}{R_r} & -\frac{1}{R_r} \\ -\frac{1}{R_r} & \frac{1}{R_r} + \frac{1}{R_w} \end{bmatrix} \begin{bmatrix} T_r \\ T_w \end{bmatrix} = \begin{bmatrix} q \\ \frac{T_o}{R_w} \end{bmatrix}$$

or

$$\begin{bmatrix} \overset{o}{T_r} \\ \overset{o}{T_w} \end{bmatrix} = \begin{bmatrix} \frac{q}{C_1} \\ \frac{T_o}{C_2 R_w} \end{bmatrix} - \begin{bmatrix} \frac{1}{C_1 R_r} & -\frac{1}{C_1 R_r} \\ -\frac{1}{C_2 R_r} & \frac{1}{C_2 R_r} + \frac{1}{C_2 R_w} \end{bmatrix} \begin{bmatrix} T_r \\ T_w \end{bmatrix} \qquad (2.135)$$

The MATLAB codes for simulating this system are given below and the system results are shown in Figure 2.56.

MATLAB codes

```
% Simulation program for Thermal System
% Initialization
clear all;
C1=0.5;
C2=1.5;
```

Systems Modeling

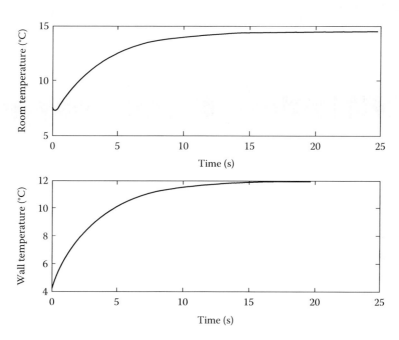

FIGURE 2.56
Temperature variation for thermal system.

```
Rr=0.5;
Rw=1.8;
Tr=8;
Tw=4;
To=3;
q=5;
A=[1/(C1*Rr)      -1/(C1*Rr)
   -1/(C2*Rr)     (1/(C2*Rr))+(1/(C2*Rw))];
B=[1/C1 0
 0 1/(C2*Rw)];
X=[Tr; Tw;];
dt=0.1;          % step size
t=0;             % Initial time
tsim=25;         % Simulation time
n=round(tsim-t)/dt;
for i=1:n
       X1(i,:)=[X' t];
   dX=-A*X+B*[q; To];
   X=X+dt*dX;
   t=t+dt;
end
Subplot(2,1,1)   % Divides the graphics window into sub windows
plot(X1(:,3),X1(:,1))
xlabel('Time (sec.)')
ylabel('Room Temperature degree C')
Subplot(2,1,2)   % Divides the graphics window into sub windows
plot(X1(:,3),X1(:,2))
xlabel('Time (sec.)')
ylabel('Wall Temperature degree C')
```

Example 2.22: Piston–Crank Mechanism

It converts, rotary motion into reciprocating motion using a piston, a connecting rod, and a crank (Figure 2.57).
Calculate and plot the position, velocity, and acceleration of the piston for two revolutions of the crank. Assume the initial condition zero.

Solution

The initial conditions are zero, i.e., at $t = 0$, $\theta = 0$, $\dot{\theta} = 0$ and $\ddot{\theta} = 0$.
The distance d_1 and h can be calculated from Figure 2.58 as

$$d_1 = r\cos\theta \quad \text{and} \quad h = r\sin\theta$$

If we know the distance h, then d_2 can be calculated as

$$d_2 = \sqrt{c^2 - h^2} = \sqrt{c^2 - r^2 \sin^2\theta}$$

The position x of piston is then given by $x = d_1 + d_2 = r\cos\theta + \sqrt{c^2 - r^2 \sin^2\theta}$
The velocity of the piston is the derivate of displacement of the piston

$$\dot{x} = -r\dot{\theta}\sin\theta + \frac{r^2 \dot{\theta} \sin 2\theta}{2\sqrt{(c^2 - r^2 \sin^2\theta)}}$$

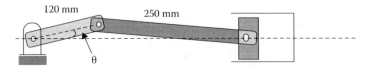

FIGURE 2.57
Piston crank mechanism.

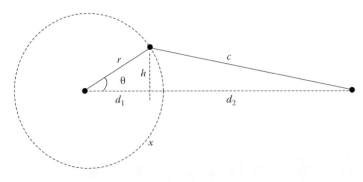

FIGURE 2.58
Motion of crank and piston.

The second derivate of displacement of the piston gives the acceleration

$$\overset{00}{x} = -r\overset{0^2}{\theta}\cos\theta + \frac{4r^2\overset{0}{\theta}\cos 2\theta(c^2 - r^2\sin^2\theta) + (r^2\overset{0}{\theta}\sin 2\theta)^2}{2\sqrt{(c^2 - r^2\sin^2\theta)}}$$

```
% Matlab codes to simulate the Piston-Crank mechanism
clear all
theta_dot=500;              % rpm
r=0.12; c=0.25;             % m
t_rev=2*pi/ theta_dot       % Time for one revolution
t=linspace(0,t_rev, 200);   % time vector
theta=theta_dot*t ;         % calculate theta at each time.
h=r*sin(theta);
d2s=c^2-r^2*sin(theta).^2;  %calculate d2 square
d2=sqrt(d2s);
x=r*cos(theta)+d2;          % calculate x for each theta
x_dot=-r*theta_dot*sin(theta)-(r^2*theta_dot*sin(2*theta)./(2*d2));
x_dot_dot=-r*theta_dot^2*cos(theta)-(4*r^2*theta_dot^2*cos(2*theta).* ...
d2s+(r^2*sin(2*theta)*theta_dot).^2)./(4*d2s.^(3/2));
subplot(2,2,1)
plot(t,x)                   % plot Position vs t
xlabel('Time(s)')
ylabel('Position(m)')
subplot(2,2,2)
plot(t,x_dot)               % plot Velocity vs t
xlabel('Time(s)')
ylabel('Velocity(m/s)')
subplot(2,2,3)
plot(t,x_dot_dot)           % plot Acceleration vs t
xlabel('Time(s)')
ylabel('Acceleration(m/s^2)')
subplot(2,2,4)
plot(t,h)                   % plot h vs t
xlabel('Time(s)')
ylabel('h(m)')
```

When the above written MATLAB code is run, it generates the plots shown in Figure 2.59. The figure shows that the velocity of the piston is zero at the end points (bottom dead center and top dead center) of the travel range where the piston changes the direction of motion. The acceleration is maximum when the piston is at the right end.

Example 2.23: Airplane Deceleration with Brake Parachute

An airplane uses a brake parachute and other means of braking as it slows down on the runway after landing as shown in Figure 2.60. Its acceleration is given by $a = -0.0045 v^2 - 3$ m/s^2. Consider an airplane with a velocity of 300 km/h that opens its parachute and starts deceleration at $t = 0$ s.
Determine

(a) Velocity as function of time from $t = 0$ until airplane stops.
(b) Distance that airplane travels as a function of time.

Also write a MATLAB program to plot the distance traveled and the velocity vs. time for the system shown in Figure 2.60.

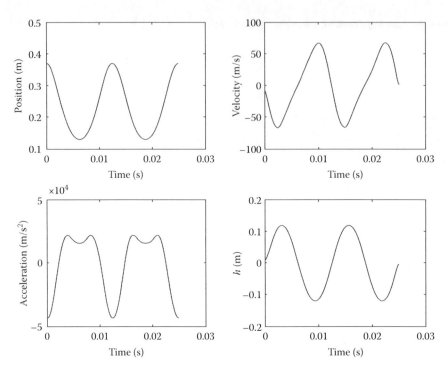

FIGURE 2.59
Position, velocity, acceleration, and *h* vs. time.

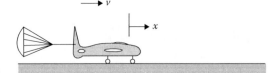

FIGURE 2.60
Airplane slowdown by brake parachute.

```
% Matlab codes to simulate the airplane braking using parachute
clear all
v=300;              % velocity (km/h)
tsim=100            % step size
npoint=10000;
dt=tsim/npoint;
acc=0;d=0;
t=0;
for i=1:npoint
   x1(i,:)=[acc,v, d, t];
   acc=-0.0045*(v*1000/3600)^2-3;      %m/s^2
   v=v+dt*acc;
   d=d+dt*v;
   t=t+dt;
   if (v<=0); break; end
end
t1=x1(:,4);
subplot(2,2,1)
plot(t1,x1(:,3))             % plot Position vs t
```

Systems Modeling

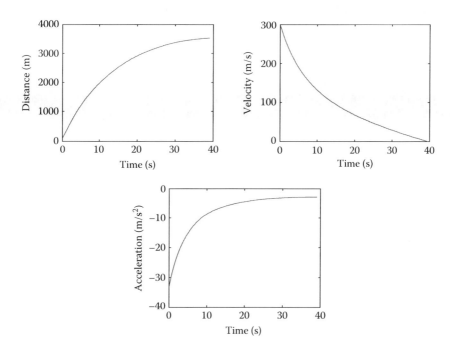

FIGURE 2.61
Simulation results of airplane problem.

```
xlabel('Time(s)')
ylabel('Distance(m)')
subplot(2,2,2)
plot(t1,x1(:,2))              % plot Velocity vs t
xlabel('Time(s)')
ylabel('Velocity(m/s)')
subplot(2,2,3)
plot(t1,x1(:,1))              % plot Acceleration vs t
xlabel('Time(s)')
ylabel('Acceleration(m/s^2)')
disp('time required in stopping the airplane')
```

The results of MATLAB program are shown in Figure 2.61 and time for stopping is 39.35 s.

2.8 Review Questions

1. Define a system model.
2. What is the need of a system model?
3. How do system models expedite the design and development of systems.
4. What are model validation and model verification?
5. How do we validate and verify a given model?
6. How can we classify models?

7. What are the steps involved in model development?
8. State the fundamental axiom of the system discipline.
9. What do you understand by physical systems?
10. Give examples of complementary terminal variables—through and across variables in the context of typical physical systems (electrical, magnetic, mechanical (transnational/rotational), pneumatic, and thermal systems).
11. Are derivatives and integrals of through and across variables respectively through and across variables or otherwise? Give suitable examples in support of your assertion.
12. State the component postulate of physical system theory.
13. a. Why are physical models required in system development.
    b. Mention the objective in using mathematical models.
    c. Give generic description of the terminal characteristics of the following two-terminal components: dissipater, delay, storage (or accumulator), and driver (or generator).
14. Figure 2.62 shows a simple mechanical system consisting of a mass, a spring, and a damper. Derive a mathematical model for the system, determine the transfer function, and draw the block diagram.
15. Consider the system of two springs of zero mass, a dashpot, and a mass as shown in Figure 2.63. Derive a mathematical model for the system.
16. Three springs of zero mass with the same stiffness constant are connected in series. Derive an expression for the equivalent spring stiffness constant.
17. Repeat the question 16, if the springs are connected in parallel.
18. Figure 2.64 shows a simple mechanical system. Derive an expression for the mathematical model for the system.

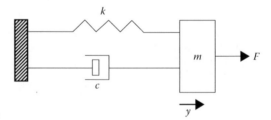

FIGURE 2.62
Simple mechanical system for Exercise 14.

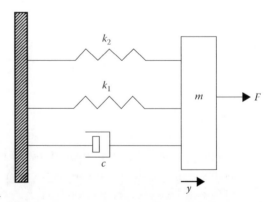

FIGURE 2.63
System of two massless springs for Exercise 15.

Systems Modeling

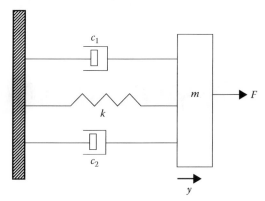

FIGURE 2.64
Simple mechanical system for Exercise 18.

19. Figure 2.65 shows a rotational mechanical system. Derive an expression for the mathematical model for the system.
20. Two rotational springs are connected in parallel. Derive an expression for the equivalent spring stiffness constant.
21. Figure 2.66 shows a simple system with a gear train. Derive an expression for the mathematical model for the system.
22. A simple electrical circuit is shown in Figure 2.67. Derive an expression for the mathematical model for the system.
23. Figure 2.68 shows an electrical circuit. Use Kirchhoff's laws to derive the mathematical model for the system.
24. A liquid level system is shown in Figure 2.69 where q_i and q_o are the inflow and outflow rates, respectively. The system has two fluid resistances, R_1 and R_2, in series. Derive an expression for the mathematical model for the system.

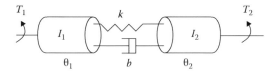

FIGURE 2.65
Simple mechanical system for Exercise 19.

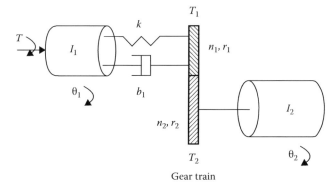

Gear train

FIGURE 2.66
Simple system with a gear train for Exercise 21.

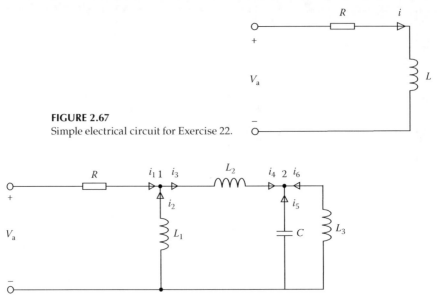

FIGURE 2.67
Simple electrical circuit for Exercise 22.

FIGURE 2.68
Electrical circuit for Exercise 23.

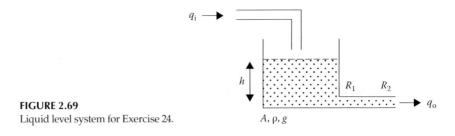

FIGURE 2.69
Liquid level system for Exercise 24.

2.9 Bibliographical Notes

The basic concepts of mathematical modeling have been well introduced in many books. Some of the important books worth to be mentioned here are written by Maki (1973), Burkhardt (1981), Berr et al. (1987), Murray (1987), Gershenfeld (1998), Ludmilla A. U. (2001), and Mark (2007). "Learn by doing" approach for mathematical modeling is given by Bender (2000). The mathematical modeling approach is very common for scientists and engineers. Some of the science and engineering problems have been modeled and explained by Peierls (1980), Lin and Segel (1988) and Hritonenko et al. (2003). Rutherford has written a good book to explain system's modeling from a chemical engineering viewpoint and explained the principles and pitfalls to all mathematical modeling of physical systems (Aris, 1999). The usefulness of mathematical modeling was nicely demonstrated for health services systems by D'Agostino et al. (1984) and for industrial systems by Cumberbatch and Fitt (2001).

3

Formulation of State Space Model of Systems

> Imagination is more important than Knowledge.
>
> **Albert Einstein**

> The purpose of science is not to analyze or describe but to make useful models of the world. A model is useful if it allows us to get use out of it.
>
> **Edward de Bono**

> The system of nature, of which man is a part, tends to be self-balancing, self-adjusting, self-cleansing. Not so with technology.
>
> **E. F. Schumacher**

3.1 Physical Systems Theory

The key to physical systems theory lies not in the notions of analogies but in the concept of a linear graph. To system analysts and designers, the linear graph can be considerably more than simply a collection of oriented line segments. The availability of high-speed computers has revolutionized the concept of what is essential in the mathematical modeling of components and systems.

If the computer is used for obtaining solutions, then t-domain models are much more convenient than the generally accepted (ω-domain or s-domain) models around which many of the existing analysis and design procedures, particularly in electrical engineering have been built. Since the digital computer inherently works best with a set of normal form of differential equations; the state space model has gravitated to the forefront.

As measured by the criteria of the efficient computer implementation and general applicability to both linear and nonlinear systems, the state space model is quickly emerging as the most useful model for analysis, synthesis, control, and optimization of engineering systems and certain socioeconomic systems and processes.

3.2 System Components and Interconnections

Consider the system of interacting components shown in Figure 3.1. Conceptually, black boxes, i.e., closed regions $\#B_j$, $j = 1, 2, \ldots, n$ may be regarded as representing components, i.e., meaningful units of a physical system or the various socially or economically meaningful sector or fundamental parts of a socioeconomic system.

These components are considered to have behavioral characteristics of their own, and their actions and behavior are transmitted to other components at points of contact (A, B, ..., K) between them, which thus represent interfaces. Each component is said to have a terminal, after the electrical usage, corresponding to each of its interfaces with other components. Terminal variables are values (numeric or nonnumeric) of component behavioral quantities measured at these terminals. The accuracy and frequency with which we record the chosen quantities is referred to as the space–time resolution level or simply, resolution level.

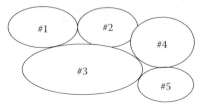

FIGURE 3.1
Conceptual system of interacting components.

The first notion of the linear graph pertinent to system modeling is that a single oriented line segment or an edge as shown in Figure 3.2 represents the measurement of two variables with respect to two distinct terminals on a component. More component terminals may be considered by introducing additional edge.

Mapping identifies a pair of complementary variables. Normally it is possible to associate two distinct terminal variables with each pair of component terminals in a physical system. The variables $X(t)$ is designated as the across variable. Since, it normally exhibits different values at terminals "a" and "b." The second variables $Y(t)$ is called the through variable. Since, it exhibits the same value at terminals "a" and "b." The arrow on the edge in Figure 3.2 also allows sign variation in the variables to be treated in a uniform manner. The concept of instantaneous oriented measurements is used as the basis for correlating the mathematical model with the physical or socioeconomic system in each case. Particular choice of complementary variables (i.e., across and through variables) depends on the type of system being modeled and a consistent choice enables the analysis of even mixed systems (electric, mechanical, pneumatic, magnetic, etc.) containing components of several types. Table 2.1 gives examples of systems and complementary variables.

It is interesting to note that the sign convention provides useful results, as far as power and energy are concerned. Specifically, these rules ensure that whenever a component instantaneously receives (or releasing) energy with respect to the pair of terminals then the product $X(t) \cdot Y(t)$ will be positive (or negative).

Each pair of complementary variables is also defined operationally, i.e., the variables are defined in terms of a specific set of operations (i.e., instrumentation) for assigning a numerical value to variables. So far as the mathematical development of the discipline of the physical systems theory is concerned, the pairs of complementary variables $X(t)$ and $Y(t)$ characterizing each of the physical and nonphysical processes are regarded as mathematically undefined. They are conceptually and operationally defined as the basis for characterizing a given process. All other variables of the system including power and energy are mathematically defined in terms of these complementary variables. Terminal equations describing the component behavior are equations which relate these to sets of

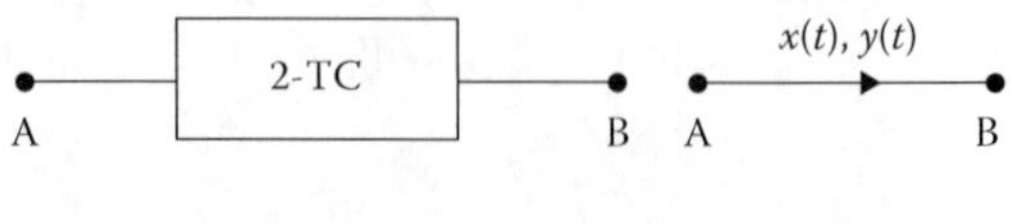

FIGURE 3.2
Edge representing two measurements on a 2T component.

Formulation of State Space Model of Systems

complementary terminal variable for the components of a given system, considered at least conceptually and in isolation.

The complementary terminal variables are so chosen that at any instant of time, the product of Fundamental across variable and Fundamental through variable is equal to power delivered to the component terminal pair, provided associated sign convention is followed. The power for n-terminal component can be written mathematically as a vector product of across and through variables:

$$P(t) = X(t) \cdot Y^T(t) = Y(t) \cdot X^T(t)$$
$$= x_1(t)y_1(t) + x_2(t)y_2(t) + \cdots + x_n(t)y_n(t) \quad (3.1)$$

where
Across variable vector $X(t) = [x_1(t), x_2(t), \ldots, x_n(t)]$
Through variable vector $Y(t) = [y_1(t), y_2(t), \ldots, y_n(t)]$

Energy of n-terminal component may be calculated as the integration of power over a time interval $t_0 \leq t \leq t_1$:

$$\int E = \frac{1}{2} \int_{t=0}^{t=t_1} P(t) dt \quad (3.2)$$

An exception to the choice of complementary terminal variables is the thermal system, because the units consistent with energy are temperature and entropy, where entropy is nonmeasurable.

Isomorphism between the component terminal graphs and physical measurements of few selected systems are shown in Figure 3.3.

In general, for actually establishing the component model beyond selecting a terminal graph, it is required to select one set S_1 of $(n - 1)$ terminal variables as independent variable functions of time and the remaining set S_2 of $(n - 1)$ terminal variables as dependent variable functions of time. The only requirement on the sets S_1 and S_2 is that each contains $(n - 1)$ equations. The mapping shows the time dependent variables in S_2 as a function of the variables in S_1.

3.3 Computation of Parameters of a Component

Consider a three-way hydraulic coupler shown in Figure 3.4. It is now desired to compute a_{11}, a_{12}, etc.

To obtain the parameters, instruments and drivers (for excitation) are connected as shown in Figure 3.5. The terminal equations for the hydraulic coupler are given by

$$\begin{bmatrix} P_1(t) \\ P_2(t) \end{bmatrix} = \begin{bmatrix} a_{11} & a_{12} \\ a_{21} & a_{22} \end{bmatrix} \begin{bmatrix} \overset{o}{g}_1(t) \\ \overset{o}{g}_2(t) \end{bmatrix} \quad (3.3)$$

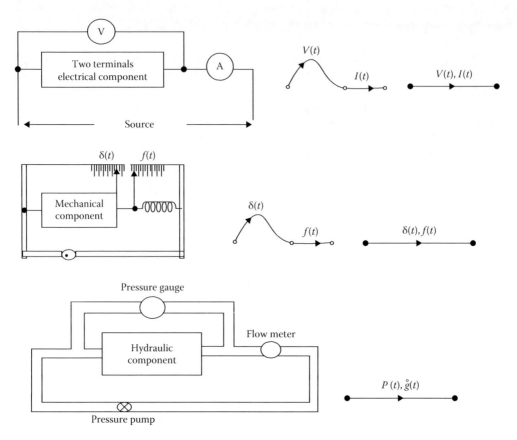

FIGURE 3.3
Various system components and their measurements.

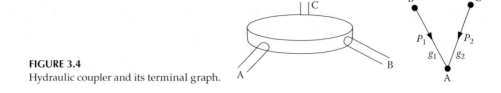

FIGURE 3.4
Hydraulic coupler and its terminal graph.

Pumps 1 and 2 are used to develop pressures P_1 and P_2, which in turn will cause respective flow rates $\overset{o}{g}_1(t)$ and $\overset{o}{g}_2(t)$. Close the valve V_2 to make $\overset{o}{g}_2(t) = 0$, and substituting in Equation 3.3 we obtain

$$\begin{bmatrix} P_1(t) \\ P_2(t) \end{bmatrix} = \begin{bmatrix} a_{11} \\ a_{21} \end{bmatrix} \overset{o}{g}_1(t) \qquad (3.4)$$

The parameters

$$a_{11} = \frac{P_1(t)}{\overset{o}{g}_1(t)} \quad \text{and} \quad a_{21} = \frac{P_2(t)}{\overset{o}{g}_1(t)}$$

Formulation of State Space Model of Systems

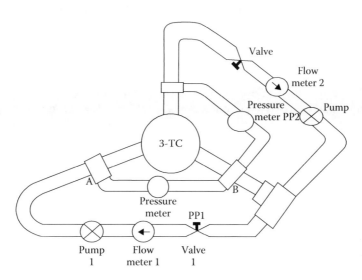

FIGURE 3.5
Schematic diagram for the measurement of open-circuit and short-circuit parameters.

Now open the valve V_2 and close valve V_1. This will make $g_1(t) = 0$ and, substituting in Equation 3.4 we get

$$\begin{bmatrix} P_1(t) \\ P_2(t) \end{bmatrix} = \begin{bmatrix} a_{12} \\ a_{22} \end{bmatrix} \overset{o}{g_2}(t) \tag{3.5}$$

This gives

$$a_{12} = \frac{P_1(t)}{\overset{o}{g_2}(t)} \quad \text{and} \quad a_{22} = \frac{P_2(t)}{\overset{o}{g_2}(t)}$$

It may be noticed that by setting one column of "AG" matrix at a time to zero, the elements of the nonzero column may be computed. Similarly, for an n-terminal component too, by setting all columns of the $[A]$ but one to zero in each test, a_{ij} of that column may be computed.

It may be noticed, in the above test, that

1. The terminal Equation 3.4 has been chosen in the impedance form similar to $V = ZI$ in an electrical system. Here Z is impedance, i.e., parameter by virtue of which the component impede the flow of current or fluid.
2. a_{ij}s have been computed by open circuiting all except one—flow rates between terminal pairs. Thus, these parameters are known as open circuit parameters, and the matrix A is known as an open circuit parametric matrix and the equation is called an open circuit parametric equation or admittance equation.

The equation may be expressed as

$$P_1(t) = a_{11} \overset{o}{g_1}(t) \tag{3.6}$$

$$P_2(t) = a_{21} \overset{o}{g}_1(t) \qquad (3.7)$$

In these equations $\overset{o}{g}_1(t)$ is the independent variable (i.e., it can be chosen arbitrarily). In Equations 3.6 and 3.7, flow rate $\overset{o}{g}_1(t)$ in the terminal pair AB develops a pressure $P_1(t)$ between the terminal pair AB itself and a pressure $P_2(t)$ between the terminal pair BC. The pressure $P_1(t)$ across the same terminal pair AB is by virtue of its impedance parameter a_{11} and a_{21}. If the flow rate $g_1(t)$ is equal to unity, then $P_1(t) = a_{11}$. This parameter a_{11} is called driving point impedance.

Similarly, flow rate $g_1(t)$ in the terminal pair AB causes a pressure of $P_2(t)$ across another terminal pair BC, as given by Equation 3.7. If we inject another flow rate $g_1(t)$ equal to unity then $a_{21} = P_2(t)$. This parameter a_{21} is called transfer impedance.

In general,

a_{ii} = open circuit driving point impedance
a_{ij} = open circuit transfer impedance

The equation $X(t) = AY(t)$ is called the open circuit parametric equation and coefficient matrix A is an open circuit parametric matrix.

Short circuit parameters

The hydraulic coupler of Figure 3.4 may be represented mathematically in admittance also called short circuit parametric form as:

$$Y(t) = GX(t) \qquad (3.8)$$

where $G = \dfrac{1}{A}$

in which the across variables are independent variables and the through variables are dependent variables. The component terminal equation in this form, then, becomes

$$\begin{bmatrix} Y_1(t) \\ Y_2(t) \end{bmatrix} = \begin{bmatrix} G_{11} & G_{12} \\ G_{21} & G_{22} \end{bmatrix} \begin{bmatrix} X_1(t) \\ X_2(t) \end{bmatrix} \quad \text{or}$$

$$\begin{bmatrix} \overset{o}{g}_1(t) \\ \overset{o}{g}_2(t) \end{bmatrix} = \begin{bmatrix} G_{11} & G_{12} \\ G_{21} & G_{22} \end{bmatrix} \begin{bmatrix} P_1(t) \\ P_2(t) \end{bmatrix} \qquad (3.9)$$

To compute G_{ij} we perform a short circuit test. First remove pump PP_2 and replace it by a short circuit. The Equation 3.9 after substituting $P_2 = 0$ becomes:

$$\begin{bmatrix} \overset{o}{g}_1(t) \\ \overset{o}{g}_2(t) \end{bmatrix} = \begin{bmatrix} G_{11} \\ G_{21} \end{bmatrix} [P_1(t)] \qquad (3.10)$$

$$\begin{bmatrix} G_{11} \\ G_{21} \end{bmatrix} = \frac{1}{P_1(t)} \begin{bmatrix} \overset{o}{g}_1(t) \\ \overset{o}{g}_2(t) \end{bmatrix} \quad (3.11)$$

From Equation 3.11 G_{11} and G_{21} can be calculated by making pressure $P_2(t) = 0$ (using short circuit test). Similarly one can calculate G_{21} and G_{22} by making $P_1(t) = 0$.

$$\begin{bmatrix} \overset{o}{g}_1(t) \\ \overset{o}{g}_2(t) \end{bmatrix} = \begin{bmatrix} G_{12} \\ G_{22} \end{bmatrix} [P_2(t)] \quad (3.12)$$

$$\begin{bmatrix} G_{12} \\ G_{22} \end{bmatrix} = \frac{1}{P_2(t)} \begin{bmatrix} \overset{o}{g}_1(t) \\ \overset{o}{g}_2(t) \end{bmatrix} \quad (3.13)$$

[G] parameters are thus computed by the short circuit in which one pump (across driver) is replaced by a SC element at a time, and hence these are called short circuit parameters.

If we short circuit PP_2 i.e. $P_2(t) = 0$ and inject a pressure $P_1(t) = 1$ between the terminals AB, then equation gives

$G_{11} = g_1(t)$: Flow rate measured between the terminal pair AB

$G_{12} = g_2(t)$: Flow rate measured between terminal pair BC used by injection of $P_1(t) = 1$ between AB

G_{ii}—Driving point admittance by the virtue of this component admits or causes a flow between terminal pairs

G_{ij}—Transfer admittance by the short circuit form of parametric equation for representing the component

3.4 Single Port and Multiport Systems

Consider a system containing fundamental two-terminal components like spring, mass, dampers or resistance, inductance, and capacitance only. These components have one terminal for input and one terminal for the output as shown in Figure 3.6.

The behavior of the system component can be mathematically written in any one of the following forms:

1. Impedance parameters

$$X(t) = A \quad Y(t) \quad (3.14)$$

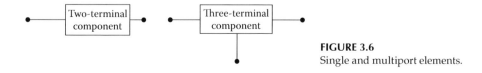

FIGURE 3.6
Single and multiport elements.

2. Admittance parameters

$$Y(t) = G \quad X(t) \quad (3.15)$$

3. Hybrid parameters

$$\begin{bmatrix} X_1(t) \\ Y_2(t) \end{bmatrix} = \begin{bmatrix} h_{11} & h_{12} \\ h_{21} & h_{22} \end{bmatrix} \begin{bmatrix} Y_1(t) \\ X_2(t) \end{bmatrix} \quad (3.16)$$

4. ABCD or cascade parameters

$$\begin{bmatrix} X_1(t) \\ Y_1(t) \end{bmatrix} = \begin{bmatrix} A & B \\ C & D \end{bmatrix} \begin{bmatrix} Y_2(t) \\ X_2(t) \end{bmatrix} \quad (3.17)$$

Applicability of these terminal equations depends upon the type of analysis, input and output parameters of interest, and ease with which these quantities may be measured. Some of the linear multiterminal components are described in the following section.

3.4.1 Linear Perfect Couplers

Couplers are components that connect two or more systems which are similar but of varying parameters. For example ideal gear couples two rotating systems at different speeds; ideal transformer couples two electric systems at different voltage levels; and ideal lever couples two mechanical systems at different forces; etc.

The ideal or perfect couplers are those coupling devices which do not consume energy. The terminal equations for ideal couplers may be written as

$$\begin{bmatrix} x_1(t) \\ y_2(t) \end{bmatrix} = \begin{bmatrix} 0 & h_{12} \\ -h_{21}^T & 0 \end{bmatrix} \begin{bmatrix} y_1(t) \\ x_2(t) \end{bmatrix} \quad (3.18)$$

For example, the terminal equation of a transformer (ref. Figure 3.7) is given by

$$\begin{bmatrix} v_1(t) \\ i_2(t) \end{bmatrix} = \begin{bmatrix} 0 & n \\ -n & 0 \end{bmatrix} \begin{bmatrix} i_1(t) \\ v_2(t) \end{bmatrix} \quad (3.19)$$

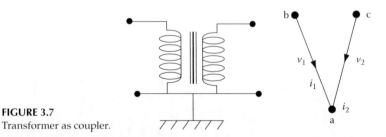

FIGURE 3.7
Transformer as coupler.

Formulation of State Space Model of Systems

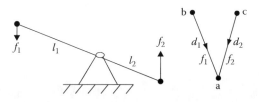

FIGURE 3.8
Lever for coupling mechanical systems and its terminal graph.

The terminal equation is given below and the terminal graph for ideal lever is illustrated in Figure 3.8.

$$\begin{bmatrix} d_1(t) \\ f_2(t) \end{bmatrix} = \begin{bmatrix} 0 & -l_1/l_2 \\ l_1/l_2 & 0 \end{bmatrix} \begin{bmatrix} f_1(t) \\ d_2(t) \end{bmatrix} \qquad (3.20)$$

There are no losses in ideal couplers and hence, power at input side and output side are the same, or in the true sense net power consumed is equal to zero.

$$P = x_1(t)\, y_1(t) + x_2(t)\, y_2(t)$$

or

$$P = \begin{bmatrix} y_1(t) & x_2(t) \end{bmatrix} \begin{bmatrix} x_1(t) \\ y_2(t) \end{bmatrix}$$

Substituting coupler terminal equation in this,

$$P = \begin{bmatrix} y_1(t) & x_2(t) \end{bmatrix} \begin{bmatrix} 0 & h_{12} \\ h_{12}^T & 0 \end{bmatrix} \begin{bmatrix} y_1(t) \\ x_2(t) \end{bmatrix} = 0$$

It is not possible to characterize a perfect coupler in open circuit or short circuit form of terminal equations. Since X_1 is reduced to zero in a coupler and then X_2 is zero.

Gyrator

Gyrator terminal equation is skewed symmetric in X explicit or Y explicit form only.

$$\begin{bmatrix} Y_1(t) \\ Y_2(t) \end{bmatrix} = \begin{bmatrix} 0 & G_{12} \\ -G_{12}^T & 0 \end{bmatrix} \begin{bmatrix} X_1(t) \\ X_2(t) \end{bmatrix} \qquad (3.21)$$

Notice that the coefficient matrix has the same form as that for ideal couplers. For ideal gyrator also, the net power is zero.

TABLE 3.1

Two-Terminal Components and Their Terminal Equations

Components	Terminal Equation in t-Domain	Terminal Equation in s-Domain	Schematic Diagrams
1. Electrical system components $v(t)$—Voltage across the component, $i(t)$—current through the component			
Resistance	$v(t) = R \cdot i(t)$	$V(s) = R \cdot I(s)$	R
Inductance	$v(t) = R \cdot di(t)/dt$	$V(s) = Ls \cdot I(s) - L \cdot i(0+)$	L
Capacitance	$i(t) = C \cdot dv(t)/dt$	$I(s) = Cs \cdot V(s) + v(0+)/s$	C
2. Translatory motion mechanical system $v(t)$—Linear velocity, $f(t)$—force acting on the component			
Damper	$f(t) = B \cdot v(t)$	$F(s) = B \cdot V(s)$	B
Spring	$df(t)/dt = K \cdot v(t)$	$sF(s) = K \cdot V(s) - f(0+)$	L
Mass	$f(t) = M \cdot dv(t)/dt$	$f(s) = Ms \cdot V(s) - M \cdot v(s)$	M
3. Rotary mechanical system $w(t)$—Angular velocity, $T(t)$—torque acting			
Rotational damper	$T(t) = B \cdot w(t)$	$T(s) = B \cdot W(s)$	B
Shaft stiffness	$dT(t)/dt = K \cdot w(t)$	$sT(s) = K \cdot W(s) - T(0+)$	K
Moment of inertia	$T(t) = J \cdot dw(t)/dt$	$T(s) = Js \cdot W(s) - J \cdot w(0+)$	I
4. Hydraulic system components $P(t)$—Pressure difference, $g(t)$—fluid flow rate			
Hydraulic resistance	$P(t) = R \cdot g(t)$	$P(s) = R \cdot G(s)$	R
Hydraulic capacitance	$g(t) = C \cdot dP(t)/dt$	$G(s) = Cs \cdot P(s) + sP(0+)$	C
5. Thermal system components $T(t)$—Temperature difference, $Q(t)$—heat flow rate			
Thermal conductance	$T(t) = Q(t)/G$	$T(s) = Q(s)/G$	R
Thermal capacitance	$Q(t) = C \cdot dT(t)/dt$	$Q(s) = C[sT(s) - T(0+)]$	C

Formulation of State Space Model of Systems

3.4.2 Summary of Two-Terminal and Multiterminal Components (Table 3.1)
3.4.3 Multiterminal Components

Multiterminal Components	Terminal Equation	
1. Vacuum tube	$\begin{bmatrix} i_g \\ v_p \end{bmatrix} = \begin{bmatrix} 0 & 0 \\ g_m & g_p \end{bmatrix} \begin{bmatrix} v_g \\ i_p \end{bmatrix}$	
2. Transformer	$\begin{bmatrix} v_2 \\ i_1 \end{bmatrix} = \begin{bmatrix} 0 & n \\ -n & 0 \end{bmatrix} \begin{bmatrix} i_2 \\ v_1 \end{bmatrix}$	
3. Transistor	$\begin{bmatrix} v_1 \\ i_2 \end{bmatrix} = \begin{bmatrix} h_{11} & h_{12} \\ h_{21} & h_{22} \end{bmatrix} \begin{bmatrix} i_1 \\ v_2 \end{bmatrix}$	
4. Iron core auto transformer	$\begin{bmatrix} v_1 \\ i_2 \end{bmatrix} = \begin{bmatrix} R_1 + L_1 d/dt & n_{12} \\ -n_{21} & 0 \end{bmatrix} \begin{bmatrix} i_1 \\ v_2 \end{bmatrix}$	
5. Gear train	$\begin{bmatrix} T_1 \\ w_2 \end{bmatrix} = \begin{bmatrix} B_1 + J_1 d/dt & n_{12} \\ -n_{21} & 0 \end{bmatrix} \begin{bmatrix} w_1 \\ T_2 \end{bmatrix}$	
6. Differential gear	$\begin{bmatrix} T_1 \\ T_2 \\ w_3 \end{bmatrix} = \begin{bmatrix} B_1 + J_1 d/dt & B_1 + J_1 d/dt & n_{12} \\ B_1 + J_1 d/dt & B_1 + J_1 d/dt & n_{23} \\ -n_{13} & -n_{23} & 0 \end{bmatrix} \begin{bmatrix} w_1 \\ w_2 \\ T_3 \end{bmatrix}$	
7. Rigid lever	$\begin{bmatrix} f_1 \\ v_2 \\ v_3 \end{bmatrix} = \begin{bmatrix} B_1 + M_1 d/dt & M_{11} & M_{13} \\ -M_{21} & 0 & 0 \\ -M_{31} & 0 & 0 \end{bmatrix} \begin{bmatrix} w_1 \\ w_2 \\ T_3 \end{bmatrix}$	
8. Rack and gear	$\begin{bmatrix} T_1 \\ v_2 \end{bmatrix} = \begin{bmatrix} B_1 + J_1 d/dt & n_{12} \\ -n_{21} & 0 \end{bmatrix} \begin{bmatrix} v_1 \\ T_2 \end{bmatrix}$	
9. Take up spool	$\begin{bmatrix} T_1 \\ v_2 \end{bmatrix} = \begin{bmatrix} B_1 + J_1 d/dt & -r \\ r & 0 \end{bmatrix} \begin{bmatrix} w_1 \\ f_2 \end{bmatrix}$	
10. Dancer pulley	$\begin{bmatrix} f_1 \\ f_2 \\ f_3 \end{bmatrix} = \begin{bmatrix} B + M d/dt & 1 & B + M d/dt \\ -1 & 0 & 2 \\ -(B + M d/dt) & -2 & B + M d/dt \end{bmatrix} \begin{bmatrix} w_1 \\ w_2 \\ w_3 \end{bmatrix}$	
11. Hydraulic reservoir	$\begin{bmatrix} P_1 \\ P_2 \end{bmatrix} = \begin{bmatrix} C_1 & R_{12} d/dt \\ R_{12} d/dt & R_2 d/dt \end{bmatrix} \begin{bmatrix} w_1 \\ f_2 \end{bmatrix}$	

(continued)

(continued)

Multiterminal Components	Terminal Equation
12. Electric potentiometer	$\begin{bmatrix} f_1 \\ V_2 \end{bmatrix} = \begin{bmatrix} B_1\,d/dt + M_1\,d^2/dt^2 & 0 \\ E/\delta_m & R\delta_1/\delta_m \end{bmatrix} \begin{bmatrix} \delta_1 \\ i_2 \end{bmatrix}$
13. Pneumatic bellows	$\begin{bmatrix} f_1 \\ q_2 \end{bmatrix} = \begin{bmatrix} K_1 & K_{12} \\ K_{12} & 0 \end{bmatrix} \begin{bmatrix} \delta_1 \\ i_2 \end{bmatrix}$
14. Tachometer	$\begin{bmatrix} T_1 \\ V_2 \end{bmatrix} = \begin{bmatrix} -K & B_1\,d/dt + M_1\,d^2/dt^2 \\ R + L\,d/dt & K\,d/dt \end{bmatrix} \begin{bmatrix} i_2 \\ \delta_1 \end{bmatrix}$

3.5 Techniques of System Analysis

Unless systematic methods of analysis are available to relate the characteristics of the proposed system to characteristics of components, one has no resource other than to build the system and try it. When the number and cost of the components are low, as in the development of a simple amplifier, "Build and Try" procedure is frequently the most expedient. However such a design would hardly be acceptable in the design of guided missile. Mathematical analysis methods are used to evaluate systematically the merits of a proposed system. Mathematical analysis techniques only extend the engineers' ability to explore the consequences of systems resulting from their imagination, and digital computers extend it even further.

In this chapter, we are primarily concerned with a discipline for developing mathematical models of interacting components of systems. These models are sets of simultaneous algebraic and/or differential equations showing the interdependence of sets of variables that characterize the observable behavior of the system. These equations are developed systematically from an identified system structure, i.e.,

1. Mathematical models of the components in the form of terminal equations
2. A mathematical model of their interconnection pattern in the form of interconnection equations, i.e., f-cutset and f-circuit equations

In this sense the system model reflects the system structure explicitly as a feature that is absolutely essential to sensitivity, reliability, stability, and other simulation studies as well as to all design and synthesis objectives. The system structure is very essential to study, especially if somebody wants to improve the system performance.

Various modeling techniques are available in the literature to model a given system. The selection of a suitable modeling and simulation technique for obtaining component model

Formulation of State Space Model of Systems

system models and their solution is very essential. Some of these techniques along with their merits and limitation are briefly presented below.

3.5.1 Lagrangian Technique

It is one of the oldest techniques of system analysis found in the area of mechanics. Lagrange in 1970 formalized what is known today as Lagrange's equation. In this technique, an energy function is associated with each component of the system. System energy functions are obtained by simply adding the energy of the components of the components. To obtain mathematical equations describing the characteristics of the system, it is necessary to take partial derivatives of system energy functions successively with respect to a set of variables called generalized coordinates.

3.5.2 Free Body Diagram Method

In this technique a schematic sketch of each rigid body along with a set of reference arrows is used to establish a mathematical relationship between through variables associated with the components. This technique cannot provide a systematic procedure for establishing a relation between across variables and through variables (forces and displacements both are required for computing the work, power and energy).

3.5.3 Linear Graph Theoretic Approach

This is based upon electrical network theory and proved to be the most effective tool. To establish mathematical characteristics of components and systems, instrumentation of one kind or other is involved, i.e., it is necessary to relate physical observations to real number systems.

The force/torque equation for mechanical systems may be written as

$$\sum f[k] = 0; \quad k = 1, 2, 3, \ldots, n \tag{3.22}$$

$$\text{or} \quad \sum T[k] = 0; \quad k = 1, 2, 3, \ldots, n \tag{3.23}$$

Similarly the current and voltage equations for electrical system may be written from Kirchoff's current law as

$$\sum i[k] = 0; \quad k = 1, 2, 3, \ldots, n \tag{3.24}$$

or Kirchoff's voltage law as

$$\sum v[k] = 0; \quad k = 1, 2, 3, \ldots, n \tag{3.25}$$

From the above discussion, it is clear that the algebraic sum of the through variables is equal to zero at a particular node.

3.6 Basics of Linear Graph Theoretic Approach

The basic definitions related with graph theoretic approach are given below, which are relevant to physical system modeling.

1. **Edge**

   An oriented line segment together with its distinct end point is called an *edge* as shown in Figure 3.9. Edge and element are generally synonyms.

2. **Vertex**

   A vertex or node is an end point of an edge. (A and B in Figure 3.9 are vertices.)

3. **Incidence**

   A vertex and an edge are incident with each other if the vertex is an end point of the edge.

4. **Linear graph**

   An oriented linear graph is defined as a set of oriented line segments (or edges) no two of which have a point in common that is not a vertex.

5. **Subgraph**

   A subgraph is any subset of the edges of the graph. A subgraph is in itself a graph.

6. **Isomorphism**

   Two graphs G and G′ are isomorphic (or concurrent) if there is one to one correspondence between the edges of G and G′ and which preserves the incidence relationship as shown in Figure 3.10.

7. **Edge sequence**

   If the edges of a graph or a subgraph can be ordered such that each edge has a vertex in common with the preceding edge (in the ordered sequence) and the other vertex in common with the succeeding edge it is called an edge sequence of a graph or subgraph. In Figure 3.10 *"adebfd"* is an edge sequence.

8. **Multiplicity**

   The number of times an edge appears in the edge sequence is the multiplicity of the edge in the given edge sequence. In the above example of edge sequence the multiplicity of *"a"* is 1 and the multiplicity of *"d"* is 2, and so on.

FIGURE 3.9
An edge.

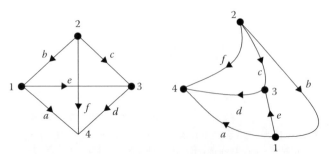

FIGURE 3.10
Isomorphism.

9. Edge train

If each edge of an edge sequence has a multiplicity one, then that edge sequence is an edge train. In Figure 3.10 a, d, c is an edge train.

10. Degree of vertex

The degree of vertex is the number of edges connected at that vertex. The degree of each vertex in Figure 3.10 is 3.

11. Path

If the degree of each internal vertex of an edge train is 2 and the degree of each terminal vertex is 1 the edge train is a path (a, d, c is a path between vertices 1 and 2).

12. Connected graph

A graph G is connected if there exists a path between any two vertices of the graph.

13. Complement of subgraph

The complement of a subgraph "S" of a graph "G" is the subgraph remaining in G when the edges of "S" are removed.

14. Separate part

A connected subgraph of a graph is said to be a separate part if it contains no vertices in common with its complement.

15. Tree

A tree is a connected subgraph of a connected graph, which contains no vertices in common with its complement. It has no circuit and contains all vertices.

Properties of tree
1. A subgraph is a tree if there exists only one path between any pair of vertices in it.
2. Every graph has at least one tree.
3. Every tree has at least two end vertices. The end vertex is a vertex at which only one edge is connected.
4. If a tree contains v vertices, then it has $(v - 1)$ edges.
5. The rank of a tree is $(v - 1)$. This is also the rank of the graph.

16. Circuit or loop

If an edge train is closed and has no terminal vertices, then that edge train is called a circuit or loop.

17. Branch

An element or edge of a tree of a connected graph is called a branch.

18. Cotree

A cotree is complement of the tree of a connected graph.

19. Chord or link

An edge of a cotree of a connected graph is called a chord or link.

20. Terminal graph

It is an oriented line segment that indicates the location of pairs of measuring instruments and the polarities of their connection.

21. System graph

System graph is the collection of component terminal graphs obtained by uniting the vertices of the terminal graphs in one to one correspondence with the union of component interfaces (terminals) in the actual system.

22. Lagrangian tree

Lagrangian tree of a graph is a tree whose branches are rooted at a single node (terminal) which is usually the reference node.

23. Forest

If the graph "G" of a system is in more than one part collection, then the tree of that multipart graph "G" is known as a forest.

24. Coforest

Complement of the forest of a multipart graph is known as a coforest.

25. Fundamental circuit (f-circuit)

The f-circuit is defined as a circuit containing exactly one and only one chord of the complement of the formulation tree. This chord is known as defining chord of the f-circuit.

Let "v" be the number of vertices
  "e" be the number of elements
  "p" is the number of separate parts in a given graph G

The number of f-circuit in a connected graph "G" = number of cotree chords
If G is a connected graph, it means $p = 1$, then

  number of tree branches = $(v - 1)$ and
  number of cotree chords = $(e - v + 1)$

else the number of branches of a forest = $(v - p)$ and

  number of chords of coforest = $(e - v + p)$

26. Rank of a graph

The rank of a graph "G" is the number of branches in tree i.e., $(v - p)$.

27. Nullity of a graph

The nullity of a graph is the number of chords in cotree i.e., $(e - v + p)$.

28. Cutset

A cutset of a connected graph "G" is defined as a set of edges "C" having the following properties:

1. When the set "C" is deleted the graph is exactly in two parts.
2. No subset of "C" has the property (1).
3. Removal of a cutset "C" reduces the rank of "G" by one.

If the vertices of a connected graph are divided into two mutual exclusive groups and contain exactly one branch of a given tree, it is called a fundamental cutset.

29. Fundamental cutset (f-cutset)

If the vertices of a connected graph is divided into two mutually exclusive groups and contains exactly one branch of a given tree, it is called a fundamental cutset.

Formulation of State Space Model of Systems

A fundamental cutset contains exactly one and only one branch of the formulation tree and the branch is called defining branch of the f-cutset.

The number of f-cutset in a connected graph "G" = number of tree branches = $(v-1)$.

30. **Incidence set**

An incidence set is defined as the set of all edges incident to any given vertex. When the incidence set is removed from a graph, the graph may be left in two or more than two parts.

3.7 Formulation of System Model for Conceptual System

Consider a conceptual system shown in Figure 3.11. Write f-cutset equation and f-circuit equations to formulate the system equations.

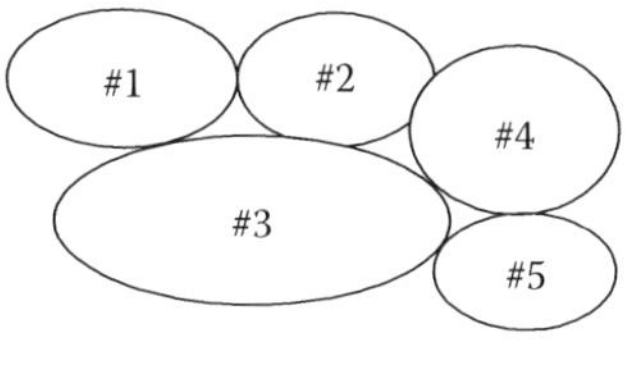

Component #1

It is a two-terminal component as shown in Figure 3.12 along with its terminal graph. Let the component terminal equation be written as

$$f_1(x_1(t), y_1(t), \overset{o}{x}_1(t), \overset{o}{y}_1(t), t) = 0 \tag{3.26}$$

Component #2

It is a three-terminal component as shown in Figure 3.13 with its two edges terminal graph. Terminal equation for this component may be written as

$$f_2(x_2(t), y_2(t), \overset{o}{x}_2(t), \overset{o}{y}_2(t), x_3(t), y_3(t), \overset{o}{x}_3(t), \overset{o}{y}_3(t), t) = 0 \tag{3.27}$$

$$f_3(x_2(t), y_2(t), \overset{o}{x}_2(t), \overset{o}{y}_2(t), x_3(t), y_3(t), \overset{o}{x}_3(t), \overset{o}{y}_3(t), t) = 0 \tag{3.28}$$

FIGURE 3.11
Conceptual system of interacting components.

Component #3

It is a four-terminal component as shown in Figure 3.14. The terminal equations may be written as follows:

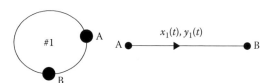

FIGURE 3.12
Two-terminal component and its terminal graph.

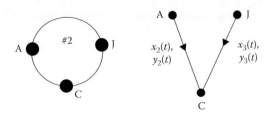

FIGURE 3.13
Three-terminal component and its terminal graph.

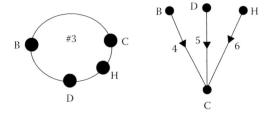

FIGURE 3.14
Four-terminal component and its terminal graph.

$$f_4(x_4(t), y_4(t), \overset{o}{x}_4(t), \overset{o}{y}_4(t), x_5(t), y_5(t), \overset{o}{x}_5(t), \overset{o}{y}_5(t), x_6(t), y_6(t), \overset{o}{x}_6(t), \overset{o}{y}_6(t), t) = 0 \quad (3.29)$$

$$f_5(x_4(t), y_4(t), \overset{o}{x}_4(t), \overset{o}{y}_4(t), x_5(t), y_5(t), \overset{o}{x}_5(t), \overset{o}{y}_5(t), x_6(t), y_6(t), \overset{o}{x}_6(t), \overset{o}{y}_6(t), t) = 0 \quad (3.30)$$

$$f_6(x_4(t), y_4(t), \overset{o}{x}_4(t), \overset{o}{y}_4(t), x_5(t), y_5(t), \overset{o}{x}_5(t), \overset{o}{y}_5(t), x_6(t), y_6(t), \overset{o}{x}_6(t), \overset{o}{y}_6(t), t) = 0 \quad (3.31)$$

Component #4

It is a three-terminal component (Figure 3.15). It has three edges in terminal graph and three-terminal equations:

$$f_7(x_7(t), y_7(t), \overset{o}{x}_7(t), \overset{o}{y}_7(t), x_8(t), y_8(t), \overset{o}{x}_8(t), \overset{o}{y}_8(t), t) = 0 \quad (3.32)$$

$$f_8(x_7(t), y_7(t), \overset{o}{x}_7(t), \overset{o}{y}_7(t), x_8(t), y_8(t), \overset{o}{x}_8(t), \overset{o}{y}_8(t), t) = 0 \quad (3.33)$$

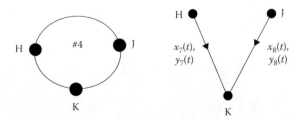

FIGURE 3.15
Three-terminal component and its terminal graph.

Formulation of State Space Model of Systems

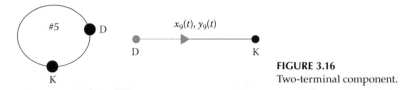

FIGURE 3.16
Two-terminal component.

Component #5

It is a two-terminal component, which may be specified by one edge in the terminal graph as shown in Figure 3.16 and one terminal equation as given below:

$$f_9(x_9(t), y_9(t), \overset{o}{x}_9(t), \overset{o}{y}_9(t), t) = 0 \tag{3.34}$$

Equations 3.26 through 3.34 represent the complete information about all the five components.

3.7.1 Fundamental Axioms

The fundamental axiom is a mathematical model of a component that characterize the behavior of that component of a system as an independent entity and does not depend on how the system component is interconnected with other components to form a system.

It implies that various components can be removed literally or conceptually from the remaining system, and can be studied in isolation to establish the model with their characteristics. This is a tool of science to make a system theory universal: analysts can go as far as they wish in breaking a system in such building blocks that are sufficiently simple to model, and can thus identify the structure that requires attention.

3.7.2 Component Postulate

The pertinent performance characteristics of each n-terminal component in an identified system structure are completely specified by a set of $(n-1)$ equations and $2(n-1)$ oriented complementary variables (across and through) identified by $(n-1)$ arbitrary chosen edges.

3.7.3 System Postulate

The system graph is formed by uniting component terminal graph in one to one correspondence with actual interconnections (interrelations) of the components in the system as shown in Figure 3.17.

These component interactions in a system may be mathematically expressed with two interconnection postulates:

1. f-cutset postulate
2. f-circuit postulate

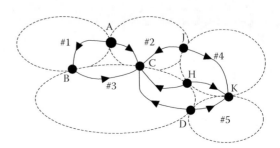

FIGURE 3.17
System graph of the conceptual system.

3.7.3.1 Cutset Postulate

The oriented sum of the through measurements $Y_i(t)$ implied by the elements of the given f-cutset is zero at any instant of time.

Mathematically, f-cutset equation may be written as

$$\sum_{k=1}^{e} a_{km} Y_k(t) = 0 \quad \text{for any } m = 1, 2, 3, \ldots, (v-1) \tag{3.35}$$

where
$a_{km} = 0$ if element k is not an element of the f-cutset m
$a_{km} = 1$ if element k is an element of the cutset m and its orientation is the same as that of its defining branch
$a_{km} = -1$ otherwise

In general, f-cutset equation in matrix form is $A \cdot Y(t) = 0$.

3.7.3.1.1 F-Cutset Equation for Conceptual System

A system graph of the previously discussed conceptual system may be drawn by combining all the terminal graphs in one to one correspondence of the components as shown in Figure 3.18. It has a tree consisting of edges (branches) 1, 2, 3, 5, 6, 8.

In the system graph

Number of vertices $v = 7$
Number of edges $e = 9$
Number of tree branches $= v - 1 = 6$
Number of chords of the cotree $= e - v + 1 = 3$

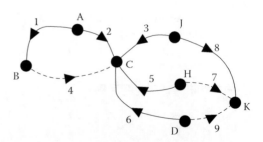

FIGURE 3.18
Tree of given system graph.

Formulation of State Space Model of Systems

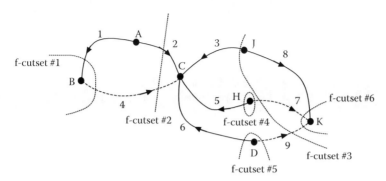

FIGURE 3.19
Different f-cutsets for conceptual system graph.

In system model formulation, we consider only the f-cutsets, as they are unique and give minimum number of cutsets. Corresponding to each tree branch a separate f-cutset is identified as shown in Figure 3.19.

1. Fundamental cutset #1 for defining branch #1 is a nodal cutset and the f-cutset equation may be written by taking the direction of the defining branch as the reference direction, which is going into the cutset and the direction of chord #4 is going out from the cutset. Therefore, $y_1(t)$ is positive and $y_4(t)$ is negative in the f-cutset equation as given by

$$y_1(t) - y_4(t) = 0 \tag{3.36}$$

2. Fundamental cutset #2 consists of a defining branch #2 and a chord #4. The f-cutset equation may be written for f-cutset #2 by considering the defining branch direction as the reference direction, which is going out of the cutset and the direction of chord #4 is also going out from the cutset same as the direction of defining branch. Hence, both the through variables in f-cutset equation are positive.

$$y_2(t) + y_4(t) = 0 \tag{3.37}$$

3. Fundamental cutset #3 is defined for defining branch #3. In the f-cutset equations, there are three through variables, because the cutset intersects three edges—one is the defining branch and the remaining two are chords. The direction of the defining branch is always considered positive and looks in the directions of remaining chords with respect to that particular defining branch. Hence, the cutset equation may be written as

$$y_3(t) - y_7(t) - y_9(t) = 0 \tag{3.38}$$

4. Fundamental cutset #4 for defining branch #5, which cuts branch #5 and chord #7. The direction of branch #5 is going out from the cutset and chord direction is also out from the cutset. Hence, both the through variables are positive in the f-cutset equation.

$$y_5(t) + y_7(t) = 0 \tag{3.39}$$

5. Fundamental cutset #5 which is a nodal cutset has defining branch #6 and chord #9. The direction of defining branch is going out from the cutset and the direction of chord #9 is also going out from the cutset.

$$y_6(t) + y_9(t) = 0 \qquad (3.40)$$

6. Fundamental cutset #6 is also a nodal cutset for the defining branch #8. The f-cutset equation is

$$y_7(t) + y_8(t) + y_9(t) = 0 \qquad (3.41)$$

Equations 3.36 through 3.41 can be written in a matrix form, keeping the tree branches in the leading position followed by the chords in the trailing position as

$$\begin{matrix} 1 & 2 & 3 & 5 & 6 & 8 & 4 & 7 & 9 \end{matrix}$$

$$\begin{bmatrix} 1 & 0 & 0 & 0 & 0 & 0 & -1 & 0 & 0 \\ 0 & 1 & 0 & 0 & 0 & 0 & 1 & 0 & 0 \\ 0 & 0 & 1 & 0 & 0 & 0 & 0 & -1 & -1 \\ 0 & 0 & 0 & 1 & 0 & 0 & 0 & 1 & 0 \\ 0 & 0 & 0 & 0 & 1 & 0 & 0 & 0 & 1 \\ 0 & 0 & 0 & 0 & 0 & 1 & 0 & 1 & 1 \end{bmatrix} \begin{bmatrix} y_1(t) \\ y_2(t) \\ y_3(t) \\ y_5(t) \\ y_6(t) \\ y_8(t) \\ y_4(t) \\ y_7(t) \\ y_9(t) \end{bmatrix} = \begin{bmatrix} 0 \\ 0 \\ 0 \\ 0 \\ 0 \\ 0 \\ 0 \\ 0 \\ 0 \end{bmatrix} \qquad (3.42)$$

where $y_3(t), y_5(t), y_6(t), y_8(t)$ form $Y_b(t)$ and $y_4(t), y_7(t), y_9(t)$ form $Y_c(t)$; U and A_c are the partitions.

$$A_{(v-1) \times e} Y(t)_{(v-1) \times 1} = 0$$

where matrix $A = [U \quad A_c]$

In the above equation matrix A may be partitioned as identity matrix (U) of size $(v - 1) \times (v - 1)$ and a submatrix A_c is called the f-cutset coefficient matrix and of size $(v - 1) \times (e - v + 1)$. The f-cutset equation may be written as

$$Y_b(t) = -A_c Y_c(t) \qquad (3.43)$$

Through variables are those which obey the relationship property of the above equation, i.e., the branch through variables are uniquely specified by the chord through variables. Therefore, the chord through variables are independent (primary) variables. The cutset equations are the continuity equations in hydraulic system, force equations in mechanical systems, and Kirchoff's current law in electrical systems.

3.7.3.2 Circuit Postulate

The oriented sum of the across measurements $Xi(t)$ implied by the elements of the given f-circuit is zero at any instant of time.

Mathematically, it may be written as given in Equation 3.44.

$$\sum_{k=1}^{e} b_{km} X_k(t) = 0 \quad \text{for any } m = 1, 2, 3, \ldots, (v-1) \tag{3.44}$$

where
$b_{km} = 0$ if element k is not an element of the f-circuit m
$b_{km} = 1$ if element k is an element of the circuit m and its orientation is same as that of its defining chord
$b_{km} = -1$ otherwise

In general, f-circuit equation in matrix form is $B. X(t) = 0$.

3.7.3.2.1 F-Circuit Equations for Conceptual System

From Figure 3.20 the f-circuit equations for the system graph may be obtained for defining chords as follows.

1. Fundamental circuit #1 consists of defining chord #4 and tree branches 1, 2. The direction of defining chord in the f-circuit is considered as reference direction and the directions of the tree branches in that particular circuit seen with respect to the defining chord. Hence, the across variable for defining chord $x_4(t)$ is positive and across variables for tree branches as $x_2(t)$ is negative and $x_1(t)$ positive. The f-circuit equation for f-circuit #1 is

$$x_1(t) - x_2(t) + x_4(t) = 0 \tag{3.45}$$

2. Fundamental circuit #2 has defining chord #7 and tree branches 3, 5, 8. The directions of across variables of tree branches 3, 5 and 8 with respect to defining chord #7 are positive and negative, respectively. The f-circuit equation may be written for the edges #3, #5, #7, and #8 as

$$x_3(t) - x_5(t) + x_7(t) - x_8(t) = 0 \tag{3.46}$$

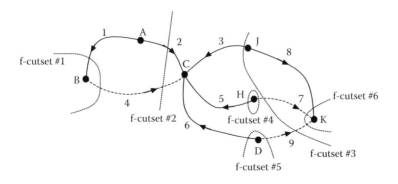

FIGURE 3.20
Fundamental circuits for given conceptual system graph.

3. Finally fundamental circuit #3 has defining chord #9 and tree branches #3, #8, #6. In this case the direction of the across variable for tree branch #6 is opposite to the direction of defining chord #9 and all the remaining tree branches have the direction same as that of the defining chord:

$$x_3(t) - x_8(t) - x_6(t) + x_9(t) = 0 \tag{3.47}$$

The above written f-circuit Equations 3.45 through 3.47 may be rearranged in the matrix form with tree branches in the leading portion and chords in the trailing portion of the matrix.

$$\overset{1\quad 2\quad 3\quad 5\quad 6\quad 8\quad 4\quad 7\quad 9}{\begin{bmatrix} 1 & -1 & 0 & 0 & 0 & 0 & 1 & 0 & 0 \\ 0 & 0 & 1 & -1 & 0 & -1 & 0 & 1 & 0 \\ 0 & 0 & 1 & 0 & -1 & -1 & 0 & 0 & 1 \end{bmatrix}} \begin{bmatrix} x_1(t) \\ x_2(t) \\ x_3(t) \\ x_5(t) \\ x_6(t) \\ x_8(t) \\ x_4(t) \\ x_7(t) \\ x_9(t) \end{bmatrix} \begin{matrix} \\ \\ x_b(t) \\ \\ \\ \\ x_c(t) \\ \end{matrix} = \begin{bmatrix} 0 \\ 0 \\ 0 \\ 0 \\ 0 \\ 0 \\ 0 \\ 0 \\ 0 \end{bmatrix} \tag{3.48}$$

$$B_{(e-v+1) \times e} \, X(t)_{(e-v+1) \times 1} = 0$$

where matrix $B = [B_b \quad U]$
The above equation may be partitioned as

$$[B_b \quad U] \begin{bmatrix} X_b(t) \\ X_c(t) \end{bmatrix} = 0$$

Submatrix B_b matrix is called the f-circuit coefficient matrix. The size of B_b matrix is $(e - v + 1) \times (v - 1)$.

The above equation can be written as

$$X_c(t) = -B_b X_b(t) \tag{3.49}$$

It is observed from Equation 3.49 that the chord across variables can be completely specified in terms of branch across variables. Hence, chord across variables are dependent (secondary) variables and branch across variables are independent (primary) variables. The primary variable vector for any given system consists of branch across variables and chord

Formulation of State Space Model of Systems

through variables $[X_b, Y_c]$ and the secondary variable vector contains branch through variables and chord across variables $[Y_b, X_c]$.

It is observed that for the conceptual system under consideration the coefficient matrices of f-cutset equation and f-circuit equation are negative transposes to each other.

$$A_c = -B_b^T \quad \text{or} \quad B_b = -A_c^T$$

This relation is not by chance, it is due to the **Orthogonal Property** of the f-cutset and f-circuit matrices.

3.7.3.2.2 Orthogonality of Circuit and Cutset Matrices

Each cutset vector (α_i) of a connected graph G is orthogonal to each circuit vector β_j of G, i.e., their product vanishes. For example, consider the first row α_i of f-cutset equation and also the first row of f-circuit equation β_j.

$$\alpha_i = [1\,0\,0\,0\,0\,0\,-1\,0\,0] \quad \text{and} \quad \beta_j = [1\,-1\,0\,0\,0\,0\,1\,0\,0]$$

$$\alpha_i(j)\,\beta_j^T = 0 \quad \text{or} \quad \beta_j\,\alpha_i(j)^T = 0$$

3.7.3.3 Vertex Postulate

The oriented sum of through measurements $Y_i(t)$ implied by the elements incident to a given vertex is zero at any instant of time.

Mathematically, it may be expressed as

$$\sum C_{km} y_k(t) = 0 \quad \text{for any } m = 1, 2, 3, \ldots, v \tag{3.50}$$

where
$C_{km} = 0$ if the element is incident to the vertex m
$C_{km} = 1$ if the element is incident to the vertex m and its orientation is away from m
$C_{km} = -1$ otherwise

Notice that the number of incidence sets is equal to the number of vertices v. But the number of f-cutset is equal to $(v - 1)$, i.e., equal to the number of branches.

3.8 Formulation of System Model for Physical Systems

The procedure outlined so far for obtaining the system model may be summarized in the following steps.

Step #1
Identify the components or the subsystems of the system. For each component identify the terminals and terminal variables, i.e., the across and through variables and the appropriate measuring system to measure these quantities.

Step #2

For each *n*-terminal component draw its terminal graph and write the terminal equation. The number of edges in the terminal graph and the number of terminal equations for *n*-terminal component are equal to $(n - 1)$.

Step #3

Combine the component terminal graph in one to one correspondence with the interconnections of the components in the actual systems to obtain the system graph "G." Identify the formulation tree for the system graph "G" according to the topological restrictions.

Step # 4

Write f-cutset and f-circuit equations. Then substitute terminal equations of components to get state model.

Linear perfect coupler

The coupler is a device which couples two similar systems. For example, to couple two rotary motion mechanical systems gears are used. Similarly, to couple translatory motion mechanical systems levers are used. In electrical and electronics systems transformers and transistors are used respectively. For hydraulic systems hydraulic couplers are used, and so on.

If there are two different types of systems coupled together, then that coupling device is called transducer. For example, techo-generator couples mechanical system to electrical system, pressure transducer converts mechanical pressure into electrical quantities, and so on. The general representation of a coupler or transducer is shown in Figure 3.21.

These coupling devices always need a little energy for their operations. But for simplification of the model we always assume that the coupling device is perfect and linear. A perfect or ideal coupler does not consume any energy.

$$\begin{bmatrix} x_1 \\ y_2 \end{bmatrix} = \begin{bmatrix} h_{11} & 0 \\ 0 & -h_{11}^T \end{bmatrix} \begin{bmatrix} y_1 \\ x_2 \end{bmatrix}$$

$$h_{11} = \frac{N_2}{N_1} = \frac{w_1}{w_2} = \frac{T_2}{T_1}$$

$$P(t) = x_1(t)y_1(t) + x_2(t)y_2(t)$$

$$= \begin{bmatrix} y_1(t) & x_2(t) \end{bmatrix} \begin{bmatrix} x_1(t) \\ y_2(t) \end{bmatrix}$$

$$= 0$$

Hence, there is no energy loss and it is said to be perfect coupler.

FIGURE 3.21
Coupling device for systems.

Formulation of State Space Model of Systems

Pseudo power or quasi power

When the across variables of a system are multiplied by the through variables of another system having the same system graph, the product is known as the pseudo power or quasi power of the system. The quasi power of systems can be proved equal to zero even when measuring the variables at two different time instants.

Example 3.1

Determine the pseudo power for the mechanical system shown in Figure 3.22 and the electrical system shown in Figure 3.23.

Determine $\sum_{k=1}^{5} \dot{\delta}_k(t) i_k(t+\tau) = \dot{\delta}_1 i_1 + \dot{\delta}_2 i_2 + \dot{\delta}_3 i_3 + \dot{\delta}_4 i_4 + \dot{\delta}_5 i_5$

$\dot{\delta}_k(t)$ = derivative of displacement of kth element at time t
$i_k(t+\tau)$ = current in kth element at time $(t+\tau)$

for both systems.

Solution

The system graphs for mechanical and electrical systems are shown in Figures 3.24 and 3.25, respectively. The measurements are taken in electrical system at time "t," and the measurements are taken in mechanical system at time "$t + \tau$."

Coefficient circuit matrix $B_f = [B_b \; U]$

$$\dot{\delta}_e = \begin{bmatrix} \dot{\delta}_b \\ \dot{\delta}_c \end{bmatrix}$$

$$\begin{bmatrix} \dot{\delta}_1 \\ \dot{\delta}_2 \\ \dot{\delta}_3 \\ \dot{\delta}_4 \\ \dot{\delta}_5 \end{bmatrix} = -\begin{bmatrix} 1 & 0 \\ 0 & 1 \\ -1 & 1 \\ -1 & 1 \\ 0 & -1 \end{bmatrix} \begin{bmatrix} \dot{\delta}_1 \\ \dot{\delta}_2 \end{bmatrix} \dot{\delta}_b$$

Coefficient cutset matrix $C_f = [U \; A_c]$

$$i_e(t+\tau) = \begin{bmatrix} i_b \\ i_c \end{bmatrix}$$

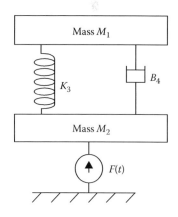

FIGURE 3.22
Mechanical system.

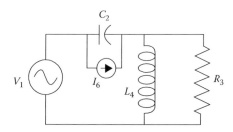

FIGURE 3.23
Electrical system.

$$\begin{bmatrix} i_1 \\ i_2 \\ i_3 \\ i_4 \\ i_5 \end{bmatrix} = \begin{bmatrix} 1 & 1 & 0 \\ -1 & -1 & 1 \\ 1 & 0 & 0 \\ 0 & 1 & 0 \\ 0 & 0 & 1 \end{bmatrix} \begin{bmatrix} i_3 \\ i_4 \\ i_5 \end{bmatrix} i_c$$

Now

$$\sum_{k=1}^{5} \dot{\delta}_k(t) i_k(t+\tau) = \dot{\delta}_1 i_1 + \dot{\delta}_2 i_2 + \dot{\delta}_3 i_3 + \dot{\delta}_4 i_4 + \dot{\delta}_5 i_5$$

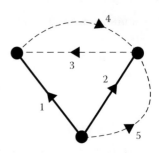

FIGURE 3.24
System graph for mechanical system.

$$= \begin{bmatrix} \dot{\delta}_1(t) & \dot{\delta}_2(t) \end{bmatrix} \begin{bmatrix} 1 & 0 \\ 0 & 1 \\ -1 & 1 \\ -1 & 1 \\ 0 & -1 \end{bmatrix}^T \begin{bmatrix} 1 & 1 & 0 \\ -1 & -1 & 1 \\ 1 & 0 & 0 \\ 0 & 1 & 0 \\ 0 & 0 & 1 \end{bmatrix} \begin{bmatrix} i_3(t+\tau) \\ i_4(t+\tau) \\ i_5(t+\tau) \end{bmatrix}$$

$$= \begin{bmatrix} \dot{\delta}_1(t) & \dot{\delta}_2(t) \end{bmatrix} \begin{bmatrix} 0 & 0 & 0 \\ 0 & 0 & 0 \end{bmatrix} \begin{bmatrix} i_3(t+\tau) \\ i_4(t+\tau) \\ i_5(t+\tau) \end{bmatrix} = 0$$

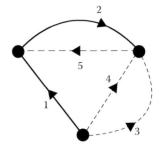

FIGURE 3.25
System graph for electrical system.

Therefore the quasi power is zero. This shows the law of conservation of energy.

Symbolic proof

For system I

$$X_{Ie}(t) = \begin{bmatrix} U \\ -B_b \end{bmatrix} [X_{Ib}(t)]$$

$$Y_{Ie}(t) = \begin{bmatrix} -A_c \\ U \end{bmatrix} [Y_{Ic}(t)]$$

For system II

$$X_{IIe}(t+\tau) = \begin{bmatrix} U \\ -B_b \end{bmatrix} [X_{IIb}(t+\tau)]$$

$$Y_{IIe}(t+\tau) = \begin{bmatrix} -A_c \\ U \end{bmatrix} [Y_{IIc}(t+\tau)]$$

Formulation of State Space Model of Systems

$$Q = \sum X_I(t) Y_{II}(t+\tau)$$

$$= [X_{lb}(t)]^T \begin{bmatrix} U \\ -B_b \end{bmatrix}^T \begin{bmatrix} -A_c \\ U \end{bmatrix} [Y_{llc}(t+\tau)]$$

$$= [X_{lb}(t)]^T [-A_c - B_b^T] [Y_{llc}(t+\tau)]$$

$$= 0$$

3.9 Topological Restrictions

3.9.1 Perfect Coupler

The terminal equations for a perfect coupler may be represented as

$$\begin{bmatrix} x_1(t) \\ y_2(t) \end{bmatrix} = \begin{bmatrix} 0 & n \\ -n & 0 \end{bmatrix} \begin{bmatrix} y_1(t) \\ x_2(t) \end{bmatrix} \quad (3.51)$$

$$x_1 = n x_2$$

$$y_2 = -n y_1$$

From Equation 3.51, it is quite clear that all through variables or all across variables cannot be explicitly placed on either side of the equation. Hence, the terminal equation of the given component is called the hybrid form, i.e., the across variable of the first edge is dependent on the across variable of the other edge and the through variable of the second edge is dependent on the through variable of the first. Therefore, all the edges of a coupler cannot be placed in a tree or a cotree. It is necessary to place one edge in the tree and the other edge in a cotree.

3.9.2 Gyrator

The terminal equation of a gyrator is

$$\begin{bmatrix} x_1(t) \\ x_2(t) \end{bmatrix} = \begin{bmatrix} 0 & n \\ -n & 0 \end{bmatrix} \begin{bmatrix} y_1(t) \\ y_2(t) \end{bmatrix} \quad (3.52)$$

From the above mentioned terminal equation, it is very clear that the across variables or the through variables may be placed explicitly on either side of the equation. Hence, all the edges of a gyrator terminal graph can be placed in a tree or in a cotree, i.e., there are no topological restrictions.

3.9.3 Short Circuit Element ("A" Type)

Instruments such as ammeter, flow meter, torque meter, or any through variables measuring device as shown in Figure 3.26 whose impedance is negligibly small and in ideal situation it is zero. Hence, the drop in across variable in these components is always zero under all operating conditions. These components are considered as short circuit components. They are always placed in tree branches because across variable is specified (i.e., zero). They can be treated as across drivers whose value is always zero.

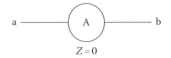

FIGURE 3.26
Through variable measuring device.

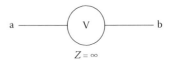

FIGURE 3.27
Across variable measuring device.

3.9.4 Open Circuit Element ("B" Type)

Instruments such as voltmeter, velocity meter or speedometer, pressure gauge, or any other across variable measuring device (ref. Figure 3.27) whose impedance is considerably large and in ideal situation it is infinite. Hence, the value of through variable in these components is always negligibly small under all operating conditions. These components are considered as open circuit components. They are always placed in cotree chords because the through variable is specified (i.e., zero). They can be treated as through drivers whose value is always zero.

3.9.5 Dissipater Type Elements

The terminal equation of dissipater type elements may be written in algebraic form explicitly with across variables on one side or with through variables on one side of the equation. Hence there is no topological restriction for these types of elements. They may be placed either in tree branches or cotree chords.

3.9.6 Delay Type Elements

The terminal equation of delay type elements may be written as

$$x(t) = b \frac{dy(t)}{dt} \qquad (3.53)$$

In this equation, the across variable is dependent upon the through variable. In other words the through variable is a primary variable and the across variable is a secondary variable. Hence, these elements are placed in the cotree.

3.9.7 Accumulator Type Elements

The terminal equation of accumulator or storage type elements may be written as in Equation 3.54.

$$y(t) = C \frac{dx(t)}{dt} \qquad (3.54)$$

Formulation of State Space Model of Systems

In this equation, the through variable is dependent upon the across variable. In other words the across variable is the primary variable and the through variable is the secondary variable. Hence, these elements are placed in the tree.

3.9.8 Across Drivers

The terminal equation for across driver is

$$x(t): \text{ specified} \tag{3.55}$$

In these elements the across variable is the independent variable (primary variable). Hence, these elements are always placed in tree branches.

3.9.9 Through Drivers

The terminal equation for through driver is

$$y(t): \text{ specified} \tag{3.56}$$

In these elements the through variable is the independent variable (primary variable). Hence, these elements are always placed in cotree chords.

Example 3.2

Draw a maximal formulation tree and write f-cutset and f-circuit equations for the given hydraulic system shown in Figure 3.28.

Step 1 System identification

The given system contains the following components as shown in Table 3.2.

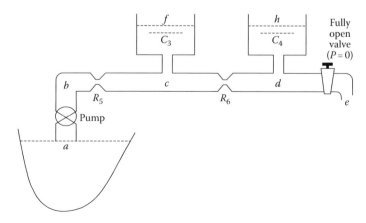

FIGURE 3.28
Hydraulic system.

TABLE 3.2

Components of Hydraulic System along with Their Terminal Equations and Terminal Graphs

Components	Diagram	Terminal Equation	Terminal Graph
1. Pressure pump		P_1: Specified	$a \xrightarrow{P_1} b$
2. Fully open valve		Pressure drop in valve $P = 0$: Specified	$d \xrightarrow{P_2} e$
2. Storage tank		$\overset{\circ}{g}(t) = C\dfrac{dP(t)}{dt}$	$c \xrightarrow{C_3} f$ $d \xrightarrow{C_4} h$
3. Pipe resistance		$\overset{\circ}{g}(t) = GP(t)$ or $P(t) = R\overset{\circ}{g}(t)$	$b \xrightarrow{R_5} c$ $c \xrightarrow{R_6} d$

Step 2 Draw the system graph and formulation tree

The system graph for the given hydraulic system may be obtained from the union of terminal graphs of all the components in one to one correspondence of these components connected in the system as shown in Figure 3.29. Once the system graph is ready, we can define the maximally formulated tree according to the topological restrictions as follows:

1. Start with across drivers, i.e., pressure pump (P_1) and fully open valve (P_2).
2. After the selection of across drivers into tree, the next priority goes to storage type elements (hydraulic tanks). So C_3 can be considered as tree branch, but C_4 cannot be taken into tree, otherwise it will form a closed loop.
3. Hence, with the elements P_1, P_2, and C_3 the tree is complete, i.e., all the vertices are connected and there is no closed loop, as shown in Figure 3.30.

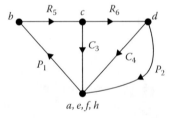

FIGURE 3.29
System graph for hydraulic system.

Step 3 Write the f-cutset equation

1. Fundamental cutset #1 contains one and only one tree branch, i.e., P_1 and the rest of the chord, i.e., R_5. The f-cutset equation for this cutset is

$$\overset{\circ}{g}_1(t) - \overset{\circ}{g}_5(t) = 0 \qquad (3.57)$$

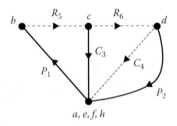

FIGURE 3.30
Formulation tree.

Formulation of State Space Model of Systems

2. Fundamental cutset #2 has P_2 as defining branch and the rest are chords as C_4 and R_6. The f-cutset equation for cutset #2 is written as

$$\overset{o}{g}_2(t) + \overset{o}{g}_4(t) - \overset{o}{g}_6(t) = 0 \tag{3.58}$$

3. Fundamental cutset #3 has been formed by defining branch C_3 and chords as R_5 and R_6. The f-cutset equation for cutset #3 is written as

$$\overset{o}{g}_3(t) - \overset{o}{g}_5(t) + \overset{o}{g}_6(t) = 0 \tag{3.59}$$

Finally, f-cutset equation may be written in the matrix as

$$\begin{matrix} & 1 & 2 & 3 & 4 & 5 & 6 \\ & \begin{bmatrix} 1 & 0 & 0 & 0 & -1 & 0 \\ 0 & 1 & 0 & 1 & 0 & -1 \\ 0 & 0 & 1 & 0 & -1 & 1 \end{bmatrix} \\ & U & & & A_c & & \end{matrix} \begin{bmatrix} \overset{o}{g}_1(t) \\ \overset{o}{g}_2(t) \\ \overset{o}{g}_3(t) \\ \overset{o}{g}_4(t) \\ \overset{o}{g}_5(t) \\ \overset{o}{g}_6(t) \end{bmatrix} \begin{matrix} y_b(t) \\ \\ \\ y_c(t) \end{matrix} = \begin{bmatrix} 0 \\ 0 \\ 0 \end{bmatrix} \tag{3.60}$$

Alternatively, the coefficient matrix of f-cutset equation may be obtained from the incidence matrix.

1. Write the incidence matrix for the given system graph as

$$\text{nodes} \downarrow \begin{matrix} & \text{edges} \longrightarrow \\ & \begin{matrix} 1 & 2 & 3 & 4 & 5 & 6 \end{matrix} \\ \begin{matrix} b \\ c \\ d \\ a,e,f,h \end{matrix} & \begin{bmatrix} -1 & 0 & 0 & 0 & 1 & 0 \\ 0 & 0 & 1 & 0 & -1 & 1 \\ 0 & 1 & 0 & 1 & 0 & -1 \\ 1 & -1 & -1 & -1 & 0 & 0 \end{bmatrix} \end{matrix}$$

2. Remove the last row related with reference node "a." The reduced incidence matrix is

$$\text{nodes} \begin{matrix} & \text{edges} \\ & \begin{matrix} 1 & 2 & 3 & 4 & 5 & 6 \end{matrix} \\ \begin{matrix} b \\ c \\ d \end{matrix} & \begin{bmatrix} -1 & 0 & 0 & 0 & 1 & 0 \\ 0 & 0 & 1 & 0 & -1 & 1 \\ 0 & 1 & 0 & 1 & 0 & -1 \end{bmatrix} \end{matrix}$$

3. Perform elementary row operations to get unit matrix in the leading portion of the matrix and the trailing portion will be A_c matrix. The unity matrix in the leading portion may be obtained by the following operations:
   i. Multiply the first row with −1
   ii. Exchange the last two rows
   iii. Multiply the last row by −1

Finally, the f-cutset coefficient matrix is obtained as given below

$$\begin{array}{c} \text{edges} \\ \begin{array}{cccccc} 1 & 2 & 3 & 4 & 5 & 6 \end{array} \\ \text{nodes} \begin{array}{c} b \\ c \\ d \end{array} \begin{bmatrix} 1 & 0 & 0 & 0 & -1 & 0 \\ 0 & 1 & 0 & 1 & 0 & -1 \\ 0 & 0 & 1 & 0 & -1 & 1 \end{bmatrix} \end{array}$$

Step 4 Write f-circuit equation
1. The f-circuit #1 consists of defining chord $\#C_4$ and a branch P_2. The f-circuit equation may be written as

$$-P_2(t) + P_4(t) = 0 \tag{3.61}$$

2. The f-circuit #2 consists of defining chord $\#R_5$ and rest of the branches such as P_1 and C_3. The f-circuit equation may be written as

$$P_1(t) + P_3(t) + P_5(t) = 0 \tag{3.62}$$

3. The f-circuit #3 formed by defining chord $\#R_6$ and rest of the branches such as C_3 and P_2. The f-circuit equation may be given by

$$P_1(t) - P_3(t) + P_6(t) = 0 \tag{3.63}$$

Finally, f-circuit equation may be written in the matrix as

$$\begin{array}{c} \begin{array}{cccccc} 1 & 2 & 3 & 4 & 5 & 6 \end{array} \\ \begin{bmatrix} 0 & -1 & 0 & 1 & 0 & 0 \\ 1 & 0 & 1 & 0 & 1 & 0 \\ 0 & 1 & -1 & 0 & 0 & 1 \end{bmatrix} \begin{bmatrix} P_1(t) \\ P_2(t) \\ P_3(t) \\ P_4(t) \\ P_5(t) \\ P_6(t) \end{bmatrix} \begin{array}{c} x_b(t) \\ \\ \\ \\ x_c(t) \end{array} = \begin{bmatrix} 0 \\ 0 \\ 0 \end{bmatrix} \tag{3.64}$$

$$\underbrace{}_{B_b} \quad \underbrace{}_{U}$$

Alternatively, B_b matrix may be determined from the property of orthogonality, i.e.,

$$B_b = -A_c^T$$

Formulation of State Space Model of Systems

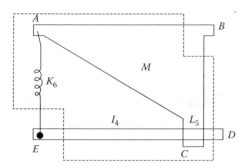

FIGURE 3.31
Subassembly of a mechanical system.

Example 3.3

Develop a state model of the subassembly shown in Figure 3.31 as a four-terminal (E, D, B, G) component and discuss the solution for an arbitrary set of terminal excitation.

The subassembly has external terminals E, D, and B from where it can be interfaced with the rest of the system as shown in the terminal graph Figure 3.32. It also has internal terminals A, C, and reference point G.

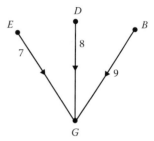

FIGURE 3.32
Terminal graph of given subassembly.

1. Recognition of the system component
   a. Mass
      It is a four-terminal component whose terminal graph is shown in Figure 3.33 and the terminal equation is given below:

$$\begin{bmatrix} f_1(t) \\ f_2(t) \\ v_3(t) \end{bmatrix} = \begin{bmatrix} M_{11}\dfrac{d}{dt} & M_{12}\dfrac{d}{dt} & 0 \\ M_{21}\dfrac{d}{dt} & M_{22}\dfrac{d}{dt} & k_{23} \\ 0 & -k_{23} & 0 \end{bmatrix} \begin{bmatrix} v_1(t) \\ v_2(t) \\ f_3(t) \end{bmatrix} \quad (3.65)$$

   b. Lever
      The terminal equation for lever as a three-terminal component is given below and the terminal graph is shown in Figure 3.34.

$$\begin{bmatrix} v_4(t) \\ f_5(t) \end{bmatrix} = \begin{bmatrix} 0 & -k_{45} \\ k_{45} & 0 \end{bmatrix} \begin{bmatrix} f_4(t) \\ v_5(t) \end{bmatrix} \quad (3.66)$$

      where $k_4 = l_4/l_5$

   c. Spring
      The terminal equation for spring as a three-terminal component

$$\dfrac{d}{dt}f_6(t) = k_6 v_6(t) \quad (3.67)$$

The terminal graph for spring is shown in Figure 3.35.

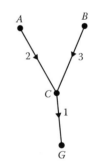

FIGURE 3.33
Terminal graph for rigid mass.

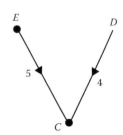

FIGURE 3.34
Terminal graph for lever.

2. Drawing the system graph
   Combine terminal graph of all components in one to one correspondence to get a system graph as shown in Figure 3.36.
3. Checking the existence and uniqueness of the solution for system model

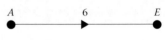

FIGURE 3.35
Terminal graph for spring.

Theorem #1(a)

The necessary condition for existence and uniqueness of the solution for the given system is that there exist two trees such that the direct sum terminal equations (DSTE) are given by

$$C\frac{d}{dt}\psi^*(t) = P^*\psi^*(t) + Q^*Z_s(t)$$

$$Z_p(t) = M^*\psi^*(t) + N^*Z_s(t)$$

(3.68)

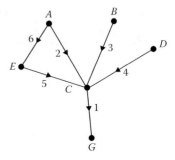

FIGURE 3.36
System graph for subassembly.

$\psi^*(t)$ = state variable vector
$Z_s(t)$ = secondary variable vector
$Z_p(t)$ = primary variable vector (across variables of a tree and through variables of a cotree)

$$Z_p(t) = \begin{bmatrix} X_{T_1} \\ Y_{T_2} \end{bmatrix} \quad \text{and} \quad Z_s(t) = \bar{Z}_p(t) = \begin{bmatrix} Y_{T_1} \\ X_{T_2} \end{bmatrix}$$

such that the matrix $(\lambda c - P^*)$ should be nonsingular.

The system graph shown in Figure 3.36 is augmented by external drivers connected in the same way as the desired terminal graph shown in Figure 3.32.

The augmented system graph for the given subassembly is shown in Figure 3.37 and the maximally formulated tree for the graph consists of branches 1, 2, 3, 4, 7′ and cotree 5, 6, 8′, 9′. Out of three external drivers connected to the subassembly, two (8′ and 9′) are in the cotree and one (7′) in the tree. Hence, the external driver 7′ must be an across driver and the remaining two external drivers 8′ and 9′ must be through drivers.

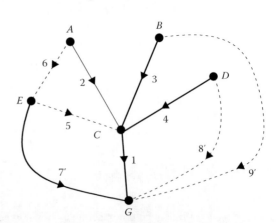

FIGURE 3.37
Augmented system graph of given subassembly.

Formulation of State Space Model of Systems

For given subassembly the DSTE can be represented by Equation 3.69.

$$\begin{bmatrix} M_{11} & M_{12} & 0 \\ M_{21} & M_{22} & 0 \\ 0 & 0 & 1 \end{bmatrix} \frac{d}{dt} \begin{bmatrix} v_1 \\ v_2 \\ f_6 \end{bmatrix} = [0] \begin{bmatrix} v_1 \\ v_2 \\ f_6 \end{bmatrix} + \begin{bmatrix} 1 & 0 & 0 & 0 & 0 & 0 \\ 0 & 1 & -k_{23} & 0 & 0 & 0 \\ 0 & 0 & 0 & 0 & 0 & k_6 \end{bmatrix} \begin{bmatrix} f_1 \\ f_2 \\ f_3 \\ f_4 \\ v_5 \\ v_6 \end{bmatrix} \quad (3.69)$$

$$C \frac{d\psi^*(t)}{dt} = P^* \psi^*(t) + QZ_s(t)$$

The primary variable vector may be expressed in terms of secondary variable vector and state vector:

$$\begin{bmatrix} v_1 \\ v_2 \\ v_3 \\ v_4 \\ f_5 \\ \dot{f_6} \end{bmatrix} = \begin{bmatrix} 1 & 0 & 0 \\ 0 & 1 & 0 \\ 0 & -k_{23} & 0 \\ 0 & 0 & 0 \\ 0 & 0 & 0 \\ 0 & 0 & 1 \end{bmatrix} \begin{bmatrix} v_1 \\ v_2 \\ f_6 \end{bmatrix} + \begin{bmatrix} 0 & 0 & 0 & 0 & 0 & 0 \\ 0 & 0 & 0 & 0 & 0 & 0 \\ 0 & 0 & 0 & 0 & 0 & 0 \\ 0 & 0 & 0 & 0 & -k_{45} & 0 \\ 0 & 0 & 0 & -k_{45} & 0 & 0 \\ 0 & 0 & 0 & 0 & 0 & 0 \end{bmatrix} \begin{bmatrix} f_1 \\ f_2 \\ f_3 \\ f_4 \\ v_5 \\ v_6 \end{bmatrix} \quad (3.70)$$

$$Z_p(t) = M\Psi(t) + NZ_s(t)$$

Continue from the previous class

$$\frac{d}{dt}\begin{bmatrix} v_1 \\ v_2 \\ f_6 \end{bmatrix} = \begin{bmatrix} \beta_{11} & \beta_{12} & 0 \\ \beta_{21} & \beta_{22} & 0 \\ 0 & 0 & 1 \end{bmatrix}[Q^*][Z_s(t)]$$

$$= \begin{bmatrix} \beta_{11} & \beta_{12} & -\beta_{12}k_{23} & 0 & 0 & 0 \\ \beta_{21} & \beta_{22} & -\beta_{22}k_{23} & 0 & 0 & 0 \\ 0 & 0 & 0 & 0 & 0 & k_6 \end{bmatrix} \begin{bmatrix} f_1 \\ f_2 \\ f_3 \\ f_4 \\ v_5 \\ v_6 \end{bmatrix}$$

Theorem #1(b)

If a linear time invariant system having a graph with e edges has a complete and unique solution for all $2e$ variables corresponding to the edges of the graph, then there exist two trees T_1 and T_2 (T_1 and T_2 may be same) in graph G such that the direction terminal equation can be written in the s-domain matrix form as

$$\begin{bmatrix} X_{b_1}(s) \\ Y_{c_1}(s) \end{bmatrix} = \begin{bmatrix} H_{11}(s) & H_{12}(s) \\ H_{21}(s) & H_{22}(s) \end{bmatrix} \begin{bmatrix} Y_{b_2}(s) \\ X_{c_2}(s) \end{bmatrix} + \begin{bmatrix} f_b(s) \\ f_c(s) \end{bmatrix} \quad (3.71)$$

where $X_{b_1}(s)$ corresponding to the edges of T_1 and $Y_{c_1}(s)$ corresponding to the edges of the complement of tree T_1, $Y_{b_2}(s)$ corresponding to the edges of T_2 and $X_{c_2}(s)$ corresponding to the edges of the complement of tree T_2, $f_b(s)$ is the initial condition for the branch elements and $f_c(s)$, initial condition for the chord elements.

Now for the given system we write the f-cutset, the f-circuit, and the constraint equation in Equations 3.72 through 3.74, respectively.

f-cutset equation

$$y_b(t) = -A_c Y_c(t)$$

$$\begin{bmatrix} f_1 \\ f_2 \\ f_3 \\ f_4 \\ f_{7'} \end{bmatrix} = \begin{bmatrix} -1 & 1 & 1 & 1 \\ 0 & 1 & 0 & 0 \\ 0 & 0 & 0 & 1 \\ 0 & 0 & 1 & 0 \\ -1 & -1 & 0 & 0 \end{bmatrix} \begin{bmatrix} f_5 \\ f_6 \\ f_{8'} \\ f_{9'} \end{bmatrix} \quad (3.72)$$

f-circuit equation

$$x_c(t) = -B_b X_b(t)$$

$$\begin{bmatrix} v_5 \\ v_6 \\ v_{8'} \\ v_{9'} \end{bmatrix} = -\begin{bmatrix} 1 & 0 & 0 & 0 & -1 \\ -1 & -1 & 0 & 0 & 1 \\ -1 & 0 & 0 & -1 & 0 \\ -1 & 0 & -1 & 0 & 0 \end{bmatrix} \begin{bmatrix} v_1 \\ v_2 \\ v_3 \\ v_4 \\ v_{7'} \end{bmatrix} \quad (3.73)$$

The constraint equation

$$\begin{bmatrix} f_1 \\ f_2 \\ f_3 \\ f_4 \\ v_5 \\ v_6 \end{bmatrix} = \begin{bmatrix} 0 & 0 & 1 \\ 0 & 0 & 1 \\ 0 & 0 & 0 \\ 0 & 0 & 0 \\ -1 & 0 & 0 \\ 1 & 1 & 0 \end{bmatrix} \begin{bmatrix} v_1 \\ v_2 \\ f_6 \end{bmatrix} + \begin{bmatrix} 0 & 1 & 1 \\ 0 & 0 & 0 \\ 0 & 0 & 1 \\ 0 & 1 & 0 \\ 1 & 0 & 0 \\ 1 & 0 & 0 \end{bmatrix} \begin{bmatrix} v_{7'} \\ f_{8'} \\ f_{9'} \end{bmatrix} + \begin{bmatrix} 0 & 0 & -1 \\ 0 & 0 & 0 \\ 0 & 0 & 0 \\ 0 & 0 & 0 \\ 0 & 0 & 0 \\ 0 & 0 & 0 \end{bmatrix} \begin{bmatrix} v_3 \\ v_4 \\ f_5 \end{bmatrix} \quad (3.74)$$

Formulation of State Space Model of Systems

From the constraint equation we try to eliminate the last matrix that is not required for representing the system behavior. In this example we use the following relation for lever to eliminate the last matrix from the constraint equation.

$$f_5 = k_{45}f_4 \quad \text{and} \quad f_4 = f_{8'}$$

Using this relation we get

$$\begin{bmatrix} f_1 \\ f_2 \\ f_3 \\ f_4 \\ v_5 \\ v_6 \end{bmatrix} = \begin{bmatrix} 0 & 0 & 1 \\ 0 & 0 & 1 \\ 0 & 0 & 0 \\ 0 & 0 & 0 \\ -1 & 0 & 0 \\ 1 & 1 & 0 \end{bmatrix} \begin{bmatrix} v_1 \\ v_2 \\ f_6 \end{bmatrix} + \begin{bmatrix} 0 & 1+k_{45} & 1 \\ 0 & 0 & 0 \\ 0 & 0 & 1 \\ 0 & 1 & 0 \\ 1 & 0 & 0 \\ 1 & 0 & 0 \end{bmatrix} \begin{bmatrix} v_{7'} \\ f_{8'} \\ f_{9'} \end{bmatrix} \quad (3.75)$$

Now the state space model can be written as given in Equation 3.76.

$$\frac{d}{dt}\begin{bmatrix} v_1 \\ v_2 \\ f_6 \end{bmatrix} = \begin{bmatrix} \beta_{11} & \beta_{12} & -\beta_{12}k_{23} & 0 & 0 & 0 \\ \beta_{21} & \beta_{22} & -\beta_{22}k_{23} & 0 & 0 & 0 \\ 0 & 0 & 0 & 0 & 0 & k_6 \end{bmatrix} \left\{ \begin{bmatrix} 0 & 0 & 1 \\ 0 & 0 & 1 \\ 0 & 0 & 0 \\ 0 & 0 & 0 \\ -1 & 0 & 0 \\ 1 & 1 & 0 \end{bmatrix} \begin{bmatrix} v_1 \\ v_2 \\ f_6 \end{bmatrix} + \begin{bmatrix} 0 & 1+k_{45} & 1 \\ 0 & 0 & 0 \\ 0 & 0 & 1 \\ 0 & 1 & 0 \\ 1 & 0 & 0 \\ 1 & 0 & 0 \end{bmatrix} \begin{bmatrix} v_{7'} \\ f_{8'} \\ f_{9'} \end{bmatrix} \right\}$$

(3.76)

State space modeling

The state variable of a dynamic system may be defined as a minimal set of numbers $x_1(t)$, $x_2(t)$, ..., $x_n(t)$ of which, if values are specified at any time instant t_0, together with the knowledge of the input to the system for $t_0 \leq \tau \leq t$ the evolution of state and the output of the system for $t_0 \leq \tau \leq t$ are completely defined.

System has m – input and p – output, the system is called multi-input multi-output (MIMO) system and is shown in Figure 3.38 with its state model. Input to a system can be divided into three groups:

1. Control input which enables influence to be exerted on the system output
2. Disturbance input which can be directly measured but not controlled
3. Disturbance input which can neither be directly measured nor controlled

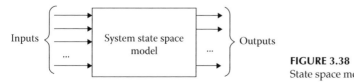

FIGURE 3.38
State space model for MIMO system.

Consider an example of modeling of automobile motion on the road; these three inputs are as follows:

1. The fuel input to the engine is controlled input.
2. The road gradient is a disturbance and can be measured directly, but cannot be controlled.
3. The wind gusts are disturbances which cannot be measured or controlled.

$$\text{Input vector } U(t) = \begin{bmatrix} u_1(t) \\ u_2(t) \\ \ldots \\ u_m(t) \end{bmatrix}, \text{ output vector } Y(t) = \begin{bmatrix} y_1(t) \\ y_2(t) \\ \ldots \\ y_p(t) \end{bmatrix}, \text{ and}$$

$$\text{state variable vector } X(t) = \begin{bmatrix} x_1(t) \\ x_2(t) \\ \ldots \\ x_n(t) \end{bmatrix}.$$

First-order differential equation in state variable form for a given system may be given as

$$\overset{0}{x_1}(t) = f_1(x_1(t), x_2(t), \ldots, x_n(t), u_1(t), u_2(t), \ldots, u_n(t), t)$$
$$\overset{0}{x_2}(t) = f_2(x_1(t), x_2(t), \ldots, x_n(t), u_1(t), u_2(t), \ldots, u_n(t), t)$$
$$\ldots$$
$$\overset{0}{x_n}(t) = f_n(x_1(t), x_2(t), \ldots, x_n(t), u_1(t), u_2(t), \ldots, u_n(t), t)$$

where $f(t) = \begin{bmatrix} f_1 \\ f_2 \\ \ldots \\ f_n \end{bmatrix}$ is a function operator.

The system output equations may also be written in the following form:

$$y_1(t) = g_1(x_1(t), x_2(t), \ldots, x_n(t), u_1(t), u_2(t), \ldots, u_n(t), t)$$
$$y_2(t) = g_2(x_1(t), x_2(t), \ldots, x_n(t), u_1(t), u_2(t), \ldots, u_n(t), t)$$
$$\ldots$$
$$Y_p(t) = g_p(x_1(t), x_2(t), \ldots, x_n(t), u_1(t), u_2(t), \ldots, u_n(t), t)$$

where $g(t) = \begin{bmatrix} g_1 \\ g_2 \\ \ldots \\ g_p \end{bmatrix}$ is a function operator for outputs.

Formulation of State Space Model of Systems

This can be represented by canonical form as a function of state variable vector and input vector as

$$\frac{dx_i(t)}{dt} = f_i\left[t, x_1(t), x_1(t), \ldots, x_n(t), u_1(t), u_2(t), \ldots, u_m(t), t\right] \quad (3.77)$$

Since state variables completely define systems, dynamic behavior output can be expressed as

$$y_k(t) = g_k(t, x_1(t), x_1(t), \ldots, x_n(t), u_1(t), u_2(t), \ldots, u_m(t), t) \quad (3.78)$$

$y_k(t)$ – system output variables, where $k = 1, 2, \ldots, P$.

The combination of state equation 3.77 and output equation 3.78 is called "state space model" of the system. These equations are nonlinear if f_i and g_k are nonlinear functions. The state space model is time variant if these functions f_i and g_k contain explicit function of time.

These equations may be written for linear time invariant system as

$$\overset{0}{X}(t) = AX(t) + BU(t) \quad \text{and} \quad Y(t) = CX(t) + DU(t)$$

where coefficient A, B, C, and D are constant matrices, and the elements of these matrices are time independent.

The above described state model for time variant system may be written as

$$\overset{0}{X}(t) = A(t)X(t) + B(t)U(t) \quad \text{and} \quad Y(t) = C(t)X(t) + D(t)U(t)$$

where coefficient $A(t)$, $B(t)$, $C(t)$, and $D(t)$ are not constant matrices, but the elements of these matrices change with time.

The order of these coefficient matrices is given below:

Coefficient Matrix	Order of Matrix
A	$n \times n$
B	$n \times m$
C	$p \times n$
D	$P \times m$

Consider the following equations

$$\frac{dx(t)}{dt} = 5t\sqrt{x(t)} + 5u(t) \quad \text{Nonlinear and time variant equation}$$

$$\frac{dx(t)}{dt} = 5\sqrt{x(t)} + 5u(t) \quad \text{Nonlinear and time invariant equation}$$

$$\frac{dx(t)}{dt} = 5x(t) + 5u(t) \quad \text{Linear and time invariant equation}$$

$$\frac{dx(t)}{dt} = 5tx(t) + 5u(t) \quad \text{Linear and time variant equation}$$

State variable vector contains variables related with the energy storing (or dynamic) element. The number of state variables equals to number of branch capacitance and the number of chord inductances, i.e., the total number of dynamic elements for nondegenerate system.

3.10 Development of State Model of Degenerative System

In any system if we are unable to keep all capacitors in branches and all inductors in chords, then the system is called degenerate system.

Example 3.4

Consider an electrical system as shown in Figure 3.39. Draw its system graph and f-tree. Also mention the number of degeneracy in the given electrical system and due to which components.

SOLUTION

Step 1   Identify the components in the system

In an electrical system, it is relatively easy to identify the components and their interactions as compared to other systems. The components identified for this system are shown in Table 3.3.

Step 2   Draw the system graph

Combine all the terminal graphs of all the components of the electrical systems in one to one correspondence to get the system graph as shown in Figure 3.40.

Step 3   Draw f-tree

Once the system graph is ready, follow the procedure given below to draw f-tree:

1. Place all across drivers in tree T_1.
2. Draw T_2 as the union of tree T_1 and terminal graphs of accumulators in such a way that there is no closed loop (i.e., T_2 = "T_1" U "C-Type components").
3. Expand tree T_2 to get tree T_3 such that T_3 = "T_2" U "terminal graph of algebraic component" and there is no closed loop.
4. Now consider the terminal graph of delay type elements and draw tree T_4 such as T_4 = "T_3" U "terminal graph of delay type component."
5. Finally, complete the tree if any node of the graph is not connected using through drivers. Hence, f-tree T_5 = "T_4" U "terminal graph of through drivers."

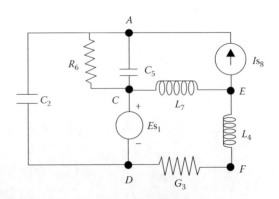

FIGURE 3.39
Electrical system.

Formulation of State Space Model of Systems

TABLE 3.3

Different Components in Given Electrical System

Type of Components	Components	Schematic Diagram	Terminal Graph
1. Across drivers	Voltage source	—◯—	C —→ D, Es_1
2. Accumulators	Capacitors C_2 and C_5	—‖—	A —→ C, C_5 A —→ D, C_2
3. Algebraic type	Resistor G_3 Conductor R_6	—⟋⟍⟋⟍—	F —→ D, G_3 A —→ C, R_6
4. Delay type	Inductor L_4 and L_7	—⟪⟫—	E —→ F, L_4 E —→ C, L_7
5. Through driver	Current source	—▶—	E —→ A, Is_8

In this procedure, at any stage if the insertion of any component terminal graph forms closed loops, then do not consider that component into the tree. That component will be a part of the cotree. The intermediate trees that develop to form the final f-tree T_4 for the given system are shown in Figure 3.41. It is very clear from the above discussion that when we consider accumulator type components like C_2 and C_5 for tree development, C_2 can be placed into the tree but C_5 forms a closed loop. Hence, C_5 cannot be kept into the tree. Similarly, for delay type components, L_4 and L_7 cannot be placed into the cotree. If we try to keep them into the cotree, then all the nodes are not connected and the tree cannot be completed. Hence, it is necessary to consider L_4 into the tree.

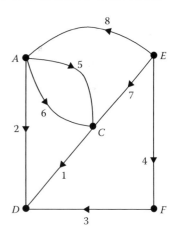

FIGURE 3.40
System graph for electrical system.

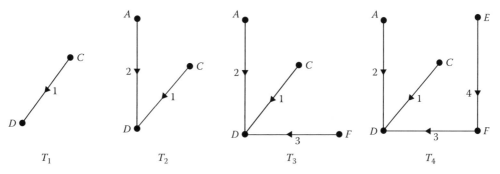

FIGURE 3.41
Development of tree for degenerate electrical system.

To get final tree T_4 there are two violations of topological restrictions, first for C_5 and second for L_4. Hence, the number of degeneracies is equal to two.

Number of degeneracies = Number of accumulator components in the cotree
+ number of components in tree

The number of state equations for degenerate system is always less than the nondegenerate system.

The number of state equation for degenerate system
= Total number of dynamic elements − number of degenerate elements

3.10.1 Development of State Model for Degenerate System

1. Let P be vector of primary variables and S be vector of secondary variables for any given system.
   These variables may be grouped according to the type of elements as shown in Table 3.4.
2. Write the interconnection (f-cutset and f-circuit) equations for the given system:

f-cutset equations $\begin{bmatrix} U & A_c \end{bmatrix} \begin{bmatrix} Y_b(t) \\ Y_c(t) \end{bmatrix} = 0$

f-circuit equation $\begin{bmatrix} B_b & U \end{bmatrix} \begin{bmatrix} X_b(t) \\ X_c(t) \end{bmatrix} = 0$

Combine the f-cutset and f-circuit equations; we get

$$\begin{bmatrix} B_b & 0 & 0 & U \\ 0 & A_c & U & 0 \end{bmatrix} \begin{bmatrix} X_b(t) \\ Y_c(t) \\ Y_b(t) \\ X_c(t) \end{bmatrix} = \begin{bmatrix} 0 \\ 0 \end{bmatrix}$$

TABLE 3.4

Grouping of Components according to Their Type

Components	Primary Variables	Secondary Variable	Terminal Equation
Drivers	P_0	S_0	P_0: Specified
Nondegenerative dynamic elements	P_1	S_1	$\dfrac{dP_1}{dt} = D_1 S_1$
Degenerative dynamic elements	P_2	S_2	$P_2 = D_2 \dfrac{dS_2}{dt}$
Algebraic elements	P_3	S_3	$P_3 = D_3 S_3$

Formulation of State Space Model of Systems

$$\begin{bmatrix} X_c(t) \\ Y_b(t) \end{bmatrix} = -\begin{bmatrix} B_b & 0 \\ 0 & A_c \end{bmatrix}\begin{bmatrix} X_b(t) \\ Y_c(t) \end{bmatrix} \quad (3.79)$$

$\begin{bmatrix} X_c(t) \\ Y_b(t) \end{bmatrix}$ —Secondary variable vector "P"

$\begin{bmatrix} X_b(t) \\ Y_c(t) \end{bmatrix}$ —Primary variable vector "S"

Equation 3.79 may be written as $S(t) = \Phi P(t)$

$$\begin{bmatrix} S_0(t) \\ S_1(t) \\ S_2(t) \\ S_3(t) \end{bmatrix} = \begin{bmatrix} \phi_{00} & \phi_{01} & \phi_{02} & \phi_{03} \\ \phi_{10} & \phi_{11} & \phi_{12} & \phi_{13} \\ \phi_{20} & \phi_{21} & \phi_{22} & \phi_{23} \\ \phi_{30} & \phi_{31} & \phi_{32} & \phi_{33} \end{bmatrix}\begin{bmatrix} P_0(t) \\ P_1(t) \\ P_2(t) \\ P_3(t) \end{bmatrix} \quad (3.80)$$

In Equation 3.80 the values of submatrices Φ_{22}, Φ_{23}, and Φ_{32} are always zero.

3. Start with terminal equations of nondegenerative dynamic elements:

$$\frac{dP_1(t)}{dt} = D_1 S_1(t)$$

Substitute S_1 from Equation 3.80:

$$= D_1\left(\Phi_{10}P_0 + \Phi_{11}P_1 + \Phi_{12}P_2 + \Phi_{13}P_3\right) \quad (3.81)$$

4. Consider the terminal equation for algebraic elements

$$P_3 = D_3 S_3$$
$$P_3 = D_3(\Phi_{30}P_0 + \Phi_{31}P_1 + \Phi_{32}P_2 + \Phi_{33}P_3)$$
$$P_3 = D_3(\Phi_{30}P_0 + \Phi_{31}P_1) + D_3\Phi_{33}P_3$$
$$(U + D_3\Phi_{33})P_3 = D_3(\Phi_{30}P_0 + \Phi_{31}P_1)$$
$$P_3 = (U + D_3\Phi_{33})^{-1}D_3(\Phi_{30}P_0 + \Phi_{31}P_1) \quad (3.82)$$

5. Let us consider the terminal equation for degenerative dynamic elements

$$P_2 = D_2 \frac{dS_2}{dt}$$

$$P_2 = D_2 \frac{d}{dt}(\phi_{20}P_0 + \phi_{21}P_1 + \phi_{22}P_2 + \phi_{23}P_3)$$

$$P_2 = D_2 \frac{d}{dt}(\phi_{20}P_0 + \phi_{21}P_1) \quad \text{because } \Phi_{22}, \Phi_{23} \text{ are zero} \tag{3.83}$$

Substituting Equations 3.82 and 3.83 in Equation 3.81, we get

$$\frac{dP_1(t)}{dt} = D_1\left[\phi_{10}P_0 + \phi_{11}P_1 + \phi_{12}D_2\frac{d}{dt}(\phi_{20}P_0 + \phi_{21}P_1) + \phi_{13}(U + D_3\phi_{33})^{-1}(\phi_{30}P_0 + \phi_{31}P_1)\right]$$

$$= \Big[D_1\phi_{10}P_0 + D_1\phi_{11}P_1 + D_1\phi_{12}D_2\phi_{20}\frac{d}{dt}P_0 + D_1\phi_{12}D_2\phi_{21}\frac{d}{dt}P_1$$
$$+ D_1\phi_{13}(U + D_3\phi_{33})^{-1}\phi_{30}P_0 + D_1\phi_{13}(U + D_3\phi_{33})^{-1}\phi_{31}P_1\Big)\Big]$$

$$[U - D_1\phi_{12}D_2\phi_{21}]\frac{d}{dt}P_1 = \{D_1\phi_{10} + D_1\phi_{13}(U + D_3\phi_{33})^{-1}\phi_{30}\}P_0$$
$$+ \{D_1\phi_{11} + D_1\phi_{13}(U + D_3\phi_{33})^{-1}\phi_{31}\}P_1 + D_1\phi_{12}D_2\phi_{20}\frac{d}{dt}P_0$$

$$\frac{d}{dt}P_1 = [U - D_1\phi_{12}D_2\phi_{21}]^{-1}\Big[\{D_1\phi_{10} + D_1\phi_{13}(U + D_3\phi_{33})^{-1}\phi_{30}\}P_0$$
$$+ \{D_1\phi_{11} + D_1\phi_{13}(U + D_3\phi_{33})^{-1}\phi_{31}\}P_1 + D_1\phi_{12}D_2\phi_{20}\frac{d}{dt}P_0\Big]$$

Example 3.5

Consider a degenerative electrical system as shown in Figure 3.42. Develop its state space model.

1. Identify the components and draw its system graph as shown in Figure 3.43.
2. Group the components and write the terminal equations.
    a. Drivers: P_0—Specified

$$\begin{bmatrix}X_1\\Y_7\end{bmatrix} = \text{specified} \tag{3.84}$$

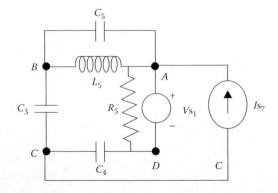

FIGURE 3.42
Electrical system.

b. Nondegenerative dynamic elements:

$$\frac{d}{dt}P_1 = D_1 S_1$$

$$\frac{d}{dt}\begin{bmatrix} X_2 \\ X_3 \\ Y_6 \end{bmatrix} = \begin{bmatrix} 1/C_2 & 0 & 0 \\ 0 & 1/C_3 & 0 \\ 0 & 0 & 1/L_6 \end{bmatrix}\begin{bmatrix} Y_2 \\ Y_3 \\ X_6 \end{bmatrix} \quad (3.85)$$

c. Degenerative dynamic elements:

$$P_2 = D_2 \frac{d}{dt} S_2 \quad (3.86)$$

$$Y_4 = C_4 \frac{d}{dt} X_4$$

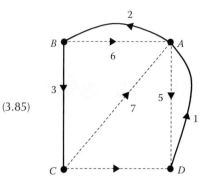

FIGURE 3.43
Tree of given electrical system.

d. Algebraic elements:

$$P_3 = D_3 S_3$$

$$Y_5 = \frac{1}{R_5} X_5 \quad (3.87)$$

3. Interconnecting equations:

$$\begin{array}{c} \\ S_0 \\ \\ \\ S_1 \\ \\ \\ S_2 \\ S_3 \end{array}\begin{bmatrix} Y_1 \\ X_7 \\ Y_2 \\ Y_3 \\ X_6 \\ X_4 \\ X_5 \end{bmatrix} = \begin{array}{cccccccc} 1 & 7 & 2 & 3 & 6 & 4 & 5 & \\ \begin{bmatrix} 0 & 0 & 0 & 0 & 0 & 1 & 1 \\ 0 & 0 & -1 & -1 & 0 & 0 & 0 \\ 0 & 1 & 0 & 0 & 1 & 1 & 0 \\ 0 & 1 & 0 & 0 & 0 & 1 & 0 \\ 0 & 0 & -1 & 0 & 0 & 0 & 0 \\ -1 & 0 & -1 & -1 & 0 & 0 & 0 \\ -1 & 0 & 0 & 0 & 0 & 0 & 0 \end{bmatrix}\begin{bmatrix} X_1 \\ Y_7 \\ X_2 \\ X_3 \\ Y_6 \\ Y_4 \\ Y_5 \end{bmatrix}\begin{array}{c} P_0 \\ \\ \\ P_1 \\ \\ P_2 \\ P_3 \end{array} \end{array} \quad (3.88)$$

4. Starting with 2

$$\frac{d}{dt}\begin{bmatrix} X_2 \\ X_3 \\ Y_6 \end{bmatrix} = \begin{bmatrix} 1/C_2 & 0 & 0 \\ 0 & 1/C_3 & 0 \\ 0 & 0 & 1/L_6 \end{bmatrix}\begin{bmatrix} Y_2 \\ Y_3 \\ X_6 \end{bmatrix}$$

$$= \begin{bmatrix} \frac{1}{C_2} & 0 & 0 \\ 0 & \frac{1}{C_3} & 0 \\ 0 & 0 & \frac{1}{L_6} \end{bmatrix}\left\{ \begin{bmatrix} 0 & 1 \\ 0 & 1 \\ 0 & 0 \end{bmatrix}\begin{bmatrix} X_1 \\ Y_2 \end{bmatrix} + \begin{bmatrix} 0 & 0 & 1 \\ 0 & 0 & 0 \\ -1 & 0 & 0 \end{bmatrix}\begin{bmatrix} X_1 \\ X_3 \\ Y_6 \end{bmatrix} + \begin{bmatrix} 1 \\ 1 \\ 0 \end{bmatrix}[Y_4] + \begin{bmatrix} 0 \\ 0 \\ 0 \end{bmatrix}[Y_5] \right\} \quad (3.89)$$

5. Eliminating Y_4 from Equation 3.89

$$Y_4 = C_4 \frac{d}{dt} X_4$$

$$= C_4 \frac{d}{dt}\left[[-1 \quad 0] \begin{bmatrix} X_1 \\ Y_7 \end{bmatrix} + [-1 \quad -1 \quad 0] \begin{bmatrix} X_2 \\ X_3 \\ Y_6 \end{bmatrix} + [0]Y_4 + [0]Y_5 \right]$$

$$= C_4 \frac{d}{dt}\left[[-1 \quad 0] \begin{bmatrix} X_1 \\ Y_7 \end{bmatrix} + [-1 \quad -1 \quad 0] \begin{bmatrix} X_2 \\ X_3 \\ Y_6 \end{bmatrix} \right] \quad (3.90)$$

6. Substituting in Equation 3.90 into 2nd row partitioned Equation 3.88

$$\begin{bmatrix} Y_2 \\ Y_3 \\ X_6 \end{bmatrix} = \begin{bmatrix} 0 & 1 \\ 0 & 1 \\ 0 & 0 \end{bmatrix} \begin{bmatrix} X_1 \\ Y_7 \end{bmatrix} + \begin{bmatrix} 0 & 0 & 1 \\ 0 & 0 & 0 \\ -1 & 0 & 0 \end{bmatrix} \begin{bmatrix} X_2 \\ X_3 \\ Y_6 \end{bmatrix} + \begin{bmatrix} 1 \\ 1 \\ 0 \end{bmatrix} \left\{ [-C_4 \quad 0] \frac{d}{dt} \begin{bmatrix} X_1 \\ Y_7 \end{bmatrix} + [-C_4 \quad -C_4 \quad 0] \frac{d}{dt} \begin{bmatrix} X_2 \\ X_3 \\ Y_6 \end{bmatrix} \right\}$$

$$= \begin{bmatrix} 0 & 1 \\ 0 & 1 \\ 0 & 0 \end{bmatrix} \begin{bmatrix} X_1 \\ Y_7 \end{bmatrix} + \begin{bmatrix} 0 & 0 & 1 \\ 0 & 0 & 0 \\ -1 & 0 & 0 \end{bmatrix} \begin{bmatrix} X_2 \\ X_3 \\ Y_6 \end{bmatrix} + C_4 \begin{bmatrix} -1 & 0 \\ -1 & 0 \\ 0 & 0 \end{bmatrix} \frac{d}{dt} \begin{bmatrix} X_1 \\ Y_7 \end{bmatrix} + C_4 \begin{bmatrix} -1 & -1 & 0 \\ -1 & -1 & 0 \\ 0 & 0 & 0 \end{bmatrix} \frac{d}{dt} \begin{bmatrix} X_2 \\ X_3 \\ Y_6 \end{bmatrix}$$

$$\frac{d}{dt}\begin{bmatrix} X_2 \\ X_3 \\ Y_6 \end{bmatrix} = \begin{bmatrix} 1/C_2 & 0 & 0 \\ 0 & 1/C_3 & 0 \\ 0 & 0 & 1/L_6 \end{bmatrix} \left\{ \begin{bmatrix} 0 & 1 \\ 0 & 1 \\ 0 & 0 \end{bmatrix} \begin{bmatrix} X_1 \\ Y_7 \end{bmatrix} + \begin{bmatrix} 0 & 0 & 1 \\ 0 & 0 & 0 \\ -1 & 0 & 0 \end{bmatrix} \begin{bmatrix} X_2 \\ X_3 \\ Y_6 \end{bmatrix} + C_4 \begin{bmatrix} -1 & 0 \\ -1 & 0 \\ 0 & 0 \end{bmatrix} \frac{d}{dt}\begin{bmatrix} X_1 \\ Y_7 \end{bmatrix} + C_4 \begin{bmatrix} -1 & -1 & 0 \\ -1 & -1 & 0 \\ 0 & 0 & 0 \end{bmatrix} \frac{d}{dt}\begin{bmatrix} X_2 \\ X_3 \\ Y_6 \end{bmatrix} \right\}$$

$$= \begin{bmatrix} 0 & 1/C_2 \\ 0 & 1/C_3 \\ 0 & 0 \end{bmatrix} \begin{bmatrix} X_1 \\ Y_7 \end{bmatrix} + \begin{bmatrix} 0 & 0 & 1/C_2 \\ 0 & 0 & 0 \\ -1/L_6 & 0 & 0 \end{bmatrix} \begin{bmatrix} X_2 \\ X_3 \\ Y_6 \end{bmatrix} + \begin{bmatrix} -\frac{C_4}{C_2} & 0 \\ -\frac{C_4}{C_3} & 0 \\ 0 & 0 \end{bmatrix} \frac{d}{dt}\begin{bmatrix} X_1 \\ Y_7 \end{bmatrix} + \begin{bmatrix} -\frac{C_4}{C_2} & -\frac{C_4}{C_2} & 0 \\ -\frac{C_4}{C_3} & -\frac{C_4}{C_3} & 0 \\ 0 & 0 & 0 \end{bmatrix} \frac{d}{dt}\begin{bmatrix} X_2 \\ X_3 \\ Y_6 \end{bmatrix}$$

Formulation of State Space Model of Systems

$$\Rightarrow \begin{bmatrix} 1+x & x & 0 \\ y & 1+y & 0 \\ 0 & 0 & 1 \end{bmatrix} \frac{d}{dt}\begin{bmatrix} X_2 \\ X_3 \\ Y_6 \end{bmatrix} = \begin{bmatrix} 0 & 1/C_2 \\ 0 & 1/C_3 \\ 0 & 0 \end{bmatrix}\begin{bmatrix} X_1 \\ Y_7 \end{bmatrix} + \begin{bmatrix} 0 & 0 & 1/C_2 \\ 0 & 0 & 0 \\ -1/L_6 & 0 & 0 \end{bmatrix}\begin{bmatrix} X_2 \\ X_3 \\ Y_6 \end{bmatrix} + \begin{bmatrix} -x & 0 \\ -y & 0 \\ 0 & 0 \end{bmatrix}\frac{d}{dt}\begin{bmatrix} X_1 \\ Y_7 \end{bmatrix}$$

where $x = -\dfrac{C_4}{C_2}$ and $y = -\dfrac{C_4}{C_3}$.

Computing inverse matrix,

$$\begin{bmatrix} 1+x & x & 0 \\ y & 1+y & 0 \\ 0 & 0 & 1 \end{bmatrix}^{-1} = \frac{1}{(1+x)(1+y)-xy}\begin{bmatrix} 1+y & -y & 0 \\ -x & 1+x & 0 \\ 0 & 0 & (1+x)(1+y)-xy \end{bmatrix}$$

$$= \frac{1}{1+x+y}\begin{bmatrix} 1+y & -y & 0 \\ -x & 1+x & 0 \\ 0 & 0 & 1+x+y \end{bmatrix}$$

$$\frac{d}{dt}\begin{bmatrix} X_2 \\ X_3 \\ Y_6 \end{bmatrix} = \frac{1}{1+x+y}\left\{\begin{bmatrix} 0 & \dfrac{1+y}{C_2} - \dfrac{y}{C_3} \\ 0 & \dfrac{1+y}{C_3} - \dfrac{x}{C_2} \\ 0 & 0 \end{bmatrix}\begin{bmatrix} X_1 \\ Y_7 \end{bmatrix} + \begin{bmatrix} 0 & 0 & \dfrac{1+y}{C_2} \\ 0 & 0 & -\dfrac{x}{C_2} \\ -\dfrac{(1+x+y)}{L_6} & 0 & 0 \end{bmatrix}\begin{bmatrix} X_2 \\ X_3 \\ Y_6 \end{bmatrix}\right.$$

$$\left. + \begin{bmatrix} (1+y)(-x)+y^2 & 0 \\ x^2+(1+x)(y) & 0 \\ 0 & 0 \end{bmatrix}\frac{d}{dt}\begin{bmatrix} X_1 \\ Y_7 \end{bmatrix}\right\} \qquad (3.91)$$

3.10.2 Symbolic Formulation of State Model for Nondegenerative Systems

1. Write terminal equations for system components explicitly in primary and secondary variables. Elements are grouped as source, dynamic, and algebraic elements.

   a. Sources: across and through drivers

   $$\begin{bmatrix} x_1(t) \\ y_6(t) \end{bmatrix} \text{—Specified function of time}$$

   b. Dynamic elements: Branch capacitors and chord inductors

   $$\frac{d}{dt}\begin{bmatrix} x_2(t) \\ y_5(t) \end{bmatrix} = \begin{bmatrix} D_2 & 0 \\ 0 & D_5 \end{bmatrix}\begin{bmatrix} y_2(t) \\ x_5(t) \end{bmatrix}$$

   where $D_2 = \dfrac{1}{C_2}$ and $D_5 = \dfrac{1}{L_5}$.

c. Algebraic elements: Branch and chord resistors.

$$\begin{bmatrix} x_3(t) \\ y_4(t) \end{bmatrix} = \begin{bmatrix} D_3 & 0 \\ 0 & D_4 \end{bmatrix} \begin{bmatrix} y_3(t) \\ x_4(t) \end{bmatrix}$$

where $D_3 = R_3$ and $D_4 = \dfrac{1}{R_4}$.

2. Write interconnection equations.

   a. F-cutset equations

$$\begin{bmatrix} u & 0 & 0 & a_{11} & a_{12} & a_{13} \\ 0 & u & 0 & a_{21} & a_{23} & a_{23} \\ 0 & 0 & u & a_{31} & a_{32} & a_{33} \end{bmatrix} \begin{bmatrix} y_1(t) \\ y_2(t) \\ y_3(t) \\ y_4(t) \\ y_5(t) \\ y_6(t) \end{bmatrix} = \begin{bmatrix} 0 \\ 0 \\ 0 \end{bmatrix}$$

   b. F-circuit equation

$$\begin{bmatrix} b_{11} & b_{12} & b_{13} & u & 0 & 0 \\ b_{21} & b_{22} & b_{23} & 0 & u & 0 \\ b_{31} & b_{32} & b_{33} & 0 & 0 & u \end{bmatrix} \begin{bmatrix} x_1(t) \\ x_2(t) \\ x_3(t) \\ x_4(t) \\ x_5(t) \\ x_6(t) \end{bmatrix} = \begin{bmatrix} 0 \\ 0 \\ 0 \end{bmatrix}$$

The second rows of f-cutset and f-circuit equations may be written as

$$y_2(t) + a_{21} y_4(t) + a_{22} y_5(t) + a_{23} y_6(t) = 0$$

$$b_{21} x_1(t) + b_{22} x_2(t) + b_{23} x_3(t) + x_5(t) = 0$$

These equations may be written in matrix form as

$$\begin{bmatrix} y_2(t) \\ x_5(t) \end{bmatrix} = \begin{bmatrix} 0 & -a_{22} \\ -b_{22} & 0 \end{bmatrix} \begin{bmatrix} x_2(t) \\ y_5(t) \end{bmatrix} + \begin{bmatrix} 0 & -a_{23} \\ -b_{21} & 0 \end{bmatrix} \begin{bmatrix} x_1(t) \\ y_6(t) \end{bmatrix} + \begin{bmatrix} 0 & -a_{21} \\ -b_{23} & 0 \end{bmatrix} \begin{bmatrix} x_3(t) \\ y_4(t) \end{bmatrix} \quad (3.92)$$

3. Start with terminal equation of dynamic elements

$$\dfrac{d}{dt} \begin{bmatrix} x_2(t) \\ y_5(t) \end{bmatrix} = \begin{bmatrix} D_2 & 0 \\ 0 & D_5 \end{bmatrix} \begin{bmatrix} y_2(t) \\ x_5(t) \end{bmatrix} \quad (3.93)$$

Formulation of State Space Model of Systems

4. Substituting Equation 3.92 in Equation 3.93

$$\frac{d}{dt}\begin{bmatrix}x_2(t)\\y_5(t)\end{bmatrix}=\begin{bmatrix}D_2 & 0\\0 & D_5\end{bmatrix}\left\{\begin{bmatrix}0 & -a_{22}\\-b_{22} & 0\end{bmatrix}\begin{bmatrix}x_2(t)\\y_5(t)\end{bmatrix}+\begin{bmatrix}0 & -a_{23}\\-b_{21} & 0\end{bmatrix}\begin{bmatrix}x_1(t)\\y_6(t)\end{bmatrix}+\begin{bmatrix}0 & -a_{21}\\-b_{23} & 0\end{bmatrix}\begin{bmatrix}x_3(t)\\y_4(t)\end{bmatrix}\right\}$$

(3.94)

5. The first row of f-cutset and the first row of f-circuit equations may be written as

$$y_1(t)+a_{11}y_4(t)+a_{12}y_5(t)+a_{13}y_6(t)=0$$

$$b_{11}x_1(t)+b_{12}x_2(t)+b_{13}x_3(t)+x_4(t)=0$$

These equations can be written in the matrix form as

$$\begin{bmatrix}y_1(t)\\x_4(t)\end{bmatrix}=\begin{bmatrix}0 & -a_{12}\\-b_{12} & 0\end{bmatrix}\begin{bmatrix}x_2(t)\\y_5(t)\end{bmatrix}+\begin{bmatrix}0 & -a_{13}\\-b_{11} & 0\end{bmatrix}\begin{bmatrix}x_1(t)\\y_6(t)\end{bmatrix}+\begin{bmatrix}0 & -a_{11}\\-b_{13} & 0\end{bmatrix}\begin{bmatrix}x_3(t)\\y_4(t)\end{bmatrix}$$

(3.95)

6. Substitute terminal equation for algebraic elements

$$\begin{bmatrix}y_1(t)\\x_4(t)\end{bmatrix}=\begin{bmatrix}0 & -a_{12}\\-b_{12} & 0\end{bmatrix}\begin{bmatrix}x_2(t)\\y_5(t)\end{bmatrix}+\begin{bmatrix}0 & -a_{13}\\-b_{11} & 0\end{bmatrix}\begin{bmatrix}x_1(t)\\y_6(t)\end{bmatrix}+\begin{bmatrix}0 & -a_{11}\\-b_{13} & 0\end{bmatrix}\begin{bmatrix}D_3 & 0\\0 & D_4\end{bmatrix}\begin{bmatrix}y_3(t)\\x_4(t)\end{bmatrix}$$

7. The third row of f-cutset equation and the first row of f-circuit equation may be written as

$$y_3(t)+a_{31}y_4(t)+a_{32}y_5(t)+a_{33}y_6(t)=0$$

$$b_{11}x_1(t)+b_{12}x_2(t)+b_{13}x_3(t)+x_4(t)=0$$

These equations can be written in the matrix form as

$$\begin{bmatrix}y_3(t)\\x_4(t)\end{bmatrix}=\begin{bmatrix}0 & -a_{32}\\-b_{12} & 0\end{bmatrix}\begin{bmatrix}x_2(t)\\y_5(t)\end{bmatrix}+\begin{bmatrix}0 & -a_{33}\\-b_{11} & 0\end{bmatrix}\begin{bmatrix}x_1(t)\\y_6(t)\end{bmatrix}+\begin{bmatrix}0 & -a_{31}\\-b_{13} & 0\end{bmatrix}\begin{bmatrix}x_3(t)\\y_4(t)\end{bmatrix}$$

Substitute this equation in the terminal equation of algebraic elements:

$$\begin{bmatrix}x_3(t)\\y_4(t)\end{bmatrix}=\begin{bmatrix}D_3 & 0\\0 & D_4\end{bmatrix}\left\{\begin{bmatrix}0 & -a_{32}\\-b_{12} & 0\end{bmatrix}\begin{bmatrix}x_2(t)\\y_5(t)\end{bmatrix}+\begin{bmatrix}0 & -a_{33}\\-b_{11} & 0\end{bmatrix}\begin{bmatrix}x_1(t)\\y_6(t)\end{bmatrix}+\begin{bmatrix}0 & -a_{31}\\-b_{13} & 0\end{bmatrix}\begin{bmatrix}x_3(t)\\y_4(t)\end{bmatrix}\right\}$$

(3.96)

$$\begin{bmatrix} x_3(t) \\ y_4(t) \end{bmatrix} = \begin{bmatrix} 0 & -D_3 a_{32} \\ -D_4 b_{12} & 0 \end{bmatrix} \begin{bmatrix} x_2(t) \\ y_5(t) \end{bmatrix} + \begin{bmatrix} 0 & -D_3 a_{33} \\ -D_4 b_{11} & 0 \end{bmatrix} \begin{bmatrix} x_1(t) \\ y_6(t) \end{bmatrix} + \begin{bmatrix} 0 & -D_3 a_{31} \\ -D_4 b_{13} & 0 \end{bmatrix} \begin{bmatrix} x_3(t) \\ y_4(t) \end{bmatrix}$$

$$\begin{bmatrix} u & -D_3 a_{31} \\ -D_4 b_{13} & u \end{bmatrix} \begin{bmatrix} x_3(t) \\ y_4(t) \end{bmatrix} = \begin{bmatrix} 0 & -D_3 a_{32} \\ -D_4 b_{12} & 0 \end{bmatrix} \begin{bmatrix} x_2(t) \\ y_5(t) \end{bmatrix} + \begin{bmatrix} 0 & -D_3 a_{33} \\ -D_4 b_{11} & 0 \end{bmatrix} \begin{bmatrix} x_1(t) \\ y_6(t) \end{bmatrix}$$

$$\begin{bmatrix} x_3(t) \\ y_4(t) \end{bmatrix} = \begin{bmatrix} u & -D_3 a_{31} \\ -D_4 b_{13} & u \end{bmatrix}^{-1} \left\{ \begin{bmatrix} 0 & -D_3 a_{32} \\ -D_4 b_{12} & 0 \end{bmatrix} \begin{bmatrix} x_2(t) \\ y_5(t) \end{bmatrix} + \begin{bmatrix} 0 & -D_3 a_{33} \\ -D_4 b_{11} & 0 \end{bmatrix} \begin{bmatrix} x_1(t) \\ y_6(t) \end{bmatrix} \right\}$$

(3.97)

8. Substituting Equation 3.97 in Equation 3.95 and then in Equation 3.94

$$\frac{d}{dt}\begin{bmatrix} x_2(t) \\ y_5(t) \end{bmatrix} = \begin{bmatrix} D_2 & 0 \\ 0 & D_5 \end{bmatrix}\begin{bmatrix} 0 & -a_{22} \\ -b_{22} & 0 \end{bmatrix}\begin{bmatrix} x_2(t) \\ y_5(t) \end{bmatrix} + \begin{bmatrix} D_2 & 0 \\ 0 & D_5 \end{bmatrix}\begin{bmatrix} 0 & -a_{23} \\ -b_{21} & 0 \end{bmatrix}\begin{bmatrix} x_1(t) \\ y_6(t) \end{bmatrix}$$

$$+ \begin{bmatrix} 0 & -a_{21} \\ -b_{23} & 0 \end{bmatrix}\begin{bmatrix} u & -D_3 a_{31} \\ -D_4 b_{13} & u \end{bmatrix}^{-1} \left\{ \begin{bmatrix} 0 & -D_3 a_{32} \\ -D_4 b_{12} & 0 \end{bmatrix}\begin{bmatrix} x_2(t) \\ y_5(t) \end{bmatrix} + \begin{bmatrix} 0 & -D_3 a_{33} \\ -D_4 b_{11} & 0 \end{bmatrix}\begin{bmatrix} x_1(t) \\ y_6(t) \end{bmatrix} \right\}$$

Simplifying results in the form

$$\frac{d}{dt}\begin{bmatrix} x_2(t) \\ y_5(t) \end{bmatrix} = [P]\begin{bmatrix} x_1(t) \\ y_6(t) \end{bmatrix} + [Q]\begin{bmatrix} x_1(t) \\ y_6(t) \end{bmatrix} \quad (3.98)$$

9. Output equation may be written as

$$\begin{bmatrix} x_1(t) \\ y_6(t) \\ x_2(t) \\ y_5(t) \\ x_3(t) \\ y_4(t) \end{bmatrix} = \begin{bmatrix} 0 & 0 \\ 0 & 0 \\ u & 0 \\ 0 & u \\ r_{11} & r_{12} \\ r_{13} & r_{14} \end{bmatrix}\begin{bmatrix} x_2(t) \\ y_5(t) \end{bmatrix} + \begin{bmatrix} u & 0 \\ 0 & u \\ 0 & 0 \\ 0 & 0 \\ s_{11} & s_{12} \\ s_{13} & s_{14} \end{bmatrix}\begin{bmatrix} x_1(t) \\ y_6(t) \end{bmatrix} \quad (3.99)$$

Example 3.6

Obtain the state space model of nondegenerative system as shown in Figure 3.44 and its system graph is shown in Figure 3.45.

1. Terminal equations for different components
    a. *Sources:*

$$\begin{bmatrix} X_1 \\ Y_6 \end{bmatrix} = \begin{bmatrix} 1 \\ 1 \end{bmatrix} 3\cos(t): \text{ Specified} \quad (3.100)$$

Formulation of State Space Model of Systems

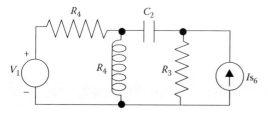

FIGURE 3.44
Nondegenerate electrical system.

b. *Dynamic elements*:

$$\frac{d}{dt}\begin{bmatrix} X_2 \\ Y_5 \end{bmatrix} = \begin{bmatrix} \frac{1}{C_2} & 0 \\ 0 & \frac{1}{L_5} \end{bmatrix} \begin{bmatrix} Y_2 \\ X_5 \end{bmatrix} \quad (3.101)$$

c. *Algebraic equations*:

$$\begin{bmatrix} X_3 \\ Y_4 \end{bmatrix} = \begin{bmatrix} R_3 & 0 \\ 0 & \frac{1}{R_4} \end{bmatrix} \begin{bmatrix} Y_3 \\ X_4 \end{bmatrix} \quad (3.102)$$

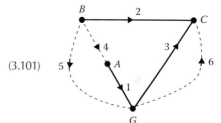

FIGURE 3.45
System graph for electrical system.

2. Interconnecting equations
   a. f-circuit equation

$$\begin{bmatrix} 1 & -1 & 1 & 1 & 0 & 0 \\ 0 & -1 & 1 & 0 & 1 & 0 \\ 0 & 0 & -1 & 0 & 0 & 1 \end{bmatrix} \begin{bmatrix} X_1 \\ X_2 \\ X_3 \\ X_4 \\ X_5 \\ X_6 \end{bmatrix} = \begin{bmatrix} 0 \\ 0 \\ 0 \end{bmatrix} \quad (3.103)$$

b. f-cutset equations

$$\begin{bmatrix} 1 & 0 & 0 & -1 & 0 & 0 \\ 0 & 1 & 0 & 1 & 1 & 0 \\ 0 & 0 & 1 & -1 & -1 & 1 \end{bmatrix} \begin{bmatrix} Y_1 \\ Y_2 \\ Y_3 \\ Y_4 \\ Y_5 \\ Y_6 \end{bmatrix} = \begin{bmatrix} 0 \\ 0 \\ 0 \end{bmatrix} \quad (3.104)$$

Note
Terminal equations have primary variable represented in terms of secondary variables. Interconnecting equations have secondary variables in terms of primary variables.

3. The second row of the f-circuit equation may be written as $X_5 = X_2 - X_3$ and the second row of the f-cutset equation $Y_2 = -Y_4 - Y_5$.

Rewriting these equations in matrix form in terms of state vector and input

$$\begin{bmatrix} Y_2 \\ X_5 \end{bmatrix} = \begin{bmatrix} 0 & -1 \\ 1 & 0 \end{bmatrix} \begin{bmatrix} X_2 \\ Y_5 \end{bmatrix} + \begin{bmatrix} 0 & 0 \\ 0 & 0 \end{bmatrix} \begin{bmatrix} X_1 \\ Y_6 \end{bmatrix} + \begin{bmatrix} 0 & -1 \\ 1 & 0 \end{bmatrix} \begin{bmatrix} X_3 \\ Y_4 \end{bmatrix}$$

Substituting the above equation in the terminal equation of dynamic elements (Equation 3.101),

$$\frac{d}{dt}\begin{bmatrix} X_2 \\ Y_5 \end{bmatrix} = \begin{bmatrix} \frac{1}{C_2} & 0 \\ 0 & \frac{1}{L_5} \end{bmatrix} \left\{ \begin{bmatrix} 0 & -1 \\ 1 & 0 \end{bmatrix} \begin{bmatrix} X_2 \\ Y_5 \end{bmatrix} + \begin{bmatrix} 0 & -1 \\ -1 & 0 \end{bmatrix} \begin{bmatrix} X_3 \\ Y_4 \end{bmatrix} \right\}$$

$$\frac{d}{dt}\begin{bmatrix} X_2 \\ Y_5 \end{bmatrix} = \begin{bmatrix} 0 & \frac{-1}{C_2} \\ \frac{1}{L_5} & 0 \end{bmatrix} \begin{bmatrix} X_2 \\ Y_5 \end{bmatrix} + \begin{bmatrix} 0 & \frac{-1}{C_2} \\ \frac{-1}{L_5} & 0 \end{bmatrix} \begin{bmatrix} X_3 \\ Y_4 \end{bmatrix} \qquad (3.105)$$

4. To eliminate $\begin{bmatrix} X_3 \\ Y_4 \end{bmatrix}$ use interconnection equations and terminal equations as follows:

   a. The last row of f-cutset equation $Y_3 = +Y_4 + Y_5 - Y_6$
   b. The first row of f-circuit equation $X_4 = -X_1 + X_2 + X_3$

These equations may be written in matrix form as

$$\begin{bmatrix} Y_3 \\ X_4 \end{bmatrix} = \begin{bmatrix} 0 & 1 \\ -1 & 0 \end{bmatrix} \begin{bmatrix} X_2 \\ Y_5 \end{bmatrix} + \begin{bmatrix} 0 & -1 \\ -1 & 0 \end{bmatrix} \begin{bmatrix} X_1 \\ Y_6 \end{bmatrix} + \begin{bmatrix} 0 & -1 \\ 1 & 0 \end{bmatrix} \begin{bmatrix} X_3 \\ Y_4 \end{bmatrix}$$

Substitute this equation in the terminal equation of algebraic elements (Equation 3.102)

$$\begin{bmatrix} X_3 \\ Y_4 \end{bmatrix} = \begin{bmatrix} R_3 & 0 \\ 0 & \frac{1}{R_4} \end{bmatrix} \left\{ \begin{bmatrix} 0 & 1 \\ 1 & 0 \end{bmatrix} \begin{bmatrix} X_2 \\ Y_5 \end{bmatrix} + \begin{bmatrix} 0 & -1 \\ -1 & 0 \end{bmatrix} \begin{bmatrix} X_1 \\ Y_6 \end{bmatrix} + \begin{bmatrix} 0 & 1 \\ 1 & 0 \end{bmatrix} \begin{bmatrix} X_3 \\ Y_4 \end{bmatrix} \right\}$$

$$\begin{bmatrix} 1 & -R_3 \\ -\frac{1}{R_4} & 1 \end{bmatrix} \begin{bmatrix} X_3 \\ Y_4 \end{bmatrix} = \begin{bmatrix} 0 & R_3 \\ \frac{1}{R_4} & 0 \end{bmatrix} \begin{bmatrix} X_2 \\ Y_5 \end{bmatrix} + \begin{bmatrix} 0 & -R_3 \\ -\frac{1}{R_4} & 0 \end{bmatrix} \begin{bmatrix} X_1 \\ Y_6 \end{bmatrix}$$

$$\begin{bmatrix} X_3 \\ Y_4 \end{bmatrix} = \begin{bmatrix} 1 & -R_3 \\ -\frac{1}{R_4} & 1 \end{bmatrix}^{-1} \left\{ \begin{bmatrix} 0 & R_3 \\ \frac{1}{R_4} & 0 \end{bmatrix} \begin{bmatrix} X_2 \\ Y_5 \end{bmatrix} + \begin{bmatrix} 0 & -R_3 \\ -\frac{1}{R_4} & 0 \end{bmatrix} \begin{bmatrix} X_1 \\ Y_6 \end{bmatrix} \right\} \qquad (3.106)$$

Substituting Equation 3.106 in Equation 3.105,

$$\frac{d}{dt}\begin{bmatrix} X_2 \\ Y_5 \end{bmatrix} = \begin{bmatrix} 0 & \frac{-1}{C_2} \\ \frac{1}{L_5} & 0 \end{bmatrix}\begin{bmatrix} X_2 \\ Y_5 \end{bmatrix} + \begin{bmatrix} 0 & \frac{-1}{C_2} \\ \frac{-1}{L_5} & 0 \end{bmatrix}\left(\begin{bmatrix} 1 & -R_3 \\ -\frac{1}{R_4} & 1 \end{bmatrix}^{-1}\left\{\begin{bmatrix} 0 & R_3 \\ \frac{1}{R_4} & 0 \end{bmatrix}\begin{bmatrix} X_2 \\ Y_5 \end{bmatrix}\right.\right.$$

$$\left.\left. + \begin{bmatrix} 0 & -R_3 \\ -\frac{1}{R_4} & 0 \end{bmatrix}\begin{bmatrix} X_1 \\ Y_6 \end{bmatrix}\right\}\right)$$

If system parameters are $C_2 = 1/3$, $L_5 = 2$, $R_3 = 4$, and $R_4 = 1$, the state model will be

$$\frac{d}{dt}\begin{bmatrix} X_2 \\ Y_5 \end{bmatrix} = \begin{bmatrix} \frac{-3}{5} & \frac{-3}{5} \\ 0.1 & 0.2 \end{bmatrix}\begin{bmatrix} X_2 \\ Y_5 \end{bmatrix} + \begin{bmatrix} \frac{-3}{5} & \frac{-12}{5} \\ \frac{-2}{5} & \frac{2}{5} \end{bmatrix}\begin{bmatrix} X_1 \\ Y_6 \end{bmatrix} \qquad (3.107)$$

5. To determine output equations, all primary variables are considered for output variables, i.e., all branch across variables and all chord voltages:

$$\begin{bmatrix} X_2 \\ Y_6 \\ X_2 \\ Y_6 \\ X_2 \\ Y_6 \end{bmatrix} = \begin{bmatrix} 0 & 0 \\ 0 & 0 \\ 1 & 0 \\ 0 & 1 \\ 0.8 & 0.8 \\ 0.2 & 0.8 \end{bmatrix}\begin{bmatrix} X_2 \\ Y_5 \end{bmatrix} + \begin{bmatrix} 1 & 0 \\ 0 & 1 \\ 0 & 0 \\ 0 & 0 \\ 0.8 & -0.8 \\ -0.2 & -0.8 \end{bmatrix}\begin{bmatrix} X_1 \\ Y_6 \end{bmatrix} \qquad (3.108)$$

3.10.3 State Model of System with Multiterminal Components

Consider a system containing multiterminal components as shown in Figure 3.46 and its system graph in Figure 3.47.

Terminal equations:

$$C_1 \frac{dv_1(t)}{dt} = i_1(t)$$

$$\begin{bmatrix} v_4(t) \\ v_2(t) \end{bmatrix} = \begin{bmatrix} 0 & 0 \\ 0 & -\gamma_p - (\mu+1)R_u \end{bmatrix}\begin{bmatrix} i_4(t) \\ i_2(t) \end{bmatrix} + \begin{bmatrix} 1 \\ -\mu \end{bmatrix}[E(t)]$$

$$L_3 \frac{d}{dt}i_3(t) = v_3(t)$$

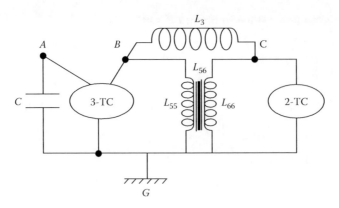

FIGURE 3.46
System containing multiterminal component.

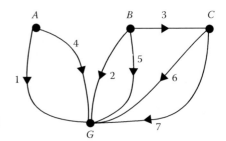

FIGURE 3.47
System graph for given system containing multiterminal component.

$$\frac{d}{dt}\begin{bmatrix} L_{55} & L_{56} \\ L_{56} & L_{66} \end{bmatrix}\begin{bmatrix} i_5(t) \\ i_6(t) \end{bmatrix} = \begin{bmatrix} v_5(t) \\ v_6(t) \end{bmatrix}$$

$$i_7(t) = 0$$

Interconnection equations:

$$P_0 = \begin{bmatrix} v_4 \\ i_7 \end{bmatrix} - \text{drivers}, \quad P_1 = \begin{bmatrix} i_5 \\ i_6 \end{bmatrix} - \text{Transformer}$$

$$P_2 = \begin{bmatrix} i_1 \\ v_3 \end{bmatrix} - \text{Dynamic elements}, \quad P_3 = [v_2] - \text{Algebraic elements}$$

$$\begin{bmatrix} i_4 \\ v_7 \\ \overline{v_5} \\ \overline{v_6} \\ \overline{v_1} \\ i_3 \\ \overline{i_2} \end{bmatrix} = \begin{bmatrix} \begin{array}{cc|cc|cc|c} 4 & 7 & 5 & 6 & 1 & 3 & 2 \\ \hline 0 & 0 & 0 & 0 & -1 & 0 & 0 \\ 0 & 0 & 0 & 0 & 0 & -1 & 1 \\ \hline 0 & 0 & 0 & 0 & 0 & 0 & 1 \\ 0 & 0 & 0 & 0 & 0 & -1 & 1 \\ \hline 1 & 0 & 0 & 0 & 0 & 0 & 0 \\ 0 & 1 & 0 & 1 & 0 & 0 & 0 \\ \hline 0 & -1 & -1 & -1 & 0 & 0 & 0 \end{array} \end{bmatrix} \begin{bmatrix} v_4 \\ i_7 \\ \overline{i_5} \\ i_6 \\ \overline{i_1} \\ v_3 \\ \overline{v_2} \end{bmatrix}$$

Formulation of State Space Model of Systems

Starting from terminal equation of group 2, 1, is dynamic elements

$$\begin{bmatrix} L_{55} & L_{56} \\ L_{56} & L_{66} \end{bmatrix} \frac{d}{dt} \begin{bmatrix} i_5 \\ i_6 \end{bmatrix} = \begin{bmatrix} v_5 \\ v_6 \end{bmatrix} = \left\{ \begin{bmatrix} 0 & 0 \\ 0 & -1 \end{bmatrix} \begin{bmatrix} i_1 \\ v_3 \end{bmatrix} + \begin{bmatrix} 1 \\ 1 \end{bmatrix} [v_2] \right\} \text{ from interconnecting equations}$$

$$= \left\{ \begin{bmatrix} 0 & 0 \\ 0 & -1 \end{bmatrix} \left\{ \begin{bmatrix} C_1 & 0 \\ 0 & L_3 \end{bmatrix} \frac{d}{dt} \begin{bmatrix} v_1 \\ i_3 \end{bmatrix} \right\} + \begin{bmatrix} 1 \\ 1 \end{bmatrix} \{ [-a]i_2 + [-\mu E] \} \right\}$$

where $a = \dfrac{1}{P} + (\mu + 1) R_{11}$

Eliminating P_2, or in fact, i_2, with the help of interconnecting equations,

$$\begin{bmatrix} v_1 \\ i_3 \end{bmatrix} = \begin{bmatrix} 1 & 0 \\ 0 & 1 \end{bmatrix} \begin{bmatrix} v_4 \\ i_7 \end{bmatrix} + \begin{bmatrix} 0 & 0 \\ 0 & 1 \end{bmatrix} \begin{bmatrix} i_5 \\ i_6 \end{bmatrix}$$

Eliminating i_2,

$$i_2 = \begin{bmatrix} 0 & -1 \end{bmatrix} \begin{bmatrix} v_1 \\ i_7 \end{bmatrix} + \begin{bmatrix} -1 & -1 \end{bmatrix} \begin{bmatrix} i_5 \\ i_6 \end{bmatrix}$$

$$\begin{bmatrix} L_{55} & L_{56} \\ L_{56} & L_{66} \end{bmatrix} \frac{d}{dt} \begin{bmatrix} i_5 \\ i_6 \end{bmatrix} = \begin{bmatrix} 0 & 0 \\ 0 & -1 \end{bmatrix} \left\{ \begin{bmatrix} C_1 & 0 \\ 0 & L_3 \end{bmatrix} \frac{d}{dt} \left\{ \begin{bmatrix} 1 & 0 \\ 0 & 1 \end{bmatrix} \begin{bmatrix} v_4 \\ i_7 \end{bmatrix} + \begin{bmatrix} 0 & 0 \\ 0 & 1 \end{bmatrix} \begin{bmatrix} i_5 \\ i_6 \end{bmatrix} \right\} \right\}$$

$$+ \begin{bmatrix} 1 \\ 1 \end{bmatrix} \left\{ -a \left[\begin{bmatrix} 0 & -1 \end{bmatrix} \begin{bmatrix} v_4 \\ i_7 \end{bmatrix} + \begin{bmatrix} -1 & -1 \end{bmatrix} \begin{bmatrix} i_5 \\ i_6 \end{bmatrix} \right] + [-\mu E] \right\}$$

$$\begin{bmatrix} 0 & 0 \\ 0 & -L_3 \end{bmatrix} \frac{d}{dt} \begin{bmatrix} v_4 \\ i_7 \end{bmatrix} + \begin{bmatrix} 0 & 0 \\ 0 & -L_3 \end{bmatrix} \frac{d}{dt} \begin{bmatrix} i_5 \\ i_6 \end{bmatrix} + \begin{bmatrix} -a \\ -a \end{bmatrix} \begin{bmatrix} 0 & -1 \end{bmatrix} \begin{bmatrix} v_4 \\ i_7 \end{bmatrix} + \begin{bmatrix} -1 & -1 \end{bmatrix} \begin{bmatrix} i_5 \\ i_6 \end{bmatrix} + \begin{bmatrix} -\mu E \\ -\mu E \end{bmatrix}$$

$$\begin{bmatrix} L_{55} & L_{56} \\ L_{56} & L_{66} + L_3 \end{bmatrix} \frac{d}{dt} \begin{bmatrix} i_5 \\ i_6 \end{bmatrix} = \underbrace{\begin{bmatrix} 0 & 0 \\ 0 & -L_3 \end{bmatrix} \frac{d}{dt} \begin{bmatrix} v_4 \\ i_7 \end{bmatrix} + \begin{bmatrix} 0 & -a \\ 0 & -a \end{bmatrix} \begin{bmatrix} v_4 \\ i_7 \end{bmatrix}}_{=0} + \begin{bmatrix} -a & -a \\ -a & -a \end{bmatrix} \begin{bmatrix} i_5 \\ i_6 \end{bmatrix} + \begin{bmatrix} -\mu E \\ -\mu E \end{bmatrix}$$

Given that $i_7 = 0$ and $\dfrac{d}{dt} i_7 = 0$

$$\frac{d}{dt} \begin{bmatrix} i_5 \\ i_6 \end{bmatrix} = \begin{bmatrix} L_{55} & L_{56} \\ L_{56} & L_{66} + L_3 \end{bmatrix} \begin{bmatrix} -a & -a \\ -a & -a \end{bmatrix} \begin{bmatrix} i_5 \\ i_6 \end{bmatrix} + \begin{bmatrix} -\mu \\ -\mu \end{bmatrix} [E]$$

Example 3.7

Develop a state model for the given mechanical system.

1. Identify various components in the system as shown in Table 3.5.
   The system graph for the mechanical system shown in Figure 3.48 may be drawn by combining all the terminal graphs of each individual terminal graph of system components as in Figure 3.49.
2. Terminal equations for different components of the system.
   For the given mechanical system $Y(t) = f(t)$ and $X(t) = \overset{o}{\delta}(t)$, and the subscript represents the element number.
   a. *Source*: Force driver

$$Y_6(t): \text{specified} \qquad (3.109)$$

   b. *Dynamic elements*:

$$\frac{d}{dt}\begin{bmatrix} X_1 \\ Y_5 \end{bmatrix} = \begin{bmatrix} \frac{1}{M_1} & 0 \\ 0 & \frac{1}{K_5} \end{bmatrix} \begin{bmatrix} Y_1 \\ X_5 \end{bmatrix} \qquad (3.110)$$

   c. *Algebraic equations*:

$$X_2 = \frac{1}{B_2} Y_2 \qquad (3.111)$$

TABLE 3.5

Different Components in Given Mechanical System

Type of Components	Components	Schematic Diagram	Terminal Graph
1. Accumulators	Mass M_1	Mass M_1	A — M_1 — G
2. Algebraic type	Dashpot B_2	(dashpot symbol)	C — B_2 — G
3. Coupler	Lever	(lever symbol)	A, B, C
4. Delay type	Spring K_5	(spring symbol)	C — K_5 — G
5. Through driver	Force driver	(force driver symbol)	B — $F_6(t)$ — G

Formulation of State Space Model of Systems

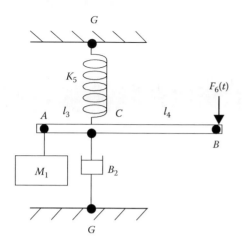

FIGURE 3.48
Mechanical system.

d. *Coupler*

$$\begin{bmatrix} X_3 \\ Y_4 \end{bmatrix} = \begin{bmatrix} 0 & n \\ -n & 0 \end{bmatrix} \begin{bmatrix} Y_3 \\ X_4 \end{bmatrix}$$

3. Interconnecting equations
   a. *f-circuit equation*

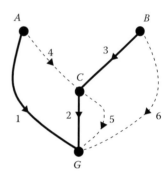

FIGURE 3.49
System graph and f-tree consisting of branches 1, 2, and 3.

$$\begin{bmatrix} -1 & 1 & 0 & | & 1 & 0 & 0 \\ 0 & -1 & 0 & | & 0 & 1 & 0 \\ 0 & -1 & -1 & | & 0 & 0 & 1 \end{bmatrix} \begin{bmatrix} X_1 \\ X_2 \\ X_3 \\ X_4 \\ X_5 \\ X_6 \end{bmatrix} = \begin{bmatrix} 0 \\ 0 \\ 0 \end{bmatrix} \quad (3.112)$$

b. *f-cutset equations*

$$\begin{bmatrix} 1 & 0 & 0 & | & 1 & 0 & 0 \\ 0 & 1 & 0 & | & -1 & 1 & 1 \\ 0 & 0 & 1 & | & 0 & 0 & 1 \end{bmatrix} \begin{bmatrix} Y_1 \\ Y_2 \\ Y_3 \\ Y_4 \\ Y_5 \\ Y_6 \end{bmatrix} = \begin{bmatrix} 0 \\ 0 \\ 0 \end{bmatrix} \quad (3.113)$$

4. Combine f-circuit and cutset equations for dynamic elements, i.e., the second row of the f-circuit equation
   $X_5 = X_2$ and the first row of the f-cutset equation $Y_1 = -Y_4$.
   Rewriting these equations in matrix form in terms of state vector and input

$$\begin{bmatrix} Y_1 \\ X_5 \end{bmatrix} = \begin{bmatrix} 0 & 0 \\ 1 & 0 \end{bmatrix} \begin{bmatrix} X_2 \\ Y_5 \end{bmatrix} + \begin{bmatrix} -1 \\ 0 \end{bmatrix} Y_4$$

Substituting the above equation into terminal equation of dynamic elements,

$$\frac{d}{dt}\begin{bmatrix} X_1 \\ Y_5 \end{bmatrix} = \begin{bmatrix} \frac{1}{M_1} & 0 \\ 0 & \frac{1}{K_5} \end{bmatrix} \left\{ \begin{bmatrix} 0 & 0 \\ 1 & 0 \end{bmatrix} \begin{bmatrix} X_2 \\ Y_5 \end{bmatrix} + \begin{bmatrix} -1 \\ 0 \end{bmatrix} Y_4 \right\}$$

$$\frac{d}{dt}\begin{bmatrix} X_1 \\ Y_5 \end{bmatrix} = \begin{bmatrix} 0 & 0 \\ \frac{1}{K_5 B_2} & 0 \end{bmatrix} \begin{bmatrix} Y_2 \\ Y_5 \end{bmatrix} + \begin{bmatrix} \frac{-1}{M_1} \\ 0 \end{bmatrix} Y_4 \quad (3.114)$$

5. Substitute terminal equation of lever $Y_4 = -nY_3$ and algebraic element $X_2 = \frac{1}{B_2} Y_2$ in the above equation:

$$\frac{d}{dt}\begin{bmatrix} X_1 \\ Y_5 \end{bmatrix} = \begin{bmatrix} 0 & 0 \\ \frac{1}{K_5 B_2} & 0 \end{bmatrix} \begin{bmatrix} Y_2 \\ Y_5 \end{bmatrix} + \begin{bmatrix} \frac{-n}{M_1} \\ 0 \end{bmatrix}, \quad (3.115)$$

From the last row of the f-cutset equation, we may write $Y_3 = -Y_6$ and $Y_2 = Y_4 - Y_5 - Y_6$

$$\frac{d}{dt}\begin{bmatrix} X_1 \\ Y_5 \end{bmatrix} = \begin{bmatrix} 0 & 0 \\ \frac{1}{K_5 B_2} & \frac{-1}{K_5 B_2} \end{bmatrix} \begin{bmatrix} X_1 \\ Y_5 \end{bmatrix} + \begin{bmatrix} \frac{n}{M_1} \\ \frac{n-1}{K_5 B_2} \end{bmatrix} Y_6 \quad (3.116)$$

3.10.4 State Model for Systems with Time Varying and Nonlinear Components

Such systems contain components that have terminal equations as either linear differential (algebraic) equations with time varying coefficients or nonlinear differential equations, whose complete solutions can be found by numerical (iterative) methods.

One reason for this is that complete solutions for such networks are not easily found by analytical methods, and so it becomes necessary to use numerical computations. Another reason is that it is appropriate to study the stability of such systems through Lia punov's method, which requires that systems be characterized by their state equations.

Although we have concerned ourselves only with simple systems whose components have terminal relations that are linear algebraic or differential equations with constant coefficient, it is reasonable to point out briefly how our methods might be extended to include more complex situations.

The state space equation may be written for time variant and nonlinear system in the following form:

$$\frac{d}{dt}[P_1] = [A(t)][P_1] + [B(t)][F(t)] \quad (3.117)$$

Usually matrices $A(t)$ and $B(t)$ were time invariant and linear.

Formulation of State Space Model of Systems

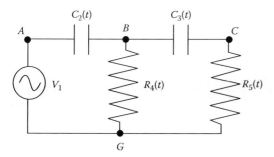

FIGURE 3.50
Electrical system.

Example 3.8

Consider the simple electrical system as shown in Figure 3.50. In this system, let us assume that the components of given electrical system are time dependent and expressed mathematically as

Capacitances

$$C_2(t) = C_{2m}(t)\sin(\omega t)$$

$$C_3(t) = C_{3m}e^{-\lambda t}$$

Resistances

$$R_4(t) = R_{4m}(t)\cos(2\omega t)$$

$$R_5(t) = R_{5m}e^{-\sigma t}$$

To formulate the state equations, we draw the system graph as shown in Figure 3.51, and write the terminal equations:

$$v_1(t) = E(t) \tag{3.118}$$

$$\frac{d}{dt}\begin{bmatrix} C_{2m}\sin(\omega t) & 0 \\ 0 & C_{3m}e^{-\lambda t} \end{bmatrix}\begin{bmatrix} v_2(t) \\ v_3(t) \end{bmatrix} = \begin{bmatrix} i_2(t) \\ i_3(t) \end{bmatrix} \tag{3.119}$$

$$\begin{bmatrix} R_{4m}\cos(2\omega t) & 0 \\ 0 & R_{5m}e^{-\sigma t} \end{bmatrix}\begin{bmatrix} i_4(t) \\ i_5(t) \end{bmatrix} = \begin{bmatrix} v_4(t) \\ v_5(t) \end{bmatrix} \tag{3.120}$$

The cutset equations are

$$\begin{bmatrix} i_2(t) \\ i_3(t) \end{bmatrix} = \begin{bmatrix} 1 & 1 \\ 0 & 1 \end{bmatrix}\begin{bmatrix} i_4(t) \\ i_5(t) \end{bmatrix} \tag{3.121}$$

and the fundamental circuit equations are

$$\begin{bmatrix} v_4(t) \\ v_5(t) \end{bmatrix} = \begin{bmatrix} -1 & 0 \\ -1 & -1 \end{bmatrix}\begin{bmatrix} v_2(t) \\ v_3(t) \end{bmatrix} + \begin{bmatrix} 1 \\ 1 \end{bmatrix}v_1(t) \tag{3.122}$$

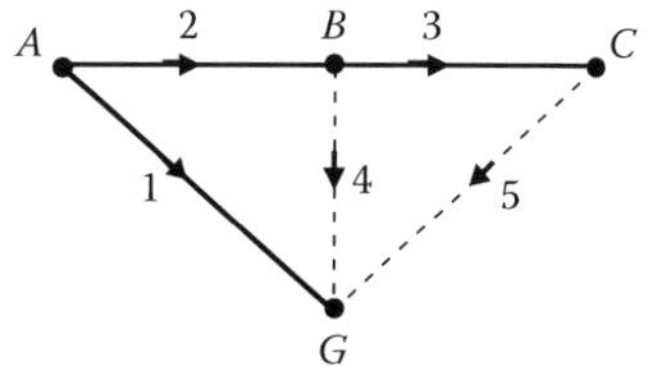

FIGURE 3.51
System graph for system.

Substituting the cutset Equation 3.121 into Equation 3.119, we get

$$\frac{d}{dt}\begin{bmatrix} C_{2m}\sin(\omega t) & 0 \\ 0 & C_{3m}e^{-\lambda t} \end{bmatrix}\begin{bmatrix} v_2(t) \\ v_3(t) \end{bmatrix} = \begin{bmatrix} 1 & 1 \\ 0 & 1 \end{bmatrix}\begin{bmatrix} i_4(t) \\ i_5(t) \end{bmatrix} \quad (3.123)$$

Now substitute the fundamental circuit equation (Equation 3.122) into Equation 3.120,

$$\begin{bmatrix} R_{4m}\cos(2\omega t) & 0 \\ 0 & R_{5m}e^{-\sigma t} \end{bmatrix}\begin{bmatrix} i_4(t) \\ i_5(t) \end{bmatrix} = \begin{bmatrix} -1 & 0 \\ -1 & -1 \end{bmatrix}\begin{bmatrix} v_2(t) \\ v_3(t) \end{bmatrix} + \begin{bmatrix} 1 \\ 1 \end{bmatrix} v_1(t) \quad (3.124)$$

After simplifying Equation 3.124

$$\begin{bmatrix} i_4(t) \\ i_5(t) \end{bmatrix} = \begin{bmatrix} \dfrac{-1}{R_{4m}\cos(2\omega t)} & 0 \\ \dfrac{-1}{R_{5m}e^{-\sigma t}} & \dfrac{-1}{R_{5m}e^{-\sigma t}} \end{bmatrix}\begin{bmatrix} v_2(t) \\ v_3(t) \end{bmatrix} + \begin{bmatrix} \dfrac{1}{R_{4m}\cos(2\omega t)} \\ \dfrac{1}{R_{5m}e^{-\sigma t}} \end{bmatrix} v_1(t) \quad (3.125)$$

Substituting Equation 3.125 into Equation 3.123, we obtain

$$\frac{d}{dt}\begin{bmatrix} C_{2m}\sin(\omega t) & 0 \\ 0 & C_{3m}e^{-\lambda t} \end{bmatrix}\begin{bmatrix} v_2(t) \\ v_3(t) \end{bmatrix} = \begin{bmatrix} 1 & 1 \\ 0 & 1 \end{bmatrix}\left\{\begin{bmatrix} \dfrac{-1}{R_{4m}\cos(2\omega t)} & 0 \\ \dfrac{-1}{R_{5m}e^{-\sigma t}} & \dfrac{-1}{R_{5m}e^{-\sigma t}} \end{bmatrix}\begin{bmatrix} v_2(t) \\ v_3(t) \end{bmatrix} + \begin{bmatrix} \dfrac{1}{R_{4m}\cos(2\omega t)} \\ \dfrac{1}{R_{5m}e^{-\sigma t}} \end{bmatrix} v_1(t)\right\} \quad (3.126)$$

or

$$\frac{d}{dt}\begin{bmatrix} C_{2m}\sin(\omega t) & 0 \\ 0 & C_{3m}e^{-\lambda t} \end{bmatrix}\begin{bmatrix} v_2(t) \\ v_3(t) \end{bmatrix} = \begin{bmatrix} \left(\dfrac{-1}{R_{4m}\cos(2\omega t)} - \dfrac{1}{R_{5m}e^{-\sigma t}}\right) & \dfrac{-1}{R_{5m}e^{-\sigma t}} \\ \dfrac{-1}{R_{5m}e^{-\sigma t}} & \dfrac{-1}{R_{5m}e^{-\sigma t}} \end{bmatrix}\begin{bmatrix} v_2(t) \\ v_3(t) \end{bmatrix} + \begin{bmatrix} \left(\dfrac{1}{R_{4m}\cos(2\omega t)} + \dfrac{1}{R_{5m}e^{-\sigma t}}\right) \\ \dfrac{1}{R_{5m}e^{-\sigma t}} \end{bmatrix} v_1(t) \quad (3.127)$$

Formulation of State Space Model of Systems

Now, we differentiate the left-hand side of Equation 3.127, and obtain

$$\begin{bmatrix} C_{2m}\omega\cos(\omega t) & 0 \\ 0 & -\lambda C_{3m}e^{-\lambda t} \end{bmatrix} \begin{bmatrix} v_2(t) \\ v_3(t) \end{bmatrix} + \begin{bmatrix} C_{2m}\sin(\omega t) & 0 \\ 0 & C_{3m}e^{-\lambda t} \end{bmatrix} \frac{d}{dt}\begin{bmatrix} v_2(t) \\ v_3(t) \end{bmatrix}$$

$$= \begin{bmatrix} \dfrac{-1}{R_{4m}\cos(2\omega t)} - \dfrac{1}{R_{5m}e^{-\sigma t}} & \dfrac{-1}{R_{5m}e^{-\sigma t}} \\ \dfrac{-1}{R_{5m}e^{-\sigma t}} & \dfrac{-1}{R_{5m}e^{-\sigma t}} \end{bmatrix} \begin{bmatrix} v_2(t) \\ v_3(t) \end{bmatrix} + \begin{bmatrix} \dfrac{1}{R_{4m}\cos(2\omega t)} + \dfrac{1}{R_{5m}e^{-\sigma t}} \\ \dfrac{1}{R_{5m}e^{-\sigma t}} \end{bmatrix} v_1(t) \quad (3.128)$$

Rearrange Equation 3.128, by combining the first term in the left with the first term on the right. This yields

$$\begin{bmatrix} C_{2m}\sin(\omega t) & 0 \\ 0 & C_{3m}e^{-\lambda t} \end{bmatrix} \frac{d}{dt}\begin{bmatrix} v_2(t) \\ v_3(t) \end{bmatrix}$$

$$= \begin{bmatrix} -C_{2m}\omega\cos(\omega t) + \left(\dfrac{-1}{R_{4m}\cos(2\omega t)} - \dfrac{1}{R_{5m}e^{-\sigma t}}\right) & \dfrac{-1}{R_{5m}e^{-\sigma t}} \\ \dfrac{-1}{R_{5m}e^{-\sigma t}} & \lambda C_{3m}e^{-\lambda t} + \dfrac{-1}{R_{5m}e^{-\sigma t}} \end{bmatrix} \begin{bmatrix} v_2(t) \\ v_3(t) \end{bmatrix}$$

$$+ \begin{bmatrix} \left(\dfrac{1}{R_{4m}\cos(2\omega t)} + \dfrac{1}{R_{5m}e^{-\sigma t}}\right) \\ \dfrac{1}{R_{5m}e^{-\sigma t}} \end{bmatrix} v_1(t) \quad (3.129)$$

Finally, we obtain the state equations by inverting the coefficient matrix on the left of Equation 3.129, and substituting for $v_1(t)$ from Equation 3.118. The result is

$$\frac{d}{dt}\begin{bmatrix} v_2(t) \\ v_3(t) \end{bmatrix} = \begin{bmatrix} -\cot(\omega t) + \dfrac{1}{C_{2m}\sin(\omega t)}\left(\dfrac{-1}{R_{4m}\cos(2\omega t)} - \dfrac{1}{R_{5m}e^{-\sigma t}}\right) & \dfrac{-1}{C_{2m}\sin(\omega t)R_{5m}e^{-\sigma t}} \\ \dfrac{-1}{R_{5m}C_{3m}e^{-\lambda t-\sigma t}} & \lambda + \dfrac{-1}{R_{5m}C_{3m}e^{-\lambda t-\sigma t}} \end{bmatrix} \begin{bmatrix} v_2(t) \\ v_3(t) \end{bmatrix}$$

$$+ \begin{bmatrix} \dfrac{1}{R_{4m}C_{2m}\sin(\omega t)\cos(2\omega t)} + \dfrac{1}{R_{5m}C_{2m}\sin(\omega t)e^{-\sigma t}} \\ \dfrac{1}{R_{5m}C_{3m}e^{-\lambda t-\sigma t}} \end{bmatrix} E(t) \quad (3.130)$$

As explained, the only difference in the procedure from the time invariant case is the coefficient matrix of the state variables in Equation 3.130. Clearly, time invariant systems are merely special cases of time-varying ones; the differentiation involved with constant coefficient is trivial.

When nonlinear components are encountered, the situation is much more complicated. It is not always clear whether or not there exist a solution for a set of nonlinear equations, and even if one does exist, there is seldom any analytical method for finding it. Matrix methods cannot be used to handle sets of nonlinear equations.

Nevertheless, we can always eliminate the chord across variables and branch through variables from a set of nonlinear terminal equations; this can be done by simple substitution because the cutset and fundamental circuit equations are linear.

Let us assume that the terminal equations of a given nonlinear system are, at worst, first-order differential equations. If we can locate all the first-order across drivers in the forest, and all the first-order through drivers in the coforest, the situation is considerably simplified. If we assume in addition that the differential terminal relations can be written explicitly in the derivatives of the first-order elements, we can proceed to eliminate all the branch currents and chord voltages through the cutset and circuit equations.

3.11 Solution of State Equations

In this section, we shall study methods for solution of the state equations from which the system transient response can then be obtained. Let us first review the classical method of solution by considering a first-order scalar differential equation,

$$\frac{dx}{dt} = ax; \quad x(0) = x_0 \tag{3.131}$$

Equation 3.131 has the solution

$$x(t) = e^{at} x_0$$

$$= \left(1 + at + \frac{1}{2!}a^2 t^2 + \cdots + \frac{1}{i!}a^i t^i\right) x_0 \tag{3.132}$$

Let us now consider the state equation

$$\overset{0}{x}(t) = Ax(t); \quad x(0) = x_0 \tag{3.133}$$

which represents a homogeneous unforced linear system with constant coefficients.

By analogy with the scalar case, we assume a solution of the form

$$x(t) = a_0 + a_1 t + a_2 t^2 + \cdots + a_i t^i + \cdots$$

where a_i are vector coefficients.

By substituting the assumed solution into Equation 3.133 we get

$$a_1 + 2a_2 t + 3a_3 t^2 + \cdots = A(a_0 + a_1 t + a_2 t^2 + \cdots)$$

Formulation of State Space Model of Systems

The comparison of vector coefficients of equal powers of t yields

$$a_1 = A a_0$$

$$a_2 = \frac{1}{2} A a_1 = \frac{1}{2!} A^2 a_0$$

$$a_2 = \frac{1}{i!} A^i a_0$$

In the assumed solution, equating $x(t = 0) = x_0$, we find that the solution $x(t)$ is thus found to be

$$x(t) = \left(I + At + \frac{1}{2!} A^2 t^2 + \cdots + \frac{1}{i!} A^i t^i + \cdots \right) x_0$$

Each of the terms inside the brackets in an $n \times n$ matrix. Because of the similarity of the entity inside the brackets with a scalar exponential of Equation 3.132, we call it a matrix exponential, which may be written as

$$e^{At} = I + At + \frac{1}{2!} A^2 t^2 + \cdots + \frac{1}{i!} A^i t^i + \cdots \tag{3.134}$$

The solution $x(t)$ can now be written as

$$x(t) = e^{At} x_0 \tag{3.135}$$

From Equation 3.135 it is observed that the initial state x_0 at $t = 0$ is driven to a state $x(t)$ at time t. This transition in state is carried out by the matrix exponential e^{At}. Because of this property, e^{At} is known as *state transition matrix* and is denoted by $\Phi(t)$.

Let us now determine the solution of the nonhomogenous state equation (forced system).

$$x(t) = Ax(t) + Bu(t); \quad x(0) = x_0 \tag{3.136}$$

Rewrite this equation in the form

$$\frac{dx}{dt} - Ax(t) = Bu(t)$$

Multiplying both sides by e^{-At}, we can write

$$e^{-At} \left[\frac{dx(t)}{dt} - Ax(t) \right] = \frac{d}{dt} \left[e^{-At} x(t) \right]$$

$$= e^{-At} Bu(t)$$

Integrating both sides with respect to t between the limits 0 and t, we get

$$e^{-At}x(t) - x(0) = \int_0^t e^{-A\tau}Bu(\tau)d\tau$$

Now premultiplying both sides by e^{At}, we have

$$x(t) = e^{-At}x(0) + \int_{t_0}^t e^{-A(t-\tau)}Bu(\tau)d\tau \qquad (3.137)$$

If the initial state is known at $t = t_0$ rather than $t = 0$, Equation 3.137 becomes

$$x(t) = e^{-A(t-t_0)}x(t_0) + \int_{t_0}^t e^{-A(t-\tau)}Bu(\tau)d\tau \qquad (3.138)$$

In Equation 3.138, the total response can be expressed as the sum of zero input and zero state components. For linear systems, the zero input and zero state components obey the principle of superposition with respect to each of their respective causes.

Properties of state transition matrix

1. $\phi(0) = e^{A0} = I$
2. $\phi(t) = e^{At} = (e^{-At})^{-1} = [\phi(-t)]^{-1}$

   or

$$\phi^{-1}(t) = \phi(-t)$$

   or

$$\phi(t_1 + t_2) = e^{-A(t_1+t_2)} = e^{-At_1}e^{-At_2}$$
$$= \phi(t_1)\,\phi(t_2) = \phi(t_2)\,\phi(t_1)$$

Example 3.9

Obtain the time response of the following system:

$$\frac{d}{dt}\begin{bmatrix} x_1 \\ x_2 \end{bmatrix} = \begin{bmatrix} 1 & 0 \\ 1 & 1 \end{bmatrix}\begin{bmatrix} x_1 \\ x_2 \end{bmatrix} + \begin{bmatrix} 1 \\ 1 \end{bmatrix}u$$

where $u(t)$ is a unit step occurring at $t = 0$ and $x^T(0) = \begin{bmatrix} 1 & 0 \end{bmatrix}$

SOLUTION

We have in this case

$$A = \begin{bmatrix} 1 & 0 \\ 1 & 1 \end{bmatrix};\quad B = \begin{bmatrix} 1 \\ 1 \end{bmatrix}$$

Formulation of State Space Model of Systems

The state transition matrix $\Phi(t)$ is given by

$$e^{At} = I + At + \frac{1}{2!}A^2t^2 + \cdots + \frac{1}{i!}A^i t^i + \cdots$$

Substituting values of A and collecting terms, we get

$$e^{At} = \begin{bmatrix} 1+t+0.5t^2+\cdots & 0 \\ t+t^2+\cdots & 1+t+0.5t^2+\cdots \end{bmatrix}$$

$$= \begin{bmatrix} \phi_{11} & \phi_{12} \\ \phi_{21} & \phi_{22} \end{bmatrix}$$

The terms ϕ_{11} and ϕ_{22} are easily recognized as series expansion of e^{At}. To recognized ϕ_{21}, more terms of the infinite series should be evaluated. In fact ϕ_{21} is te^t.

We have the state transition matrix as

$$\phi(t) = \begin{bmatrix} e^t & 0 \\ te^t & e^t \end{bmatrix}$$

The time response of the system is given by

$$x(t) = \phi(t)\left[x_0 + \int_0^t \phi(-\tau) Bu\, d\tau \right]$$

Now with $u = 1$,

$$\phi(-\tau)Bu = \begin{bmatrix} e^{-\tau} & 0 \\ -\tau e^{-\tau} & e^{-\tau} \end{bmatrix}\begin{bmatrix} 1 \\ 1 \end{bmatrix} = \begin{bmatrix} e^{-\tau} \\ e^{-\tau(1-\tau)} \end{bmatrix}$$

Therefore

$$\int_0^t \phi(-\tau)Bu\, d\tau = \begin{bmatrix} \int_0^t e^{-\tau} d\tau \\ \int_0^t e^{-\tau(1-\tau)} d\tau \end{bmatrix} = \begin{bmatrix} 1-e^{-t} \\ te^{-t} \end{bmatrix}$$

Then the solution $x(t)$ is given by

$$x(t) = \begin{bmatrix} e^t & 0 \\ te^t & e^t \end{bmatrix}\left\{ \begin{bmatrix} 1 \\ 0 \end{bmatrix} + \begin{bmatrix} 1-e^t \\ te^t \end{bmatrix} \right\}$$

$$= \begin{bmatrix} 2e^t - 1 \\ 2te^t \end{bmatrix}$$

Computation of state transition matrix

1. *Machine computation:*

$$e^{At} = I + At + \frac{1}{2!}A^2t^2 + \cdots + \frac{1}{i!}A^it^i + \cdots$$

2. *Diagonalization:*
   The matrix having elements only on its diagonal is known as diagonal matrix. The inverse of a diagonal matrix is obtained by inspection, e.g.,

$$\begin{bmatrix} \alpha & 0 & 0 \\ 0 & \beta & 0 \\ 0 & 0 & \gamma \end{bmatrix}^{-1} = \begin{bmatrix} \frac{1}{\alpha} & 0 & 0 \\ 0 & \frac{1}{\beta} & 0 \\ 0 & 0 & \frac{1}{\gamma} \end{bmatrix}$$

The state model having elements of coefficient matrix A only on its diagonal is called canonical state model, which is obtained by parallel decomposition of a transfer function. However, any general state model can be transformed into a canonical form using the process of diagonalization.

Consider the state equation

$$\frac{dx}{dt} = AX$$

where A is a nondiagonal matrix

$$D = P^{-1}AP$$

where D is a diagonal matrix in which $\lambda_1, \lambda_2, \lambda_3, \ldots, \lambda_n$ are diagonals:

$$P = \begin{bmatrix} 1 & 1 & \cdots & 1 \\ \lambda_1 & \lambda_2 & \cdots & \lambda_n \\ \lambda_1^2 & \lambda_2^2 & \cdots & \lambda_n^2 \\ \vdots & \vdots & \cdots & \vdots \\ \lambda_1^n & \lambda_2^n & \cdots & \lambda_n^n \end{bmatrix}_{n*n}$$

There are n number of roots

$$e^{Dt} = \begin{bmatrix} e^{\lambda_1 t} & 0 & 0 & \cdots & 0 \\ 0 & e^{\lambda_2 t} & 0 & \cdots & 0 \\ 0 & 0 & e^{\lambda_3 t} & \cdots & 0 \\ 0 & 0 & 0 & \cdots & 0 \\ 0 & 0 & 0 & \cdots & e^{\lambda_n t} \end{bmatrix}$$

Formulation of State Space Model of Systems

So now

$$e^{At} = Pe^{Dt}P^{-1}$$

where matrix P is known as Vandermonde's matrix.

Example 3.10

The matrix A is given. Find the state transition matrix.

$$A = \begin{bmatrix} 2 & 0 \\ 0 & 1 \end{bmatrix}$$

The characteristic equation is

$$|\lambda I - A| = \begin{bmatrix} \lambda - 2 & 0 \\ 0 & \lambda - 1 \end{bmatrix}$$

$$|\lambda I - A| = (\lambda - 1)(\lambda - 2) = 0$$

$\lambda = 1, 2$, i.e., $\lambda_1 = 1, \lambda_2 = 2$

$$P = \begin{bmatrix} 1 & 1 \\ 1 & 2 \end{bmatrix} \text{ and } P^{-1} = \frac{1}{2-1}\begin{bmatrix} 2 & -1 \\ -1 & 1 \end{bmatrix}$$

$$P^{-1} = \begin{bmatrix} 2 & -1 \\ -1 & 1 \end{bmatrix} \text{ and } D = \begin{bmatrix} 1 & 0 \\ 0 & 2 \end{bmatrix}$$

$$e^{At} = Pe^{Dt}P^{-1} = \begin{bmatrix} 1 & 1 \\ 1 & 2 \end{bmatrix}\begin{bmatrix} e^t & 0 \\ 0 & e^{2t} \end{bmatrix}\begin{bmatrix} 2 & -1 \\ -1 & 1 \end{bmatrix}$$

$$\text{Then } = \begin{bmatrix} e^t & e^{2t} \\ e^t & 2e^{2t} \end{bmatrix}\begin{bmatrix} 2 & -1 \\ -1 & 1 \end{bmatrix}$$

$$\begin{bmatrix} 2e^t - e^{2t} & -e^t + 2e^{2t} \\ 2e^t - 2e^{2t} & -e^t + 2e^{2t} \end{bmatrix}$$

3. *Laplace Method*:

By the Laplace transform

$$e^{At} = L^{-1}[SI - A]^{-1}$$

Example 3.11

Find the state transition matrix for given matrix A:

$$A = \begin{bmatrix} 0 & 1 \\ 0 & -2 \end{bmatrix}$$

$$[SI - A] = \begin{bmatrix} S & -1 \\ 0 & S+2 \end{bmatrix}$$

$$[SI - A]^{-1} = \frac{1}{S(S+2)} \begin{bmatrix} S+2 & 1 \\ 0 & S \end{bmatrix}$$

$$= \begin{bmatrix} \frac{1}{S} & \frac{1}{S(S+2)} \\ 0 & \frac{1}{S+2} \end{bmatrix}$$

Taking inverse Laplace transforms

$$L^{-1}\left\{\frac{1}{S}\right\} = 1$$

and $\quad L^{-1}\left\{\dfrac{1}{S+2}\right\} = e^{-2t}$

$$\frac{1}{S(S+2)} = \frac{A}{S} + \frac{B}{S+2}$$

$$1 = A(S+2) + B(S)$$

$$B = -\frac{1}{2} \quad \text{and} \quad A = \frac{1}{2}$$

$$\frac{1}{S(S+2)} = \frac{1}{2}\left(\frac{1}{S} - \frac{1}{S+2}\right)$$

$$L^{-1}\left(\frac{1}{S(S+2)}\right) = \frac{1}{2}(1 - e^{-2t})$$

$$[SI - A]^{-1} = \begin{bmatrix} 1 & \frac{1}{2}(1 - e^{-2t}) \\ 0 & e^{-2t} \end{bmatrix} = e^{At}$$

4. *Calay Hamilton Method*:

In Calay Hamilton method

$$e^{At} = \begin{bmatrix} 1 & \lambda_1 & \lambda_1^2 & \cdots & \cdots & \lambda_1^{n-1} & e^{\lambda_1 t} \\ 1 & \lambda_2 & \lambda_2^2 & \cdots & \cdots & \lambda_2^{n-1} & e^{\lambda_2 t} \\ \vdots & \vdots & \vdots & \vdots & \vdots & \vdots & \vdots \\ \vdots & \vdots & \vdots & \vdots & \vdots & \vdots & \vdots \\ 1 & \lambda_m & \lambda_m^2 & \cdots & \cdots & \lambda_m^{n-1} & e^{\lambda_n t} \\ I & A & A^2 & \cdots & \cdots & A^{n-1} & e^{At} \end{bmatrix}_{(n+1)xn}$$

Formulation of State Space Model of Systems

Example 3.12

Find the state transition matrix for given coefficient matrix A

$$A = \begin{bmatrix} 0 & 1 \\ 0 & -2 \end{bmatrix}$$

$$\lambda I - A = \begin{bmatrix} \lambda & -1 \\ 0 & \lambda+2 \end{bmatrix} = 0$$

$$\lambda_1 = 0, \quad \lambda_2 = -2$$

$$\begin{bmatrix} 1 & 0 & e^{0t} \\ 1 & -2 & e^{-2t} \\ I & A & e^{At} \end{bmatrix} = 0$$

Characteristic equation is

$$1[-2e^{At} - Ae^{-2t}] + 1[A + 2I] = 0$$

or $-2e^{At} - Ae^{-2t} + A + 2I = 0$

$$e^{At} = \frac{1}{2}(A + 2I - Ae^{-2t})$$

$$= \frac{1}{2}\begin{bmatrix} 0 & 1 \\ 0 & -2 \end{bmatrix} + \frac{1}{2}\begin{bmatrix} 2 & 0 \\ 0 & 2 \end{bmatrix} - \frac{1}{2}\begin{bmatrix} 0 & e^{-2t} \\ 0 & -2e^{-2t} \end{bmatrix}$$

$$= \begin{bmatrix} 1 & 0.5(1-e^{-2t}) \\ 0 & e^{-2t} \end{bmatrix}$$

Example 3.13

Solve the state equation $\dfrac{d}{dt}\begin{bmatrix} x_1 \\ x_2 \end{bmatrix} = \begin{bmatrix} 0 & 1 \\ 0 & -2 \end{bmatrix}\begin{bmatrix} x_1 \\ x_2 \end{bmatrix} + \begin{bmatrix} 0 \\ 6 \end{bmatrix} u$ and also determine the output from the output equation:

$$y = \begin{bmatrix} 1 & 0 \end{bmatrix}\begin{bmatrix} x_1 \\ x_2 \end{bmatrix}$$ and initial conditions

$$\begin{bmatrix} x_1(0) \\ x_2(0) \end{bmatrix} = \begin{bmatrix} 1 \\ 1 \end{bmatrix}$$

From Example 3.12 the matrix e^{At} is

$$e^{At} = \begin{bmatrix} 1 & \dfrac{1}{2}(1-e^{-2t}) \\ 0 & e^{-2t} \end{bmatrix}$$

Solution of state equation is $x(t) = e^{At}x(0) + \int_0^\tau e^{A(t-\tau)}Bu(\tau)d\tau$

$$= \begin{bmatrix} 1 & \frac{1}{2}(1-e^{-2t}) \\ 0 & e^{-2t} \end{bmatrix} \begin{bmatrix} 1 \\ 1 \end{bmatrix} + [u] \int_0^t \begin{bmatrix} 1 & \frac{1}{2}(1-e^{-2(t-\tau)}) \\ 0 & e^{-2(t-\tau)} \end{bmatrix} \begin{bmatrix} 0 \\ 6 \end{bmatrix} d\tau$$

$$= \begin{bmatrix} 1.5(1-e^{-2t}) \\ e^{-2t} \end{bmatrix} + \begin{bmatrix} 3(t-1.5e^{0t}) \\ 3e^{0t} \end{bmatrix} - \begin{bmatrix} -1.5e^{2t} \\ 3e^{-2t} \end{bmatrix}$$

$$x(t) = \begin{bmatrix} 3t - e^{-2t} + 1.5e^{2t} \\ 3 + e^{-2t} + 3e^{-2t} \end{bmatrix}$$

Example 3.14

Find the solution of state equation.

$$\frac{d}{dt}\begin{bmatrix} x_1 \\ x_2 \\ x_3 \end{bmatrix} = \begin{bmatrix} 1 & 0 & 0 \\ 0 & 1 & 0 \\ 0 & 0 & -2 \end{bmatrix} \begin{bmatrix} x_1 \\ x_2 \\ x_3 \end{bmatrix} + \begin{bmatrix} 1 \\ 0 \\ 0 \end{bmatrix}[u(t)]$$

and $y(t) = \begin{bmatrix} 1 & 0 & 0 \end{bmatrix} x(t)$

$$[SI - A] = \begin{bmatrix} S-1 & 0 & 0 \\ 0 & S-1 & 0 \\ 0 & 0 & S+2 \end{bmatrix}$$

$$[SI - A]^{-1} = \frac{1}{(S-1)(S-1)(S+2)} \begin{bmatrix} (S-1)(S+2) & 0 & 0 \\ 0 & (S-1)(S+2) & 0 \\ 0 & 0 & (S-1)^2 \end{bmatrix}$$

$$[SI - A]^{-1} = \begin{bmatrix} \frac{1}{(S-1)} & 0 & 0 \\ 0 & \frac{1}{(S-1)} & 0 \\ 0 & 0 & \frac{1}{(S+2)} \end{bmatrix}$$

Taking inverse Laplace transform

$$L^{-1}[SI - A]^{-1} = \begin{bmatrix} e^t & 0 & 0 \\ 0 & e^t & 0 \\ 0 & 0 & e^{-2t} \end{bmatrix}$$

Formulation of State Space Model of Systems

i.e., $e^{At} = \begin{bmatrix} e^t & 0 & 0 \\ 0 & e^t & 0 \\ 0 & 0 & e^{-2t} \end{bmatrix}$

The solution of state equation is $x(t) = e^{At}x(0) + \int_0^{\tau} e^{A(t-\tau)}Bu(\tau)d\tau$

$$x(t) = \begin{bmatrix} e^t \\ e^t \\ e^{-2t} \end{bmatrix} + \int_0^t \begin{bmatrix} e^{(t-\tau)} \\ 0 \\ 0 \end{bmatrix} u(t-\tau)d\tau$$

$$x(t) = \begin{bmatrix} e^t \\ e^t \\ e^{-2t} \end{bmatrix} + u(t)\begin{bmatrix} -e^{(t-\tau)} \\ 0 \\ 0 \end{bmatrix}_0^t$$

$$x(t) = \begin{bmatrix} e^t \\ e^t \\ e^{-2t} \end{bmatrix} - (1-e^t)u(t), \quad y(t) = \begin{bmatrix} 1 & 0 & 0 \end{bmatrix}\left\{ \begin{bmatrix} e^t \\ e^t \\ e^{-2t} \end{bmatrix} - \begin{bmatrix} 1-e^t \\ 0 \\ 0 \end{bmatrix} u(t) \right\}$$

$$y(t) = e^t - (1 - e^t)u(t)$$

Euler's method

Consider the equation

$$\frac{dy}{dx} = f(x, y) \tag{3.139}$$

given that $y(x_0) = y_0$. Its curve of solution through $P(x_0, y_0)$ is shown dotted in Figure 3.52. Now we have to find the ordinate of any other point on this curve.

Let us divide x-axis into n subintervals each of width h so that h is quite small. In the interval 0 to x_0 we approximate the curve by the tangent at P. Then y_1 may be calculated as

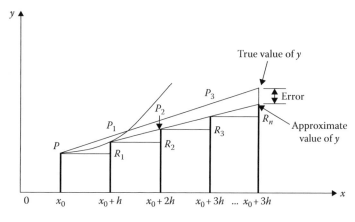

FIGURE 3.52
Solution of state model.

$$y_1 = y_0 + PR_1 \tan \theta$$

$$= y_0 + h\left(\frac{dy}{dx}\right)_p = y_0 + hf(x_0, y_0)$$

Let P_1Q_1 be the curve of solution of Equation 3.139 through P_1 and let its tangent at P_1 meet the ordinate through $P_2(x_0 + 2h, y_2)$, then

$$y_2 = y_1 + hf(x_0 + h, y_1) \tag{3.140}$$

Repeating this process n times we finally reach on an approximation, given by

$$y_n = y_{n-1} + hf(x_0 + (n-1)h, y_{n-1})$$

Note: In Euler's method, we approximate the curve of solution by the tangent in each interval, i.e., by a sequence of short lines. Unless h is small, the error is bound to be quite significant. This sequence of lines may also deviate considerably from the curve of solution. As such, the method is very slow and hence there is a modification of this method which is given in the next section.

Example 3.15

Using Euler's method, find an approximate value of y corresponding to $x = 1$, given that $\frac{dy}{dx} = x + y$ and $y = 1$ when $x = 0$.

We take $n = 10$ and $h = 0.1$ which is sufficiently small. The various calculations are arranged in Table 3.6.

Thus the required approximate value of $y = 3.18$.

Modified Euler's Method:

In the Euler's method, the curve of solution in the interval is approximated by the tangent at P (Figure 3.52) such that at P_1 we have

$$y_1 = y_0 + hf(x_0, y_0) \tag{3.141}$$

TABLE 3.6

Iterative Solution for First-Order Differential Equation

x	y	Mean Slope (y')	New Value of y
0.0	1.00	1.00	1.00 + 0.1(1.00) = 1.10
0.1	1.10	1.20	1.10 + 0.1(1.20) = 1.22
0.2	1.22	1.42	1.22 + 0.1(1.42) = 1.36
0.3	1.36	1.66	1.36 + 0.1(1.66) = 1.53
0.4	1.53	1.93	1.53 + 0.1(1.93) = 1.72
0.5	1.72	2.22	1.72 + 0.1(2.22) = 1.94
0.6	1.94	2.54	1.94 + 0.1(2.54) = 2.19
0.7	2.19	2.89	2.19 + 0.1(2.89) = 2.48
0.8	2.48	3.29	2.48 + 0.1(3.29) = 2.81
0.9	2.81	6.71	2.81 + 0.1(3.71) = 3.18
1.0	3.18		

Formulation of State Space Model of Systems

Then the slope of the curve of solution through P_1, i.e., $\left(\dfrac{dy}{dx}\right)_{P_1} = f(x_0 + h, y_1)$ is computed and the tangent at P_1 is drawn. Now we find a better approximation $y_1^{(1)}$ of $y(x_0 + h)$ by taking the slope of the curve as the mean of the slopes of the tangents at P and P_1, i.e.,

$$y_1^{(1)} = y_0 + \dfrac{h}{2}\left[f(x_0, y_0) + f(x_0 + h, y_1)\right] \tag{3.142}$$

As the slope of the tangent at P_1 is not known, we take y_1 as found in Equation 3.141 by Euler's method and insert it on R.H.S. of Equation 3.142 to obtain the first modified value of $y_1^{(1)}$.
Again Equation 3.142 is used and an even better value $y_1^{(2)}$ is found as

$$y_1^{(2)} = y_0 + \dfrac{h}{2}\left[f(x_0, y_0) + f(x_0 + h, y_1^{(1)})\right]$$

We repeat this step till consecutive values of y agree. This is then taken as the starting point for the next interval.

Once y_1 is obtained to the desired degree of accuracy, y_2 may be calculated by

$$y_2 = y_1 + hf(x_0 + h, y_1)$$

and a better approximation $y_2^{(1)}$ is obtained from Equation 3.142

$$y_2^{(1)} = y_1 + \dfrac{h}{2}\left[f(x_0 + h, y_1) + f(x_0 + 2h, y_2)\right]$$

Repeat this step until y_2 becomes stationary. Then we proceed to calculate y_3 as above and so on.

Example 3.16

Using modified Euler's method find an approximate value of y when $x = 0.3$, given that $\dfrac{dy}{dx} = x + y$ and $y = 1$ when $x = 0$. Taking $h = 0.1$, the calculations are shown in Table 3.7.
Hence, $y(0.3) = 1.4004$ approximately.

TABLE 3.7
Iterative Solution for First-Order Differential Equation

x	$x + y = y'$	Mean Slope	New Value of y
0.0	0 + 1	—	1.00 + 0.1(1.00) = 1.10
0.1	0.1 + 1.1	½(1 + 1.2)	1.00 + 0.1(1.1) = 1.11
0.1	0.1 + 1.11	½(1 + 1.21)	1.00 + 0.1(1.105) = 1.1105
0.1	0.1 + 1.1105	½(1 + 1.2105)	1.00 + 0.1(1.1052) = 1.1105
0.1	1.2105	—	1.1105 + 0.1(1.2105) = 1.2316
0.2	0.2 + 1.2136	½(1.12105 + 1.4316)	1.1105 + 0.1(1.3211) = 1.2426
0.2	0.2 + 1.2426	½(1.2105 + 1.4426)	1.1105 + 0.1(1.3266) = 1.2432
0.2	0.2 + 1.2432	½(1.2105 + 1.4432)	1.1105 + 0.1(1.3268) = 1.2432
0.2	1.4432	—	1.2432 + 0.1(1.4432) = 1.3875
0.3	0.3 + 1.3875	½(1.4432 + 1.6875)	1.2432 + 0.1(1.5654) = 1.3997
0.3	0.3 + 1.3997	½(1.4432 + 1.6997)	1.2432 + 0.1(1.5715) = 1.4003
0.3	0.3 + 1.4003	½(1.4432 + 1.7003)	1.2432 + 0.1(1.5718) = 1.4004
0.3	0.3 + 1.4004	½(1.4432 + 1.7004)	1.2432 + 0.1(1.5718) = 1.4004

Adams–Bashforth method

The Euler's and Runge–Kutta methods are examples of single step methods because they use only the most recent value, x^k, to compute the solution point x^{k+1}. Once the solution has been progressed over several steps, it is possible to use past values of $x(t)$ to construct a polynomial approximation to $f(t, x(t))$. This approximation can then be integrated over the interval $[t^k, t^{k+1}]$ and the result added to x^k to produce x^{k+1}. This is the basic idea behind the multistep methods.

To generate a multistep method using Adams–Bashforth method we start by approximating $f(t, x(t))$ using a Newton backward difference interpolating polynomial of degree m. For $m = 3$, this corresponds to the following cubic polynomial:

$$f[t, x(t)] \approx f^k + s\Delta f^{k-1} + \frac{s(s+1)}{2}\Delta^2 f^{k-2} + \frac{s(s+1)(s+2)}{6}\Delta^3 f^{k-3}$$

where $s = \dfrac{t - t^k}{h}$.

Example 3.17

Given $\dfrac{dy}{dx} = f(x, y)$ and initial conditions are y_0 at x_0, we compute

$$y_{-1} = y(x_0 - h), \quad y_{-2} = y(x_0 - 2h), \quad y_{-3} = y(x_0 - 3h)$$

by Taylor's series or Euler's method or Runge–Kutta method.

Next we calculate

$$f_{-1} = f(x_0 - h, y_{-1}), \quad f_{-2} = f(x_0 - 2h, y_{-2}), \quad f_{-3} = f(x_0 - 3h, y_{-3})$$

Then to find y_1, we substitute Newton's backward interpolation formula

$$f(x, y) = f_0 + n\Delta f_0 + \frac{n(n+1)}{2}\Delta^2 f_0 + \frac{n(n+1)(n+2)}{6}\Delta^3 f_0$$

in

$$y_1 = y_0 + \int_{x_0}^{x_0+h} f(x, y)\,dx \tag{3.143}$$

$$y_1 = y_0 + \int_{x_0}^{x_0+h} \left(f_0 + n\Delta f_0 + \frac{n(n+1)}{2}\Delta^2 f_0 + \frac{n(n+1)(n+2)}{6}\Delta^3 f_0 \right) dx$$

Substitute $x = x_0 + nh$ and $dx = h\,dn$

$$y_1 = y_0 + h\int_{0}^{1} \left(f_0 + n\Delta f_0 + \frac{n(n+1)}{2}\Delta^2 f_0 + \frac{n(n+1)(n+2)}{6}\Delta^3 f_0 \right) dn$$

$$= y_0 + h\left(f_0 + \frac{1}{2}\Delta f_0 + \frac{5}{12}\Delta^2 f_0 + \frac{3}{8}\Delta^3 f_0 + \cdots \right)$$

Neglecting fourth and higher order differences and expressing Δf_0, $\Delta^2 f_0$, and $\Delta^3 f_0$ in terms of function values, we get

$$y_1 = y_0 + \frac{h}{24}(55f_0 - 59f_{-1} + 37f_{-2} - 9f_{-3} + \cdots) \tag{3.144}$$

This is called Adam–Bashfourth predictor formula. In this method, to find y_1, we have to find $f_1 = f(x_0 + h, y_1)$. Then to find a better value of y_1, we derive a corrector formula by substituting Newton's backward formula at f_1, i.e.,

$$f(x,y) = f_1 + n\Delta f_1 + \frac{n(n+1)}{2}\Delta^2 f_1 + \frac{n(n+1)(n+2)}{6}\Delta^3 f_1 + \cdots$$

in Equation 3.143

$$y_1 = y_0 + \int_{x_0}^{x_1}\left(f_0 + n\Delta f_0 + \frac{n(n+1)}{2}\Delta^2 f_0 + \frac{n(n+1)(n+2)}{6}\Delta^3 f_0\right)dx$$

Substitute $x = x_1 + nh$ and $dx = h\,dn$

$$y_1 = y_0 + h\int_{-1}^{0}\left(f_0 + n\Delta f_0 + \frac{n(n+1)}{2}\Delta^2 f_0 + \frac{n(n+1)(n+2)}{6}\Delta^3 f_0 + \cdots\right)dn$$

$$y_1 = y_0 + h\left(f_1 - \frac{1}{2}\Delta f_1 - \frac{1}{12}\Delta^2 f_1 - \frac{1}{24}\Delta^3 f_1 - \cdots\right)$$

Neglecting fourth and higher order differences and expressing Δf_1, $\Delta^2 f_1$, and $\Delta^3 f_1$ in terms of function values, we obtain

$$y_1 = y_0 + \frac{h}{24}(9f_1 + 19f_0 - 5f_{-1} + f_{-2} + \cdots) \tag{3.145}$$

which is called Adam–Moulton corrector formula.

Then an improved value of f_1 is calculated and again the corrector equation (Equation 3.145) is applied to find a still better value y_1. This step is repeated till y_1 remains unchanged, and then we proceed to calculate y_2 as above.

Example 3.18

Given $dy/dx = x^2(1 + y)$ and $y(1) = 1$, $y(1.1) = 1.233$, $y(1.2) = 1.548$, $y(1.3) = 1.979$, evaluate by Adam–Bashfourth method.

$$\text{Here } f(x, y) = x^2(1 + y)$$

Starting values of the Adam–Bashforth method with $h = 0.1$ are

$$x = 1.0, \quad y_{-3} = 1.000, \quad f_{-3} = (1.0)^2((1 + 1.000) = 2.000$$

$$x = 1.1, \quad y_{-2} = 1.233, \quad f_{-2} = 2.702$$

$$x = 1.2, \quad y_{-1} = 1.548, \quad f_{-1} = 3.669$$
$$x = 1.3, \quad y_0 = 1.979, \quad f_0 = 5.035$$

Using predictor,

$$y_1 = y_0 + \frac{h}{24}(55f_0 - 59f_{-1} + 37f_{-2} - 9f_3 + \cdots)$$

$$x = 1.4, \quad y_1 = 2.573, \quad f_1 = 7.004$$

Using the corrector,

$$y_1 = y_0 + \frac{h}{24}(9f_1 + 19f_0 - 5f_{-1} + f_{-2} + \cdots)$$

$$= 2.575$$

Hence, $y(1.4) = 2.575$

3.12 Controllability

A system is said to be controllable at time t_0, if it is possible by means of an unconstrained control vector to transfer the system from any initial state $x(t_0)$ to any other state in a finite interval of time.

$$x(0) = \begin{bmatrix} B & AB & A^2B & \cdots & A^{n-1}B \end{bmatrix} \begin{bmatrix} \beta_0(t) \\ \beta_1(t) \\ \beta_2(t) \\ \vdots \\ \beta_{n-1}(t) \end{bmatrix}$$

where matrix A is of size $n \times n$ and B of size $n \times 1$.

If rank of $\begin{bmatrix} B & AB & A^2B & \cdots & A^{n-1}B \end{bmatrix} \geq n$ then system is controllable.

Example 3.19

Find the state of the system for which the state equation is given below:

$$\begin{bmatrix} \overset{\circ}{x_1} \\ \overset{\circ}{x_2} \end{bmatrix} = \begin{bmatrix} -1 & 0 \\ 0 & -2 \end{bmatrix} \begin{bmatrix} x_1 \\ x_2 \end{bmatrix} + \begin{bmatrix} 2 \\ 5 \end{bmatrix} u(t)$$

$$n = 2$$

From the above equation matrix [B AB] is

$$[B \quad AB] = \begin{bmatrix} 2 & -2 \\ 5 & -10 \end{bmatrix}$$

Because $A = \begin{bmatrix} -1 & 0 \\ 0 & -2 \end{bmatrix}$, $B = \begin{bmatrix} 2 \\ 5 \end{bmatrix}$ and

$$AB = \begin{bmatrix} -1 & 0 \\ 0 & -2 \end{bmatrix} \begin{bmatrix} 2 \\ 5 \end{bmatrix}$$

$$= \begin{bmatrix} -2 \\ -10 \end{bmatrix}$$

Rank of [B AB] = 2 = n.
Hence, the system states are controllable.

3.13 Observability

A system is said to be observable at time t_0 if, with the system in state $x(t_0)$, it is possible to determine this state from the observation of the output over a finite time interval.

$$y(t) = \begin{bmatrix} \alpha_0(t) & \alpha_1(t) & \alpha_2(t) & \cdots & \alpha_{n-1}(t) \end{bmatrix} \begin{bmatrix} C \\ CA \\ CA^2 \\ \cdots \\ CA^{n-1} \end{bmatrix}$$

where matrix A is of size $n \times n$ and matrix C is of size $n \times 1$.

If rank of $\begin{bmatrix} C & CA & CA^2 & \cdots & CA^{n-1} \end{bmatrix} \geq n$ then system is observable.

Example 3.20

Find the state of the system of which state equation is given below:

$$\begin{bmatrix} \overset{\circ}{x_1} \\ \overset{\circ}{x_2} \end{bmatrix} = \begin{bmatrix} 1 & 1 \\ -2 & -1 \end{bmatrix} \begin{bmatrix} x_1 \\ x_2 \end{bmatrix} + \begin{bmatrix} 0 \\ 1 \end{bmatrix} u(t)$$

and $y = \begin{bmatrix} 1 & 0 \end{bmatrix} \begin{bmatrix} x_1 \\ x_2 \end{bmatrix}$

From above equation matrix $\begin{bmatrix} C \\ CA \end{bmatrix}$ is $\begin{bmatrix} C \\ CA \end{bmatrix} = \begin{bmatrix} 1 & 0 \\ 1 & 1 \end{bmatrix}$

Because $A = \begin{bmatrix} 1 & 1 \\ -2 & -1 \end{bmatrix}$, $C = \begin{bmatrix} 1 & 0 \end{bmatrix}$ and $CA = \begin{bmatrix} 1 & 1 \end{bmatrix}$

Rank of $\begin{bmatrix} C \\ CA \end{bmatrix} = 2 = n$. Hence, the system states are observable.

3.14 Sensitivity

The parameters of a system may have a tendency to vary due to changing environment conditions and this variation in parameters affects the desired performance of a system. The use of feedback in a system reduces the effect of parameter variations. The term sensitivity in relation to systems gives an assessment of the system performance as affected due to parameter variations.

Let the variable in a system which changes its value be A such as the output, and this change is considered due to parameter variations of element K such as gain or feedback, then the system sensitivity is expressed as

$$\text{Sensitivity} = \frac{\text{Percentage change in } A}{\text{Percentage change in } K}$$

In mathematical terms sensitivity is written as

$$S_K^A = \frac{\partial A/A}{\partial K/K}$$

The notation S_K^A denotes sensitivity of variable A with respect to parameter K.

So it is preferable that the sensitivity function S_K^A should be minimum.

The sensitivity function for the overall transfer function $M(s)$ with respect to variation in $G(s)$ is written as

$$S_G^M = \frac{\partial M(s)/M(s)}{\partial G(s)/G(s)} \quad \text{or} \quad S_G^M = \frac{G(s)}{M(s)} \frac{\partial M(s)}{\partial G(s)} \tag{3.146}$$

For the open loop control system which is shown in Figure 3.53, the overall transfer function is

$$M(s) = \frac{C(s)}{R(s)} = G(s) \quad \text{or} \quad \frac{M(s)}{G(s)} = 1 \tag{3.147}$$

Formulation of State Space Model of Systems

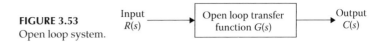

FIGURE 3.53
Open loop system.

Differentiating $M(s)$ with respect to $G(s)$

$$\frac{\partial M(s)}{\partial G(s)} = 1 \qquad (3.148)$$

Substituting Equations 3.147 and 3.148 in Equation 3.146 we get

$$S_G^M = \frac{G(s)}{M(s)} \frac{\partial M(s)}{\partial G(s)} = 1 \qquad (3.149)$$

For the closed loop system shown in Figure 3.54, the overall transfer function is

$$M(s) = \frac{G(s)}{1 + G(s)H(s)} \qquad (3.150)$$

Differentiating Equation 3.150 with respect to $G(s)$

$$\frac{\partial M(s)}{\partial G(s)} = \frac{[1 + G(s)H(s)] - G(s)H(s)}{[1 + G(s)H(s)]^2}$$

or

$$= \frac{1}{[1 + G(s)H(s)]^2} \qquad (3.151)$$

Substituting Equations 3.150 and 3.151 in Equation 3.146, we get

$$S_G^M = \frac{G(s)}{M(s)} \frac{\partial M(s)}{\partial G(s)}$$

$$= \frac{G(s)}{G(s)/(1 + G(s)H(s))} \frac{1}{[1 + G(s)H(s)]^2}$$

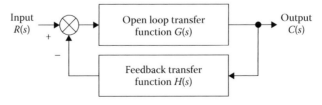

FIGURE 3.54
Closed loop feedback system.

$$S_G^M = \frac{1}{[1+G(s)H(s)]} \qquad (3.152)$$

Comparing sensitivity functions shown in Equations (4) and 3.152 it is observed that the sensitivity of the overall transfer function with respect to forward path transfer function in the case of closed loop system is reduced by a factor $[1 + G(s)H(s)]$ as compared to open loop system.

Sensitivity of overall transfer function $M(s)$ with respect to feedback path transfer function. $H(s)$ is written as

$$S_H^M = \frac{H(s)}{M(s)} \frac{\partial M(s)}{\partial H(s)}$$

For closed loop control system is

$$M(s) = \frac{G(s)}{1+G(s)H(s)}$$

Differentiating the above equation with respect to $H(s)$

$$\frac{\partial M(s)}{\partial H(s)} = \frac{[G(s)]^2}{[1+G(s)H(s)]^2}$$

Substituting the second and third equations in the first, we get

$$S_H^M = \frac{G(s)H(s)}{1+G(s)H(s)}$$

Comparing sensitivity function given in Equations 3.149 and 3.152 it is concluded that a closed loop control system is more sensitive to variation in feedback path parameters than the variations in forward path parameters, therefore, the specifications of feedback elements in a closed loop control system should be more rigid as compared to that of forward path elements.

3.15 Liapunov Stability

The general state equation for a nonlinear system can be expressed as

$$\overset{o}{x} = f(x(t), u(t), t) \qquad (3.153)$$

Formulation of State Space Model of Systems

A variety of methods have therefore been devised which yield information about the stability and domain of stability of Equation 3.153 without restoring to its complete solution.

For an autonomous system the technique of investigating stability by linearizing about a singular point gives information only about stability in the small range. It, therefore, has very limited practical use. Merely looking for existence of limit cycles (self-oscillation) apart from being cumbersome yields limited information. Approximation technique of describing function in determining system stability has some merit but the answer may not always be depended upon.

Therefore, exact approaches for stability investigation wherever possible must be used. Liapunov's direct method and Popov's method are two such approaches which provide us with qualitative aspects of system behavior.

Basic stability theorems

The direct method of Liapunov is based on the concept of energy and the relation of stored energy with system stability. Consider an autonomous physical system described as

$$\overset{\circ}{x} = f(x(t))$$

and let $x(x(t_0), t)$ be a solution. Further, let $V(x)$ be the total energy associated with the system. If the derivative $dV(x)/dx$ is negative for all $x(x(t_0), t)$ except the equilibrium point, then it follows that energy of the system decreases as t increases, and finally the system will reach the equilibrium point. This holds because energy is nonnegative function of system state which reaches a minimum only if the system motion stops.

The Liapunov's method in its simplest form is given by the following theorem.

Theorem 1: Consider the system

$$\overset{\circ}{x} = f(x(t)), \quad f(0) = 0$$

Suppose there exists a scalar function $V(x)$ which, for some real number $\varepsilon > 0$, satisfies the following properties for all X in the region $\|X\| \le \varepsilon$:

1. $V(x) > 0; X \ne 0$.
2. $V(0) = 0$.
3. $V(x)$ has continuous derivatives with respect to all components of X.
4. $\dfrac{dV}{dt} \le 0$ (i.e., dV/dt is negative semi-definite scalar function).

Then the system is stable at the origin.

Theorem 2: If the property (4) of Theorem 1 is replaced with $dV/dt < 0$, $x \ne 0$ (i.e., dV/dt is negative definite scalar function), then the system is asymptotically stable.

It is intuitively obvious that since a continuous V function, $V > 0$ except at $X = 0$, satisfies the condition $dV/dt < 0$, we expect that X will eventually approach the origin. We shall avoid the rigorous proof of this theorem.

Theorem 3: If all the conditions of Theorem 2 hold and in addition,

$$V(X) \to \infty \text{ as } \|X\|$$

Then the system is asymptotically stable in-the-large at the origin.

Liapunov functions

The determination of stability via Liapunov's direct method centers around the choice of a positive definite function $V(x)$ called the Liapunov function.

Theorem 4: Consider the system

$$\overset{o}{x} = f(x(t)), \quad f(0) = 0$$

Suppose there exists a scalar function $W(x)$ which, for some real number $\varepsilon > 0$, satisfies the following properties for all X in the region $\|X\| \leq \varepsilon$:

5. $W(x) > 0; X \neq 0$.
6. $W(0) = 0$.
7. $W(x)$ has continuous derivatives with respect to all components of X.
8. $\dfrac{dW}{dt} \geq 0$.

Then the system is unstable at the origin.

3.16 Performance Characteristics of Linear Time Invariant Systems

When the mathematical model of a physical system is solved for various input conditions, the result represents the dynamic response of the system. The mathematical model of a system is linear if it obeys the *principal of superposition and homogeneity*. This principal implies that if a system model has responses $y_1(t)$ and $y_2(t)$ to any two inputs $x_1(t)$ and $x_2(t)$, respectively, then the system response to the linear combination of these inputs $\alpha_1 x_1(t) + \alpha_2 x_2(t)$ is given by the linear combination of the individual outputs, i.e., $\alpha_1 y_1(t) + \alpha_2 y_2(t)$ where α_1 and α_2 are constants.

Mathematical models of most physical systems are characterized by differential equations. A mathematical model is linear, if the differential equation describing it has coefficients, which are either functions only of the independent variable or are constants. If the coefficients of describing differential equations are functions of time (the independent variable), then the mathematical model is linear time-varying. On the other hand, if the coefficients of the describing differential equations are constants, the model is linear time invariant.

Formulation of State Space Model of Systems   187

3.17 Formulation of State Space Model Using Computer Program (SYSMO)

System state model (SYSMO) is a digital computer program for analysis of lumped linear time invariant systems without any restriction with regard to the "type" of elements or their interconnection (as far as they are "legal").

3.17.1 Preparation of the Input Data

1. Replace all the elements of the given network by their equivalent circuits in Table 3.1 using the SYSMO standard elements given in Table 3.10. The fictitious element A and B can be placed wherever needed but always such that the element A is in series whereas the element B is in parallel with the other one-port elements in the network.
2. Set up an oriented linear graph of the network. If the resulting graph is unconnected join its separate parts by coalescence of vertices in such a way that every two parts have just one vertex in common.
3. Number the edges and vertices of the graph in an ascending way starting at one. Orientation of each edge has to be related to the orientation of the corresponding network element and to the sign of its nominal value.

The input data for each edge consist of the following information given in Table 3.8.
The nominal and initial values have to be given in a compatible system of physical units.

3.17.2 Algorithm for the Formulation of State Equations

In SYSMO the system of the state and output equations in normal form is not given explicitly but is derived algorithmically through the following steps:

1. The incidence matrix A is set up and its columns are rearranged according to the type of the elements which are represented in Table 3.9.
2. The $(n-1)$ rows of the incidence matrix A are eliminated by Gauss–Jordan reduction, modified in such a way that if at the ith step all entries in the ith column

TABLE 3.8

Element Information for SYSMO

S. No.	Type of Elements	Description
1.	EL	Number of the edge
2.	ND1	Number of the node incident with the "tail" of the edge
3.	ND2	Number of the node incident with the "arrow" of the edge
4.	CPL	Number of the edge to which the given edge is coupled (nonzero in case of dependent sources and mutual inductance only)
5.	TYP	Type of the element represented by the edge (according to Table 3.1)
6.	VAL	Nominal value of the element represented by the edge
7.	INC	Initial value of the variable conserved by the element represented by the edge

TABLE 3.9

Priority Level for Different Elements in Incidence Matrix

Type of Elements	Description
E	Independent across driver
A	Short circuit elements (through variable measuring devices)
V	Dependent across drivers
C	Accumulator type elements
R	Dissipater (algebraic elements)
L	Delay type elements
I	Dependent through drivers
B	Open circuit elements (across variable measuring devices)
J	Independent through drivers

below and including the *i*th row are zero, the *i*th column is interchanged with the nearest column *j* in *A* with a nonzero entry in its *i*th row. Thus the resultant matrix is a fundamental cutset matrix [U Q]. Columns of the unit submatrix *U* correspond to the edges of the normal tree.

3. The interchange of the matrix columns in Step 2 is checked so that it can be used for topological diagnostic of the given network. If when searching for a column with a nonzero entry in its *i*th row

   a. The column *i* corresponds either to E, A, or V—the network is not topologically consistent (it contains a circuit made of across variable sources and/or short circuit elements A).

   b. The column *j* corresponds either to I, B, or J—the network is not topologically consistent (it contains a cutset made of through variable sources and/or open circuit elements B).

   c. No column *j* with a nonzero entry in its *i*th row can be found—the graph of the network is not connected [see instruction (2) for the data preparation].

4. Let us denote the network primary variables (across variables of tree branches and through variables of cotree chords) by upper case letters and the secondary variables (branch through variables and chord across variables) by lower case letters. Then according to f-cutset and f-circuit equations

$$\begin{bmatrix} y_b \\ x_c \end{bmatrix} = \begin{bmatrix} 0 & -Q \\ Q^T & 0 \end{bmatrix} \begin{bmatrix} x_b \\ y_c \end{bmatrix} \qquad (3.154)$$

where *Q* is the matrix derived in the Step 2. *Q* is its transpose and the whole middle square matrix is the combined matrix.

5. The rows and columns of the combined matrix are rearranged according to the types of the elements so as to get a matrix *T* which in the partitioned form fully describes the topological relations among the network variables in the following way:

Formulation of State Space Model of Systems

$$\begin{bmatrix} w_1 \\ w_2 \\ w_3 \end{bmatrix} = \begin{bmatrix} T_{11} & T_{12} & T_{13} \\ T_{21} & T_{22} & T_{23} \\ T_{31} & T_{32} & T_{33} \end{bmatrix} \begin{bmatrix} W_1 \\ W_2 \\ W_3 \end{bmatrix}$$

where

$$W_1^T = \begin{bmatrix} X_{bc} & Y_{cL} & X_{bL} & Y_{cc} \end{bmatrix}^T$$

$$w_1^T = \begin{bmatrix} y_{bc} & x_{cL} & y_{bL} & x_{cc} \end{bmatrix}^T$$

are the vectors of the primary and secondary system dynamic variables

$$W_2^T = \begin{bmatrix} X_{bR} & Y_{cR} & X_{bX} & Y_{cY} \end{bmatrix}^T \quad \text{and}$$

$$w_2^T = \begin{bmatrix} y_{bR} & x_{cR} & y_{bA} & x_{cR} \end{bmatrix}^T$$

are the vectors of the primary and secondary static variables and $W_3^T = \begin{bmatrix} X_{bE} & Y_{cL} \end{bmatrix}^T$ is the vector of the specified variables. As $X_{bA} = Y_{cB} = 0$, these vectors do not have to be considered further.

6. The terminal equations for memoryless network elements

$$\begin{bmatrix} X_{bR} \\ Y_{cR} \end{bmatrix} = \begin{bmatrix} R & 0 \\ 0 & G \end{bmatrix} \begin{bmatrix} y_{bR} \\ x_{cR} \end{bmatrix} \quad \text{and}$$

$$\begin{bmatrix} X_{bX} \\ Y_{cY} \end{bmatrix} = \begin{bmatrix} r & m \\ a & g \end{bmatrix} \begin{bmatrix} y_{bA} \\ x_{cB} \end{bmatrix}$$

can be combined so as to give

$$W_2 = P_2 w_2 \tag{3.155}$$

where the matrix P_2 of the memoryless network parameters

$$P_2 = \begin{vmatrix} R^T & 0 & & \\ 0 & G^T & & \\ & & r & m \\ & & a & g \end{vmatrix}$$

G is a matrix of reciprocal values of chord resistors.

7. The second row of Equation given in step-5 yields

$$w_2 = T_{21} W_1 + T_{22} W_2 + T_{23} W_3$$

After substituting (2) into the last equation premultiplied by P_2 we obtain

$$(U - P_2 T_{22})W_2 = P_2 T_{22} W_1 + P_2 T_{23} W_3 \tag{3.156}$$

Equation 3.156 can be solved with respect to the vector of primary variables of memoryless elements W_2 by Gauss–Jordan reduction that will give the equation

$$W_2 = K W_1 + K W_3 \tag{3.157}$$

8. The Gauss–Jordan reduction in Step 7 simultaneously carries out the diagnostic of algebraic consistency and determinacy of the given network. If the matrix $(U - P_2 T_{22})$ is not of its full rank the program has to be terminated at this point.
9. The terminal equations of the memory elements are given as

$$\begin{vmatrix} y_{bC} \\ Y_{CC} \end{vmatrix} = \begin{vmatrix} C_b & 0 \\ 0 & C_C \end{vmatrix} \frac{d}{dt} \begin{vmatrix} X_{bc} \\ x_{cc} \end{vmatrix}$$

and

$$\begin{vmatrix} X_{bL} \\ x_{CL} \end{vmatrix} = \begin{vmatrix} L_{bb} & L_{bC} \\ L_{cb} & L_{CC} \end{vmatrix} \frac{d}{dt} \begin{vmatrix} y_{bL} \\ Y_{CL} \end{vmatrix}$$

The subscripts b and c refer to the branches and chords of the normal tree, respectively. These two matrix equations combined together

$$\begin{bmatrix} y_{bC} \\ x_{CL} \\ X_{bL} \\ Y_{CC} \end{bmatrix} = \begin{bmatrix} C_b & & & \\ & L_{CC} & L_{bc} & \\ & L_{cb} & L_{bb} & \\ & & & C_{CC} \end{bmatrix} \frac{d}{dt} \begin{bmatrix} x_{bC} \\ y_{CL} \\ Y_{bL} \\ X_{CC} \end{bmatrix}$$

can be simply written as

$$\begin{bmatrix} w_{11} \\ w_{12} \end{bmatrix} = P_1 \frac{d}{dt} \begin{bmatrix} W_{11} \\ W_{12} \end{bmatrix} \tag{3.158}$$

where the vectors

$$W_{11} = \begin{bmatrix} X_{bC} & Y_{cL} \end{bmatrix}^T \quad \text{and} \quad w_{11} = \begin{bmatrix} Y_{bC} & X_{cL} \end{bmatrix}^T$$

$$W_{12} = \begin{bmatrix} X_{bL} & Y_{cC} \end{bmatrix}^T \quad \text{and} \quad w_{12} = \begin{bmatrix} Y_{bL} & X_{cC} \end{bmatrix}^T$$

are just parts of the vectors W and w from.

Formulation of State Space Model of Systems

10. Substitution of Equation 4 into the first equation of (1)

$$w_1 = T_{11}W_1 + T_{12}W_2 + T_{13}W_3$$

yields

$$w_1 = (T_{11} + T_{12}K_1)W_1 + (T_{13} + T_{12}K_3)W_3$$

or simply

$$w_1 = MW_1 + NW_3 \tag{3.159}$$

Partitioning Equation 3.159 in the same way as the vectors W and w have been partitioned in Equation 3.158 and interchanging the vectors W_2 and w_2 in the last equation yield

$$\begin{bmatrix} U & -M_{12} \\ 0 & -M_{22} \end{bmatrix} \begin{bmatrix} w_{11} \\ W_{12} \end{bmatrix} = \begin{bmatrix} M_{11} & 0 \\ M_{21} & -U \end{bmatrix} \begin{bmatrix} W_{11} \\ w_{12} \end{bmatrix}$$

After substitution of (5) into this equation we obtain a system of preliminary state equations

$$\begin{bmatrix} U & -M_{12} \\ 0 & -M_{22} \end{bmatrix} \frac{d}{dt} \begin{bmatrix} w_{11} \\ W_{12} \end{bmatrix} = \begin{bmatrix} M_{11} & 0 \\ M_{21} & -U \end{bmatrix} \begin{bmatrix} W_{11} \\ w_{12} \end{bmatrix} + NW_3$$

11. The preliminary state equations

$$SX = AX + BU$$

and output equations

$$y = RX + CX + DU$$

would be in the final form if S were an identity matrix. Since this is generally not so, a procedure suggested by reducing the partitioned matrix $[S\ A\ B]$ to row echelon form. Three possibilities are now recognized:

a. S becomes an identity matrix. In this case S is of full rank and the final state equations are obtained.

b. S and A contain at least one zero row. Unless the same row of B is also zero the network has no solution because an "illegal" interconnection of independent sources is indicated.

c. S but not A contains at least one zero row. It indicates that the order of complexity of the network is smaller than the number of memory elements. In this case the last row expresses the equation

$$0 = a_n X^T + b_n U^T$$

implying a dependence among variables. One may be chosen for elimination and the relation above used to remove a column from A and C matrices. Differentiating the expression gives

$$0 = a_n^T \dot{X} + b_n^T \dot{U}$$

which is used to remove the corresponding element from $\dot{X}$ and hence a column from R and S. Derivatives of input signals may thus arise during state variable elimination. When these relations have been exhausted, a smaller matrix $[S \ A \ B]$ is again reduced to row echelon form and the process repeated until an identity matrix is obtained for S or all state variables disappear.

Alternatively

$$S\dot{X}_1 = A_1 X_1 + B_1 U$$

$$S = LU = LDU +$$

where $L = \begin{bmatrix} 1 & & 0 \\ & 1 & \\ 1 & & 1 \end{bmatrix}$ = nonsingular

and $U = \begin{bmatrix} u_1 & & & U \\ & u_2 & & \\ & & u_r & \\ & 0 & & 0 \end{bmatrix}$ = singular

$D = \begin{bmatrix} 1 & & & & 0 \\ & 1 & & & \\ & & 1 & & \\ & & & 1 & \\ & 0 & & & 0 \end{bmatrix}$ = singular

Let $X_1 = U^{-1} \dot{X} + X$

$$SU^{-1} \dot{X} + SX = A_1 U^{-1} \dot{X} + A_1 X + B_1 U$$

or $\quad L^{-1} S U^{-1} \dot{X} + L^{-1} S X = L^{-1} A_1 U^{-1} \dot{X} + L^{-1} A_1 X + L^{-1} B_1 U$

or $\quad D\dot{X} = PX + QU \quad$ where $P = L^{-1} A_1 U^{-1} + \ $ and $Q = L^{-1} B_1$

$$\begin{bmatrix} I_r & 0 \\ 0 & 0_{n-r} \end{bmatrix} \frac{d}{dt} \begin{bmatrix} x_1 \\ x_2 \end{bmatrix} = \begin{bmatrix} P_{11} & P_{12} \\ P_{21} & P_{22} \end{bmatrix} \begin{bmatrix} x_1 \\ x_2 \end{bmatrix} + \begin{bmatrix} Q_1 \\ Q_2 \end{bmatrix} U$$

Formulation of State Space Model of Systems

TABLE 3.10

SYSMO Standard Elements

Elements/Terminal Equation	EL	ND1	ND2	CPL	TYP	VAL	INC
Independent across driver $X(t)$ = specified	i	P	q	—	E	e	—
Independent through driver $Y(t)$ = specified	i	P	q	—	J	j	—
Across dependent across driver $Xi(t) = r\, Xj(t)$	i	P	q	j	V	r	—
Through dependent across driver $Xi(t) = m\, Yj(t)$	i	p	q	j	V	m	—
Across dependent through driver $Yi(t) = a\, Xj(t)$	i	p	q	j	I	a	—
Through dependent through driver $Yi(t) = g\, Yj(t)$	i	p	q	j	I	g	—
Dissipator $X(t) = R\, Y(t)$	i	p	q	—	R	R	—
Impedance of zero value/short circuit element/Through variable measuring device $X(t) = 0$	i	p	q	—	A	0	—
Impedance of infinite value/Open circuit element/Across variable measuring device $Y(t) = 0$	i	p	q	—	B	inf	—
Capacitor $Y(t) = c\, dX(t)/dt$	i	p	q	—	C	c	$Xc(0+)$
Inductance $X(t) = L\, dY(t)/dt$	i	p	q	—	L	l	$Y_l(0+)$
Mutual inductance	i	p	q	j	M	ml	$Xc(0+)$

The Table 3.10 shows the standard elements used in system model.

Assuming P_{22} is nonsingular

$$X_2 = -P_{22}^{-1}P_{21}X_1 - P_{22}^{-1}Q2U$$

or

$$X_1 = AX_1 + BU$$

where $A = [P_{11} - P_{12}P_{22}^{-1}P_{21}]$ and $B = [Q_1 - P_{12}P_{22}^{-1}Q_2]$.

Example 3.21

The schematic diagram of a field controlled dc motor is shown in Figure 3.55 and its terminal graph in Figure 3.56. Develop its state model for two inputs such as load torque and field excitation.

The detailed equivalent circuit for the system shown in Figure 3.55 is given in Figure 3.57. The first part of equivalent circuit shows the field circuit of generator, which is consisting of dc supply, field resistance, and field inductance. The second part of the system contains generator armature components such as armature resistance, armature inductance, and generated emf in the armature. The third part represents the armature equivalent circuit of motor, which has armature resistance, armature inductance, and back emf induced in the armature. Finally, the last part shows the mechanical load containing damper, inertia, and load torque as through driver.

From the equivalent circuit the multipart system graph may be drawn as shown in Figure 3.58. Different type of elements in the given system

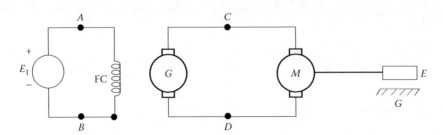

FIGURE 3.55
Schematic diagram for dc motor–generator set.

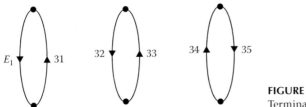

FIGURE 3.56
Terminal graph for dc motor–generator set.

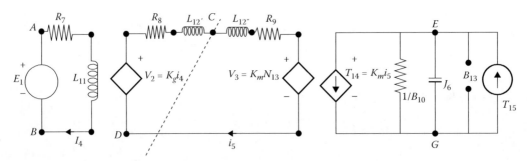

FIGURE 3.57
Equivalent circuit for given dc motor–generator set.

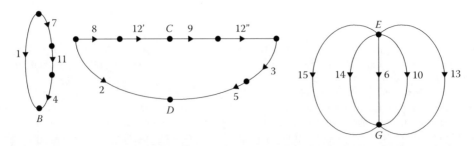

FIGURE 3.58
Multipart system graph.

Formulation of State Space Model of Systems

S. No.	Elements	Element Type	Element Numbers
1.	Independent across drivers	E	1
2.	Dependent across drivers	X	2, 3
3.	Short circuit elements	SC	4, 5
4.	Accumulators	C	6
5.	Dissipater elements	R	7, 8, 9, 10
6.	Delay type elements	L	11, 12', 12"
7.	Open circuit elements	OC	13
8.	Dependent through drivers	Y	14
9.	Independent through drivers	J	15

From the equivalent circuit of dc motor–generator set it is very clear that the two delay elements 6 and 8 are in series and can be combined together as $L_{12} = L_{12'} + L_{12''}$. After combining these delay elements, the multipart graph may be rooted at a common node to obtain a connected graph as shown in Figure 3.59. Covalence of nodes B, D, and G is one, because cutset equations remains unchanged and no property of system graph is violated.

For this connected graph of Figure 3.59, draw the f-tree consisting of edges related to E-type element, short circuit elements (SC), dependent across drivers (X), accumulators (C), and resistors (R) to complete it.

Hence, f-tree has elements (branches)—1, 2, 3, 4, 5, 6, 7, 8, 9 and
Cotree elements (chords)—10, 11, 12, 13, 14, 15

In this system, total number of dynamic element = number of "C" type elements + number of "L" type elements

$$= 1 + 2 = 3$$

Hence the total number of state equations equals to 3 (i.e., the total number of dynamic elements in nondegenerate system).

Step 1 Write the incidence matrix as given in Table 3.11.

Step 2 Rearrange the incidence matrix according to the topological restrictions. In this case, we will get the same matrix as shown in Table 3.11, because we have already taken the elements as per the topological restrictions.

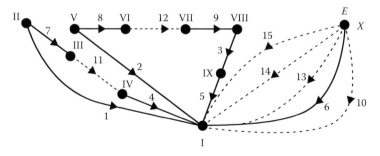

FIGURE 3.59
Connected graph for given dc motor–generator set.

TABLE 3.11

Incidence Matrix for Given System Graph

	E	X		SC	C		R			L		Y	OC	J
	1	2	3	4	5	6	7	8	9	10	11	12	13	14	15
I	−1	−1	0	−1	−1	−1	0	0	0	−1	0	0	−1	−1	−1
II	1	0	0	0	0	0	1	0	0	0	0	0	0	0	0
III	0	0	0	0	0	0	−1	0	0	0	1	0	0	0	0
IV	0	0	0	1	0	0	0	0	0	0	−1	0	0	0	0
V	0	1	0	0	0	0	0	1	0	0	0	0	0	0	0
VI	0	0	0	0	0	0	0	−1	0	0	0	1	0	0	0
VII	0	0	0	0	0	0	0	0	1	0	0	−1	0	0	0
VIII	0	0	1	0	0	0	0	0	−1	0	0	0	0	0	0
IX	0	0	−1	0	1	0	0	0	0	0	0	0	0	0	0
X	0	0	0	0	0	1	0	0	0	1	0	0	1	1	1

TABLE 3.12

Reduced Incidence Matrix for Given System Graph

	E	SC		X	C		R			L		Y	OC	J
	1	2	3	4	5	6	7	8	9	10	11	12	13	14	15
II	1	0	0	0	0	0	1	0	0	0	0	0	0	0	0
III	0	0	0	0	0	0	−1	0	0	0	1	0	0	0	0
IV	0	0	0	1	0	0	0	0	0	0	−1	0	0	0	0
V	0	1	0	0	0	0	0	1	0	0	0	0	0	0	0
VI	0	0	0	0	0	0	0	−1	0	0	0	1	0	0	0
VII	0	0	0	0	0	0	0	0	1	0	0	−1	0	0	0
VIII	0	0	1	0	0	0	0	0	−1	0	0	0	0	0	0
IX	0	1	−1	0	1	0	0	0	0	0	0	0	0	0	0
X	0	0	0	0	0	1	0	0	0	1	0	0	1	1	1

Step 3 Delete a row related with the reference node (i.e., node I) to get a reduced incidence matrix as shown in Table 3.12.

Step 4 Obtain the unity matrix in the leading portion of the reduced incidence matrix using Gauss–Jordan elimination method as shown in Table 3.13. The trailing portion of the table contains f-cutset coefficient matrix (Q or A_c).

Step 5 Grouping of elements
Secondary variables of

1. Dynamic elements ("C" type and "L" type) $w_1 = [i_6 \ V_{11} \ v_{12}]$
2. Dissipaters ("R" type) $w_2 = [i_7 \ i_8 \ i_9 \ \overset{o}{\theta}_{10}]$
3. Drivers (across and through drivers) $w_3 = [i_1 \ i_2 \ i_3 \ i_4 \ i_5 \ \overset{o}{\theta}_{13} \ \overset{o}{\theta}_{14} \ \overset{o}{\theta}_{15}]$

Formulation of State Space Model of Systems

TABLE 3.13

Coefficient Matrix of f-Cutset Equation

	E	SC		X		C		R			L		Y	OC	J
	1	2	3	4	5	6	7	8	9	10	11	12	13	14	15
II + III	1	0	0	0	0	0	0	0	0	0	1	0	0	0	0
V + VI	0	1	0	0	0	0	0	0	0	0	0	1	0	0	0
VII + VIII	0	0	1	0	0	0	0	0	0	0	0	−1	0	0	0
IV	0	0	0	1	0	0	0	0	0	0	−1	0	0	0	0
VII + VIII + IX	0	0	0	0	1	0	0	0	0	0	0	−1	0	0	0
X	0	0	0	0	0	1	0	0	1	0	0	1	1	1	1
(−1) * III	0	0	0	0	0	0	1	0	0	0	−1	0	0	0	0
(−1) * VI	0	0	0	0	0	0	0	1	0	0	0	−1	0	0	0
VII	0	0	0	0	0	0	0	0	1	0	0	−1	0	0	0

Step 6 Write a combined constraint equation

$$\begin{bmatrix} i_1 \\ i_2 \\ i_3 \\ i_4 \\ i_5 \\ T_6 \\ i_7 \\ i_8 \\ i_9 \\ \overset{o}{\theta}_{10} \\ v_{11} \\ v_{12} \\ \overset{o}{\theta}_{13} \\ \overset{o}{\theta}_{14} \\ \overset{o}{\theta}_{15} \end{bmatrix} = \begin{bmatrix} 0 & 0 & 0 & 0 & 0 & 0 & 0 & 0 & 0 & 0 & -1 & 0 & 0 & 0 & 0 \\ 0 & 0 & 0 & 0 & 0 & 0 & 0 & 0 & 0 & 0 & 0 & -1 & 0 & 0 & 0 \\ 0 & 0 & 0 & 0 & 0 & 0 & 0 & 0 & 0 & 0 & 0 & 1 & 0 & 0 & 0 \\ 0 & 0 & 0 & 0 & 0 & 0 & 0 & 0 & 0 & 0 & 1 & 0 & 0 & 0 & 0 \\ 0 & 0 & 0 & 0 & 0 & 0 & 0 & 0 & 0 & 0 & 0 & 1 & 0 & 0 & 0 \\ 0 & 0 & 0 & 0 & 0 & 0 & 0 & 0 & 0 & -1 & 0 & 0 & -1 & -1 & -1 \\ 0 & 0 & 0 & 0 & 0 & 0 & 0 & 0 & 0 & 0 & 1 & 0 & 0 & 0 & 0 \\ 0 & 0 & 0 & 0 & 0 & 0 & 0 & 0 & 0 & 0 & 0 & 1 & 0 & 0 & 0 \\ 0 & 0 & 0 & 0 & 0 & 0 & 0 & 0 & 0 & 0 & 0 & 1 & 0 & 0 & 0 \\ \hline 0 & 0 & 0 & 0 & 0 & 1 & 0 & 0 & 0 & 0 & 0 & 0 & 0 & 0 & 0 \\ 1 & 0 & 0 & -1 & 0 & 0 & -1 & 0 & 0 & 0 & 0 & 0 & 0 & 0 & 0 \\ 0 & 1 & -1 & 0 & -1 & 0 & 0 & -1 & -1 & 0 & 0 & 0 & 0 & 0 & 0 \\ 0 & 0 & 0 & 0 & 0 & 1 & 0 & 0 & 0 & 0 & 0 & 0 & 0 & 0 & 0 \\ 0 & 0 & 0 & 0 & 0 & 1 & 0 & 0 & 0 & 0 & 0 & 0 & 0 & 0 & 0 \\ 0 & 0 & 0 & 0 & 0 & 1 & 0 & 0 & 0 & 0 & 0 & 0 & 0 & 0 & 0 \end{bmatrix} \begin{bmatrix} v_1 \\ v_2 \\ v_3 \\ v_4 \\ v_5 \\ \overset{o}{\theta}_6 \\ v_7 \\ v_8 \\ v_9 \\ \hline T_{10} \\ i_{11} \\ i_{12} \\ T_{13} \\ T_{14} \\ T_{15} \end{bmatrix}$$

Step 7 Rearranged the combined constraint equation

$$\begin{bmatrix} T_6 \\ v_{11} \\ v_{12} \\ i_7 \\ i_8 \\ i_9 \\ \overset{o}{\theta}_{10} \\ i_1 \\ i_2 \\ i_3 \\ i_4 \\ i_5 \\ \overset{o}{\theta}_{13} \\ \overset{o}{\theta}_{14} \\ \overset{o}{\theta}_{15} \end{bmatrix} = \begin{bmatrix} & 6 & 11 & 12 & 7 & 8 & 9 & 10 & 1 & 2 & 3 & 4 & 5 & 13 & 14 & 15 \\ 6 & 0 & 0 & 0 & 0 & 0 & 0 & -1 & 0 & 0 & 0 & 0 & 0 & -1 & -1 & -1 \\ 11 & 0 & 0 & 0 & -1 & 0 & 0 & 0 & 1 & 0 & 0 & -1 & 0 & 0 & 0 & 0 \\ 12 & 0 & 0 & 0 & 0 & -1 & -1 & 0 & 0 & 1 & -1 & 0 & -1 & 0 & 0 & 0 \\ 7 & 0 & 1 & 0 & 0 & 0 & 0 & 0 & 0 & 0 & 0 & 0 & 0 & 0 & 0 & 0 \\ 8 & 0 & 0 & 1 & 0 & 0 & 0 & 0 & 0 & 0 & 0 & 0 & 0 & 0 & 0 & 0 \\ 9 & 0 & 0 & 1 & 0 & 0 & 0 & 0 & 0 & 0 & 0 & 0 & 0 & 0 & 0 & 0 \\ 10 & 1 & 0 & 0 & 0 & 0 & 0 & 0 & 0 & 0 & 0 & 0 & 0 & 0 & 0 & 0 \\ 1 & 0 & -1 & 0 & 0 & 0 & 0 & 0 & 0 & 0 & 0 & 0 & 0 & 0 & 0 & 0 \\ 2 & 0 & 0 & -1 & 0 & 0 & 0 & 0 & 0 & 0 & 0 & 0 & 0 & 0 & 0 & 0 \\ 3 & 0 & 0 & 1 & 0 & 0 & 0 & 0 & 0 & 0 & 0 & 0 & 0 & 0 & 0 & 0 \\ 4 & 0 & 1 & 0 & 0 & 0 & 0 & 0 & 0 & 0 & 0 & 0 & 0 & 0 & 0 & 0 \\ 5 & 0 & 0 & 1 & 0 & 0 & 0 & 0 & 0 & 0 & 0 & 0 & 0 & 0 & 0 & 0 \\ 13 & 1 & 0 & 0 & 0 & 0 & 0 & 0 & 0 & 0 & 0 & 0 & 0 & 0 & 0 & 0 \\ 14 & 1 & 0 & 0 & 0 & 0 & 0 & 0 & 0 & 0 & 0 & 0 & 0 & 0 & 0 & 0 \\ 15 & 1 & 0 & 0 & 0 & 0 & 0 & 0 & 0 & 0 & 0 & 0 & 0 & 0 & 0 & 0 \end{bmatrix} \begin{bmatrix} \overset{o}{\theta}_6 \\ i_{11} \\ i_{12} \\ v_7 \\ v_8 \\ v_9 \\ T_{10} \\ v_1 \\ v_2 \\ v_3 \\ v_4 \\ v_5 \\ T_{13} \\ T_{14} \\ T_{15} \end{bmatrix}$$

Consider w_2 row of the above equation

$$\begin{bmatrix} i_7 \\ i_8 \\ i_9 \\ \overset{o}{\theta}_{10} \end{bmatrix} = \begin{bmatrix} 0 & 1 & 0 \\ 0 & 0 & 1 \\ 0 & 0 & 1 \\ 1 & 0 & 0 \end{bmatrix} \begin{bmatrix} \overset{o}{\theta}_6 \\ i_{11} \\ i_{12} \end{bmatrix} \qquad (3.160)$$

Step 8 Write the terminal equation for memoryless elements

$$\begin{bmatrix} v_7 \\ v_8 \\ v_9 \\ T_{10} \end{bmatrix} = \begin{bmatrix} R_7 & 0 & 0 & 0 \\ 0 & R_8 & 0 & 0 \\ 0 & 0 & R_9 & 0 \\ 0 & 0 & 0 & B_{10} \end{bmatrix} \begin{bmatrix} i_7 \\ i_8 \\ i_9 \\ \overset{o}{\theta}_{10} \end{bmatrix} \qquad (3.161)$$

Substitute Equation 3.160 in Equation 3.161.

$$\begin{bmatrix} v_7 \\ v_8 \\ v_9 \\ T_{10} \end{bmatrix} = \begin{bmatrix} R_7 & 0 & 0 & 0 \\ 0 & R_8 & 0 & 0 \\ 0 & 0 & R_9 & 0 \\ 0 & 0 & 0 & B_{10} \end{bmatrix} \begin{bmatrix} 0 & 1 & 0 \\ 0 & 0 & 1 \\ 0 & 0 & 1 \\ 1 & 0 & 0 \end{bmatrix} \begin{bmatrix} \overset{o}{\theta}_6 \\ i_{11} \\ i_{12} \end{bmatrix} \qquad (3.162)$$

Step 9 Consider w_1 row of the matrix equation

Formulation of State Space Model of Systems

$$\begin{bmatrix} T_6 \\ v_{11} \\ v_{12} \end{bmatrix} = \begin{bmatrix} 0 & 0 & 0 & -1 \\ -1 & 0 & 0 & 0 \\ 0 & -1 & -1 & 0 \end{bmatrix} \begin{bmatrix} v_7 \\ v_8 \\ v_9 \\ T_{10} \end{bmatrix} + \begin{bmatrix} 0 & 0 & 0 & 0 & 0 & -1 & -1 & -1 \\ 1 & 0 & 0 & -1 & 0 & 0 & 0 & 0 \\ 0 & 1 & -1 & 0 & -1 & 0 & 0 & 0 \end{bmatrix} \begin{bmatrix} v_1 \\ v_2 \\ v_3 \\ v_4 \\ v_5 \\ T_{13} \\ T_{14} \\ T_{15} \end{bmatrix} \quad (3.163)$$

Substitute Equation 3.162 in Equation 3.163, we will get

$$\begin{bmatrix} T_6 \\ v_{11} \\ v_{12} \end{bmatrix} = \begin{bmatrix} 0 & 0 & 0 & -1 \\ -1 & 0 & 0 & 0 \\ 0 & -1 & -1 & 0 \end{bmatrix} \begin{bmatrix} R_7 & 0 & 0 & 0 \\ 0 & R_8 & 0 & 0 \\ 0 & 0 & R_9 & 0 \\ 0 & 0 & 0 & B_{10} \end{bmatrix} \begin{bmatrix} 0 & 1 & 0 \\ 0 & 0 & 1 \\ 0 & 0 & 1 \\ 1 & 0 & 0 \end{bmatrix} \begin{bmatrix} \overset{o}{\theta}_6 \\ i_{11} \\ i_{12} \end{bmatrix}$$

$$+ \begin{bmatrix} 0 & 0 & 0 & 0 & 0 & -1 & -1 & -1 \\ 1 & 0 & 0 & -1 & 0 & 0 & 0 & 0 \\ 0 & 1 & -1 & 0 & -1 & 0 & 0 & 0 \end{bmatrix} \begin{bmatrix} v_1 \\ v_2 \\ v_3 \\ v_4 \\ v_5 \\ T_{13} \\ T_{14} \\ T_{15} \end{bmatrix} \quad (3.164)$$

Step 10 Write the terminal equation for memory elements

$$\begin{bmatrix} T_6 \\ v_{11} \\ v_{12} \end{bmatrix} = \begin{bmatrix} J_6 & 0 & 0 \\ 0 & L_{11} & 0 \\ 0 & 0 & L_{12} \end{bmatrix} \frac{d}{dt} \begin{bmatrix} \overset{o}{\theta}_6 \\ i_{11} \\ i_{12} \end{bmatrix} \quad (3.165)$$

Substitute Equation 3.165 in Equation 3.164.

$$\begin{bmatrix} J_6 & 0 & 0 \\ 0 & L_{11} & 0 \\ 0 & 0 & L_{12} \end{bmatrix} \frac{d}{dt} \begin{bmatrix} \overset{o}{\theta}_6 \\ i_{11} \\ i_{12} \end{bmatrix} = \begin{bmatrix} 0 & 0 & 0 & -1 \\ -1 & 0 & 0 & 0 \\ 0 & -1 & -1 & 0 \end{bmatrix} \begin{bmatrix} R_7 & 0 & 0 & 0 \\ 0 & R_8 & 0 & 0 \\ 0 & 0 & R_9 & 0 \\ 0 & 0 & 0 & B_{10} \end{bmatrix} \begin{bmatrix} 0 & 1 & 0 \\ 0 & 0 & 1 \\ 0 & 0 & 1 \\ 1 & 0 & 0 \end{bmatrix} \begin{bmatrix} \overset{o}{\theta}_6 \\ i_{11} \\ i_{12} \end{bmatrix}$$

$$+\begin{bmatrix} 0 & 0 & 0 & 0 & 0 & -1 & -1 & -1 \\ 1 & 0 & 0 & -1 & 0 & 0 & 0 & 0 \\ 0 & 1 & -1 & 0 & -1 & 0 & 0 & 0 \end{bmatrix}\begin{bmatrix} v_1 \\ v_2 \\ v_3 \\ v_4 \\ v_5 \\ T_{13} \\ T_{14} \\ T_{15} \end{bmatrix} \qquad (3.166)$$

In Equation 3.166 voltages $v_4 = v_5 = 0$ and $T_{13} = 0$

$$\begin{bmatrix} J_6 & 0 & 0 \\ 0 & L_{11} & 0 \\ 0 & 0 & L_{12} \end{bmatrix}\frac{d}{dt}\begin{bmatrix} \overset{o}{\theta}_6 \\ i_{11} \\ i_{12} \end{bmatrix} = \begin{bmatrix} 0 & 0 & 0 & -1 \\ -1 & 0 & 0 & 0 \\ 0 & -1 & -1 & 0 \end{bmatrix}\begin{bmatrix} R_7 & 0 & 0 & 0 \\ 0 & R_8 & 0 & 0 \\ 0 & 0 & R_9 & 0 \\ 0 & 0 & 0 & B_{10} \end{bmatrix}\begin{bmatrix} 0 & 1 & 0 \\ 0 & 0 & 1 \\ 0 & 0 & 1 \\ 1 & 0 & 0 \end{bmatrix}\begin{bmatrix} \overset{o}{\theta}_6 \\ i_{11} \\ i_{12} \end{bmatrix}$$

$$+\begin{bmatrix} 0 & 0 & 0 & -1 & -1 \\ 1 & 0 & 0 & 0 & 0 \\ 0 & 1 & -1 & 0 & 0 \end{bmatrix}\begin{bmatrix} v_1 \\ v_2 \\ v_3 \\ T_{14} \\ T_{15} \end{bmatrix}$$

Now substitute the terminal equations of dependent drivers

$$\overset{o}{\theta}_{13} = \overset{o}{\theta}_6, \quad v_3 = -k_m\overset{o}{\theta}_6, \quad v_2 = k_g i_{11} = k_g i_4, \quad T_{14} = k_m i_5 = k_m i_{12}$$

$$\begin{bmatrix} J_6 & 0 & 0 \\ 0 & L_{11} & 0 \\ 0 & 0 & L_{12} \end{bmatrix}\frac{d}{dt}\begin{bmatrix} \overset{o}{\theta}_6 \\ i_{11} \\ i_{12} \end{bmatrix} = \begin{bmatrix} 0 & 0 & 0 & -1 \\ -1 & 0 & 0 & 0 \\ 0 & -1 & -1 & 0 \end{bmatrix}\begin{bmatrix} R_7 & 0 & 0 & 0 \\ 0 & R_8 & 0 & 0 \\ 0 & 0 & R_9 & 0 \\ 0 & 0 & 0 & B_{10} \end{bmatrix}\begin{bmatrix} 0 & 1 & 0 \\ 0 & 0 & 1 \\ 0 & 0 & 1 \\ 1 & 0 & 0 \end{bmatrix}\begin{bmatrix} \overset{o}{\theta}_6 \\ i_{11} \\ i_{12} \end{bmatrix}$$

$$+\begin{bmatrix} 0 & 0 & 0 & -1 & -1 \\ 1 & 0 & 0 & 0 & 0 \\ 0 & 1 & -1 & 0 & 0 \end{bmatrix}\begin{bmatrix} v_1 \\ k_g i_{11} \\ -k_m\overset{o}{\theta}_6 \\ k_m i_{12} \\ T_{15} \end{bmatrix}$$

Formulation of State Space Model of Systems

Combine the similar terms to get the final state space model as

$$\frac{d}{dt}\begin{bmatrix} \overset{o}{\theta_6} \\ i_{11} \\ i_{12} \end{bmatrix} = \begin{bmatrix} \dfrac{-B_{10}}{J_6} & 0 & \dfrac{k_m}{J_6} \\ 0 & \dfrac{-R_7}{L_{11}} & 0 \\ \dfrac{-k_m}{L_{12}} & \dfrac{k_g}{L_{12}} & \dfrac{-(R_8+R_9)}{L_{12}} \end{bmatrix} \begin{bmatrix} \overset{o}{\theta_6} \\ i_{11} \\ i_{12} \end{bmatrix} + \begin{bmatrix} 0 & \dfrac{-1}{J_6} \\ \dfrac{1}{L_{11}} & 0 \\ 0 & 0 \end{bmatrix} \begin{bmatrix} V_1 \\ T_{15} \end{bmatrix}$$

Example 3.22

Develop the state space model for the given mechanical system shown in Figure 3.60 using MATLAB® Program SYSMO.

Output of SYSMO Program

WELCOME TO SYSMO
A PACKAGE FOR STATE SPACE EQUATIONS FORMULATION

The package formulates the state space equations once the interconnections within the system, the nature of each component, and its terminal equations are provided.

The formulation is applicable to linear, lumped parameter, bilateral, and time invariant systems. It is applicable to electrical, mechanical, pneumatic, hydraulic, or a combination of two or more of the above types of system.

IMPORTANT:

  i. If the system is unconnected, connect each separation at one node only.
  ii. For each controlled source, create an element [an "Impedance", either of zero value or of infinite value] the value of whose variable controls the dependent source.
  iii. Number the nodes and elements in an unambiguous manner.

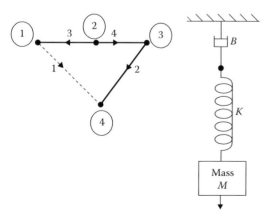

FIGURE 3.60
Mechanical system with its system graph.

Press any key to continue.....
The following type of elements are recognized

    i. Independent across driver: Type 1
    ii. Dependent across driver: Type 2
    iii. Impedance of zero value: Type 3
    iv. Capacitor type: Type 4
    v. Resistance type: Type 5
    vi. Inductance type: Type 6
    vii. Impedance of infinite value: Type 7
    viii. Dependent through driver: Type 8
    ix. Independent through driver: Type 9

The maximum allowable number of elements is 25.

What is the total number of nodes? 4

What is the total number of elements? 4

Please give the element data in the following order:

element no.; positivenode; negativenode; coupling; type;

Start giving the individual element data please

Element no.? 1

The positive node? 1

The negative node? 4

The element no. to which it is coupled [if uncoupled then print "0"]? 0

The element type? 5

Value of the element? 1

Please give the value of the initial conditions? 0

Next Please?

Element no.? 2

The positive node? 1

The negative node? 2

The element no. to which it is coupled [if uncoupled then print "0"]? 0

The element type? 6

Value of the element? 1

Please give the value of the initial conditions? 0

Next Please?

Element no.? 3

The positive node? 2

The negative node? 3

The element no. to which it is coupled [if uncoupled then print "0"]? 0

The element type? 4

Value of the element? 1

Please give the value of the initial conditions? 0

Next Please?

Element no.? 4

Formulation of State Space Model of Systems 203

The positive node? 3
The negative node? 4
The element no. to which it is coupled [if uncoupled then print "0"]? 0
The element type? 9
If the source is step/sinusoidal/cosinusoidal then
write 1/2/3–> 1
give the value of the step function source-1
n, the total no. of nodes

no_of_nodes =

    4

g, the total number of elements

g =

    4

The incidence matrix is

incidence =

1	1	0	0
0	-1	1	0
0	0	-1	1
-1	0	0	-1

Press any key to continue.....
Please wait....
The rearranged incidence matrix is

incidence =

0	1	1	0
1	0	-1	0
-1	0	0	1
0	-1	0	-1

Press any key to continue.....
Please wait....
Incidence matrix reduced to f-cutset matrix

incidence =

1	0	0	-1
0	1	0	1
0	0	1	-1

Press any key to continue.....
The Q Matrix is

qa =

    1
   -1
    1

The negative transpose of Q matrix is

qt =

  -1 1 -1

Press any key to continue.....

The combined constraint equations matrix is

combined =

  0 0 0 1
  0 0 0 -1
  0 0 0 1
  -1 1 -1 0

Press any key to continue.....

Please wait...

W, the rearranged combined constraint equation matrix is

w =

  0 0 0 1
  0 0 0 1
  0 0 0 -1
  -1 -1 1 0

Press any key to continue.....

The number of nondegenerate elements,

xw11 =

  1

The number of degenerate elements,

xw12 =

  1

The number of memoryless elements,

xw2 =

  1

The T.E. matrix, P2 of the memoryless elements is

tememless =

  1

Press any key to continue.....

The various coefficient matrices of the middle constraint eqn. are as follows

The t21 matrix is

t21 =

  0 0

the t22 matrix is

t22 =

  0

Formulation of State Space Model of Systems

and the t23 matrix is

t23 =

  -1

Press any key to continue.....
Premultiplying the entire equation by P2, the
T.E. of algebraic elements,
we get
The p2t21 matrix is

p2t21 =

  0 0

The p2t22 matrix

p2t22 =

  0

The p2t23 matrix

p2t23 =

  -1

Press any key to continue.....
After multiplying the middle equation by the inverse of the
matrix of terminal equations of the memoryless elements, the equation
reduces to W2 = K1 * W1 + K3 * W3 where

the k1 matrix is

k1 =

  0 0

and the k3 matrix is

k3 =

  -1

Press any key to continue.....
p1 : the terminal equations matrix of dynamic elements

p1 =

  1 0
  0 1

Press any key to continue.....
the various coefficient matrices of first constraint eqn. are:
t12 matrix

t12 =

  0
  0

t11 matrix

t11 =

  0 0
  0 0

t13 matrix

t13 =

  1
  1

Press any key to continue.....

After manipulation the top constraint equation takes the form

w1 = M1 * W1 + M3 * W3

where M1 is

m1 =

  0 0
  0 0

and M3 is

m3 =

  1
  1

Press any key to continue.....

The preliminary state equation is of the form

*S2 * dX/dt = s3 *X + m3 * U*

where S2 matrix is

s2 =

  1 0
  0 0

Press any key to continue.....

and s3 matrix is

s3 =

  0 0
  0 -1

and m3 matrix is

m3 =

  1
  1

Press any key to continue.....
welcome
welcome

Formulation of State Space Model of Systems

welcome2

welcome3

welcome4

The state space model is

dx/dt = s3 * x + m3 *u + m4 du/dt

dx/dt = s3 * x + m3 *u + m4 du/dt

s3 =

   0

Press any key to continue.....

and

m3 =

   1

and

u =

   1

&

m4 =

   1

and du/dt

derivative = 0

Model development is finished

MATLAB codes

```
% program for simulation
x=[0;0]
s3=0
m3=[-1,-1]
u=1
m4=-1
dt=.01
tsim=10
t=0
n=round((tsim-t/dt)
for I=1:n
    x1(I,:)=[t,x']
    c=s3*x
    e=m3*u
    dx=(c+e)*dt
    x=x+dx
    t=t+dt
end
plot(x1(0,1),x1(:,2:3))
```

3.18 Review Questions

1. Mention the fundamental variables for electrical, mechanical, hydraulic, and economic systems.
2. Are derivatives and integrals of through and across variables respectively through and across variables or otherwise? Give suitable examples in support of your answer.
3. State the component postulate of physical system theory.
4. State the interconnection postulates in the form of cutset and circuit postulates.
5. Write fundamental cutset and fundamental circuit equations in the matrix form with the branch chord partitioning as implied by the selected formulation tree of the system graph.
6. What is a maximally selected formulation tree?
7. Derive the principle of conservation of energy in a closed physical system on the basis of the fundamental cutset and circuit equations.
8. Consider two different physical systems with the same topology (i.e., interconnection pattern or system graph). Let $X_I(t)$ and $Y_I(t)$ represent the across variable vector and the through variable vector of the first physical system while $X_{II}(t)$ and $Y_{II}(t)$ represent the respective column vectors for the second physical system at time $(t + t_1)$. Prove that Quassi Power is

$$X_I^T(t)Y_{II}(t+t_1) = 0 \quad \text{and} \quad Y_I^T(t)X_{II}(t+t_1) = 0$$

9. State the terminal characteristics together with topological restrictions, if any, for the following multilateral/multiport components:

    Perfect couplers,

    Gyrators,

    Electromechanical transducers (2-port dc generator, 2-port dc motor).
10. Draw the equivalent circuits for the multiterminal components in question 9 in terms of appropriate two-terminal components and dependent drivers (connected sources).
11. The definition of a system depends not only on the physical entity involved but also on the purpose of the investigation. Comment with particular reference to the system and its environment.
12. State the determinateness theorem for linear system in any one form giving a set of necessary conditions on the system structure under which a complete and unique solution exists.
13. Apply the theorem in question 12 to the system shown in Figure 3.61 to establish whether or not necessary conditions are satisfied for the existence of a complete and unique solution for this system. $R = 3\,\Omega$, $L = 1$ H, $C = 0.5$ pF.
14. Consider the hydraulic system in Figure 3.62 consisting of a three-terminal reservoir with the terminal graph in Figure 3.63 and the corresponding terminal characteristics as given by equation

$$dP(t)/dt = AG(t) \quad \text{where } P(t) = [P_2(t)P_3(t)]^T, \quad G(t) = [g_2(t)g_3(t)]^T$$

Formulation of State Space Model of Systems

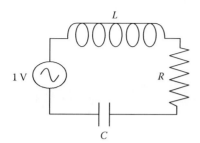

FIGURE 3.61
Series RLC circuit.

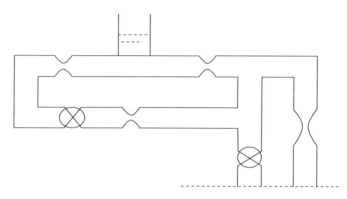

FIGURE 3.62
Hydraulic system.

$$A = [1/C2r23*d/dt$$

$$R23*d/dtr23*d/dt]$$

and $r23 = R3/2$

FIGURE 3.63
Terminal graph of three-terminal tank.

a. State the determinateness theorem in any one of its three alternative forms and apply it to the system in Figure 3.62 to determine whether necessary conditions for the existence and uniqueness of the solution of this system are satisfied or not.
b. Obtain the formulation tree(s) and the interconnection constraint equations for the system in Figure 3.62.
15. Draw the system graph and identify the tree for the given mechanical system shown in Figure 3.64.
16. Define precisely and explain with the help of a simple example in each case the following topological terms used in system theory:

   1. Terminal graph
   2. System graph
   3. Edge
   4. Path
   5. Branch
   6. Chord
   7. Tree of the graph
   8. Circuit
   9. Forest
   10. Coforest
   11. f-cutset
   12. f-circuit
   13. Isomorphism
   14. Edge-sequence
   15. Degree of vertex
   16. Multiplicity

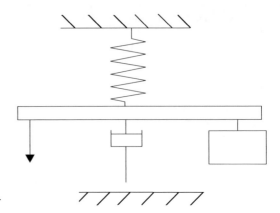

FIGURE 3.64
Mechanical system.

17. The terminal graphs for a four-terminal component are given in Figure 3.65. Give all other possible terminal graphs for this component.

18. Consider the mass and spring components and their linear graphs shown in Figure 3.66. Let a spring balance be the force meter and a calibrated scale be the displacement meter.

    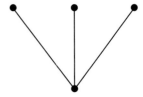

    FIGURE 3.65
    Terminal graph for four-terminal component.

    a. Show how the meters would be placed on the components to measure their terminal variables as indicated by their linear graphs. Draw two diagrams for each component, one with a displacement driver and one with a force driver. Assume that all meters read zero before measurement starts.

    b. For the mass components, assume that the calibrated scale reads −2 before measurement starts. Assuming linear operation write the terminal equations for the component.

    c. For the spring component, assume that both force meter and displacement meter read +2 before measurement starts. Sketch the terminal equation of the component.

19. Identify the two-terminal components, and draw the system graphs of the hydraulic, mechanical, and electrical systems shown in Figures 3.67 through 3.69, respectively. Also determine the rank and nullity for each graph.

20. Find out the number of possible trees of the network shown in Figure 3.70. Evaluate the determinant of $A*A^T$, where A is the reduced incidence matrix. Is this equal to the number of trees of the graph?

21. a. For the system graphs of the system shown in Figure 3.70 choose a formulation tree for each and write the fundamental circuit and cutset equations.

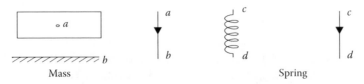

FIGURE 3.66
Mass and spring components of mechanical system.

Formulation of State Space Model of Systems

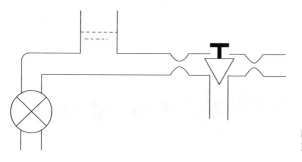

FIGURE 3.67
Hydraulic system.

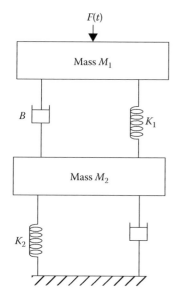

FIGURE 3.68
Mechanical system.

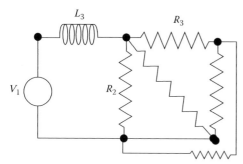

FIGURE 3.69
Electrical system.

b. How many independent circuit and cutset equations can be written for the graphs of the systems in Figure 3.70?

Let the cutset and circuit equations in part (a) be represented symbolically by the equations

$$\begin{bmatrix} B_1 & U \end{bmatrix} \begin{bmatrix} X_b \\ X_c \end{bmatrix} = 0 \quad \text{and} \quad \begin{bmatrix} U & A_1 \end{bmatrix} \begin{bmatrix} Y_b \\ Y_c \end{bmatrix} = 0$$

Show that in all cases $B_1 = -A_1^T$.

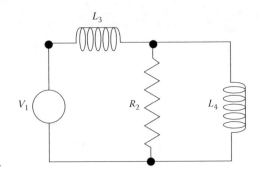

FIGURE 3.70
Electrical system.

22. Prove that for a graph
    1. $\alpha\beta^T = \alpha^T\beta = 0$
    2. $A_1 + B_1^T = 0$ or $B_1 + A_1^T = 0$
23. Choose a formulation forest (set of trees) and write the fundamental circuit and cutset equations for the graph in Figure 3.71.
24. Let the graph G of a system be connected and contain "e" elements and "v" vertices.
    a. What is the maximum number of across variables that can be arbitrarily specified?
    b. What is the maximum number of through drivers that can be arbitrarily specified?
    c. Is there any restriction as to where in the system the specified through and across drivers can be located? Explain.
    d. Why, out of all the possible circuit and cutset equations that can be written, is it convenient to use the fundamental circuit and cutset equations?
25. a. How one can maximally select a formulation tree.
    b. What are the properties of a tree?
26. State and explain the following with the help of suitable examples:
    a. Fundamental axiom
    b. Component postulate
    c. Interconnection postulates
29. a. Identify the two-terminal components of the mechanical system shown in Figure 3.72 and write their terminal equations.
    b. What are the properties of an f-tree. Draw an f-tree for the mechanical system shown in Figure 3.72.
30. Apply SYSMO algorithm to derive interconnection constraint equations for the system shown in Figure 3.72 in the following form:
$$w = TW$$

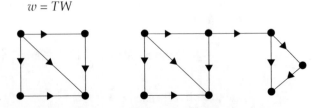

FIGURE 3.71
Unconnected multipart system graph.

Formulation of State Space Model of Systems

where
  w = secondary variable vector,
  W = primary variable vector (corresponding to a maximally selected formulation tree)
  T = combined matrix

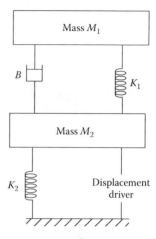

FIGURE 3.72
Mechanical system.

31. Continue question 30 to formulate a state model for the system shown in Figure 3.72 by using SYSMO MATLAB program.
32. State topological restrictions of the multiterminal/multiport components used in electrical systems.
33. How are input data prepared for SYSMO formulation?
34. What is the order of types of elements in which the columns of the incidence matrix are arranged in SYSMO formulation?
35. What is the topological diagnostic available in SYSMO formulation while applying the Gauss–Jordan reduction to the reduced incidence matrix?
36. How are interconnection constraint equations written in a combined form in SYSMO formulation?
38. How are rows and columns of the combined matrix in question 36 rearranged according to the types of elements in SYSMO formulation?
39. Formulate a state model for the system shown in Figure 3.73 using SYSMO algorithm.
42. Formulate the state model of an automobile suspension system shown in Figures 3.74 and 3.75. Consider the input is a change in displacement $x(t)$ as it goes ahead. Find the output $y(t)$ for the given input.
43. Develop the state model for the mechanical system shown in Figures 3.76 and 3.77.

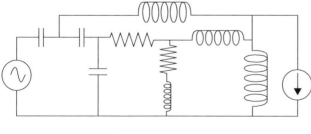

FIGURE 3.73
Electrical system.

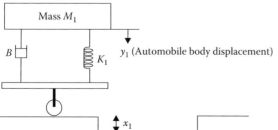

FIGURE 3.74
Schematic diagram of automobile suspension system.

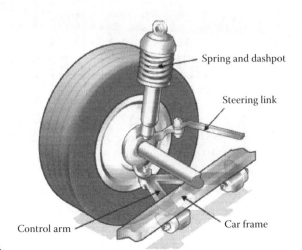

FIGURE 3.75
Automobile suspension systems.

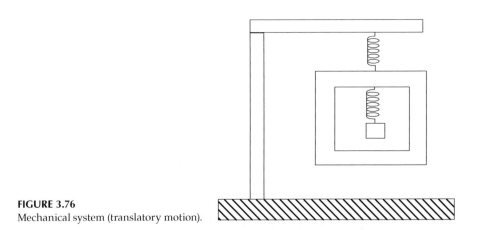

FIGURE 3.76
Mechanical system (translatory motion).

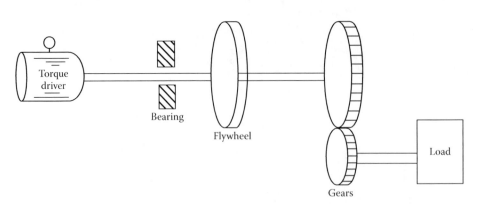

FIGURE 3.77
Mechanical system (rotating system).

Formulation of State Space Model of Systems   215

Q. Multiple Choice Questions (Bold is correct answer)

1. If there are b elements and n nodes in a system graph, the number of **f-circuits** is given
   a. b
   b. $b - n$
   c. $n - 1$
   d. **$(b - n + 1)$**

2. Which is incorrect to define "tree"?
   a. It is a set of branches which together connects all nodes
   b. The set of branches should not form a loop
   c. There cannot be more than one tree for a graph
   d. The branches of tree are called twigs

3. In a system
   a. The number of tree branches is equal to the number of links
   b. The number of tree branches cannot be equal to the number of links
   c. The number of tree branches has no relation with the number of link branches
   d. None of these

4. A cutset has
   a. Only one tree branch
   b. Always more than one tree branch
   c. Only one tree link
   d. None of these

5. Let the coefficient matrices of a state model are $A = \begin{bmatrix} 0 & 1 \\ -1 & -2 \end{bmatrix}$, $B = \begin{bmatrix} 0 \\ 1 \end{bmatrix}$, $C = \begin{bmatrix} 1 \\ 1 \end{bmatrix}$ and $D = \begin{bmatrix} 0 \\ 0 \end{bmatrix}$, which of the following is correct for the given system:
   a. Neither state controllable nor observable
   b. State controllable but not observable
   c. Not state controllable but observable
   d. Both state controllable and observable

6. The value of A in $\overset{o}{X} = Ax$ for system $\overset{oo}{y} + 2\overset{o}{y} + 3y = 0$ is
   a. $A = \begin{bmatrix} 1 & 0 \\ 2 & -1 \end{bmatrix}$
   b. $A = \begin{bmatrix} 0 & 1 \\ -1 & -2 \end{bmatrix}$
   c. $A = \begin{bmatrix} 0 & 1 \\ -2 & 1 \end{bmatrix}$
   d. $A = \begin{bmatrix} 0 & 1 \\ -3 & -2 \end{bmatrix}$

7. The size of coefficient matrices A and B in the state space model are:
   a. both square
   b. A rectangular and B square
   c. A square and B rectangular
   d. both rectangular

8. In a mechanical system with rotating mass, the time constant of the system can be decreased by
   a. Increasing the inertia of the system
   b. Output rate feedback
   c. Increasing input to the system
   d. Reducing friction

9. In the following, pick out the linear systems

   i. $\dfrac{d^2y(t)}{dt^2} + a_1 \dfrac{dy(t)}{dt} + a_2 y(t) = u(t)$

   ii. $y\dfrac{dy(t)}{dt} + a_1 y(t) = a_2 u(t)$

   iii. $2\dfrac{d^2y(t)}{dt^2} + t\dfrac{dy(t)}{dt} + t^2 y(t) = 10u(t)$

   a. (i) and (ii)   b. (i) only
   c. (i) and (iii)   d. (ii) and (iii)

10. In the following, pick out the nonlinear systems

    i. $\dfrac{d^3y(t)}{dt^3} + t^3 \dfrac{d^2y(t)}{dt^2} + t\dfrac{dy(t)}{dt} + y^2 = 20\sin(\omega t)$

    ii. $\dfrac{d^2y(t)}{dt^2} + \dfrac{1}{t}\dfrac{dy(t)}{dt} + y(t) = 4$

    iii. $\left(\dfrac{d^2y(t)}{dt^2}\right)^2 + \dfrac{dy(t)}{dt} + y(t) = 10$

    a. (i) and (iii)   b. (ii) and (iii)
    c. (i) and (ii)   d. none

11. In a linear system an input of 5 sin(ωt) produces an output of 15 cos(ωt), the output corresponding to 20 sin(ωt) will be equal to
    a. (15/4) sin(ωt)   b. 60 sin(ωt)
    c. 60 cos(ωt)   d. (15/4) cos(ωt)

12. In the following, pick out the time invariant systems:

    i. $\dfrac{dy(t)}{dt} + 5y(t) = u(t)$

    ii. $t\dfrac{d^2y(t)}{dt^2} + \dfrac{dy(t)}{dt} + 2y(t) = 10u(t)$

    iii. $\dfrac{d^2y(t)}{dt^2} + 5\dfrac{dy(t)}{dt} + 8y(t) = v(t)$

    a. (i) and (ii)   b. (i) only
    c. (i) and (iii)   d. (ii) and (iii)

13. A mass–damper spring system is given by
    $\dfrac{d^2x(t)}{dt^2} + \dfrac{dx(t)}{dt} + 0.6x(t) = f(t)$, where $f(t)$ is the external force acting on the system and x is the displacement of mass. The steady state displacement corresponding to a force of 6 Newton is given by
    a. 10 m   b. 0.36 m   **c. 6 m**   d. none of these

14. The system equation for a mechanical system is $5\dfrac{d\omega(t)}{dt} + \omega(t) = 5$, where ω is angular velocity. The solution for ω is given by
    a. $5(1 - e^{-t/5})$ b. $5e^{-t/5}$ c. $5(1 + e^{-t/5})$ d. $5(1 - e^{t/5})$

15. The mechanical time constant of a motor is 50 s. If the friction coefficient is 0.02 Nm/rad/s, the value of moment of inertia of the motor is equal to
   a. 50 kg-m²   b. $\frac{1}{50}$ kg-m²   c. **1 kg-m²**   d. 0.1 kg-m²

16. A mass–spring–damper system is:

    a. First-order system   **b. Second-order system**
    c. Third-order system   d. none

Q. Fill in the Blanks:

1. If the number of vertices in a system is 8, and then the rank of the system is _____

2. If branch k is associated with node h and oriented away from node h, and then A_a is _____

3. The relation between A and B is _____

   Answers: 1. 7   2. –1   3. $A \cdot B^T$

3.19 Bibliographical Notes

The basic concepts of systems had been described by Zadeh (1963), Kailath (1980), and Lathi (1965, 1987). Revolutionary and remarkable discoveries in modern science such as the theory of relativity (with four-dimensional space–time construct), quantum theory [with superposition of n (even infinite) number of probable states], and string theory (currently the leading and only candidate for a theory of everything, the holy grail of physics) pose formidable challenges to graph theoretic modeling approach. The latest version of string theory (M-theory) predicts there are 11 dimensions (Kaku, 2008; Satsangi, 2009).

4

Model Order Reduction

4.1 Introduction

Large-scale systems are all around and exist in diverse fields such as complex chemical processes, biomedical systems, social economic systems, transportation systems, ecological systems, electrical systems, mechanical systems, and aeronautical and astronautics systems. All these large and complex systems are difficult to model with conventional techniques. Hence, these systems can be decoupled or partitioned into suitable numbers of interconnected subsystems to reduce their complexity for modeling purposes.

In many practical situations, a fairly complex and high-order model is obtained from theoretical considerations. This complexity often makes it difficult to obtain a good understanding of the behavior of the system. These high-order models are generally represented in the state space form or the transfer function form in the time domain or the frequency domain, respectively.

The exact analysis of most higher-order systems is both tedious as well as costly, and poses a great challenge to both the system analyst and the control engineer. The preliminary design and optimization of such complex systems can often be accomplished with greater ease if a low-order model is derived, which provides a good approximation to the system.

A great deal of work has been done in the past to get better low-order models for high-order systems as evident from the comprehensive bibliography prepared by Genesio and Milaness (1976).

One of the most desirable features of such models will be simplicity while preserving the features of interest. Since the models may be developed with various aims and objectives in mind or with various view points, it is possible to have more than one model for a given system, each one satisfying some predefined objectives. Different people develop different models for the same system; sometimes the same person may develop different models for the same system, if his understanding or view point changes with time. Simplicity of the model is a crucial aspect, especially in online controls. Simpler models generally give a better feel of the original system.

Another important feature of a model is its order, which again gives a measure of its complexity. The fact is that a large selection of the first course on automatic control theory deals with second-order models of original systems.

The term reduced-order model is normally used for state space models; whereas, the term simplified model is used for frequency domain transfer function models. Reducing the order of a state model is equivalent to the search for a coarser representation of the system, which, therefore, yields a lower-dimensional state space, for the same class of inputs.

The problem of concern would be that the underlying techniques must yield a controllable and an observable reduced model.

On the other hand, decreasing the order of a transfer function does not ensure that the resulting transfer function is always physically realizable. For this purpose, the term simplification is used to designate the building of approximate transfer function dominants of lower orders than those of the original system subjected to a functional criterion.

The order reduction of a linear time-invariant system is applied in almost all fields of engineering. The use of reduced-order models for test simulations of complex systems is a lot easier than utilizing full-order models. This is due to the fact that the lower-order transfer function can be analyzed more easily. Therefore, order reduction algorithms are standard techniques in the system interdependent analysis, approximation, and simulation of models arising from interconnect and system interdependent.

There are several procedures that seek to automate the model reduction process. Suppose a high-order, linear, time-invariant model, G, is given, then the Prototype $H\infty$ model reduction problem is to find a low-order approximation, $\hat{G}$, of G such that $\|G - \hat{G}\|\infty$ is small. Consider the more difficult problem of selecting $\hat{G}$ such that the problem of selecting W_1 and W_2 such that $\|W_2(G - \hat{G})W_1\|\infty$, is small; the weighting functions, W_1 and W_2, are used to frequency-shape the model reduction error. For example, one might select the weights so that the modeling error is small in the unity gain range of frequency.

Among the various classes of model reduction techniques, *explicit moment-matching* algorithms (Pillage, Rohrer, 1990; Ratzlaff, Pillage, 1994) and Krylov-subspace-based methods (Feldmann et al., 1995; Kerns et al., 1997; Odabasioglu et al., 1998) have been most commonly employed for generating the reduced-order models of the interconnects. The computational complexity of these model order reduction techniques is primarily due to the matrix–vector products. However, these methods do not provide a provable error bound for the reduced system. Extensions of explicit moment-matching techniques have been proposed recently, to reduce the linear time-varying (LTV) as well as nonlinear dynamic systems (Roychowdhury, 1999; Peng and Pileggi, 2003).

4.2 Difference between Model Simplification and Model Order Reduction

The term reduced-order model is frequently used for time domain (state space) models; whereas, model simplification is related to the frequency domain. The reason for differentiating is due to the problem of realization. Reducing the order of a state space model is equivalent to the search for a coarser representation of the process, which, therefore, yields a lower-dimensional state space, for the same class of inputs. The problem of concern would be that the underlying technique must yield a controllable and an observable reduced-order model.

On the other hand, decreasing the order of a transfer function does not ensure that the resulting transfer function is realizable. For this purpose, the term simplification is used to designate the building of an approximate transfer function whose denominators are of lower orders than those of the original system subjected to a functional criterion.

4.3 Need for Model Order Reduction

Every physical system can be translated into a mathematical model. These mathematical models give a comprehensive description of a system in the form of higher-order differential equations. It is useful and sometimes necessary to find a lower-order model that adequately reflects the dominant characteristics of the system under consideration. Some of the reasons for model order reduction are as follows:

1. Quick and easy understanding of the system

    A higher-order model of a system possesses difficulties in its analysis, synthesis, or identification. An obvious method of dealing with such a type of system is to approximate it to a lower-order model that reflects the characteristics of the original system such as the time constant, the damping ratio, and the natural frequency.

2. Reduced computational burden

    When the order of the system model is higher, the numerical techniques need to compute the system response at the cost of computation time and memory required by digital computers.

3. Reduced hardware complexity

    Most controllers are designed on the basis of linear low-order models, which are more reliable, less costly, and easy to implement and maintain due to less hardware complexity.

4. Making feasible designs

    Reduced-order models may be effectively used in control applications like
    i. Model reference adaptive control schemes
    ii. Hierarchical control schemes
    iii. Suboptimal control
    iv. Decentralized controllers

5. Generalization

    The results studied for a simple low-order model can be easily generalized to other comparable systems.

6. To improve the methodology of computer-aided control system design.

4.4 Principle of Model Order Reduction

Both, the ordinary differential equation (ODE) systems, which arise from the spatial discretization of first-order, time-dependent partial differential equations (PDEs), and the differential algebraic equation (DAE) systems, which describe the dynamics of electrical circuits at time t, can be described by

$$\frac{d}{dt}q(x,t) + j(x,t) + Bu(t) = 0 \qquad (4.1)$$

with the difference that for "real" DAEs the partial derivative, q_x, is singular. In the case of circuit equations for example, the vector-valued functions, $q(x,t)$ and j, represent the contributions of reactive elements (such as capacitors and inductors) and of nonreactive elements (such as resistors), respectively, and all time-dependent sources are stored within $u(t)$. In case of linear or linearized models, Equation 4.1 simplifies to

$$C\dot{x} + Gx = Bu(t)$$
$$y = Lx \qquad (4.2)$$

where
- $x(t) \in R^n$ is the state vector
- $u(t) \in R^m$ is the input excitation vector
- $y(t) \in R^p$ is the output measurement vector
- $G \in R^{n \times n}$ and $C \in R^{n \times n}$ are the symmetric and sparse system matrices

and for DAEs, C is singular, $B \in R^{n \times m}$ and $L \in R^{p \times n}$ are the user-defined input and output distribution arrays, n is the dimension of the system and m and p are the numbers of inputs and outputs. The idea of model order reduction is to replace Equation 4.2 by a system of the same form but with a much smaller dimension, $r \ll n$:

$$C_r \dot{z} + G_r z = B_r u(t)$$
$$y_r = L_r z \qquad (4.3)$$

which can be solved by a suitable DAE or ODE solver and will approximate the input/output characteristics of Equation 4.2. A transition from Equation 4.2 to Equation 4.3 is formal and is done in two steps.

Step I
The transformation of the state vector, x, to the vector of generalized coordinates, z, and the truncation of a number of those generalized coordinates, which leads to some (hopefully) small error, J:

$$x = Vz + e \qquad (4.4)$$

The time-independent $V \in R^{n \times r}$ is called the transformation or projection matrix, as Equation 4.4 can be seen as the projection of the state vector onto some low-dimensional subspace defined by V. Note that the spatial and physical meaning of x is lost during such a projection.

Step II
In the second step, Equation 4.2 is multiplied from the left-hand side with another matrix, $W^T \in R^{r \times n}$, so that $C_r = W^T C V$, $G_r = W^T G V$, $B_r = W^T B$, and $L_r = EV$. Note that the number of inputs and outputs in the reduced system (Equation 4.3) is the same as in Equation 4.2.

The above principle of projection can also be applied directly to the second-order ODE systems [19], which may arise from the spatial discretization of the second-order, time-dependent PDEs (e.g., the equation of motion, which is often solved in MEMS simulation). Furthermore, it can be applied to the nonlinear system (Equation 4.1). This, however, is much more complicated and does not necessarily result in the reduction of computational time.

4.5 Methods of Model Order Reduction

The main objective of model order reduction is that the reduced-order approximation should reproduce the significant characteristics of the original system as closely as possible.

The model order reduction techniques can broadly be classified as

1. Time domain simplification techniques
    a. Dominant eigenvalue approach
    b. Aggregation method
    c. Subspace projection methods
    d. Optimal order reduction
    e. Balance realization method
    f. Singular perturbation method
    g. Hankel matrix approach
    h. Hankel–Norm model reduction
2. Frequency domain simplification techniques
    a. Continued fraction expansion (CFE) and truncation
    b. Pade approximation techniques
    c. Moment-matching method
    d. Matching frequency response
    e. Simplification using canonical form
    f. Reduction using orthogonal polynomials
    g. Reduction based on Routh stability criterion
    h. Stability equation method
    i. Reduction based on integral least square techniques
    j. Polynomial differentiation method
    k. Polynomial truncation method
    l. Factor division method

4.5.1 Time Domain Simplification Techniques

In time domain reduction techniques, the original and reduced systems are expressed in the state space form. The order of matrices, A_r, B_r, and C_r, are less than A, B, and C, and the output, Y_r, will be a close approximation to Y for specified inputs. The time domain techniques belong to either of the following categories.

4.5.1.1 Dominant Eigenvalue Approach

This category attempts to retain the dominant eigenvalues of the original system and then obtain the remaining parameters of the low-order model in such a way that its response, to a certain specified input, should approximate closely to that of a high-order system. The methods proposed by Davison (1966), Marshall (1966), Mitra (1967), and Aoki (1968) belong to this category. It has been shown that these (Hicken, 1978; Sinha, 1980) may be regarded as special cases of the aggregation method proposed by Aoki.

Davison's method consists of diagonalizing the system matrix and ignoring the large eigenvalues. In this case, the input is taken as a step function and all the eigenvalues are assumed to be distinct. This restriction, however, was removed by Chidambara (1967) and Davison (1968). Aoki (1968) took a more general approach based on aggregation. Gruca (1978) introduced delay in the output vector of the aggregated model to minimize the quality index function of the output vector. This led to improvement in the quality of the simplified aggregated model of the system without increasing the order of the state differential equation. However, the numerical difficulties and the absence of guidelines for selecting the weighting matrices in the performance index of this method were well observed by the researches. Inooka (1977) proposed a method based on combining the methods of aggregation and integral square criterion. An important variation of the dominant eigenvalues' concept was proposed by Gopal (1988) wherein the high-order system is replaced by three models, successfully representing the initial intermediate and final stages of the transient response.

The location of eigenvalues plays an important role on system stability. To understand the concept of system stability, we consider an example of a right circular cone. There may be three possible situations (equilibrium states) for a right circular cone to stand as shown in Figure 4.1, which is mentioned below.

1. The cone may stand forever on its circular base. If the cone is disturbed in this state, it would eventually return to its original state when the disturbance is removed. Hence, the cone is said to be in a stable state.
2. If the cone is standing on its apex, then the slightest disturbance will cause the cone to move farther and farther from its original equilibrium state. The cone in this case is said to be in an unstable state.
3. In the last situation, the cone is lying on its side. If the cone is disturbed in this state, it will neither go back to its original state nor continue to move. This is called the neutral equilibrium state.

Consider a state space model of a system:

$$\begin{bmatrix} \dot{x}_1 \\ \dot{x}_1 \\ \vdots \\ \vdots \\ \dot{x}_n \end{bmatrix} = \begin{bmatrix} A_{11} & A_{12} & \cdots & A_{1n} \\ A_{21} & A_{22} & \cdots & A_{2n} \\ \vdots & \vdots & \cdots & \vdots \\ A_{n1} & A_{n2} & \cdots & A_{nn} \end{bmatrix} \begin{bmatrix} x_1 \\ x_2 \\ \vdots \\ \vdots \\ x_n \end{bmatrix} + \begin{bmatrix} B_{11} & B_{12} & \cdots & B_{1m} \\ B_{21} & B_{22} & \cdots & B_{2m} \\ \vdots & \vdots & \cdots & \vdots \\ B_{n1} & B_{n2} & \cdots & B_{nm} \end{bmatrix} \begin{bmatrix} u_1 \\ u_2 \\ \vdots \\ \vdots \\ u_n \end{bmatrix}$$

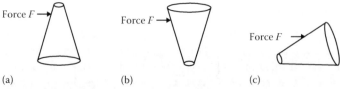

FIGURE 4.1
System stability. (a) Stable cone. (b) Unstable cone. (c) Cone in neutral equilibrium state.

$$\dot{X} = AX + BU \tag{4.5}$$

The characteristic equation is

$$|\lambda I - A| = 0 \tag{4.6}$$

The roots of this characteristic equation are λ_i, where $i = 1, 2, 3, \ldots, n$.

When the roots are in the left half of the $j\omega$-plane, then the system is stable, and when the roots are in the right half of the $j\omega$-plane, then the system is unstable, as shown in Figure 4.2.

When the roots are on the $j\omega$-axis, then the system is oscillatory. The locations of these roots decide the responses of the system. Figure 4.3 shows the locations of the characteristic roots and the step responses of a second-order system. In the transfer

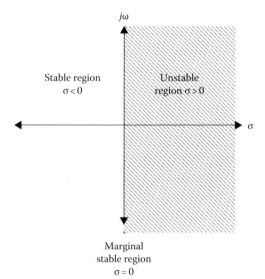

FIGURE 4.2
Characteristic roots' locations and system stability.

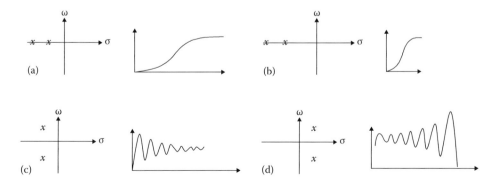

FIGURE 4.3
Effects of locations of poles on system responses. (a) Roots are real and near to imaginary plane (on left side). (b) Roots are real and far from imaginary plane (on left side). (c) Roots are complex and near to imaginary plane (on left side). (d) Roots are complex and near to imaginary plane (on right side).

function model, zeros play an important role on the response, but not on the stability of the system. The poles or eigenvalues of the system model have a great impact on the transient responses of the system. Since most system models are of a higher order, it would be useful to reduce their order by retaining the dominant eigenvalues and eliminating the insignificant eigenvalues. If the poles are on the right-hand side of the imaginary plane, the system becomes unstable, and if the eigenvalues are on the left-hand side of the imaginary plane, the system is stable. When the real value of the eigenvalues is zero, then the system response is oscillating.

From the above discussion, it is evident that as the eigenvalues go away from the imaginary ($j\omega$-) plane, the oscillations decay fast. If we go far away from the real axis, the frequency will be more. At the imaginary axis there is no decay, and hence the oscillations are of constant amplitude. If the distance D is 5–10 times greater, the fast-decaying transients can be ignored, and these eigenvalues are called insignificant or nondominant eigenvalues, as shown in Figure 4.4.

The important property of the characteristics and the eigenvalue is that they are invariant under a nonsingular transformation, such as the phase variable canonical form of transformation. In other words, when matrix A is transformed by a nonsingular transformation

$$x = Py \tag{4.7}$$

$$\bar{A} = P^{-1}AP \tag{4.8}$$

So, the characteristic equation and the eigenvalues of $\bar{A}$ are identical to those of A:

$$\begin{aligned} |sI - \bar{A}| &= |sI - A| \\ &= |sP^{-1}P - P^{-1}AP| \\ &= |P^{-1}(sI - A)P| \end{aligned} \tag{4.9}$$

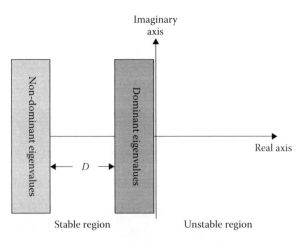

FIGURE 4.4
Difference between dominant eigenvalues and nondominant eigenvalues.

Model Order Reduction

Since the determinant of a product of matrices is equal to the product of a determinant of matrices:

$$= |P^{-1}||sI - A||P|$$

$$= |sI - A|$$

Dominant eigenvalue method

Let M be a modal matrix that is nonsingular:

$$x = My$$

Substitute this x into Equation 4.5:

$$M\overset{0}{\dot{y}} = AMy + Bu$$

$$\overset{0}{\dot{y}} = M^{-1}AMy + M^{-1}Bu$$

$$\overset{0}{\dot{y}} = \Lambda y + \eta u \qquad (4.10)$$

$$\begin{bmatrix} \overset{0}{\dot{y}_1} \\ \overset{0}{\dot{y}_2} \end{bmatrix} = \begin{bmatrix} \Lambda_1 & 0 \\ 0 & \Lambda_2 \end{bmatrix} \begin{bmatrix} y_1 \\ y_2 \end{bmatrix} + \begin{bmatrix} \eta_1 \\ \eta_2 \end{bmatrix} u$$

where
  y_1 is the dominant eigenvalue
  y_2 is the nondominant eigenvalue

$$\dot{y}_2 = \Lambda_2 y_2 + \eta_2 u$$

For long-term dynamics, $\overset{0}{\dot{y}_2} = 0$. Substitute this by the above equation:

$$y_2 \overset{.}{=} -\Lambda_2^{-1} \eta_2 u$$

$$= Du$$

$$M = \begin{bmatrix} M_0 & M_{11} \\ M_{21} & M_{22} \end{bmatrix}$$

$$\begin{bmatrix} x_1 \\ x_2 \end{bmatrix} = \begin{bmatrix} M_0 & M_{11} \\ M_{21} & M_{22} \end{bmatrix} \begin{bmatrix} y_1 \\ y_2 \end{bmatrix}$$

$$x_1 = M_0 y_1 + M_{11} y_2$$

$$y_1 \overset{.}{=} M_0^{-1} x_1 - M_0^{-1} M_{11} y_2$$

The second row of the above matrix equation:

$$x_2 = M_{21}y_1 + M_{22}y_2$$

$$x_2 = M_{21}\{M_0^{-1}x_1 - M_0^{-1}M_{11}y_2\} + M_{22}y_2$$

$$x_2 = M_{21}M_0^{-1}x_1 + \{M_{22} - M_{21}M_0^{-1}M_{11}\}y_2$$

$$\overset{0}{x_1} = A_{11}x_1 + A_{12}x_2 + B_1u$$

$$= A_{11}x_1 + A_{12}M_{21}M_0^{-1}x_1 + A_{12}\{M_{22} - M_{21}M_0^{-1}M_{11}\}Du + B_1u$$

$$= \{A_{11} + A_{12}M_{21}M_0^{-1}\}x_1 + \{A_{12}M_{22} - A_{12}M_{21}M_0^{-1}M_{11}D + B_1\}u$$

$$\overset{0}{x_1} = F_1x_1 + G_1u$$

where

$$F_1 = \{A_{11} + A_{12}M_{21}M_0^{-1}\}$$

$$G_1 = \{A_{12}M_{22}D - A_{12}M_{21}M_0^{-1}M_{11}D + B_1\}$$

4.5.1.1.1 Limitations of Eigenvalue Approach

The above-mentioned eigenvalue approaches, although useful in many applications, suffer from the following problems:

1. The computations of eigenvalues and eigenvectors may be quite formidable for a high-order system.
2. In the case when the eigenvalues of a system are close together or when the eigenvalues are not easily identified, this method obviously fails.
3. There may be considerable differences between the responses of high-order systems and low-order systems to certain inputs.

4.5.1.2 Aggregation Method

The concept of aggregation is well known to economists but less known to control engineers. The popular aggregation method is proposed by Aoki (1968). He has shown the usefulness of an aggregation matrix for designing suboptimal controllers. The key to the success of formulating the problem is the existence of an aggregation matrix, which relates the system states and the reduced-order model states. There exists a large class of aggregation matrices—the choice depends on how one selects the modes in aggregated model and model state. Then, although the formulation of the suboptimal control design problem is straightforward, to set up a suitable aggregated model for the solution is by no means simple, owing to the wide range of choices (Sandell et al., 1978; Siret et al., 1979). The main

Model Order Reduction

advantage of this method is that some internal structural properties of the original system are preserved in the reduced-order model, which is useful not only in the analysis of system but also in deriving state–feedback suboptimal control.

Let the state space equation be

$$\overset{0}{X} = AX + BU$$

Assume

$$Z(t) = CX(t)$$

where
- Z is the reduced state vector of $r \times 1$
- X is the state vector of $n \times 1$
- C is the constant vector of size $r \times n$, $r \leq n$

$$X(t) = C^{\#} Z(t)$$

where $C^{\#}$ is the left pseudo inverse of C

$$C^{\#} C = I$$

$$C^{\#}[CC^T] = IC^T$$

$$C^{\#} = [C^T[CC^T]^{-1}]$$

Pre-multiply by C in the state equation:

$$C\overset{0}{X} = CAX + CBu$$

$$C\overset{0}{X} = CAC^{\#}X + CBu$$

Reduced-order state equation:

$$\overset{0}{z}(t) = Fz(t) + Gu(t)$$

$$F = CAC^{\#}$$

$$G = CB$$

$$e(t) = Z(t) + CX(t)$$

$$\overset{0}{e}(t) = \overset{0}{Z}(t) + C\overset{0}{X}(t)$$

$$\overset{0}{e}(t) = FZ(t) + Gu(t) - C\{AX(t) + Bu(t)\}$$
$$= \{FZ(t) - FCX(t)\} + \{FCX(t) - CAX(t)\} + \{Gu(t) - CBu(t)\}$$
$$= F\{Z(t) - CX(t)\} + \{FC - CA\}X(t) + \{G - CB\}u(t)\}$$
$$FC = CA(CC^{\#}) = CA$$
$$\overset{0}{e}(t) = F\{Z(t) - C\overset{0}{X}(t)\}$$

Case I (Perfect aggregation)

If $\overset{0}{e}(0) = 0$ and $\overset{0}{e}(0) = 0$, $t \geq 0$

Case II (Aggregation is perfect in asymptotic sense)

If $\overset{0}{e}(0) \neq 0$ and $\underset{t \to 0}{\overset{0}{e}(t)} = 0$

But F is asymptotically stable.

Case III

If $FC \neq CA$
But $G = CB$.

$$\overset{0}{e}(t) = Fe(t) + (FC - CA)X(t)$$

$$\overset{0}{e}(t) = e^{FT}e(0) + \int_0^t e^{F(t-\tau)}(FC - CA)X(t)d\tau$$

Example 4.1

$$\dot{X} = AX + BU$$

$$A = \begin{bmatrix} 0 & 1 & 0 \\ 0 & 0 & 1 \\ -20 & -32 & -13 \end{bmatrix}$$

$$B = \begin{bmatrix} 3/2 \\ -3/2 \\ 9/2 \end{bmatrix}$$

Dominant eigenvalue approach

$$|\lambda I - A| = 0$$

$$\begin{bmatrix} \lambda & -1 & 0 \\ 0 & \lambda & -1 \\ 20 & 32 & \lambda+13 \end{bmatrix} = 0$$

$$\lambda = -1, -2, -10$$

$$M = \begin{bmatrix} 1 & 1 & 1 \\ \lambda_1 & \lambda_2 & \lambda_3 \\ \lambda_1^2 & \lambda_2^2 & \lambda_3^2 \end{bmatrix}$$

$$M = \begin{bmatrix} 1 & 1 & 1 \\ -10 & -2 & -1 \\ 100 & 4 & 1 \end{bmatrix}$$

$$|M| = 72$$

$$M^{-1} = 1/72 \begin{bmatrix} 2 & 3 & 1 \\ -90 & -99 & -9 \\ 60 & 96 & 8 \end{bmatrix}$$

$$X = MY$$

$$M\dot{Y} = AMY + Bu$$

$$\dot{Y} = M^{-1}AMY + M^{-1}Bu$$

$$M^{-1}AM = \begin{bmatrix} -10 & 0 & 0 \\ 0 & -2 & 0 \\ 0 & 0 & -1 \end{bmatrix}$$

$$\begin{bmatrix} \dot{Y}_1 \\ \dot{Y}_2 \\ \dot{Y}_3 \end{bmatrix} = \begin{bmatrix} -10 & 0 & 0 \\ 0 & -2 & 0 \\ 0 & 0 & -1 \end{bmatrix} Y + \begin{bmatrix} 0 \\ 1.5 \\ 0.5 \end{bmatrix} u$$

$$\begin{bmatrix} \dot{Y}_2 \\ \dot{Y}_3 \end{bmatrix} = \begin{bmatrix} -2 & 0 \\ 0 & -1 \end{bmatrix} \begin{bmatrix} Y_2 \\ Y_3 \end{bmatrix} + \begin{bmatrix} 1.5 \\ 0.5 \end{bmatrix} u$$

Aggregation method

We assume, the aggregation matrix

$$C = \begin{bmatrix} 0.2 & 1 & 0 \\ 0 & 0 & 1 \end{bmatrix}$$

$$Z(t) = CX(t)$$

$$F = CAC^T(CC^T)^{-1}$$

$$= \begin{bmatrix} -2.3 & 0 \\ 0 & -1 \end{bmatrix}$$

$$G = CB$$

$$= \begin{bmatrix} 1.0 \\ 0.5 \end{bmatrix}$$

4.5.1.2.1 Limitations of Aggregation Approach

The above-mentioned aggregation approaches suffer from the following problems:

1. They are iterative processes and the accuracy of the results of the lower-order model depends on initial values of the aggregation matrix.
2. The computation time depends on the size of the aggregation matrix and calculation of the error gradient.
3. The aggregation matrix is always rectangular. Hence, the inverse of the aggregation matrix is not possible. Its pseudo inverse is required to be determined.

4.5.1.3 Subspace Projection Method

An alternative approach to the model order reduction is based on the construction of the best invariant subspace in the state space such that the projection error is minimal. The development of such an approach is attributed to Mitra (1969), and further elaboration is due to Sinha (1975) and Siret (1975). This approach is still further developed by Parkinson (1986). The subspace projection method consists of two stages, as depicted in Figure 4.5 (Mitra, 1969).

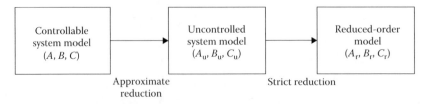

FIGURE 4.5
Stages for model order reduction using subprojection method.

4.5.1.4 Optimal Order Reduction

This group of methods is based on obtaining a model of a specified order such that its impulse or step response (or alternatively its frequency response) matches that of the original system in an optimum manner, with no restriction on the location of the eigenvalues. Such techniques aim at minimizing a selected performance criterion, which in general is a function of the error between the response of the original high-order system and its reduced-order model. The parameters of the reduced-order model are then obtained either from the necessary conditions of optimality or by means of numerical algorithms. Anderson (1967) proposed a geometric approach based on orthogonal projection, to obtain a lower-order model minimizing the integral square error in the time domain. Sinha and Pille (1971) proposed the utilization of the matrix pseudo inverse for least squares fit with the samples of the response. Other criteria for optimization have also been studied by Sinha and Bereznai (1971), and Bandler et al. (1973). Sinha and Bereznai (1971) have suggested the use of the pattern search method of Hooke and Jeeves (1961), whereas Bandler et al. (1973) have proposed gradient methods, which require less computation time but now the gradient of the objective has to be evaluated. Since the dominant part of the expression for the gradient consists of the partial derivatives of the response of the model with respect to its parameters, the additional work required is not excessive. Moreover, this expression is the same for any high-order system and any error criterion. The development of optimal order reduction is attributed to Wilson and Mishra (1979), who studied the approximation for step and impulse responses. Methods for obtaining optimum low-order models in the frequency domain have been proposed by Langholz and Bistritz (1978).

4.5.1.5 Hankel Matrix Approach

It is well known that given the Markov parameters of a system, a minimum realization in the form of the matrices, A, B, and C, can be obtained. A simple procedure is suggested by Rozsa and Sinha (1974) in which one starts with a block Hankel matrix consisting of Markov parameters. Following Ho and Kalman (1965), the block Hankel matrix, S_{ij}, is defined as

$$S_{ij} = \begin{bmatrix} J_0 & J_1 & \cdots & J_{j-1} \\ J_1 & J_2 & \cdots & J_0 \\ \cdots & \cdots & \cdots & \cdots \\ J_{j-1} & J_i & \cdots & J_{i+j-2} \end{bmatrix}$$

where $J_i = CA^iB$, for $i = 0, 1, 2, \ldots$, are called the Markov parameters of the system.

This procedure can be generalized to include the time moments in the block Hankel matrix, which is defined as

$$H_{ij}(k) = \begin{bmatrix} J_{-k} & J_{-k+1} & \cdots & J_{j-1} \\ J_{-k+1} & J_{-K+2} & \cdots & J_{-k+j} \\ \cdots & \cdots & \cdots & \cdots \\ J_{-k+i-1} & J_{-k+i} & \cdots & J_{i+j-2} \end{bmatrix}$$

where $J_{-i} = T_{i-1}$ for $i \geq 1$.

And, therefore, the term "generalized Markov parameters" will be used to include T_i. T_i are related to the time moments of the impulse response matrix through a multiplicative constant.

To determine a low-order approximation to the system, one should obtain the matrices, A_r, B_r, and C_r, so that a number of the generalized Markov parameters of the two systems are identical. The Hankel matrix, S_{ij}, is transformed into the Hermite normal form, H, as given in Rozsa and Sinha (1974). In particular, it must be noted that by matching the time moments we shall be equating the steady state responses to inputs in the form of power series (steps, ramps, parabolic functions, etc.). On the other hand, matching Markov parameters will improve the approximation in the transient portion of the response.

4.5.1.6 Hankel–Norm Model Order Reduction

In this case, the problem of model order reduction is to find a transfer matrix, $G_r(s)$, of degree $r < n$ such that the Hankel norm of the error matrix, $E(s) = G(s) - G_r(s)$, becomes $\|E(s)\| = \|G(S) - G_r(s)\|$, which will be minimized (Glover, 1984).

4.5.2 Model Order Reduction in Frequency Domain

The frequency domain model order reduction methods can be divided into three subgroups.

1. Classical reduction method (CRM): The classical reduction method is based on classical theories of mathematical approximation, such as the CFE and truncation, Pade approximation, and time moment matching. The problems of these methods are instability, non-minimum-phase behavior, and low accuracy.
2. Stability preservation method (SPM): The stability preservation method includes Routh Hurwitz approximation, dominant pole retention, reduction based on differentiation, Mihailov criterion, and factor division. The SPM suffers from a serious drawback of lack of flexibility when the reduced model does not produce a good enough approximation.
3. Stability criterion method (SCM): The stability criterion method is a mixed method in which the denominator of the reduced order model is derived by the SPM, while numerator parameters are obtained by the CRM. This improves the degree of accuracy in the low-frequency range.

4.5.2.1 Pade Approximation Method

The Pade approximation method has a number of advantages, such as computational simplicity, fitting of the initial time moments, and the steady state value of the outputs of the system and the model being the same for an input of the form, $\alpha_i t_i$. An important drawback of the methods using the Pade approximation is that the low-order model obtained may sometimes turn out to be unstable even though the original system is stable.

4.5.2.2 Continued Fraction Expansion

The CFE method was first proposed by Chen and Shieh (1968). They showed that if the CFE of a transfer function was truncated, it led to a low-order model with a step response matching closely that of the original system. The main attraction of this approach was its computational simplicity, as compared with methods described in the earlier categories. Various improvements and extensions of this approach have been presented by Chen and Shieh. Chen has extended the CFE techniques to model the reduction and the design of multivariable control systems. In the formulation of reduced-order models by using the CFE techniques, the formulation of laborious computer-oriented algorithms for expansion into various Cauer's forms and their inversion has been derived. Shieh has given a good approximation in the transient, the steady state, and the overall region of the response curve, respectively. Shieh and Goldman have shown that a mixture of Cauer's first and second forms gives an approximation for both the transient and the steady state responses.

In this method, the CFE of the transfer function is truncated to get a low-order model, with a step response matching closely that of the original system. The main attraction of this approach is its computational simplicity as compared to other methods.

Consider the nth-order system described by a transfer function:

$$G(s) = \frac{a_{21} + a_{22}s + \cdots + a_{2,m+1}s^m}{1 + a_{12}s + a_{13}s^2 + \cdots + a_{1,n+1}s^n}$$

It is desired to find a lower-order transfer function for this system. The power series expansion of e^{-st} about $s = 0$ results in

$$G(s) = \int_0^\infty g(t)e^{-st}\,dt$$

4.5.2.3 Moment-Matching Method

Another approach to model order reduction is based on matching certain time moments of the impulse response of the original system with those of the reduced-order model. The moment-matching techniques are based on equating a few lower-order moments of the model to those of the original system and no consideration is given to the remaining higher-order moments. This would preserve the low-frequency responses of the system, $G(s)$, while the transient response would not be so accurate. The simplification of higher-order systems using moments was first suggested by Paynter (1956). The method of matching time moments proposed by Gibilaro and Lees (1969), and Zakian (1973) is another interesting approach to the problem. It was shown later by Shamash (1974) that these methods are equivalent and can be classified as Pade approximations. Although, initially, these methods were developed for the SISO systems only, it is possible to extend them to the MIMO systems by matching the time or the generalized Markov parameters through partial realization (Hickin and Sinha, 1976). An important drawback of the methods using the Pade approximation is that the low-order models obtained may sometimes turn out to be unstable even if the original system is stable.

4.5.2.4 Balanced Realization-Based Reduction Method

The balanced realization model order reduction method is one of the important methods. Moore proposed a fully analytical method of model order reduction based on the balanced realization method, which drew the attention of many researchers. In this approach, the realization term "balanced" is chosen for the system such that the inputs to the state coupling and the state output coupling are weighted equally, so that those state components that are weakly coupled to both the input and the output are discarded.

Moore proposed that the natural first step in model reduction is to apply the mechanism of minimal realization using an internally balanced model (i.e., in which the controllability and observability grammians are diagonal and equal). The key computational problem is the calculation of balancing transformation and the matrices of the balanced realization. The input–output behavior of the system is not changed too much if the least controllable and/or the least observable part is deleted. This method is based on the simultaneous diagonalization of the controllability and observability grammians. A balancing transform (T_{bal}) is then used to convert the original system (A, B, C, D) to an equivalent "internally balanced" system (A_{bal}, B_{bal}, C_{bal}, D_{bal}). A low-order balanced truncated (BT) model is then obtained by eliminating the least controllable and/or the least observable part of the transformed system. Moore pointed out that the resulting BT model is also internally balanced and retains the dominant second-order modes (the square root of the eigenvalues of the product of the grammians of the original system).

A state space realization, {A, B, C, D}, of $G(s) = C(sI - A)^{-1}B + D$ is said to be the minimal realization of $G(s)$ if "A" has the smallest possible dimension, that is, if no pole zero cancellation in $G(s)$ is going to occur. The state space realization is also minimal if {A, B} is controllable and {C, A} is observable.

A system, G = {A, B, C, D}, can be decomposed into four parts, each characterized by the controllability and observability properties of the system, by means of a suitable coordinate transformation (say, $X_{new} = T^{-1}X_{old}$, to diagonalize the states). This decomposition is shown in Figure 4.6.

In Figure 4.6, c denotes controllability, o denotes observability, $\bar{c}$ denotes uncontrollability, and $\bar{o}$ denotes unobservability of the system. Since only the controllable and observable part of the system, G_{co}, contributes to the system dynamics, it is called the minimal realized model of the original system, G. The other decomposed parts that do not contribute to the system dynamics may be deleted from the transfer function, which does not affect the input–output behavior of the system.

Similarly, it may be concluded that those subsystems that are loosely connected to the input and the output may be deleted from the transfer function, and the resulting subsystem will approximate the input–output behavior of the original system. This concept of controllability and observability leads to model reduction. One of the analytical tools to remove the weakly controllable and/or observable part from the original system is Moore's balanced realization theory.

Controllability and observability grammians

For a continuous time system, G = {A, B, C, D}, the following two continuous time Lyapunov equations will be satisfied:

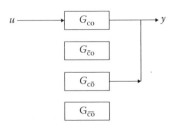

FIGURE 4.6
Controllability and observability decomposition of a system.

Model Order Reduction

$$AP + PA' + BB' = 0, \quad P = \int_0^\infty e^{\tau A} BB' e^{\tau A'} d\tau$$

$$A'Q + QA + C'C = 0, \quad Q = \int_0^\infty e^{\tau A'} C'C e^{\tau A} d\tau$$

Balanced realized models

Let P and Q be the controllability and observability grammians of the nth-order stable minimal system, $G = \{A, B, C, D\}$, which has a transfer function given by

$$G(s) = C(sI - A)^{-1} B + D$$

For the system, G, it is to be noted that the grammians are not balanced, that is, P is not equal to Q.

Let T_{bal} be the similarity transformation for balancing the system. Then the transformed system becomes

$$G_{bal} = \begin{bmatrix} A_{bal} & B_{bal} \\ C_{bal} & D_{bal} \end{bmatrix} = \begin{bmatrix} T_{bal}^{-1} A T_{bal} & T_{bal}^{-1} B \\ C T_{bal} & D \end{bmatrix}$$

Let the balanced grammians be $\Sigma = P_{bal} = Q_{bal} = (\Sigma_1, \Sigma_2)$

$$P_{bal} = T_{bal}^{-1} P (T_{bal}^{-1})' \in R^{n \times n}$$

$$Q_{bal} = T_{bal}' Q T_{bal} \in R^{n \times n}$$

Then, $\Sigma = \text{diag}(\sigma_1, \sigma_2, \ldots, \sigma_m, \sigma_{m+1}, \ldots, \sigma_n) = \text{diag}(\Sigma_1, \Sigma_2)$
where

$$\sigma_1 \geq \sigma_2, \ldots, \sigma_m \geq \sigma_{m+1}, \ldots, \geq \sigma_n$$

$$\sigma_i = \sqrt{[\lambda_i (PQ)]}$$

are the Hankel singular values of the second-order modes of the system.

The full-order balanced realized model of the system is the combination of a strong and a weak subsystem, which is given by

$$G_{bal} = \begin{bmatrix} A_{bal} & B_{bal} \\ C_{bal} & D_{bal} \end{bmatrix} = \begin{bmatrix} A_{11} & A_{12} & B_1 \\ A_{21} & A_{22} & B_2 \\ C_1 & C_2 & D \end{bmatrix} \begin{matrix} \updownarrow r \\ \updownarrow n-r \end{matrix}$$

The BT model of G of the order r is

$$G_{r\_bt} = \begin{bmatrix} A_{11} & B_1 \\ C_1 & D \end{bmatrix} \updownarrow r$$

The BT model does not preserve the DC gain of the original system.

To preserve the DC gain, the balanced residualized model or the balanced SPA (BSPA) model for a minimal system is given by

$$G_{r\_dc} = \begin{bmatrix} A_{11} - A_{12}(-\phi I + A_{22})^{-1}A_{21} & B_1 - A_{12}(-\phi I + A_{22})^{-1}B_2 \\ C_1 - C_2(-\phi I + A_{22})^{-1}A_{21} & D - C_2(-\phi I + A_{22})^{-1}B_2 \end{bmatrix}$$

where

$\Sigma = 0$ for a continuous system
$\Sigma = 1$ for a discrete system
I is an identity matrix

Properties of balanced realized models

Stability properties of subsystems:

1. Assume that Σ_1 and Σ_2 have no diagonal entries in common, then both subsystems, $\{A_i, B_i, C_i, D_i\} = 1, 2$, are asymptotically stable.
2. The reciprocal system, $\{\bar{A}, \bar{B}, \bar{C}, \bar{D}\}$ of $\{A_{bal}, B_{bal}, C_{bal}, D_{bal}\}$, given by $\bar{A} = A_{bal}^{-1}$, is $\bar{B} = B_{bal}^{-1}$, $\bar{C} = C_{bal}^{-1}$, and $\bar{D} = D_{bal}^{-1}$ also stable and balanced.

4.5.2.5 Balanced Truncation

The main idea in balanced truncation is to transform the original stable LTI system into a representation of the same size, but with the property that those states of the transformed system that are uncontrollable are also unobservable and vice versa. In this form, the system is said to be "balanced." Once this is done, one simply discards those states of the transformed system that are the least controllable (and hence the least observable), which gives rise to the term "truncation." Thus, after truncation, one has an LTI model with (typically) far fewer states than the original model.

The method of balanced truncation for model reduction of linear systems was proposed by Moore (1981) in the context of the realization theory, and is now a well-developed method of model reduction that appears in standard textbooks (Dullerud and Paganini, 2000). For linear systems, the approach requires only matrix computations, and has been very successfully used in control design. A priori error bounds in the induced two-norm are known for the error between the original and the reduced system (Glover, 1984; Enns, 1984).

The method of balanced truncation has been extensively developed for nonlinear systems by Scherpen (1993, 1996), and Scherpen and van der Schaft (1994), based on energy functions. A closed-loop approach is also presented in Pavel and Fairman (1997), and balancing methods for bilinear systems have also been developed (Al-Baiyat and Bettayeb, 1993; Al-Baiyat et al., 1994; Gray and Mesko, 1998). Computational approaches for these

Model Order Reduction

methods have been investigated, and compared with the Karhunen–Loe've method in Newman and Krishnaprasad (1998a,b).

Another important feature of a balanced realization is that the state coordinate basis is selected such that the controllability and observability grammians are both equal to some diagonal matrix, Σ, normally with the diagonal entries of Σ in the descending order. The state space representation is then called a balanced realization. The magnitudes of the diagonal entries reflect the contributions of different entries of the state vector to system responses. The state vector entries that contribute the least are associated with the smallest σ_i. Thus, we can use truncation to eliminate the unimportant state variables from the realization. The following two Lyapunov equations give the relation to the system matrices, A, B, and C, for a balanced realization:

$$A\Sigma + \Sigma A^T + BB^T = 0$$
$$\Sigma A + A^T \Sigma + C^T C = 0 \tag{4.11}$$

Hence, $\Sigma = \mathrm{diag}\,[\sigma_i]$ and $\sigma_1 > \sigma_2 > \cdots > \sigma_n$. The σ_i are termed the Hankel singular values of the original system.

Let P and Q be, respectively, the controllability and observability grammians associated with an arbitrary minimal realization, $\{A, B, C\}$, of a stable transfer function, respectively. Since P and Q are symmetric, there exist orthogonal transformations, U_c and U_o, such that

$$P = U_c S_c U_T$$

$$Q = U_o S_o U_T$$

where S_c and S_o are diagonal matrices. The matrix

$$H = S_O^{1/2} U_O^T U_C S_C^{1/2}$$

is constructed and a singular value decomposition is obtained from it:

$$H = U_H S_H V_H^T$$

Using these matrices, the balancing transformation is given by

$$T = U_O^{1/2} S_O^{-1/2} U_H S_H^{1/2}$$

The balanced realization is

$$\begin{bmatrix} A & B \\ C & D \end{bmatrix} = \begin{bmatrix} T^{-1}AT & T^{-1}B \\ TC & D \end{bmatrix}$$

Through simple manipulations, it can be confirmed that

$$T^T Q T = T^{-1} P (T^T)^{-1} = S_H$$

Consider a stable system, $G \in RH\infty$, and suppose that

$$G = \begin{bmatrix} \bar{A} & \bar{B} \\ \bar{C} & D \end{bmatrix}$$

is a balanced realization. Denoting the balanced grammians by Σ, we have

$$\bar{A}\Sigma + \Sigma \bar{A}^* + \bar{B}\bar{B}^* = 0$$
$$\Sigma \bar{A} + \bar{A}^* \Sigma + \bar{C}^*\bar{C} = 0 \tag{4.12}$$

Now, partition the balanced grammian as

$$\Sigma = \begin{bmatrix} \Sigma_1 & 0 \\ 0 & \Sigma_2 \end{bmatrix}$$

and also partition the system accordingly:

$$G = \left[\begin{array}{cc|c} A_{11} & A_{12} & B_1 \\ A_{21} & A_{22} & B_2 \\ \hline C_1 & C_2 & D \end{array} \right]$$

Then Equation 4.12 can be written in terms of their partitioned matrices as

$$\bar{A}_{11}\Sigma_1 + \Sigma_1 \bar{A}_{11}^* + \bar{B}_1\bar{B}_1^* = 0$$
$$\Sigma_1 \bar{A}_{11} + \bar{A}_{11}^* \Sigma_1 + \bar{C}_1^*\bar{C}_1 = 0 \tag{4.13}$$

$$\bar{A}_{21}\Sigma_1 + \Sigma_2 \bar{A}_{12}^* + \bar{B}_2\bar{B}_1^* = 0$$
$$\Sigma_2 \bar{A}_{21} + \bar{A}_{12}^* \Sigma_1 + \bar{C}_2^*\bar{C}_1 = 0 \tag{4.14}$$

$$\bar{A}_{22}\Sigma_2 + \Sigma_2 \bar{A}_{22}^* + \bar{B}_2\bar{B}_2^* = 0$$
$$\Sigma_2 \bar{A}_{22} + \bar{A}_{22}^* \Sigma_2 + \bar{C}_2^*\bar{C}_2 = 0 \tag{4.15}$$

By virtue of the method adopted to construct Σ, the most energetic modes of the system are in Σ_1 and the less energetic ones are in Σ_2. Thus, the system with Σ_1 is chosen as its balanced grammian would be a good approximation of the original system. Thus, the procedure to obtain a reduced-order model would be as follows:

1. Obtain the balanced realization of the system.
2. Choose an appropriate order, r, of the reduced-order model. Partition the system matrices accordingly.

Model Order Reduction

Thus, the reduced-order model is obtained as $G_r = \begin{bmatrix} \bar{A}_{11} & \bar{B}_1 \\ C_1 & D \end{bmatrix}$.

Properties of truncated systems:

Lemma 4.1: Suppose X is the solution of the Lyapunov equation:

$$A^*X + XA + Q = 0 \tag{4.16}$$

Then,

1. $\text{Re}(\lambda_i(A)) < 0$ if $X > 0$ and $Q > 0$.
2. A is stable if $X > 0$ and $Q > 0$.
3. A is stable if $X > 0$, $Q > 0$, and (Q, A) is detectable.

Proof Let λ be an eigenvalue of A and $v \neq 0$ be a corresponding eigenvector, then $Av = \lambda v$. Pre-multiply (3) by v^* and postmultiply it by v to get

$$2\,\text{Re}(\lambda(v^*Xv)) + v^*Qv = 0$$

Now, if $X > 0$ and $v^*Xv > 0$, it is clear that $\text{Re}(\lambda) < 0$ if $Q > 0$ and $\text{Re}(\lambda) < 0$ if $Q > 0$. Hence, (1) and (2) hold. To see (3), we assume $\text{Re}(\lambda) > 0$. Then, we must have $v^*Qv = 0$, that is, $Qv = 0$. This implies that λ is an unstable and unobservable mode, which contradicts the assumption that (Q, A) is detectable.

Lemma 4.2: Consider the Sylvester equation:

$$AX + XB = C \tag{4.17}$$

where $A \in F^{n \times n}$, $B \in F^{m \times m}$, and $C \in F^{n \times m}$ are given matrices. There exists a unique solution, $X \in F^{n \times m}$, if and only if $\lambda_i(A) + \lambda_j(B) \neq 0;\ i = 1, 2, \ldots, n,\ j = 1, 2, \ldots, m$.

Proof Equation 4.17 can be written as a linear matrix equation by using the Kronecker product:

$$(B^T \oplus A)\text{vec}(X) = \text{vec}(C)$$

Now that equation has a unique solution if and only if $B^T \oplus A$ is nonsingular. Since the eigenvalues of $B^T \oplus A$ have the form: $\lambda_i(A) + \lambda_j(B^T) = \lambda_i(A) + \lambda_j(B)$, the conclusion follows.

Theorem 4.1: Assume that Σ_1 and Σ_2 have no diagonal entries in common. Then both subsystems $(\mathbf{A}_{ii}, \mathbf{B}_i, \mathbf{C}_i)$, $i = 1, 2$, are asymptotically stable.

Proof It is sufficient to show that A_{11} is asymptotically stable. The proof for the stability of A_{22} is similar.

Since Σ is a balanced realization, by the properties of SVD, Σ_1 can be assumed to be positive definite without a loss of generality. Then, it is obvious that $\lambda_i(A_{11}) \leq 0$ by the lemma.

1. Assume that A_{11} is not asymptotically stable; then there exists an eigenvalue at $j\omega$ for some ω. Let V be a basis matrix for $\ker(A_{11} - j\omega I)$. Then,

$$(A_{11} - J\omega I)V = 0$$

which gives

$$V^*(A_{11}^* + J\omega I) = 0$$

Adding and subtracting $j\omega \Sigma_1$, equations (A and B) can be rewritten as

$$(\bar{A}_{11} - j\omega I)\Sigma_1 + \Sigma_1(\bar{A}_{11}^* + j\omega I) + \bar{B}_1 \bar{B}_1^* = 0$$

$$\Sigma_1(\bar{A}_{11} - j\omega I) + (\bar{A}_{11}^* + j\omega I)\Sigma_1 + \bar{C}_1^* = 0 \qquad (4.18)$$

Now, first multiply Equation 4.18 from the right by $\Sigma_1 V$ and from the left by $V^* \Sigma_1$ to obtain

$$B_1^* \Sigma_1 V = 0$$

Then multiply Equation 4.18 from the right by $\Sigma_1 V$ to get

$$(A_{11}^* - J\omega I)\Sigma_1^2 V = 0$$

It follows that the columns of $\Sigma_1^2 V$ are in $\ker(A_{11}^* - J\omega I)$. Therefore, there exists a matrix, $\bar{\Sigma}_1$, such that

$$\Sigma_1^2 V = V \bar{\Sigma}_1^2$$

Since $\bar{\Sigma}_1^2$ is the restriction of Σ_1^2 to the space spanned by V, it follows that it is possible to choose V such that $\bar{\Sigma}_1^2$ is diagonal. It is then also possible to choose $\bar{\Sigma}_1$ diagonal, such that the diagonal entries of $\bar{\Sigma}_1$ are a subset of the diagonal entries of Σ_1.

Further,

$$(\bar{A}_{21} V)\bar{\Sigma}_1^2 = \Sigma_1^2(\bar{A}_{21}V)$$

This is a Sylvester equation in $A_{21}V$. Because Σ_1^2 and $\bar{\Sigma}_2^2$ have no diagonal entries in common, it follows from Lemma 2 that

$$\bar{A}_{21} V = 0$$

Model Order Reduction

is the unique solution. Now, it can be easily seen that

$$\begin{bmatrix} A_{11} & A_{12} \\ A_{21} & A_{22} \end{bmatrix} \begin{bmatrix} V \\ 0 \end{bmatrix} = j\omega \begin{bmatrix} V \\ 0 \end{bmatrix}$$

which means that the $\bar{A}$ matrix of the original system has an eigenvalue at $j\omega$. This contradicts the fact that the original system is asymptotically stable. Therefore, $\bar{A}_{11}$ must be asymptotically stable.

Reduction of unstable systems by balanced truncation

Unstable systems cannot be directly reduced using balanced truncation. The original system is transformed through Schur transformations into

$$\begin{bmatrix} \dot{x}_1 \\ \dot{x}_2 \end{bmatrix} = \begin{bmatrix} A_- & A_c \\ 0 & A_+ \end{bmatrix} \begin{bmatrix} x_1 \\ x_2 \end{bmatrix} + \begin{bmatrix} B_- \\ B_+ \end{bmatrix} U$$

$$y = \begin{bmatrix} C_- & C_+ \end{bmatrix} \begin{bmatrix} x_1 \\ x_2 \end{bmatrix} + DU$$

where
  A_- has all its eigenvalues stable
  A_+ has all its eigenvalues unstable

This system is then converted to two systems, S_1 and S_2, with

$$S_2 \Rightarrow \dot{x}_2 = [A_+]x_2 + [B_+]U$$

$$\dot{x}_1 = [A_-]x_1 + [B_- \quad A_c]\begin{bmatrix} U \\ x_2 \end{bmatrix}$$

$$S_1 \Rightarrow y = [C_-]x_1 + [D \quad C_+]\begin{bmatrix} U \\ x_2 \end{bmatrix}$$

Now, using the above described methods, a reduced-order model can be obtained for the stable system, S_1. Let the reduced system obtained for S_1 be represented as S_1':

$$S_1' \Rightarrow \dot{z} = [A_r]z + [B_{ru} \quad B_{rx_2}]\begin{bmatrix} U \\ x_2 \end{bmatrix}$$

$$y = [C_r]z + [D_{ru} \quad D_{rx_2}]\begin{bmatrix} U \\ x_2 \end{bmatrix}$$

The reduced-order model for the original system can then be formulated as

$$\begin{bmatrix} \dot{z} \\ \dot{x}_2 \end{bmatrix} = \begin{bmatrix} A_r & B_{rx2} \\ 0 & A_+ \end{bmatrix} \begin{bmatrix} z \\ x_2 \end{bmatrix} + \begin{bmatrix} B_{rU} \\ B_+ \end{bmatrix} U$$

$$y = \begin{bmatrix} C_r & D_{rx2} \end{bmatrix} \begin{bmatrix} z \\ x_2 \end{bmatrix} + [D_{rU}]U$$

Thus, even unstable systems can be reduced using balanced truncation.

4.5.2.6 Frequency-Weighted Balanced Model Reduction

Given the original full-order model, $G \in RH\infty$, the input weighting matrix, $W_i \in RH\infty$, and the output weighting matrix, $W_o \in RH\infty$, our objective is to find a lower-order model, G_r, such that

$$\|W_o(G - G_r)W_i\|_\infty$$

is made as small as possible. Assume that G, W_i, and W_o have the following state space realizations:

$$G = \begin{bmatrix} A & B \\ C & 0 \end{bmatrix}, \quad W_i = \begin{bmatrix} A_i & B_i \\ C_i & D_i \end{bmatrix}, \quad W_o = \begin{bmatrix} A_o & B_o \\ C_o & D_o \end{bmatrix}$$

with $A \in R^{n \times n}$. Note that there is no loss of generality in assuming $D = G(\infty) = 0$, since, otherwise, it can be eliminated by replacing G_r with $D + G_r$.

Now the state space realization for the weighted transfer matrix is given by

$$W_o G W_i = \begin{bmatrix} A & 0 & BC_i & BD_i \\ B_o C & A_o & 0 & 0 \\ 0 & 0 & A_i & B_i \\ D_o C & C_o & 0 & 0 \end{bmatrix} = \begin{bmatrix} \hat{A} & \hat{B} \\ \hat{C} & 0 \end{bmatrix}$$

Let $\bar{P}$ and $\bar{Q}$ be the solutions of the following Lyapunov equations:

$$\hat{A}\bar{P} + \bar{P}\hat{A}^* + \hat{B}\hat{B}^* = 0$$

$$\bar{Q}\hat{A} + \hat{A}^*\bar{Q} + \hat{C}^*\hat{C} = 0$$

Then the input weighted grammian, $\hat{P}$, and the output weighted grammian, $\hat{Q}$, are defined by

$$\hat{P} : \begin{bmatrix} I_n & 0 \end{bmatrix} \bar{P} \begin{bmatrix} I_n \\ 0 \end{bmatrix}$$

$$\hat{Q} : \begin{bmatrix} I_n & 0 \end{bmatrix} \bar{Q} \begin{bmatrix} I_n \\ 0 \end{bmatrix}$$

It can be shown easily that $\hat{P}$ and $\hat{Q}$ can satisfy the following lower-order equations:

$$\begin{bmatrix} A & BC_i \\ 0 & A_i \end{bmatrix} \begin{bmatrix} \hat{P} & \hat{P}_{12} \\ \hat{P}_{12}^* & \hat{P}_{22} \end{bmatrix} + \begin{bmatrix} \hat{P} & \hat{P}_{12} \\ \hat{P}_{12}^* & \hat{P}_{22} \end{bmatrix} \begin{bmatrix} A & BC_i \\ 0 & A_i \end{bmatrix}^* + \begin{bmatrix} BD_i \\ B_i \end{bmatrix} \begin{bmatrix} BD_i \\ B_i \end{bmatrix}^* = 0$$

$$\begin{bmatrix} \hat{Q} & \hat{Q}_{12} \\ \hat{Q}_{12}^* & \hat{Q}_{22} \end{bmatrix} \begin{bmatrix} A & 0 \\ B_0 C & A_0 \end{bmatrix} + \begin{bmatrix} A & 0 \\ B_0 C & A_0 \end{bmatrix}^* \begin{bmatrix} \hat{Q} & \hat{Q}_{12} \\ \hat{Q}_{12}^* & \hat{Q}_{22} \end{bmatrix} + \begin{bmatrix} C^* D_0^* \\ C_0^* \end{bmatrix} \begin{bmatrix} C^* D_0^* \\ C_0^* \end{bmatrix}^* = 0$$

The computation can be further reduced if $W_i = I$ or $W_o = I$. In the case of $W_i = I$, P can be obtained from

$$\hat{P} A^* + A \hat{P} + BB^* = 0$$

while in the case of $W_o = I$, Q can be obtained from

$$\hat{Q} A^* + A^* \hat{Q} + C^* C = 0$$

Now, let T be a nonsingular matrix such that

$$T P T^* = (T^{-1})^* Q T^{-1} = \begin{bmatrix} \Sigma_1 & \\ & \Sigma_2 \end{bmatrix}$$

(i.e., balanced) with

$$\Sigma_1 = \text{diag}(\sigma_1 I_{s1}, \sigma_2 I_{s2}, \ldots, \sigma_r I_{sr})$$

$$\Sigma_2 = \text{diag}(\sigma_{r+1} I_{s1r+1}, \sigma_{r+2} I_{s1r+2}, \ldots, \sigma_n I_{sn})$$

and partition the system accordingly as

$$\begin{bmatrix} TAT^{-1} & TB \\ T^{-1}C & 0 \end{bmatrix} = \begin{bmatrix} A_{11} & A_{12} & B_1 \\ A_{21} & A_{22} & B_2 \\ \hline C_1 & C_2 & 0 \end{bmatrix}$$

Then the reduced-order model, G_r, is obtained as

$$G_r = \begin{bmatrix} A_{11} & B_1 \\ \hline C_1 & 0 \end{bmatrix}$$

Empirical balanced truncation

The empirical grammians give a quantitative method for deciding upon the importance of particular subspaces of the state space, with respect to the inputs and outputs of the system. We propose to use these for model reduction of nonlinear systems in the same way as for linear systems, find a linear change of coordinates such that the empirical grammians are balanced, and perform a Galerkin projection onto the states corresponding to the largest eigenvalues.

Since, for linear systems, the empirical grammians are exactly the usual grammians, the method is exactly the balanced truncation method when it is applied to a linear system. When applied to a nonlinear system, it requires only matrix computations, and results in a new nonlinear model.

The empirical balanced truncation gives a reduced-order model, which takes into account the input–output behavior and is directly computable from data. For linear systems, the Hankel singular values are unaffected by coordinate changes, even though the grammians matrices themselves are not coordinate invariant. For nonlinear systems, this property no longer holds.

Steady state matching

The algorithm discussed above has a basic disadvantage. Though the responses of the reduced system get closer and closer to the original system, there is no certainty that they would match at steady state. This is because of the upper bound on the approximation error that is dependent on the ignored Hankel singular values.

This steady state error occurs because the ignored states, even if not contributing much to the dynamics of the system, do contribute to its steady state. Hence, they should be considered while deriving the reduced-order model without the steady state error. This steady state error can be eliminated by modifying the reduced-order model using the concept of singular perturbations.

An example is given by the linear-parameter-varying (LPV) control design (Wu et al., 1996), in which the required optimization for control synthesis involves solving a linear matrix inequality feasibility problem where the number of variables grows as the square of the state-dimension of the system. A significant reduction in the overall computational effort required is achieved by constructing a reduced-order model via simulation of the full-order model, and then using this reduced-order model for control synthesis.

Example 4.2: Mechanical Links

Mechanical systems not only are high-order systems and hard to control, but it is difficult to develop intuition as to their behavior. One of the advantages of the procedure developed here is that it can be viewed as a selection of appropriate mode shapes on which to project the nonlinear dynamics. These mode shapes are often physically meaningful, and in this section we give examples for a simple, mildly nonlinear mechanical system.

The system is shown in Figure 4.7. It consists of five uniform rigid rods in two dimensions, connected via torsional

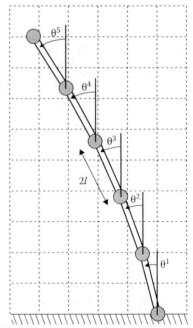

FIGURE 4.7
Mechanical system.

Model Order Reduction

springs and dampers. The lowest rod is pinned to the ground with a torsional spring, so that the system has a stable equilibrium in the upright vertical position. There is no gravity. The system has a single input, a torque about the lowest pin joint, and a single output, the horizontal displacement of the end of the last rod from the vertical symmetry axis.

The potential energy, V, of this system is

$$V = \frac{1}{2}k(\theta^1)^2 + \frac{1}{2}k\sum_{i=2}^{n}(\theta^i - \theta^{i-1})^2$$

and the kinetic energy, T, is

$$T = \frac{1}{2}\sum_{i=1}^{n}\frac{ml^2}{3}\left(\dot{\theta}^i\right)^2 + \frac{1}{2}\sum_{i=1}^{n}m\left(\left(\dot{x}^i\right)^2 + \left(\dot{y}^i\right)^2\right)$$

where x^i and y^i are the Cartesian coordinates of the center of mass of the ith rod, given by

$$x_i = \begin{cases} l\sin\theta^i & \text{if } i = 1 \\ -l\sin\theta^i - 2l\sum_{p=1}^{i-1}\sin\theta^p & \text{if } i = 2,\ldots,n \end{cases}$$

$$y_i = \begin{cases} l\cos\theta^i & \text{if } i = 1 \\ l\cos\theta^i + 2l\sum_{p=1}^{i-1}\cos\theta^p & \text{if } i = 2,\ldots,n \end{cases}$$

The Lagrangian $L = T - V$, and the equations of motion are then

$$\frac{d}{dt}\frac{\partial L}{\partial \dot{\theta}^i} - \frac{\partial L}{\partial \theta^i} = F_i$$

where F is the force term containing dissipative forces and the external force term, w:

$$F_i = \begin{cases} -b\dot{\theta}^i + w & \text{if } i = 1 \\ -b(\dot{\theta}^i - \dot{\theta}^{i-1}) & \text{if } i = 2,\ldots,n \end{cases}$$

The measurement equation is

$$z = h(\theta, \dot{\theta}) := -2l\sum_{i=1}^{n}\sin\theta^i$$

The constants are given by $b = 0.5$, $k = 3$, $m = 1$, $l = 1$, and $n = 5$.

Linearized model
We first analyze the linearization of the system about its stable equilibrium. Figure 4.8 shows the configuration parts of the first four of the ten mode shapes and the corresponding singular values of the balanced realization. The singular values have been normalized so that they sum up to one.

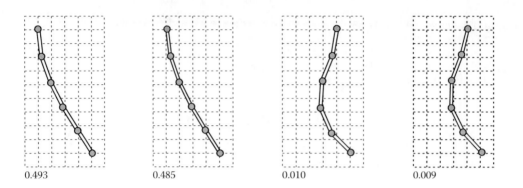

FIGURE 4.8
Balanced modes of linearization, with the singular value of each mode. (From Lall, S. et al., *Int. J. Robust Nonlinear Control*, 12, 519, 2002. With permission.)

Empirical balancing

The application of the empirical model reduction procedure to the nonlinear Lagrangian model of this system leads to a set of corresponding modes. In this case, Figure 4.9 shows these modes; we can see that the first two mode shapes have split into three.

Since this system is mechanical, the dynamics have a Lagrangian structure. In the absence of forcing and dissipation, Lagrangian systems conserve energy as well as quantities associated with the symmetries of the system. The dynamics of a mechanical system also satisfy a variational principle, and the evolution maps are simplistic transformations. All of these properties can be viewed as fundamental to a model of a mechanical system. The importance of this is evidenced in Figure 4.8, where each mode shape appears twice; this is a consequence of the underlying correspondence between configuration variables and their generalized moment. These repeated structures are captured by the linearized method, and the method in this chapter should be improved to take account of this repeated structure. One way to achieve this would be by combining the techniques and constructing the empirical grammians on the configuration space rather than the phase space. In this way, the underlying geometric structure is preserved by the reduction, and the resulting reduced-order system is itself a Lagrangian system. Similarly, taking account of symmetry (Glavaski et al., 1998; Rowley and Marsden, 2000) and the other special structure present in the original system can be of great use, leading to reduced computational requirements and producing more accurate reduced-order models.

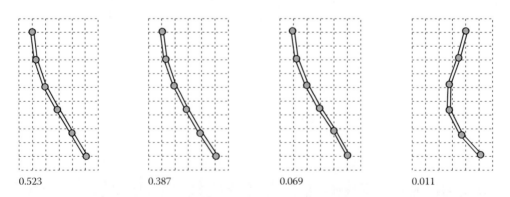

FIGURE 4.9
Empirical balanced modes of the nonlinear model, for $c_i = 0.4$. (From Lall, S. et al., *Int. J. Robust Nonlinear Control*, 12, 519, 2002. With permission.)

Model Order Reduction

Note that the method of balanced truncation is not optimal, in that, in general, it does not achieve the minimal possible error in any known norm even for linear systems. However, it is a well-used method in control, which is both intuitively motivated by the realization theory and known to perform well in engineering practice. The intended use of the methods in this chapter is to deliberately construct approximate reduced-order models motivated by the ideas of balanced truncation.

4.5.2.7 Time Moment Matching

This method is based on matching the time moments of the impulse response of the original model with those of the impulse response of the reduced-order model. The number of matched time moments determines in turn the order of the simplified/reduced model.

Consider the nth-order system described by the following transfer function:

$$G(s) = \frac{a_0 + a_1 s + a_2 s^2 + \cdots + a_m s^m}{b_0 + b_1 s + b_2 s^2 + \cdots + b_n s^n} \quad m < n \tag{4.19}$$

Dividing the numerator and the denominator of Equation 4.19 with b_0, we get

$$G(s) = \frac{a_{21} + a_{22} s + a_{23} s^2 + \cdots + a_{2,(m+1)} s^m}{1 + a_{12} s + a_{13} s^2 + \cdots + b_{1,(n+1)} s^n} \tag{4.20}$$

It is desired to find a lower-order transfer function for this system. The power series expansion of e^{-st} about $s = 0$ results in

$$G(s) = \int_0^\infty g(t) e^{-st} dt$$

$$= \int_0^\infty g(t)\{1 - st + (st)^2 + \cdots\} dt$$

$$= \int_0^\infty g(t) dt - s \int_0^\infty g(t) t \, dt + s^2 \int_0^\infty t^2 g(t) dt + \cdots$$

$$= c_0 + c_1 s + c_2 s^2 + \cdots$$

$$= \sum_{i=0}^\infty c_i s^i \tag{4.21}$$

where c_0 is the area under the given function, $g(t)$, and is known as the zeroth moment of the transient response. If $g(t)$ is normalized, then $c_1 = 1$, and c_1 represents the center of gravity, while $2c_2$ is the moment of inertia of the response about $t = 0$.

It can be shown that

$$G(S) = \sum_{i=0}^\infty M_i s^{-i} = \sum_{i=0}^\infty CA^{i-1}B \quad \text{about } s = \infty$$

where M_i ($i = 1, 2, 3\ldots$) are called the Markov parameters of the system, j.

$$G(s) = \sum_{i=0}^{\infty} c_i s^i = -\sum_{i=0}^{\infty} CA^{-(i+1)}B \quad \text{about } s = 0$$

TABLE 4.1
Routh Array

a_{11}	a_{12}	a_{13}	a_{14}	...
a_{21}	a_{22}	a_{23}		...
a_{31}	a_{32}	...		
a_{41}				

From this we may get

$$G(s) = a_{21} - a_{31}s + a_{41}s^2 - a_{51}s^3 + \cdots \quad (4.22)$$

where a_{21} is the zeros term coefficient of the numerator and the remaining coefficients are obtained from the Routh array, by using the following recursion:

$$a_{k,v} = a_{k-1,1}a_{1,v+1} - a_{k-1,v+1} \quad (4.23)$$

for $k = 3, 4, \ldots, 1n + 1$ and $v = 1, 2, \ldots, n$. Note that once a Routh array is formed based on the above Table 4.1, the jth moment can be obtained by dividing $a_{j+1,1}$, by j for $j = 0, 1, 2, \ldots$. The expansion coefficient, c_j, is then obtained by $c_j = (-1)^j a_{j+2}$.

From Equations 4.7, 4.12, and 4.13, the following equation is obtained:

$$\begin{bmatrix} c_0 \\ c_1 \\ c_2 \\ \cdots \\ c_m \\ c_{m+1} \\ c_{m+2} \\ \cdots \\ c_{m+n} \end{bmatrix} = \begin{bmatrix} 0 & 0 & 0 & 0 & 0 & 0 & 0 & 0 & 0 \\ -c_0 & 0 & & & & & & & \\ -c_1 & & & & & & & & \\ \cdots & & & & & & & & \\ -c_{m-1} & -c_{m-2} & \cdots & -c_0 & & & & & \\ -c_m & -c_{m-1} & \cdots & -c_1 & -c_0 & \cdots & 0 & 0 & 0 \\ -c_{m+1} & -c_m & \cdots & \cdots & c_m & -c_0 & 0 & 0 & 0 \\ \cdots & \cdots & \cdots & \cdots & \cdots & \cdots & \cdots & \cdots & \cdots \\ -c_{m+n-1} & -c_{m+n-2} & 0 & \cdots & \cdots & -c_1 & -c_0 & \cdots & 0 \end{bmatrix} \begin{bmatrix} a_{12} \\ a_{13} \\ a_{14} \\ \cdots \\ a_{1,n+1} \\ 0 \\ 0 \\ \cdots \\ 0 \end{bmatrix} + \begin{bmatrix} a_{21} \\ a_{22} \\ a_{23} \\ \cdots \\ a_{2,m+1} \\ 0 \\ 0 \\ \cdots \\ 0 \end{bmatrix}$$

It is an $(n + m + 1)$-dimensional vector relation, which can be rewritten as

$$\begin{bmatrix} \hat{c}_1 \\ \cdots \\ \hat{c}_2 \end{bmatrix} = \begin{bmatrix} C_{11} & | & 0 \\ \cdots & \cdots & \cdots \\ C_{21} & | & C_{22} \end{bmatrix} \begin{bmatrix} \hat{a}_1 \\ \cdots \\ 0 \end{bmatrix} + \begin{bmatrix} \hat{a}_2 \\ \cdots \\ 0 \end{bmatrix} \quad (4.24)$$

where C_{11}, C_{12}, and C_{22} are $(m + 1) \times n$, $n \times n$, and $n \times (m + 1)$ matrices, respectively, and $\hat{c}_i$, $\hat{a}_i = 1, 2$ are vectors of the $(m + 1)$th and the nth dimension defined by Equation 4.23. Partitioning the set of two equations and solving for $\hat{a}_1$ and $\hat{a}_2$, one gets

$$\hat{a}_1 = C_{21}^{-1}\hat{c}_2, \quad \hat{a}_2 = \hat{c}_1 - C_{11}\hat{a}_1 = \hat{c}_1 - C_{11}C_{21}^{-1}\hat{c}_2 \quad (4.25)$$

Model Order Reduction

Once the moments, c_j, $j = 0, 1, 2$, are determined, the coefficient matrix, C, of Equation 4.24 is defined, and $\hat{a}_1$ and $\hat{a}_2$ are obtained from Equation 4.25, the submatrix, C_{21}, is normally nonsingular and its singularity means that the given set of moments can be matched by a simpler model.

Example 4.3

Consider a third-order asymptotically stable system with the following transfer function:

$$G(s) = \frac{s^2 + 13s + 40}{s^3 + 13s^2 + 32s + 20}$$

$$= \frac{2 + 0.65s + 0.05s^2}{1 + 1.6s + 0.65s^2 + 0.05s^3}$$

Determine the second-order system using the moment-matching method.

Solution

First, a Routh array is constructed:

1	1.6	0.65	0.05
2	0.65	0.05	
2.55	1.25	0.1	
2.83	1.5575		
2.9705			

The elements in the Routh array may be calculated from $a_{k,v} = a_{k-1,1} a_{1,v+1} - a_{k-1,v+1}$

$a_{31} = 2 \times 1.6 - 0.65 = 2.55$
$a_{32} = 2 \times 0.65 - 0.05 = 1.25$
$a_{33} = 2 \times 0.05 - 0 = 0.1$
$a_{41} = 2.55 \times 1.6 - 1.25 = 2.83$
$a_{42} = 2.55 \times 0.65 - 0.1 = 1.5575$
$a_{51} = 2.83 \times 1.6 - 1.5575 = 2.9705$

which indicates that the first few expansion coefficients, c_j, for $j = 0, 1, 2$ are

$c_0 = 2$, $c_1 = -2.55$, $c_2 = 2.83$, $c_3 = -2.9705$, etc.

The denominator and numerator coefficients of the reduced-order model are obtained from Equation 4.25, that is

$$\hat{a}_1 = C_{12}^{-1}\, \hat{c}_2 = \begin{bmatrix} -c_1 & -c_0 \\ -c_2 & -c_1 \end{bmatrix}^{-1} \begin{bmatrix} -c_1 \\ c_3 \end{bmatrix}$$

$$= \begin{bmatrix} 2.55 & -2 \\ -2.83 & 2.55 \end{bmatrix}^{-1} \begin{bmatrix} 2.83 \\ -2.9705 \end{bmatrix} = \begin{bmatrix} 1.5144 \\ 0.5171 \end{bmatrix}$$

$$\hat{a}_2 = \hat{c}_1 - C_{11}^{-1}\, \hat{a}_1 = \begin{bmatrix} c_0 \\ c_1 \end{bmatrix} - \begin{bmatrix} 0 & 0 \\ -c_0 & 0 \end{bmatrix}^{-1} \begin{bmatrix} a_{12} \\ a_{13} \end{bmatrix}$$

$$= \begin{bmatrix} -2 \\ 2.55 \end{bmatrix} - \begin{bmatrix} 0 & 0 \\ -2 & 0 \end{bmatrix}^{-1} \begin{bmatrix} 1.5144 \\ 0.5171 \end{bmatrix} = \begin{bmatrix} 2 \\ 0.48 \end{bmatrix}$$

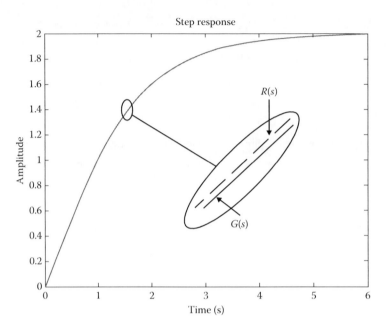

FIGURE 4.10
Comparison of responses of reduced-order model and original system model.

Hence, the second-order reduced model is

$$R(s) = \frac{a_{21} + a_{22}s}{1 + a_{12}s + a_{13}s^2} = \frac{2 + 0.48s}{1 + 1.5144s + 0.5171s^2}$$

The step responses for the original model, $G(s)$, and the reduced-order model, $R(s)$, are shown in Figure 4.10, and both responses are stable.

```
%Matlab script for getting step responses of G(s) and R(s)
clear all
num1=[1 13 40]; den1=[1 13 32 20]; % TF of original system
num2=[0.48 2]; den2=[0.5171 1.5144 1]; % TF of reduced order model
step(num1, den1) % step response of original system
hold
step(num2, den2) % step response of reduced order system
```

4.5.2.8 Continued Fraction Expansion

Consider the transfer function given below:

$$G(s) = \frac{b_0 + b_1 s + b_2 s^2 + \cdots + b_n s^{n-1}}{a_0 + a_1 s + a_2 s^2 + \cdots + a_n s^n} \qquad (4.26)$$

The order of the denominator is n. It is often convenient to let $a_n = 1$, and this may be done without a loss of generality, because a general control system is a low-pass filter in nature and, therefore, in simplification, we take care of the steady state first, and then the transient part. Hence, we start the CFE from the constant term or arrange the polynomial in the ascending powers of s, and, thus, may rewrite the transfer function as

Model Order Reduction

$$G(s) = \frac{A_{21} + A_{22}s + A_{23}s^2 + \cdots + A_{2,n}s^{n-1}}{A_{11} + A_{12}s + A_{13}s^2 + \cdots + a_{1,(n+1)}s^n} \quad (4.27)$$

The CFE of Equation 4.27, has been used to write the Routh array, as given in Table 4.2, where

TABLE 4.2
Routh Array

A_{11}	A_{12}	A_{13}	A_{14}
A_{21}	A_{22}	A_{23}	A_{24}
A_{31}	A_{32}	A_{33}	—
A_{41}	A_{42}	A_{43}	—
A_{51}	A_{52}	—	—

$$A_{j,k} = A_{j-2,k+1} - \frac{A_{j-2,1}A_{j-1,k+1}}{A_{j-1,1}} \quad (4.28)$$

$J = 3, 4\ldots$
$K = 1, 2\ldots$

In the Routh array, the first two rows are formed by the denominator and numerator coefficients of $G(s)$, and the elements of the third, the fourth, and subsequent rows can be evaluated by the Routh algorithm(s) (Chen and Shieh, 1969):

$$G(s) = \cfrac{1}{\cfrac{A_{11}}{A_{21}} + \cfrac{s}{\cfrac{A_{21}}{A_{31}} + \cfrac{s}{\cfrac{A_{31}}{A_{41}} + \cfrac{s}{\cdot}}}} = \cfrac{1}{h_1 + \cfrac{s}{h_2 + \cfrac{s}{h_3 + \cfrac{s}{\cdots}}}} \quad (4.29)$$

The simplified models with the denominator of the reduced order, r, may be derived using the first $2r$ values of quotients, h_i, and will be of the general form as

$$R(s) = \frac{A_{21}^* + A_{22}^* s + A_{23}^* s^2 + \cdots + A_{2,r}^* s^{r-1}}{A_{11}^* + A_{12}^* s + A_{13}^* s^2 + \cdots + a_{1,(r+1)}^* s^r} \quad (4.30)$$

For the second-order model considering that four scalar quotients, $h_i = 1, 2, \ldots, 4$, are given, the continued fraction inversion of Equation 4.28 is

$$R(s) = \cfrac{1}{h_1 + \cfrac{s}{h_2 + \cfrac{s}{h_3 + \cfrac{s}{\cdots}}}} = \frac{h_2 h_3 h_4 + (h_2 + h_4)s}{h_1 h_2 h_3 h_4 + (h_1 h_2 + h_1 h_4 + h_3 h_4)s + s^2} \quad (4.31)$$

Example 4.3

Consider the following transfer function:

$$G(s) = \frac{360 + 171s + 10s^2}{720 + 702s + 71s^2 + s^3}$$

Reduce the model order using the CFE method.

Solution

The Routh array may be written as given in Table 4.3, from which a continued fraction is written immediately as

$$= \cfrac{1}{2 + \cfrac{1}{\cfrac{1}{s} + \cfrac{1}{3 + \cfrac{1}{\cfrac{5}{s} + \cfrac{1}{6 + \cfrac{4}{s}}}}}}$$

Truncating the CFE up to four quotients, one obtains a second-order model as

$$R_2(s) = \frac{15 + 6s}{30 + 27s + s^2}$$

The unit step responses of the original system and the reduced-order model are shown in Figure 4.11.

The approach of Chen and Shieh to the CFE is based on the following principle: The low performance terms can be discarded and the high performance terms should be retained.

They pointed out that the first quotient in the expansion dominates the characteristics of the steady state for step function disturbances. The second and the subsequent quotients in the expansion make up the parts of the transient response. A simplified model is obtained by keeping the first few significant quotients and discarding the others. As a result, the simplified model gives a satisfactory approximation in the steady state region but not in the transient portion.

Expanding a rational transfer function into a continued fraction and inverting a continued fraction to a transfer function are two fundamentally important operations in system synthesis, system analysis, and model order reduction. Theoretically, the two iterations are trivial, one involves many divisions and the other is related to many multiplications. Practically speaking, however, when the order is high, the heavy labor of driving multiplications and divisions is unavoidable. Facing the tedious work, we naturally think of an algorithmic approach to the problem in order that we can use the digital computer to free us from drudgery.

TABLE 4.3

Routh Array for Given System

	720	702	71	1
$h_1 = \dfrac{720}{360} = 2$	360	171	10	
$h_2 = \dfrac{360}{360} = 1$	360	51	1	
$h_3 = \dfrac{360}{120} = 3$	120	9		
$h_4 = \dfrac{120}{24} = 5$	24	1		
$h_5 = \dfrac{24}{4} = 6$	4			
$h_6 = \dfrac{4}{1} = 4$	1			

Model Order Reduction

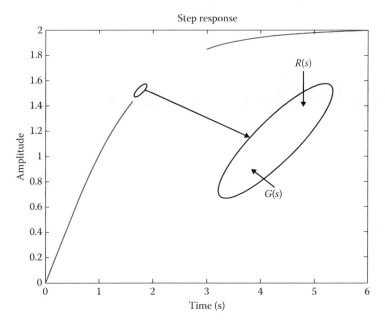

FIGURE 4.11
Comparison of responses of original system and reduced model.

Continued fraction inversion of Cauer's first form

Parathasarthy and Harpreet (1975) gave an elegant method for inverting the continued fraction given in Cauer's first form:

$$G(s) = \frac{A_{21}s^{n-1} + A_{22}s^{n-2} + \cdots + A_{2n-1}s + A_{2,n}}{s^n + A_{12}s^{n-1} + A_{13}s^{n-2} + \cdots + A_{1n}s + A_{1n+1}} \quad (4.32)$$

The corresponding first form of the CFE is

$$G(s) = \cfrac{1}{H_1 s + \cfrac{1}{H_2 + \cfrac{1}{H_3 s + \cfrac{1}{\cdots + \cfrac{1}{H_{2n}}}}}} \quad (4.33)$$

where $H_1, H_2, \ldots$ are known as partial coefficients. Without a loss of generality, the coefficient of S^n in Equation 4.32 can be taken as unity, and the numerator is at least one degree less than the denominator. The problem is to evaluate the coefficients A_{21} to A_{2n}, and A_{12} to $A_{1,n+1}$ from a knowledge of H_1 to H_{2n}, and construct the transfer function (4.33).

The Routh array is formed as in Table 4.4.

It is to be noted that $A_{11} = 1$. And the end elements of the odd rows will be constant and equal to $A_{2n+1,1}$ (the last element of the array), and those of the even rows will be zeros. The partial coefficients are related to the elements of the first column by

$$\frac{A_{p+1,1}}{A_{p,1}} = \frac{1}{H_p}, \quad p = 1, 2, \ldots, 2n; \quad H_p \neq 0 \quad (4.34)$$

TABLE 4.4

Routh Array

1	A_{12}	A_{13}	...	A_{1n}	$A_{1,n+1}$
A_{21}	A_{22}	A_{23}	...	A_{2n}	0
A_{31}	A_{32}	A_{33}	...	A_{3n}	
A_{41}	A_{42}	A_{43}	...	0	
$\vdots$					
$A_{2n-1,1}$	$A_{2n-1,2}$				
$A_{2n,1}$	0				
$A_{2n+1,1}$					

And the other elements of the Routh table are connected by the following recursive relation:

$$A_{j-2,k+1} = A_{jk} + \frac{A_{j-2,1} A_{j-1,k+1}}{A_{j-1,1}} \qquad (4.35)$$

where
$j = 2n, 2n - 1, \ldots, 3$
$k = 1, 2, \ldots, (n - 1)$

Thus, the steps for the proposed algorithm are

Step 1: Starting from the first element of the Routh table ($A_{11} = 1$) and from the knowledge of partial coefficients, write down the first column of the Routh table.

Step 2: Evaluate recursively the remaining elements of the table.

Step 3: Copy the first rows of the Routh table, which give the coefficients of the denominator and numerator polynomials of the transfer function.

Example 4.4

Let us consider

$$G(s) = \cfrac{1}{2s + \cfrac{1}{1 + \cfrac{1}{2}s + \cfrac{1}{\cfrac{8}{3} + \cfrac{1}{\cfrac{3}{2}s + \cfrac{1}{\cfrac{1}{2/3}}}}}}$$

Reduce the model order using the CFE method.

Solution

The partial coefficients are
$H_1 = 2$, $H_2 = 1$, $H_3 = 1/2$, $H_4 = 8/3$, $H_5 = 3/2$, and $H_6 = 2/3$.

Model Order Reduction

TABLE 4.5
Routh Table

		1	9/2	17/4	3/8
$H_1 = 2$	1/2	2	13/8	0	
$H_2 = 1$	1/2	1	3/8		
$H_3 = 1/2$	1	5/4	0		
$H_4 = 8/3, 3/8$	3/8				
$H_5 = 3/2, 1/4$	0				
$H_6 = 2/3, 3/8$					

The entries of the first column are

$$A_{11} = 1, \quad A_{21} = 1/2 \quad A_{31} = 1/2 \quad A_{41} = 1 \quad A_{51} = 3/8 \quad A_{61} = 2/3$$

For
$j = 6, k = 1$
$j = 5, k = 1$
$j = 4, k = 1, 2 \qquad A_{42} = 5/4$
$j = 3, k = 1, 2$

$$A_{32} = 1$$
$$A_{22} = 2 \qquad A_{23} = 13/8$$
$$A_{12} = 9/2 \qquad\qquad A_{13} = 17/4$$

By inspection

$$A_{62} = A_{43} = A_{24} = 0$$
and $\quad A_{71} = A_{52} = A_{33} = A_{14} = 3/8$

The Routh table is given in Table 4.5.
The corresponding transfer function is

$$G(s) = \frac{(1/2)s^2 + 2s + (13/8)}{s^3 + (9/2)s^2 + (17/4)s + (3/8)}$$

An improved algorithm for continued fraction inversion (Kumar and Singh, 1978)

Given the continued fraction, $G_1(s)$, in Cauer's second form and $G_2(s)$ in Cauer's first form, that is

$$G_1(s) = \cfrac{1}{H_1 + \cfrac{1}{\cfrac{H_2}{s} + \cfrac{1}{H_3 + \cfrac{1}{\cdots + \cfrac{1}{\cfrac{H_{2n}}{s}}}}}}$$

$$G_2(s) = \cfrac{1}{H_1 s + \cfrac{1}{H_2 + \cfrac{1}{H_3 s + \cfrac{1}{\cdots + \cfrac{1}{H_{2n}}}}}}$$

The problem is to determine the corresponding rational forms:

$$G_1(s) = \frac{a_0 + a_1 s + a_2 s^2 + \cdots + a_{n-1} s^{n-1}}{b_0 + b_1 s + b_2 s^2 + \cdots + b_n s^n}$$

$$G_2(s) = \frac{a_{n-1} s^{n-1} + \cdots + a_2 s^2 + a_1 s + a_0}{b_n s^n + \cdots + b_2 s^2 + b_1 s + b_0}$$

TABLE 4.6
Routh Table

A_{11}	A_{12}	A_{13}	...	A_{1n-1}	A_{1n}	1
A_{21}	A_{22}	A_{23}	...	A_{2n-1}	A_{2n}	0
A_{31}	A_{32}	A_{33}	...	A_{3n-1}		1
A_{41}	A_{42}	A_{43}	...	A_{4n-1}		0
⋮						
$A_{2n-1,1}$	1					
$A_{2n,1}$	0					
1						

A formula is proposed of the form:

$$A_{j,k} = A_{j+2,k-1} + H_j A_{j+1,k} \tag{4.36}$$

where
$j = 2n, 2n-1, \ldots, 1$
$k = 1, 2, \ldots, n$

In view of the fact that

$$\frac{A_{j,1}}{A_{j+1,1}} = H_j$$

the use of Equation 4.36 requires the following initial conditions:

$A_{j,k} = 1$ for $j + 2k = 2n + 3$
$A_{j,k} = 0$ for $j + 2k > 2n + 3$
$A_{j,0} = 0$

Using Equation 4.36 we construct the Routh table given in Table 4.6.
Using the first two rows of the above table:

$$a_0 = a'_{n-1} = A_{21}, \quad a_1 = a'_{n-2} = A_{22}, \ldots, a_{n-1} = a'_0 = A_{2n}$$
$$b_0 = b'_n = A_{11}, \quad b_1 = b'_{n-1} = A_{12}, \ldots, b_{n-1} = b'_1 = A_{1n}$$

Example 4.5

Consider the system transfer function given below:

$$G_1(s) = \cfrac{1}{2 + \cfrac{1}{\cfrac{1}{s} + \cfrac{1}{\cfrac{1}{2} + \cfrac{1}{\cfrac{8}{3s} + \cfrac{1}{\cfrac{3}{2} + \cfrac{1}{\cfrac{2}{3s}}}}}}} \tag{4.37}$$

Model Order Reduction

$$G_2(s) = \cfrac{1}{2s + \cfrac{1}{1 + \cfrac{1}{\cfrac{s}{2} + \cfrac{1}{\cfrac{8}{3} + \cfrac{1}{\cfrac{3s}{2} + \cfrac{1}{\cfrac{2}{3}}}}}}}$$

TABLE 4.7
Routh Table

8/3	12	34/3	1
4/3	16/3	13/3	0
4/3	8/3		
8/3	10/3	1	
1	1	0	
2/3	0		

Reduce the model order using improved continued fraction inversion.

Solution

Both $G_1(s)$ and $G_2(s)$ have the same set of partial coefficients:
$H_1 = 2$, $H_2 = 1$, $H_3 = 1/2$, $H_4 = 8/3$, $H_5 = 3/2$, and $H_6 = 2/3$.
Using $G_1(s)$ we can calculate the elements of the table, given in Table 4.7.
The desired rational forms of the transfer functions are

$$G_1(s) = \frac{4/3 + (16/3)s + (13/3)s^2}{8/3 + 12s + (34/3)s^2 + s^3}$$

$$G_2(s) = \frac{(4/3)s^2 + (16/3)s + 13/3}{(8/3)s^3 + 12s^2 + (34/3)s + 1}$$

This method of model order reduction is computationally better.

4.5.2.9 Model Order Reduction Based on the Routh Stability Criterion

The Routh algorithm is a computational method that develops a sequence of computed numbers from two generating rows of numbers. This method is used to check the stability of a system. The array is usually written in the following form, from the coefficients of a given polynomial:

$$P(s) = a_0 s^n + b_0 S^{n-1} + a_1 s^{n-2} + b_1 s^{n-2} + \cdots$$

Generating rows
$\quad a_0 \quad a_1 \quad a_2 \quad a_3 \quad a_4 \quad - \quad - \quad -$
$\quad b_0 \quad b_1 \quad b_2 \quad b_3 \quad - \quad - \quad - \quad -$

Computed rows
$\quad c_0 \quad c_1 \quad c_2 \quad c_3 \quad - \quad - \quad - \quad -$
$\quad d_0 \quad d_1 \quad d_2 \quad - \quad - \quad - \quad - \quad -$
$\quad e_0 \quad e_1 \quad e_2$

etc.

The third and each subsequent row are evaluated from the proceeding two rows by means of a systematic form of calculation, namely

$$c_0 = \frac{b_0 a_1 - a_0 b_1}{b_0}, \quad c_1 = \frac{b_0 a_2 - a_0 b_2}{b_0}$$

$$c_3 = \frac{b_0 a_3 - a_0 b_3}{b_0}$$

$$d_0 = \frac{c_0 b_1 - b_0 c_1}{c_0}, \quad d_1 = \frac{c_0 b_2 - b_0 c_2}{c_0}$$

There is a possibility that certain elements in the array may vanish; unless the first column number, c_0, d_0, etc., vanishes, the computation proceeds without difficulty. The array becomes roughly triangular and terminates with a row having only one element. The numbers $a_0, a_1, \ldots, b_0, b_1, \ldots$, etc., in the generating rows are normally coefficients taken either alternately from one polynomial or taken from two polynomials, the source of which depends upon the specific applications.

The Routh criterion is an efficient test that tells how many roots of a polynomial lie in the right half of the s-plane, and also gives information about the roots symmetrically located about the origin. The number of sign changes in the first column of the Routh array determines the number of roots lying in the right of the s-plane.

Let the higher-order transfer function be

$$G(s) = \frac{b_{11} s^m + b_{21} s^{m-1} + b_{12} s^{m-2} + b_{22} s^{m-3} + \cdots}{b_{11} s^n + a_{21} s^{n-1} + a_{12} s^{n-2} + a_{22} s^{n-3} + \cdots}$$

The reduced-order transfer function is constructed directly from the elements in the Routh stability arrays of the high-order numerator and denominator, as shown in Tables 4.8 and 4.9.

TABLE 4.8

Numerator Stability Array

b_{11}	b_{12}	b_{13}	b_{14}	—	—	—
b_{21}	b_{22}	b_{23}	b_{24}	—	—	—
b_{31}	b_{32}	b_{33}	—	—	—	—
b_{41}	b_{42}	b_{43}	—	—	—	—
$\vdots$						
$b_{m,1}$						
$b_{m+1,1}$						

TABLE 4.9

Denominator Stability Array

a_{11}	a_{12}	a_{13}	a_{14}	—	—	—
a_{21}	a_{22}	a_{23}	a_{24}	—	—	—
a_{31}	a_{32}	a_{33}	—	—	—	—
a_{41}	a_{42}	a_{43}	—	—	—	—
$\vdots$						
$a_{n-2,1}$	$a_{n-2,2}$					
$a_{n-1,1}$	$a_{n-1,2}$					
$a_{n,1}$						
$a_{n+1,1}$						

Model Order Reduction

The first two rows of each table consist of odd coefficients (e.g., first, fifth, etc.). The tables are computed in the conventional way by computing coefficients of successive rows by the algorithms:

$$c_{ij} = c_{i-2,j+2} - \frac{c_{i-2,1}\, c_{i-1,j*1}}{c_{i-1,1}}$$

for $i \geq 3$ and $l \leq (n - l + 3)/2$.

Suppose that the first two rows of each table are made available, then the transfer function of the system of order n can easily be reconstructed. Let us take this a little further down the table; consider now the second and third rows of each table. A transfer function may be constructed in a similar way with these coefficients with the system order reduced to $(n - 1)$. It should be noted that the effect of all coefficients of the previous two rows has been taken into consideration while computing the coefficients of the third rows. This applies equally well to all other rows computed from the previous two rows.

The transfer function thus constructed with the second and third rows of each table is given by

$$G_{n-1}(s) = \frac{b_{21}s^{m-1} + b_{31}s^{m-2} + b_{22}s^{m-3} + b_{32}s^{m-4} + \cdots}{a_{21}s^{n-1} + a_{31}s^{n-2} + a_{22}s^{n-3} + a_{32}s^{n-4} + \cdots}$$

Example 4.6

Consider an eighth-order system:

$$G(s) = \frac{35s^7 + 1086s^6 + 13285s^5 + 82402s^4 + 278376s^3 + 511812s^2 + 482964s + 194480}{s^8 + 33s^7 + 437s^6 + 3017s^5 + 11870s^4 + 27470s^3 + 37492s^2 + 28880s + 9600}$$

Reduce the system order using the Routh stability criterion from the eighth-order to the fifth- and second-orders.

Solution

The reduced-order transfer function can be immediately derived of any order k that is less than or equal to n from stability Tables 4.10 and 4.11.

TABLE 4.10

Numerator Stability Array

s^7	35	13,285	278,376	482,964 (coefficients of odd terms)
s^6	1,086	82,402	511,812	194,480 (coefficients of even terms)
s^5	10,629	261,881	476,696	—
s^4	55,645	463,108	194,480	
s^3	173,419	439,547		
s^2	322,069	194,480		
s^1	334,828			
s^0	194,480			

TABLE 4.11

Denominator Stability Array

S^8	1	437	11,870	37,492	9,600 (coefficients of even terms)
S^7	33	3,017	27,470	28,880 (coefficients of odd terms)	
S^6	346	11,037	36,617	9,600	
S^5	1,963	23,973	27,963		
S^4	6,817	31,694	9,600		
S^3	14,847	25,199			
S^2	20,124	9,600			
S^1	181,162				
S^0	9,600				

For example, the second-order model:

$$R_2(s) = \frac{334828s + 194880}{20123s^2 + 18116s + 9600}$$

and the fifth-order model:

$$R_5(s) = \frac{55645s^4 + 173419s^3 + 463108s^2 + 439547s + 194480}{1963s^5 + 6817s^4 + 23973s^3 + 31694s^2 + 27963s + 9600}$$

These models are simulated in MATLAB® and the results have been compared, as shown in Figure 4.12.

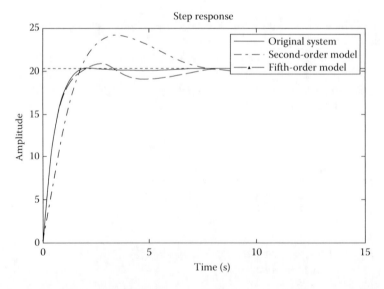

FIGURE 4.12
Model response comparison.

Model Order Reduction

```
%Matlab program for comparison of responses
clear all
Num_sys=[35 1086 13285 82402 278376 511812 482964 194480];
Den_sys=[1 33 437 3017 11870 27470 37492 28880 9600];
step(Num_sys, Den_sys)
hold
Num_2=[334828 194880];
Den_2=[20123 18116 9600];
step(Num_2, Den_2)
Num_5=[55645 173419 463108 439547 194480];
Den_5=[1963 6817 23973 31694 27963 9600];
step(Num_5, Den_5)
legend("Original System," "Second order Model," "Fifth order model")
```

4.5.2.10 Differentiation Method for Model Order Reduction

The differentiation method for model order reduction was introduced by Gutman et al. (1982). This method is based on differentiation of polynomials. The reciprocals of the numerator and denominator polynomials of the higher-order transfer functions are differentiated successively many times to yield the coefficients of the reduced-order transfer function. The reduced polynomials are reciprocated back and normalized. The straightforward differentiation is discarded because it has a drawback that zeros with large moduli tend to be better approximated than those with small moduli (Prasad et al., 1995).

Example 4.7

Consider $G(s) = \dfrac{40 + 13s + s^2}{20 + 32s + 13s^2 + s^3}$.

The reciprocal of $G(s)$ is given by

$$\overline{G(s)} = \frac{1}{s} G\left(\frac{1}{s}\right)$$

$$\overline{G(s)} = \frac{40s^2 + 13s + 1}{20s^3 + 32s^2 + 13s + 1}$$

After differentiation, the second-order model is

$$\overline{R_2(s)} = \frac{80s + 13}{60s^2 + 64s + 13}$$

Its reciprocal is given by

$$R_2(s) = \frac{80 + 13s}{60 + 64s + 13s^2}$$

Applying the steady state correction:

$$R_2(s) = \frac{3/2(80 + 13s)}{60 + 64s + 13s^2} = \frac{9.23 + 1.5s}{4.62 + 4.92s + s^2}$$

The comparison of original system responses and second-order reduced model responses has been shown in Figure 4.13.

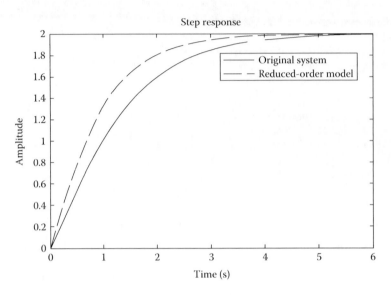

FIGURE 4.13
Comparison of results of original system and reduced-order model.

Example 4.8

Consider the following transfer function:

$$\frac{C(s)}{R(s)} = \frac{10}{(s+10)(s^2+2s+2)}$$

$s = -10, -1 + j.1, -1 - j.1$

$$\frac{C(s)}{R(s)} = \frac{10}{10(s/10+1)(s^2+2s+2)}$$

If $\dfrac{s}{10} \ll 1$, then $\dfrac{C(s)}{R(s)} = \dfrac{1}{(s^2+2s+2)}$.

The higher-order system can be represented as

$$M(s) = K \frac{1 + a_1 s + a_2 s^2 + \cdots + a_m s^m}{1 + b_1 s + b_2 s^2 + \cdots + b_n s^n}$$

The lower-order system can be represented as

$$L(s) = K \frac{1 + c_1 s + c_2 s^2 + \cdots + c_q s^q}{1 + d_1 s + d_2 s^2 + \cdots + d_p s^p}$$

where $n \geq p \geq q$.

Model Order Reduction

Approximation criterion

$$\frac{|M(j\omega)|^2}{|L(j\omega)|^2} = 1 \quad \text{for } 0 < \omega < \infty$$

The last condition implies that the amplitude characteristics of the two systems in the frequency domain, ($s = j\omega$), are similar. It is hoped that this will lead to the same time responses for the two systems.

Steps involved in the approximation:

1. Choose the appropriate order of the numerator polynomial, q, and the denominator polynomial, p, of $L(s)$.
2. Determine the coefficient c_i where $i = 1, 2, \ldots, q$, and d_j where $j = 1, 2, \ldots, p$, so that the condition in approximation criterion is approached:

$$\frac{M(s)}{L(s)} = \frac{(1 + a_1 s + a_2 s^2 + \cdots + a_m s^m)(1 + d_1 s + d_2 s^2 + \cdots + d_p s^p)}{1 + b_1 s + b_2 s^2 + \cdots + b_n s^n (1 + c_1 s + c_2 s^2 + \cdots + c_q s^q)}$$

$$= K \frac{1 + m_1 s + m_2 s^2 + \cdots + m_u s^u}{1 + l_1 s + l_2 s^2 + \cdots + l_v s^v}$$

where $u = m + p$ and $v = n + q$.

$$\frac{|M(j\omega)|^2}{|L(j\omega)|^2} = \left.\frac{M(s)M(-s)}{L(s)L(-s)}\right|_{s=j\omega}$$

The numerator and the denominator of $\frac{M(s)M(-s)}{L(s)L(-s)}$ are even polynomials of s, that is, these contain only even powers of s:

$$\frac{|M(j\omega)|^2}{|L(j\omega)|^2} = \left.\frac{1 + e_2 s^2 + e_4 s^4 + \cdots + e_{2u} s^{2u}}{1 + f_2 s^2 + f_4 s^4 + \cdots + f_{2v} s^{2v}}\right|_{s=j\omega}$$

Divide the numerator by the denominator on the right-hand side:

$$= \left.\frac{1 + e_2 s^2 + e_4 s^4 + \cdots + e_{2u} s^{2u}}{1 + f_2 s^2 + f_4 s^4 + \cdots + f_{2v} s^{2v}}\right|_{s=j\omega} - 1 = 0$$

$$\frac{(e_2 - f_2)s^2 + (e_4 - f_4)s^4 + \cdots + (e_{2u} - f_{2u})s^{2u} - \cdots - f_{2v} s^{2v}}{1 + f_2 s^2 + f_4 s^4 + \cdots + f_{2v} s^{2v}} = 0$$

For the approximation criterion:

$$e_2 = f_2, \quad e_4 = f_4, \ldots e_{2u} = f_{2u}$$

Beyond the 2uth term, all others are error terms:

$$\frac{M(s)M(-s)}{L(s)L(-s)} = \frac{(1+m_1s+m_2s^2+\cdots+m_us^u)(1-m_1s+m_2s^2+\cdots+(-1)^u m_u s^u)}{(1+l_1s+l_2s^2+\cdots+l_v s^u)(1-l_1s+l_2s^2+\cdots+(-1)^v l_v s^v)}$$

$$= \frac{1+e_2 s^2 + e_4 s^4 + \cdots + e_{2u} s^{2u}}{1+f_2 s^2 + f_4 s^4 + \cdots + f_{2v} s^{2v}}$$

$$e_2 = 2m_2 - m_1^2$$

$$e_4 = 2m_4 - 2m_3 m_1 + m_2^2$$

$$e_6 = 2m_6 - 2m_1 m_5 + 2m_2 m_4 - m_3^2$$

$$e_{2x} = \sum_{i=0}^{x-1} (-1)^i 2 m_i m_{2x-i} + (-1)^x m_x^2$$

where $x = 1, 2, \ldots, u$ and $m_2 = 1$.
Similarly, we can write

$$f_{2y} = \sum_{i=0}^{y-1} (-1)^i 2 l_i l_{2y-i} + (-1)^y l_y^2$$

where $y = 1, 2, \ldots, v$ and $l_0 = 1$.

Example 4.9

Let the open-loop transfer function be

$$G(s) = \frac{8}{s(s^2 + 6s + 1)}$$

The closed-loop transfer function is

$$M(s) = \frac{C(s)}{R(s)} = \frac{G}{1+GH}$$

$$= \frac{8}{s^3 + 6s^2 + 12s + 1}$$

$$= \frac{1}{l_3 s^3 + l_2 s^2 + l_1 s + 1}$$

where
 $l_3 = 1/8$
 $l_2 = 6/8$
 $l_1 = 12/8$

Model Order Reduction

Now, for a reduced-order model:

$$u(s) = \frac{1}{d_2 s^2 + d_1 s + 1} = \frac{1}{m_2 s^2 + m_1 s + 1}$$

$$m_1 = d_1, \quad m_2 = d_2$$

According to the approximation criterion, we can write

$$\frac{M(s)M(-s)}{L(s)L(-s)} = \frac{e_4 s^4 + e_2 s^2 + 1}{f_6 s^6 + f_4 s^4 + f_2 s^2 + \cdots + 1}$$

Comparing the coefficients of powers of s from both sides, we get

$$e_2 = f_2 \rightarrow 2d_2 - d_1^2 = 2l_2 - l_1^2$$

$$2d_2 - d_1^2 = 2\left(\tfrac{6}{8}\right) - \left(\tfrac{12}{8}\right)^2 = -0.75$$

$$e_4 = f_4 \Rightarrow d_2^2 = -2l_1 l_3 + l_2^2$$

$$d_2^2 = -2\left(\tfrac{12}{8}\right)\left(\tfrac{1}{8}\right) + \left(\tfrac{6}{8}\right)^2 = 0.1875$$

From the above equations, we get
$d_2 = 0.433$ and $d_1 = 1.271$
Hence, the reduced-order model will be

$$L(s) = \frac{1}{0.433 s^2 + 1.271 s + 1}$$

If we take the first-order reduced model:

$$L(s) = \frac{1}{ds + 1}$$

$$e_2 = f_2 \Rightarrow -d_1^2 = 2l_2 - l_1^2 = 0.75$$

$$d_1 = 0.866$$

$$\text{Hence,} \quad L(s) = \frac{1}{0.866 s + 1}$$

The Simulink® model is shown in Figure 4.14, which is simulated in MATLAB, and the results are shown in Figure 4.15.

MATLAB program

```
step(1, [.125 .75 1.5 1])
Pause
hold
step(1, [.433 1.271 1])
step(1, [.866 1])
```

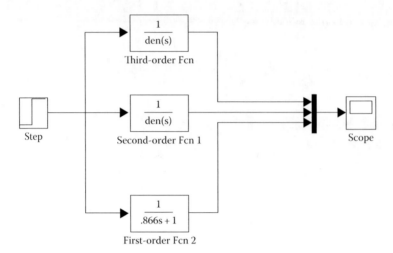

FIGURE 4.14
Simulink model to find the effect of model order reduction on system response.

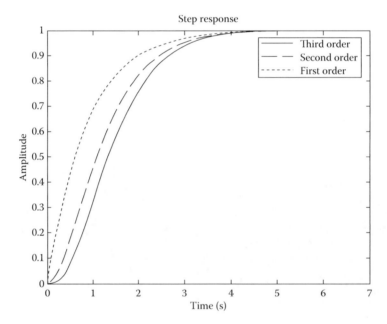

FIGURE 4.15
System responses for different model orders.

Example 4.10

Let the closed-loop transfer function be

$$M(s) = \frac{1.5 \times 10^7 K}{(1/1.5 \times 10^7 K)s^3 + (3408.3/1.5 \times 10^7 K)s^2 + (120400/1.5 \times 10^7 K)s + 1}$$

$$= \frac{1.5 \times 10^7 K}{l_3 s^3 + l_2 s^2 + l_1 s + 1}$$

Model Order Reduction

Here, $l_3 = \dfrac{1}{1.5 \times 10^7 K}$

$$l_3 = \dfrac{3408.3}{1.5 \times 10^7 K}$$

$$l_3 = \dfrac{120400}{1.5 \times 10^7 K}$$

Now, for a reduced-order model:

$$u(s) = \dfrac{1}{d_2 s^2 + d_1 s + 1} = \dfrac{1}{m_2 s^2 + m_1 s + 1}$$

$$m_1 = d_1, \quad m_2 = d_2$$

According to the approximation criterion, we can write

$$\dfrac{M(s)M(-s)}{L(s)L(-s)} = \dfrac{e_4 s^4 + e_2 s^2 + 1}{f_6 s^6 + f_4 s^4 + f_2 s^2 + \cdots + 1}$$

Comparing the coefficients of powers of s from both sides, we get

$$e_2 = f_2 \rightarrow 2d_2 - d_1^2 = 2l_2 - l_1^2$$

$$2d_2 - d_1^2 = 2\left(\dfrac{3408.3}{1.5 \times 10^7 K}\right) - \left(\dfrac{120400}{1.5 \times 10^7 K}\right)^2 \tag{4.38}$$

$$e_4 = f_4 \rightarrow 2d_2^2 = 2l_1 l_3 + l_2^2$$

$$d_2^2 = -2\left(\dfrac{120400}{1.5 \times 10^7 K}\right)\left(\dfrac{1}{1.5 \times 10^7 K}\right) + \left(\dfrac{3408.3}{1.5 \times 10^7 K}\right)^2 \tag{4.39}$$

We can solve Equations 4.38 and 4.39 for different values of K (e.g., $K = 7.5$ and 14.5).

From the response plots shown in Figures 4.16 through 4.18, we can say that for small values of K the second-order approximation is quite good. As we increase the value of K, the error increases.

Example 4.11

Consider the closed-loop transfer function:

$$M(s) = \dfrac{1}{(0.5s^2 + s + 1)(1 + Ts)}$$

$$= \dfrac{1}{(0.5Ts^3 + (T + 0.5)s^2 + (T + 1)s + 1}$$

$$M(s) = \dfrac{1}{l_3 s^3 + l_2 s^2 + l_1 s + 1}$$

where
$l_3 = 0.5T$
$l_2 = 0.5 + T$
$l_1 = 1 + T$

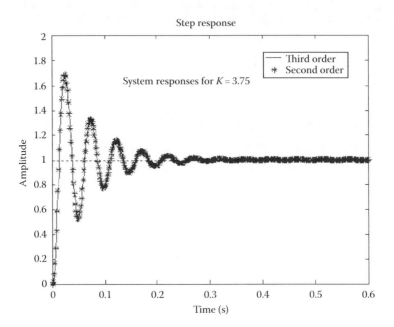

FIGURE 4.16
Systems responses for second- and third-order models, for $K = 3.75$.

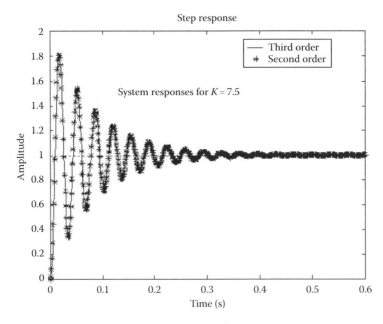

FIGURE 4.17
System responses for second- and third-order models, for $K = 7.5$.

Model Order Reduction

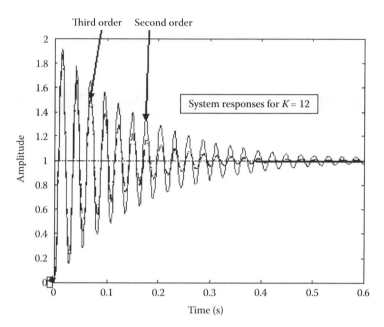

FIGURE 4.18
System responses for second- and third-order models, for $K = 12$.

Now, for a reduced-order model:

$$u(s) = \frac{1}{d_2 s^2 + d_1 s + 1} = \frac{1}{m_2 s^2 + m_1 s + 1}$$

$$m_1 = d_1, \quad m_2 = d_2$$

According to the approximation criterion, we can write

$$\frac{M(s)M(-s)}{L(s)L(-s)} = \frac{e_4 s^4 + e_2 s^2 + 1}{f_6 s^6 + f_4 s^4 + f_2 s^2 + \cdots + 1}$$

Comparing the coefficients of powers of s from both sides, we get

$$e_2 = f_2 \rightarrow 2d_2 - d_1^2 = 2l_2 - l_1^2$$

$$2d_2 - d_1^2 = 2(0.5 + T) - (1 + T)^2 \tag{4.40}$$

$$e_4 = f_4 \rightarrow 2d_2^2 = 2l_1 l_3 + l_2^2$$

$$d_2^2 = -2(1 + T)(0.5T) + (0.5 + T)^2 \tag{4.41}$$

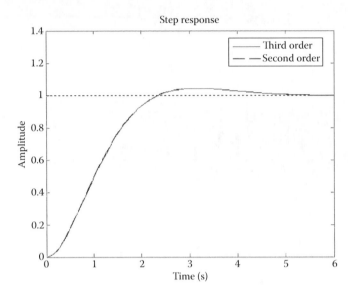

FIGURE 4.19
System responses for second- and third-order models, for $T = 0.01$.

We can solve Equations 4.40 and 4.41 for different values of T:

$$d_2 = 0.5, \quad d_1 = \sqrt{(1+T^2)}$$

From the response plots shown in Figures 4.19 through 4.21, we can say that for small values of T, the second-order approximation is quite good. As we increase the value of T, the error increases.

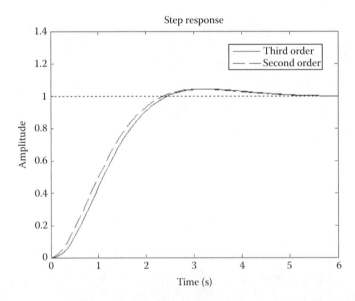

FIGURE 4.20
System responses for second- and third-order models, for $T = 0.1$.

Model Order Reduction

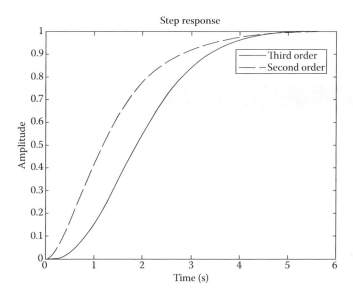

FIGURE 4.21
System responses for second- and third-order models, for $T = 1.0$.

4.6 Applications of Reduced-Order Models

Reduced-order models and reduction techniques have been widely used for the analysis and the synthesis of high-order systems. Reduced-order models are useful due to the following reasons:

1. Predicting transient response sensitivities of high-order systems using low-order models
2. Predicting dynamic errors of high-order systems using low-order equivalents
3. Control system design
4. Adaptive control using low-order models
5. Designing reduced-order estimator
6. Suboptimal control derived by simplified models
7. Providing guidelines for online interactive modeling

4.7 Review Questions

1. State the approximation criterion for reducing a model's order from given a high-order, $M(s)$, to a low-order, $L(s)$, using the dominant eigenvalue approach.
2. Consider that the open-loop transfer function of a unity feedback system is

$$G(S) = \frac{8}{s(s^2 + 6s + 12)}$$

Determine the first- and second-order models for the given third-order model. Also compare the computer simulation responses in MATLAB of these systems and check how the response of the model deviates when the order of the model is reduced.

3. Explain mathematically the dominant eigenvalue approach for model order reduction of a linear time-invariant dynamic system. Obtain a second-order reduced model for

$$\frac{d}{dt}\begin{bmatrix} X_1(t) \\ X_2(t) \\ X_3(t) \end{bmatrix} = \begin{bmatrix} -10 & 0 & 0 \\ 0 & -2 & 0 \\ 0 & 0 & -1 \end{bmatrix}\begin{bmatrix} X_1(t) \\ X_2(t) \\ X_3(t) \end{bmatrix} + \begin{bmatrix} 0 \\ 1.5 \\ 0.5 \end{bmatrix}U(t)$$

and

$$Y(t) = \begin{bmatrix} 0.2 & 0.4 & 0.6 \end{bmatrix}\begin{bmatrix} X_1(t) \\ X_2(t) \\ X_3(t) \end{bmatrix}$$

using (1) the dominant eigenvalue approach and (2) the aggregation method.

4. Develop the state space model for the electrical and electromechanical systems shown in Figures 4.22 and 4.23. Also reduce the model order using a suitable technique.

5. Take the open-loop transfer function for a closed-loop unity feedback system as

$$G(s) = \frac{(1+T_s)}{s(s+1)(s+2)(s+6)}$$

Determine the second-order model for the system and simulate it to find the effect of model order reduction on the response of this system.

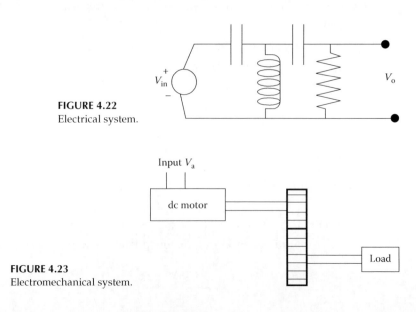

FIGURE 4.22
Electrical system.

FIGURE 4.23
Electromechanical system.

6. Obtain an approximate first-order model for the given system.
    i. Take the open-loop transfer function $G(s) = \dfrac{(1+T_s)}{(s^2+s+4)(s^3+2s^2+6)}$ for a unity feedback system. Determine approximate third-, second-, and first-order models and compare their unit step responses.
    ii. The open-loop transfer function and feedback transfer functions are given as

    $$H(s) = \dfrac{1}{s(s+1.25)(s^2+2.5s+10)}$$

    Obtain the second-order system model and compare the results for $K = 35, 40$, and 50.

4.8 Bibliographical Notes

An interesting survey work in the field of model reduction has been reported by Towill (1963), Gutman et al. (1982), and Genesio and Milaness (1976). Whatever be the approach used for model order reduction, the main objective is that the reduced-order approximation should reproduce the significant characteristics of the original system. There are many papers/books published in the area of model order reduction, but some of the interesting papers are published by Feldmann and Freund (1995), Kerns and Yang (1998), Phillips et al. (2003), and Tan and He (2007).

5
Analogous of Linear Systems

5.1 Introduction

In the analysis of linear systems, the mathematical procedure for obtaining the solutions to a given set of equations does not depend upon what physical system the equations represent. Hence, if the response of one physical system to a given excitation is determined, the responses of all other systems that can be described by the same set of equations are known for the same excitation function. Systems which are governed by the same types of equations are called analogous systems. When we deal with systems other than electrical, there are distinct advantages if we can reduce the systems under consideration to their analogous electrical circuits.

1. Once the circuit diagram of the analogous electrical system is determined, it is possible to visualize and even predict system behaviors by inspection.
2. Electrical circuit theory techniques can be applied in actual analysis of the system.
3. The solution of the set of differential equations describing a particular physical system can be directly applied to analogous systems of their type.
4. The ease of changing the parameters of electrical components, of connecting and disconnecting them in a circuit, and measuring the voltages and currents all prove invaluable in model construction and testing.
5. Since electrical or electronic systems can be built up easily, it is easier to build such a system rather than build up a mechanical or hydraulic system for experimental studies.

The two electrical analogies for mechanical systems are

1. Force–voltage analogy
2. Force–current analogy

Electric duality is a special type of analogy. The duality is shown in Figure 5.1.

5.1.1 D'Alembert's Principle

D'Alembert's principle is a slightly modified form of Newton's second law of motion and can be stated as follows:

> For any body, the algebraic sum of the externally applied forces and the forces resisting motion in any given direction is zero.

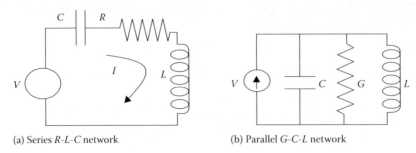

FIGURE 5.1
Dual networks.

It applies for all instants of time. A positive reference direction must be chosen. Forces acting in the reference direction are then considered as positive and those against the reference direction as negative. D'Alembert's principle is useful in writing the equations of motion for a mechanical system as Kirchoff's laws are in writing the circuit equations for an electric network.

D'Alembert's principle modified for a rotational system can be stated as follows:

> For any body, the algebraic sum of the externally applied torques and the torques resisting rotation about any axis is zero.

5.2 Force–Voltage (f–v) Analogy

The behavior of the mechanical systems can be completely predicted by what we know about the simple electrical circuits. If the force in the mechanical system is set to be analogous to voltage in the electrical system, we designate this type of analogy to force–voltage analogy. The analogous quantities of mechanical system and electrical system for f–v analogy are shown in Table 5.1.

5.2.1 Rule for Drawing f–v Analogous Electrical Circuits

Each junction in the mechanical system corresponds to a closed loop, which consists of electrical excitation sources and passive elements analogous to the mechanical driving

TABLE 5.1

Analogous Quantities for f–v Analogy

Translational Motion Mechanical System	Rotary Motion Mechanical System	Electrical System
Force, f	Torque, T	Voltage, V
Velocity, u	Angular velocity, ω	Current, I
Displacement, x	Angular displacement, Θ	Charge, q
Mass, M	Moment of inertia, J	Inductance, L
Damping coefficient, D	Viscous friction coefficient, B	Resistance, R
Compliance, K	Spring constant, K	Reciprocal of capacitance, $\dfrac{1}{C}$

Analogous of Linear Systems 279

sources and passive elements connected to the junction. All points on a rigid mass are considered as the same junction.

5.3 Force–Current (*f–i*) Analogy

From the point of view of physical interpretation, it is a natural analogy, because forces acting through mechanical elements are made to be analogous to voltage across the corresponding electrical elements and velocities across mechanical elements are made to be analogous to currents through the corresponding electrical elements. A direct consequence is that a junction in the mechanical system goes over to the analogous electrical circuits as a loop. The analogous quantities for *f–i* analogy are given in Table 5.2.

5.3.1 Rule for Drawing *f–i* Analogous Electrical Circuits

Each junction in the mechanical system corresponds to a node, which joins electrical excitation sources and passive elements analogous to the mechanical driving sources and passive elements connected to the junction. All points on a rigid mass are considered as the same junction and one terminal of the capacitance analogous to a mass is always connected to the ground.

The reason that one terminal of the capacitance analogous to a mass is always connected to the ground is that the velocity of a mass is always referred to the earth.

Example 5.1

Find the equations that describe the motion of the mechanical system of Figure 5.2 using

1. D'Alembert's principle
2. *f–v* analogy
3. *f–i* analogy

TABLE 5.2

Analogous Quantities for *f–i* Analogy

Translatory Motion Mechanical System	Rotary Motion Mechanical System	Electrical System
Force, f	Torque, T	Current, I
Velocity, u	Angular velocity, ω	Voltage, V
Linear displacement, x	Angular displacement, Θ	Magnetic flux, Φ
Mass, M	Moment of inertia, J	Capacitance, C
Damping coefficient, D	Viscous friction coefficient, B	Reciprocal of resistance, $\dfrac{1}{R}$
Compliance, K	Spring constant, K	Reciprocal of inductance, $\dfrac{1}{L}$

Solution

1. *Using d'Alembert's principle*

   It is clear that the system in Figure 5.2 is a two-coordinate system, that is, two variables, x_1 and x_2 are needed to describe the system completely.

   For mass M_1

   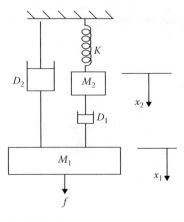

   FIGURE 5.2
   Mechanical system.

   $$M_1 \frac{du_1}{dt} + (D_1 + D_2)u_1 - D_1 u_2 = f \qquad (5.1)$$

   where $u_1 = x_1$ and $u_2 = x_2$

   For mass M_2

   $$-D_1 u_1 + M_2 \frac{du_2}{dt} + D_1 u_2 + \frac{1}{K}\left[\int_0^t u_2\, dt + x_2(0)\right] = 0 \qquad (5.2)$$

2. *Using f–v analogy*

   Corresponding to the two coordinates x_1 and x_2, the mechanical system has two junctions. Hence, we will have two loops in the f–v analogous electrical system. The analogous electrical circuit is shown in Figure 5.3.

   The equations can now be written using KVL as follows:

   $$L_1 \frac{di_1}{dt} + (R_1 + R_2)i_1 - R_1 i_2 = v$$

   $$-R_1 i_1 + L_2 \frac{di_2}{dt} + R_1 i_2 + \frac{1}{C}\left[\int_0^t i_2\, dt + q_2(0)\right] = 0 \qquad (5.3)$$

3. *Using f–i analogy*

   Corresponding to the two coordinates x_1 and x_2, the mechanical system has two junctions. Hence, we will have two independent nodes in the f–i analogous electrical system. The analogous electrical circuit is shown in Figure 5.4.

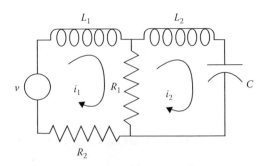

FIGURE 5.3
f–v analogy.

Analogous of Linear Systems

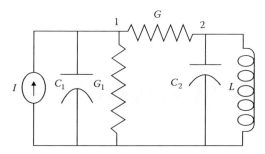

FIGURE 5.4
f–i analogy.

By applying KCL, the nodal equations can be easily written as

$$C_1 \frac{dv_1}{dt} + (G_1 + G_2)v_1 - G_1 v_2 = i$$

$$-G_1 v_1 + C_2 \frac{dv_2}{dt} + G_1 v_2 + \frac{1}{L}\left[\int_0^t v_2 dt + \phi_2(0)\right] = 0$$

(5.4)

Example 5.2: Mechanical Coupling Devices

Common mechanical coupling devices such as gears, friction wheels, and levers, also have electrical analogs. Let us first consider the pair of nonslipping friction wheels shown in Figure 5.5.

At the point of contact, P_1, on wheel 1 and P_2 on wheel 2 must have the same linear velocity because they move together, and experience equal and opposite forces(action and reaction). Since this is a rotational system, it is convenient to use angular velocities and torques. The following relations between magnitudes hold:

$$\frac{\tau_1}{\tau_2} = \frac{r_1}{r_2}$$

$$\frac{\omega_1}{\omega_2} = \frac{r_2}{r_1}$$

(5.5)

These relations are similar that exist among the voltages and currents in the primary and secondary windings of an ideal transformer. If $r_1:r_2$ is considered as the turns ratio $N_1:N_2$ of an ideal transformer, the torque will then be analogous to voltage and angular velocity to current. This forms the basis of f–v analogy shown in Figure 5.6.

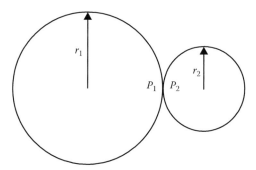

FIGURE 5.5
Friction wheel.

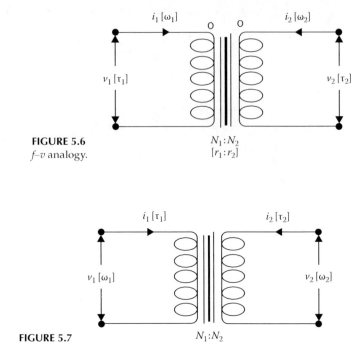

FIGURE 5.6
f–v analogy.

FIGURE 5.7
f–i analogy.

If $r_2:r_1$ is considered as the turns ratio $N_1:N_2$, we have the analogous ideal transformer based on the f–i analogy in Figure 5.7.

The reversal of current directions and voltage polarities in the secondaries of these two figures is to show the reversal of the directions of both torque and angular velocity due to coupling; it is equivalent to putting dots on opposite ends of the primary and secondary windings of the transformer. In general, it is easy to determine the relative directions of motion of the coupled mechanical elements by inspection of mechanical system without elaborating notations in the electrical circuits.

The simple lever is another type of mechanical coupling device that is analogous to the transformer. Consider the lever in Figure 5.8 which rests on a rigid fulcrum P. The lever is assumed to be massless but rigid, and its left end is connected to the ground through some mechanical

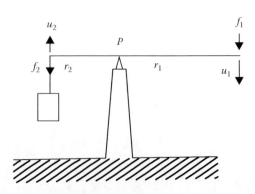

FIGURE 5.8
Translatory motion mechanical system.

Analogous of Linear Systems

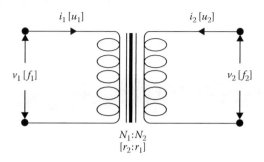

FIGURE 5.9
f–v analogy.

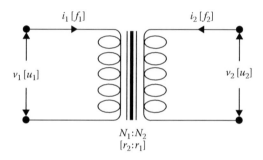

FIGURE 5.10
f–i analogy.

elements which resists motion. If a force f_1 applied to the right end makes it move with a velocity u_1, the following relations hold:

$$\frac{u_1}{u_2} = \frac{r_1}{r_2}$$

$$\frac{f_1}{f_2} = \frac{r_2}{r_1}$$
(5.6)

These equations are also similar to that of an ideal transformer. We find that the velocities of the ends of the simple lever correspond to the torques on the gears, and the forces on the lever correspond to the angular velocities of the gears. Therefore, although the electrical analog of a simple lever is also an ideal transformer, the *f–v* analog for a lever will correspond to the *f–i* analog of a pair of meshed gears, and vice versa, as shown in Figures 5.9 and 5.10.

Example 5.3

Find the *f–i* analogous electrical circuit of the mechanical system shown in Figure 5.11. Assume that the bar is rigid but massless and that the junctions are restricted to have vertical motion only.

Solution

Since this system has a lever-type coupling, the existence of an ideal transformer in the electrical analog is apparent. However, we do not have a simple lever here because the bar does not rest or pivot on a fixed fulcrum. The primary circuit in the electrical analog can be drawn without

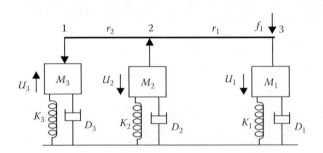

FIGURE 5.11
Mechanical system.

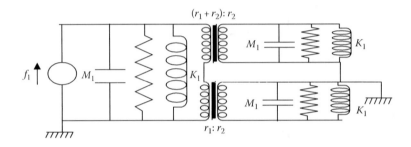

FIGURE 5.12
f–i analogy.

difficulty. By the rule of f–i analogy, we know that the primary circuit has one independent node with a current source (f) and three elements (a capacitance M_1, a resistance $1/D_1$, and an inductance K_1) connected to it, as shown in Figure 5.12.

To draw the secondary circuit, we apply the principle of superposition. First, consider junction 3 as fixed. We then have

$$\frac{u_1}{u_2} = \frac{r_1 + r_2}{r_2}$$

$$\frac{f_1}{f_2} = \frac{r_2}{r_1 + r_2} \quad (5.7)$$

$$\frac{N_1}{N_2} = \frac{r_1 + r_2}{r_2}$$

Next considering junction 2 as fixed, we have

$$\frac{u_1}{u_3} = \frac{r_1}{r_2}$$

$$\frac{f_1}{f_3} = \frac{r_2}{r_1} \quad (5.8)$$

$$\frac{N_1}{N_3} = \frac{r_1}{r_2}$$

Analogous of Linear Systems

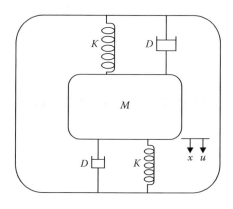

FIGURE 5.13
Schematic diagram of cushioned package system.

Example 5.4

A certain cushioned package is to be dropped from a height $h = 10$ ft. The package can be represented by the schematic diagram, as shown in Figure 5.13. Determine the motion of the mass M. Assume that the package falls onto the ground with no rebound.

$M = 2$ lb-s^2/ft
$K = 5 \times 10^{-4}$ ft/lb
$D = 40$ lb-s/ft

Solution

The *f–v* analogous electrical circuit for this mechanical system can be easily drawn as in Figure 5.14. Note that the only externally applied force on the system is the static weight Mg of the mass M. We write the equation of motion in terms of velocity u as follows:

$$M\frac{du}{dt} + 2Du + \frac{2}{K}\left[\int_0^t u\,dt + x(0)\right] = Mg \tag{5.9}$$

This equation can be solved for u, and displacement and acceleration can then be derived from it. Alternatively we can write the equation of motion in terms of displacement x:

$$M\frac{d^2x}{dt^2} + 2D\frac{dx}{dt} + \frac{2}{K}x = Mg \tag{5.10}$$

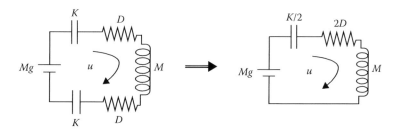

FIGURE 5.14
f–v analogy of given cushioned package system.

The initial conditions at $t = 0$ (the instant at which the package first touches the ground) are

$$x(0) = 0, \quad x'(0) = u(0) = \sqrt{2gh}$$

Taking Laplace transform of the equation of motion, taking due account to the initial conditions, yields

$$\left(Ms^2 + 2Ds + \frac{2}{K}\right)X(s) = M\sqrt{2gh} + \frac{Mg}{s}$$

$$\text{or} \quad X(s) = \frac{\sqrt{2gh}}{s^2 + 2(D/M)s + (2/MK)} + \frac{g}{s^2 + 2(D/M)s + (2/MK)} \tag{5.11}$$

Putting the numerical values, we have

$$X(s) = \frac{25.4}{(s+40s)^2 + 40^2} + 0.016\left[\frac{1}{s} - \frac{s+40}{(s+20)^2 + 40^2}\right] \tag{5.12}$$

Taking the inverse Laplace transform of the above equation, we get the displacement $x(t)$:

$$x(t) = L^{-1}[X(s)]$$

$$= \frac{25.4}{40}e^{-20t}\sin 40t + 0.016\left[1 - e^{-20t}\left(\cos 40t + \frac{1}{2}\sin 40t\right)\right]$$

$$= 0.016 + 0.627e^{-20t}\sin(40t - 1.46°) \text{ ft}$$

Both velocity and acceleration functions can be obtained from $x(t)$ by differentiation.
State model may be written from Equation 5.10:

$$\frac{d}{dt}\begin{bmatrix} v \\ f \end{bmatrix} = \begin{bmatrix} -\frac{D}{M} & \frac{-1}{M} \\ K & 0 \end{bmatrix}\begin{bmatrix} v \\ f \end{bmatrix} + \begin{bmatrix} \frac{-1}{M} \\ 0 \end{bmatrix}U$$

MATLAB® Program

```
clear all;
k = 2.1; %spring constant
d = 0.1;
m = 50.5;
x = [0; 0];
dt = 0.01;
t = 0;
tsim = 100;
n = round((tsim-t)/dt);
A = [-d/m -1/m
     k 0];
B = [-1/m
     0];
u = 1;
for i = 1:n
    if t < = .01; u = 1; else u = 0; end;
    x1(i,:) = [t x'];
```

Analogous of Linear Systems

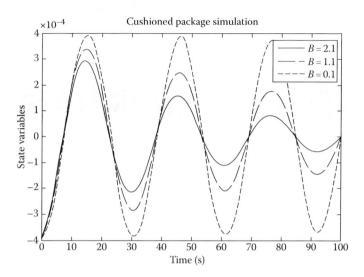

FIGURE 5.15
Cushioned package response for different values of B.

```
      dx = A*x + B*u;
      x = x + dt*dx;
      t = t + dt;
end
hold on
plot (x1(:,1), x1(:,2), 'k')

xlabel ('Time(sec.)')
ylabel('State Variables')
title('Cushioned package Simulation')
```

Results

1. Effect of damping coefficient (B) on system response has been studied when changed from 0.1 to 2.1 for $K = 2.1$ and $M = 50.5$ kg and the results are shown in Figure 5.15.
2. Effect of spring constant ($K = 0.1$–2.1) has been studied on cushioned package when droped from certain height with $B = 1.1$ and $M = 50.5$ and the simulation results are shown in Figure 5.16.
3. Lastly, the effect of change in mass (M) from 10.5 to 100.5 kg with $B = 1.1$ and $K = 2.1$ has also been studied and it is found that the velocity overshoot is more with less weight compared to higher weights, as shown in Figure 5.17.

Example 5.5

Truck–trailer system

Draw the mechanical network for the truck–trailer system shown in Figure 5.18. Draw the analogous electrical system in which force is analogous to current. Also draw force–voltage analogy.

Solution

The systematic mechanical system is shown in Figure 5.19.
Let us draw the equivalent mechanical network as shown in Figure 5.20.

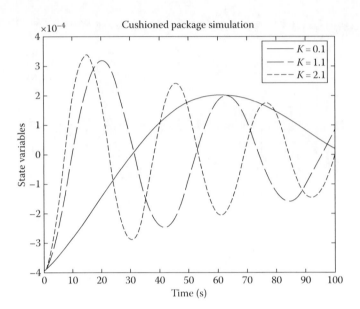

FIGURE 5.16
Cushioned package response for different values of spring constant K.

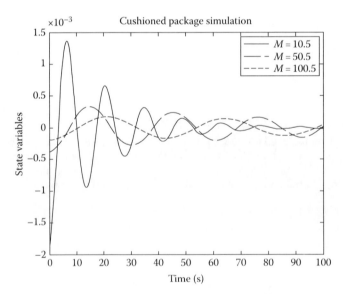

FIGURE 5.17
Cushioned package response for different values of mass (M).

Based on the f–i analogy referring to Table 5.2, the analogous electrical system is drawn as in Figure 5.21.

Similarly, based on the f–v analogy referring to Table 5.1, the analogous electrical system is drawn as in Figure 5.22.

Analogous of Linear Systems

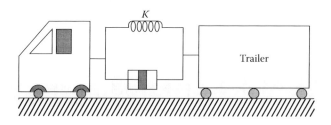

FIGURE 5.18
Truck–trailer system.

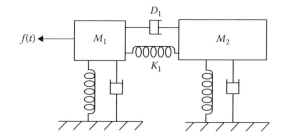

FIGURE 5.19
Systematic mechanical system.

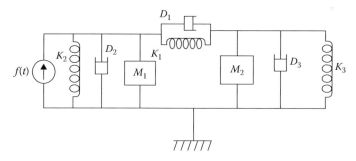

FIGURE 5.20
Equivalent mechanical network.

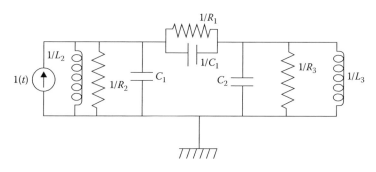

FIGURE 5.21
f–i analogy of truck–trailer system.

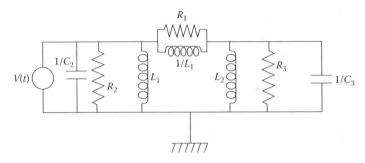

FIGURE 5.22
f–v analogy of truck–trailer system.

Example 5.6: Modeling of Log Chipper

The log chipper cuts logs into smaller chips to make paper pulp. The engine accelerates the massive chipper head to high speed. Then logs are forced against the spinning blades. The high inertia of the head forces the fluid clutch and provides smooth engagement and disengagement of the engine from the drive shaft.

- Derive a lumped model that will represent the rotational behavior of the system
- Draw system graph
- Derive a set of system equations
- Simulate the above developed model using MATLAB
- Draw its f–i and f–v analogy

Solution

The schematic diagram of log chipping system is shown in Figure 5.23. It consists of an engine, which supplies power to log chipping head through a fluid clutch. The fluid clutch engages and disengages the engine shaft with chipper head. The chipper head is supported from both the ends with bearings. The load applied to log chipper head by application of wooden log, and it depends upon the size of log and the pressure applied on it.

The lumped mechanical network for log chipping system is shown in Figure 5.24.

Step 1   Variable identification

To model the given system, it is necessary to identify the key variables of the system. The list of key variables is mentioned below.

List of key variables

J_1 inertia of chipper head
J_2 inertia of the rotating part of engine and attached shaft and clutch disk

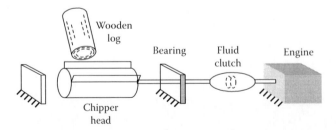

FIGURE 5.23
Schematic diagram for log chipping system.

Analogous of Linear Systems

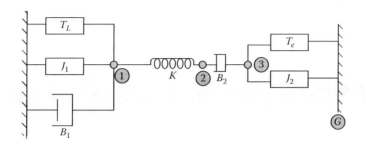

FIGURE 5.24
Lumped model of given log chipping machine.

- K shaft stiffness between 1 and 2
- T engine torque (specified)
- B_1 damper for both bearing friction
- B_2 damper for clutch
- T_L load torque due to log of chipper head
- T_e engine torque

The system graph may be drawn, as shown in Figure 5.25.

Step 2 Topological constraints

1. Across drivers (X)
   a. Independent
   b. Short circuit elements
   c. Dependent
2. Accumulator (C) type elements
3. Dissipater (R) type elements
4. Delay (L) type elements
5. Through drivers
   a. Dependent
   b. Open circuit elements
   c. Independent

The tree may also be drawn based on the topological restrictions and it is consisting of elements—J_1, J_2, and B_2. The rest of the elements belong to the complement of the tree (co-tree), which are—B_1, K, T_L, and T_e, as shown in Figure 5.26.

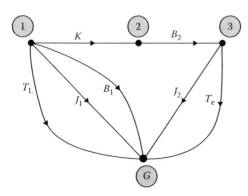

FIGURE 5.25
System graph.

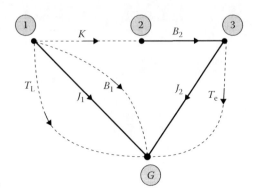

FIGURE 5.26
Formulation tree for given system graph.

Step 3   Incidence matrix

$$
\Lambda = \begin{array}{c|ccccccc}
\text{ND El} & J_1 & J_2 & B_2 & B_1 & K & T_L & T_e \\
\hline
1 & 1 & 0 & 0 & 1 & 1 & 1 & 0 \\
2 & 0 & 0 & 1 & 0 & -1 & 0 & 0 \\
3 & 0 & 1 & -1 & 0 & 0 & 0 & 1 \\
G & -1 & -1 & 0 & -1 & 0 & -1 & -1
\end{array}
$$

Observations: 1. Column wise sum equals to zero.
2. Size must be # node × # elements.

Step 4   Reduced incidence matrix

$$
A_r = \begin{array}{c|ccccccc}
\text{ND El} & J_1 & J_2 & B_2 & B_1 & K & T_L & T_e \\
\hline
1 & 1 & 0 & 0 & 1 & 1 & 1 & 0 \\
2 & 0 & 0 & 1 & 0 & -1 & 0 & 0 \\
3 & 0 & 1 & -1 & 0 & 0 & 0 & 1
\end{array}
$$

Delete the row related to reference node in the incidence matrix.

Step 5   Gauss–Jorden elimination

$$
A_r = \begin{array}{c|ccccccc}
\text{ND El} & J_1 & J_2 & B_2 & B_1 & K & T_L & T_e \\
\hline
R_1 & 1 & 0 & 0 & 1 & 1 & 1 & 0 \\
R_2 + R_3 & 0 & 1 & 0 & 0 & -1 & 0 & 1 \\
R_2 & 0 & 0 & 1 & 0 & -1 & 0 & 0
\end{array}
$$

Step 6   Combined constraint equation

F-cut-set equation $y_b = -A_c^* Y_c$ and f-circuit equation $x_c = -B_b^* X_b$

	J_1	J_2	B_2	B_1	K	T_L	T_e	
T_{J1}	0	0	0	−1	−1	−1	0	w_{J1}
T_{J2}	0	0	0	0	1	0	−1	w_{J2}
T_{B2}	0	0	0	0	1	0	0	w_{B2}
$w_{B1} =$	1	0	0	0	0	0	0	T_{B1}
w_k	1	−1	−1	0	0	0	0	T_k
w_{TL}	1	0	0	0	0	0	0	T_{TL}
w_{Te}	0	1	0	0	0	0	0	T_{Te}

Analogous of Linear Systems

Rearranged combined constraint equation

	J_1	J_2	B_2	B_1	K	T_L	T_e	
T_{J1}	0	0	−1	−1	0	−1	0	ω_{J1}
T_{J2}	0	0	1	0	0	0	−1	ω_{J2}
ω_k	1	−1	0	0	−1	0	0	T_k
$\omega_{B1} =$	1	0	0	0	0	0	0	T_{B1}
T_{B2}	0	0	1	0	0	0	0	ω_{B2}
ω_{TL}	1	0	0	0	0	0	0	T_{TL}
ω_{Te}	0	1	0	0	0	0	0	T_{Te}

(5.13)

Write middle row of matrix equation

$$\begin{bmatrix} \omega_{B1} \\ T_{B2} \end{bmatrix} = \begin{bmatrix} 1 & 0 & 0 \\ 0 & 0 & 1 \end{bmatrix} \begin{bmatrix} \omega_{J1} \\ \omega_{J2} \\ T_k \end{bmatrix}$$

(5.14)

Step 7 **Write terminal equation for memory-less elements**

$$\begin{bmatrix} T_{B1} \\ \omega_{B2} \end{bmatrix} = \begin{bmatrix} B_1 & 0 \\ 0 & 1/B_2 \end{bmatrix} \begin{bmatrix} \omega_{B1} \\ T_{B2} \end{bmatrix}$$

(5.15)

Substitute Equation 5.14 in Equation 5.15.

$$\begin{bmatrix} T_{B1} \\ \omega_{B2} \end{bmatrix} = \begin{bmatrix} B_1 & 0 \\ 0 & 1/B_2 \end{bmatrix} \begin{bmatrix} 1 & 0 & 0 \\ 0 & 0 & 1 \end{bmatrix} \begin{bmatrix} \omega_{J1} \\ \omega_{J2} \\ T_k \end{bmatrix}$$

(5.16)

Write top row of matrix Equation 5.13.

$$\begin{bmatrix} T_{J1} \\ T_{J2} \\ \omega_K \end{bmatrix} = \begin{bmatrix} 0 & 0 & -1 \\ 0 & 0 & 1 \\ 1 & -1 & 0 \end{bmatrix} \begin{bmatrix} \omega_{J1} \\ \omega_{J2} \\ T_K \end{bmatrix} + \begin{bmatrix} -1 & 0 \\ 0 & 0 \\ 0 & -1 \end{bmatrix} \begin{bmatrix} T_{B1} \\ \omega_{B2} \end{bmatrix} + \begin{bmatrix} -1 & 0 \\ 0 & -1 \\ 0 & 0 \end{bmatrix} \begin{bmatrix} T_{TL} \\ T_{Te} \end{bmatrix}$$

(5.17)

Substitute Equation 5.16 of algebraic elements into Equation 5.17.

$$\begin{bmatrix} T_{J1} \\ T_{J2} \\ \omega_K \end{bmatrix} = \begin{bmatrix} 0 & 0 & -1 \\ 0 & 0 & 1 \\ 1 & -1 & 0 \end{bmatrix} \begin{bmatrix} \omega_{J1} \\ \omega_{J2} \\ T_K \end{bmatrix} + \begin{bmatrix} -1 & 0 \\ 0 & 0 \\ 0 & -1 \end{bmatrix} \begin{bmatrix} B_1 & 0 \\ 0 & 1/B_2 \end{bmatrix} \begin{bmatrix} 1 & 0 & 0 \\ 0 & 0 & 1 \end{bmatrix} \begin{bmatrix} \omega_{J1} \\ \omega_{J2} \\ T_K \end{bmatrix} + \begin{bmatrix} -1 & 0 \\ 0 & -1 \\ 0 & 0 \end{bmatrix} \begin{bmatrix} T_{TL} \\ T_{Te} \end{bmatrix}$$

(5.18)

After simplifying Equation 5.18, we get

$$\begin{bmatrix} T_{J1} \\ T_{J2} \\ \omega_K \end{bmatrix} = \begin{bmatrix} -B_1 & 0 & 0 \\ 0 & 0 & 0 \\ 0 & 0 & -1/B_2 \end{bmatrix} \begin{bmatrix} \omega_{J1} \\ \omega_{J2} \\ T_K \end{bmatrix} + \begin{bmatrix} -1 & 0 \\ 0 & -1 \\ 0 & 0 \end{bmatrix} \begin{bmatrix} T_{TL} \\ T_{Te} \end{bmatrix}$$

(5.19)

Step 8 Write terminal equation for memory elements

$$\begin{bmatrix} T_{J1} \\ T_{J2} \\ \omega_K \end{bmatrix} = \begin{bmatrix} J_1 & 0 & 0 \\ 0 & J_2 & 0 \\ 0 & 0 & 1/K \end{bmatrix} \frac{d}{dt} \begin{bmatrix} \omega_{J1} \\ \omega_{J2} \\ T_K \end{bmatrix} \quad (5.20)$$

Substitute terminal Equation 5.20 in Equation 5.19.

$$\begin{bmatrix} J_1 & 0 & 0 \\ 0 & J_2 & 0 \\ 0 & 0 & 1/K \end{bmatrix} \frac{d}{dt} \begin{bmatrix} \omega_{J1} \\ \omega_{J2} \\ T_K \end{bmatrix} = \begin{bmatrix} -B_1 & 0 & 0 \\ 0 & 0 & 0 \\ 0 & 0 & -1/B_2 \end{bmatrix} \begin{bmatrix} \omega_{J1} \\ \omega_{J2} \\ T_K \end{bmatrix} + \begin{bmatrix} -1 & 0 \\ 0 & -1 \\ 0 & 0 \end{bmatrix} \begin{bmatrix} T_{TL} \\ T_{Te} \end{bmatrix} \quad (5.21)$$

Step 9 Final state space equation

$$\frac{d}{dt} \begin{bmatrix} \omega_{J1} \\ \omega_{J2} \\ T_K \end{bmatrix} = \begin{bmatrix} -B_1/J_1 & 0 & 0 \\ 0 & 0 & 0 \\ 0 & 0 & -K/B_2 \end{bmatrix} \begin{bmatrix} \omega_{J1} \\ \omega_{J2} \\ T_K \end{bmatrix} + \begin{bmatrix} -1/J_1 & 0 \\ 0 & -1/J_2 \\ 0 & 0 \end{bmatrix} \begin{bmatrix} T_{TL} \\ T_{Te} \end{bmatrix} \quad (5.22)$$

Simulation Program

```
%Simulation of Log Chipper Machine
clear all
clc
B1 = 0.2;
B2 = 0.1;
J1 = .2;
J2 = 3;
K = 5;
TL = -3;
Te = 4;
a = [-B1/J1 0 0; 0 0 0; 0 0 -K/B2];
b = [-1/J1 0; 0 -1/J2; 0 0];
v = [TL; Te];
x = [0;0;0];
dt = 0.01;
t = 0; tsim = 15; n = (tsim-t)/dt;
for i = 1:n
    dx = a*x + b*v;
    x = x + dt*dx;
    x1(i,:) = [t, x']
    t = t + dt;
end
plot(x1(:,1),x1(:,2))
xlabel('Time, sec.')
ylabel('state variable')
title('plot of x')
```

The simulation result for the speed of log chipper head given load is shown in Figure 5.27.

Simulink® Model

The Simulink model for the same system is shown in Figure 5.28 and the result is also shown in Figure 5.29.

Analogous of Linear Systems

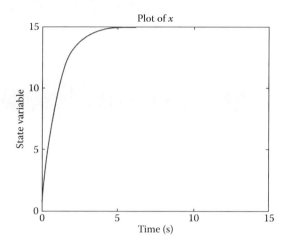

FIGURE 5.27
Speed of log chipper head.

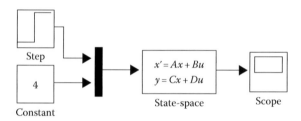

FIGURE 5.28
Simulink model for log chipping system.

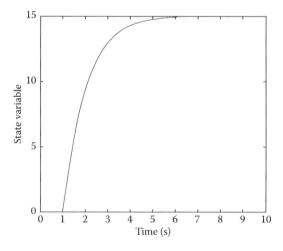

FIGURE 5.29
Simulation results of Simulink model.

To draw its electrical analogous systems, the mechanical network is helpful as shown in Figure 5.30.

Based on f–v analogous quantities mentioned in Table 5.1, the f–v analogy may be drawn as shown in Figure 5.31.

Based on f–i analogous quantities mentioned in Table 5.2, the f–i analogy may be drawn as shown in Figure 5.32.

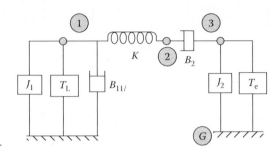

FIGURE 5.30
Mechanical network of log chipping system.

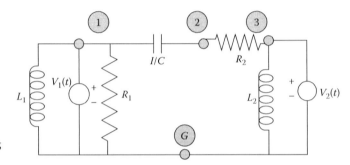

FIGURE 5.31
f–v analogy of given log chipping system.

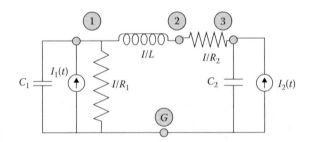

FIGURE 5.32
f–i analogy for given log chipping system.

Example 5.7

Consider a speedometer cable of an automobile shown in Figure 5.33, in which rotation of wheel turns the cable within an oil sheath. The cable turns a cup, which in turn drags a second cup by viscous friction of oil film between cups. A spring restrains rotation of second cup. An indicator needle, attached to the second cup, indicates the speed of an automobile.

Data
Cable
  Length = 1 m
  Diameter = 2 mm
  Inside diameter of sheath = 2.5 mm
  Density of cable material = 6000 kg/m³
  Shear modulus $G = 7 \times 10^{10}$

Analogous of Linear Systems

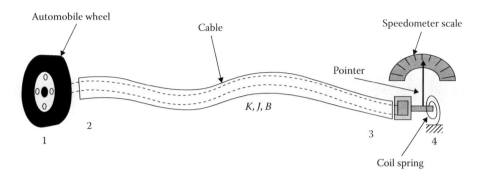

FIGURE 5.33
Speedometer cable of an automobile.

Viscosity of oil = 0.1 N-s/m^2
Thickness of oil film between cups = 0.1 mm.
Rotating cup made by aluminum
   Thickness = 0.3 mm
   Length = 0.7 cm
   Inside diameter = 2 cm
Lumped model of speedometer cable is shown in Figure 5.34.

Parameter Calculation

- Spring constants

  $K = 0.44$ N-m/rad
  $K_s = 8.8 \times 10^{-3}$ N-m/rad

- Inertia

  $J_c = 6.11 \times 10^{-4}$ kg-m^2; $J_s = 2.49 \times 10^{-4}$ kg-m^2
  $J = 2.36 \times 10^{-9}$ kg m^2

- Dampers

  $B = 6.28 \times 10^{-7}$ N-m-s/rad
  $B_c = 4.4 \times 10^{-5}$ N-m-s/rad

The lumped model of given system may be redrawn to get a mechanical network, as shown in Figure 5.35, which will help in drawing electrical analogous systems. The *f–v* and *f–i* analogies for a given mechanical system are drawn in Figures 5.36 and 5.37, respectively.

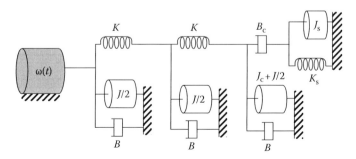

FIGURE 5.34
Lumped model of speedometer cable.

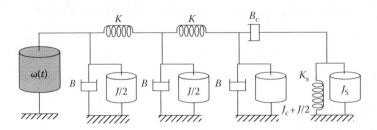

FIGURE 5.35
Mechanical network for speedometer cable.

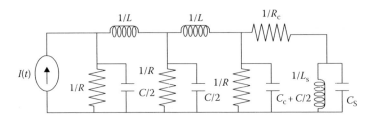

FIGURE 5.36
f–i analogy of speedometer cable.

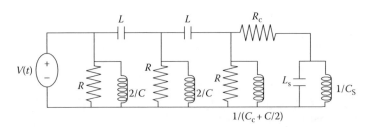

FIGURE 5.37
f–v analogy of speedometer cable.

5.4 Review Questions

1. The mechanical system shown in Figure 5.38.
    a. Draw its f–v and f–i analogy of given system.
    b. Write their system equations.
    c. Write MATLAB executable codes to simulate the given mechanical system.
    d. Analyze the results obtained.
2. Draw f–i analogy of a given spring balance shown in Figures 5.39 and 5.40.
3. Draw f–v analogy for the systems shown in Figures 5.41 through 5.43.

Analogous of Linear Systems

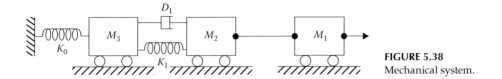

FIGURE 5.38
Mechanical system.

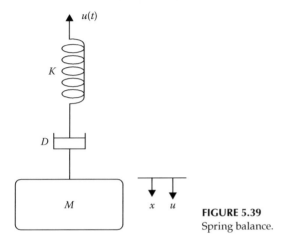

FIGURE 5.39
Spring balance.

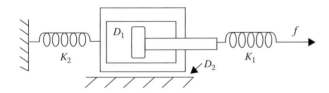

FIGURE 5.40
Mechanical system.

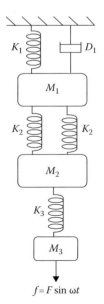

FIGURE 5.41
Mechanical system-I.

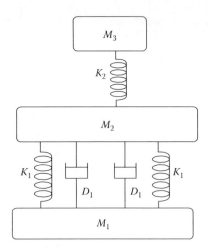

FIGURE 5.42
Mechanical system-II.

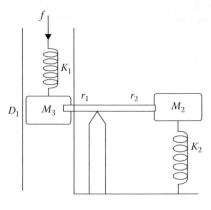

FIGURE 5.43
Lever-based system.

6

Interpretive Structural Modeling

6.1 Introduction

An individual often encounters complexity while dealing with systems. This complexity exists because of a large number of elements involved and the large number of interactions among them. These elements and interactions may be of any form, they may be the elements of a large organization or components of a large technical system. Another form of elements interactions may exist only in the mind of the individual.

A common characteristic of complex systems is that each has a structure associated with it. This structure may be obvious, as in the case of managerial organization of a cooperation or less obvious as in the case of the value structure of a decision maker. An individual or a group can make better decisions concerning systems when the structure of the system is well defined. Hence, it is desirable that some method be developed that aids in identifying the structure within the system.

Interpretive structural modeling (ISM) is one of such method. ISM refers to the systematic applications of graph theory notions such that there results a directed graph representation of complex patterns of a particular contextual relationship among a set of elements. It transforms unclear poorly articulated mental models of systems into visible and well-defined models.

The ISM process transforms unclear, poorly articulated mental models of systems into visible, well-defined models useful for many purposes. The underlying theory for ISM will be presented and discussed by nets, relations, and digraphs. ISM has its basis in mathematics, particularly in graph theory, set theory, mathematical logic, and matrix theory.

6.2 Graph Theory

Many elements in a large-scale system interact with each other. It is, therefore, desirable for us to have methods for identifying these interactions among elements. Interaction matrices and graphs are very useful ways of doing this. To develop an interaction matrix (often called self-interaction matrix or SIM), we must consider all elements and determine if a pair interacts or does not interact with respect to some contextual relation. If the influence between the two elements does not occur by means of a third element, then it is called a direct interaction or a first-order interaction. For indirect or higher order interactions, the interaction graph is the best way of displaying them.

Interaction matrix is filled with 0 and 1 only; therefore, it is called binary interaction matrix. Binary 1 is used to indicate direct interaction while 0 is used to indicate no direct interaction. SIM can always be written in triangular form. For n elements, there are $n(n-1)/2$ entries to be made in SIM. If all entries in an interaction matrix are 1 then every element is connected to rest of the elements in the system and it is represented by nondirected lines in an interaction graph. For example, SIM of typical systems shown in Tables 6.1 and 6.2 are represented by nondirected line in an interaction graph, as shown in Figures 6.1 and 6.2.

If every line segment in a graph has a direction, then the graph is called an oriented or directed graph. It can be represented by a full matrix (Tables 6.3 and 6.4) of size $n \times n$, where n is the number of elements in the system, as shown in Figure 6.3. Mathematically, it can be written as $e_{ij} = 0$, if there is no relation (interaction) from element i to element j.

$e_{ij} = 1$, if there is a relation (interaction) from element i to element j.

Sometimes, we use numbers like 1, 2, 3, 4, and 5 to represent the interaction between elements rather than binary numbers. This shows the strength of interactions such as

1. No interaction
2. Very poor interaction
3. Poor interaction
4. Moderate interaction
5. Strong interaction

Example 6.1

A simple single loop feedback system for an antenna servomechanism might appear as shown in Figure 6.4a. A flow graph for this system could also be represented, as shown in Figure 6.4b. In this flow graph, arrows show the direction of signals and power flow. Flow graph of this system is also the digraph in ISM, because ISM uses contextual relationship such as change in element **a** will lead to change in element **b**.

The SIM for the above system could be written as

$$A_i = \begin{array}{c} \\ 1 \\ 2 \\ 3 \\ 4 \\ 5 \\ 6 \end{array} \begin{array}{c} \begin{matrix} 1 & 2 & 3 & 4 & 5 & 6 \end{matrix} \\ \begin{bmatrix} 0 & 1 & 0 & 0 & 0 & 0 \\ 0 & 0 & 1 & 0 & 0 & 0 \\ 0 & 0 & 0 & 1 & 0 & 0 \\ 1 & 0 & 0 & 0 & 0 & 1 \\ 1 & 0 & 0 & 0 & 0 & 0 \\ 0 & 0 & 0 & 0 & 0 & 0 \end{bmatrix} \end{array}$$

TABLE 6.1
SIM

	A_3	A_2	A_1
A_1	0		1
A_2	1		
A_3			

TABLE 6.2
SIM

	A_4	A_3	A_2	A_1
A_1	1	0	1	
A_2	0	1		
A_3	1			
A_4				

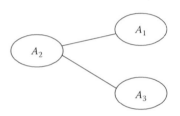

FIGURE 6.1
Nondirected interaction graph.

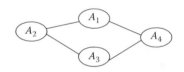

FIGURE 6.2
Nondirected interaction graph.

TABLE 6.3
SIM

	A_3	A_2	A_1
A_1	0	0	0
A_2	0	0	1
A_3	0	1	0

TABLE 6.4
SIM

	A_4	A_3	A_2	A_1
A_1	1	0	0	0
A_2	0	1	0	1
A_3	0	0	0	0
A_4	0	1	0	0

Interpretive Structural Modeling

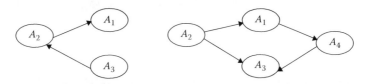

FIGURE 6.3
Directed interaction graph for SIM shown in Tables 6.3 and 6.4.

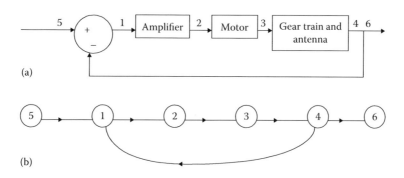

FIGURE 6.4
(a) Block diagram for simple antenna and servomechanism. (b) Flow graph for simple antenna and servomechanism.

SIM often written as and called structured self-interaction matrix (SSIM).

N	A	A	N	V	1
N	N	N	V	2	
N	N	V	3		
V	N	4			
N	5				
6					

where

0 or N—no interaction, that is, element i is not related with element j and vice versa ($e_{ij} = 0$ and $e_{ji} = 0$)

A—element j is related with element i (i.e., $e_{ji} = 1$) and element i is not related with element j (i.e., $e_{ij} = 0$)

V—element i is related with element j but element j is not related with element i ($e_{ji} = 0$ and $e_{ij} = 1$)

X—element i is related with element j and element j is also related with element i ($e_{ji} = 1$ and $e_{ij} = 1$)

If all the diagonal entries of SIM are made 1 (i.e., $e_{ii} = 1$) then it becomes reachability matrix. The reachability matrix shows that either element i is reachable from element j or not. The number of lines in the path from element i to element j is called the length of the path. In Figure 6.4b element 4 is reachable from element 2 by the length of path 2. For completeness, we define every point reachable from itself by a path length 0. Hence, we add identity matrix in A_i to get reachability matrix. The reachability matrix for the above system is equal to $(A_i + I)$.

$$(A_i + I) = \begin{matrix} & 1 & 2 & 3 & 4 & 5 & 6 \\ 1 & \begin{bmatrix} 1 & 1 & 0 & 0 & 0 & 0 \\ 2 & 0 & 1 & 1 & 0 & 0 & 0 \\ 3 & 0 & 0 & 1 & 1 & 0 & 0 \\ 4 & 1 & 0 & 0 & 1 & 0 & 1 \\ 5 & 1 & 0 & 0 & 0 & 1 & 0 \\ 6 & 0 & 0 & 0 & 0 & 0 & 1 \end{bmatrix} \end{matrix}$$

If we multiply $(A_i + I)$ with itself and if the value of any element is greater than 1 make it 1, we will get

$$(A_i + I)^2 = \begin{matrix} & 1 & 2 & 3 & 4 & 5 & 6 \\ 1 & \begin{bmatrix} 1 & 1 & 0 & 0 & 0 & 0 \\ 2 & 0 & 1 & 1 & 1 & 0 & 0 \\ 3 & 1 & 0 & 1 & 1 & 0 & 0 \\ 4 & 1 & 1 & 1 & 1 & 0 & 1 \\ 5 & 1 & 1 & 0 & 0 & 1 & 0 \\ 6 & 0 & 0 & 0 & 0 & 0 & 1 \end{bmatrix} \end{matrix}$$

If we continue this multiplication by itself, a stage is reached when successive powers of $(A + I)$ produce identical matrices:

$$(A_i + I)^3 = \begin{matrix} & 1 & 2 & 3 & 4 & 5 & 6 \\ 1 & \begin{bmatrix} 1 & 1 & 1 & 1 & 0 & 0 \\ 2 & 1 & 1 & 1 & 1 & 0 & 1 \\ 3 & 1 & 1 & 1 & 1 & 0 & 1 \\ 4 & 1 & 1 & 1 & 1 & 0 & 1 \\ 5 & 1 & 1 & 1 & 0 & 1 & 0 \\ 6 & 0 & 0 & 0 & 0 & 0 & 1 \end{bmatrix} \end{matrix}$$

$$(A_i + I)^4 = \begin{matrix} & 1 & 2 & 3 & 4 & 5 & 6 \\ 1 & \begin{bmatrix} 1 & 1 & 1 & 1 & 0 & 1 \\ 2 & 1 & 1 & 1 & 1 & 0 & 1 \\ 3 & 1 & 1 & 1 & 1 & 0 & 1 \\ 4 & 1 & 1 & 1 & 1 & 0 & 1 \\ 5 & 1 & 1 & 1 & 0 & 1 & 0 \\ 6 & 0 & 0 & 0 & 0 & 0 & 1 \end{bmatrix} \end{matrix}$$

$$(A_i + I)^5 = \begin{matrix} & 1 & 2 & 3 & 4 & 5 & 6 \\ 1 & \begin{bmatrix} 1 & 1 & 1 & 1 & 0 & 1 \\ 2 & 1 & 1 & 1 & 1 & 0 & 1 \\ 3 & 1 & 1 & 1 & 1 & 0 & 1 \\ 4 & 1 & 1 & 1 & 1 & 0 & 1 \\ 5 & 1 & 1 & 1 & 0 & 1 & 1 \\ 6 & 0 & 0 & 0 & 0 & 0 & 1 \end{bmatrix} \end{matrix}$$

Interpretive Structural Modeling

$$(A_i + I)^6 = \begin{array}{c} \\ 1 \\ 2 \\ 3 \\ 4 \\ 5 \\ 6 \end{array} \begin{array}{cccccc} 1 & 2 & 3 & 4 & 5 & 6 \\ \left[\begin{array}{cccccc} 1 & 1 & 1 & 1 & 0 & 1 \\ 1 & 1 & 1 & 1 & 0 & 1 \\ 1 & 1 & 1 & 1 & 0 & 1 \\ 1 & 1 & 1 & 1 & 0 & 1 \\ 1 & 1 & 1 & 0 & 1 & 1 \\ 0 & 0 & 0 & 0 & 0 & 1 \end{array}\right] \end{array}$$

It is clear that $(A_i + I)^5 = (A_i + I)^6 = (A_i + I)^7 \ldots$

This shows that in this system digraph has maximum length of path is 5. Element 6 is reachable from element 5 by length of path equal to 5. In general, it could be written as

$$(A_i + I)^{r-2} \neq (A_i + I)^{r-1} = (A_i + I)^r = p$$

where
$r \leq (n - 1)$
n is number of elements

The longest possible path for n elements is $(n - 1)$, and higher power of $(A_i + I)$ would merely indicate reachability through paths of length $(n - 1)$ or less. Matrix p is a transitive closure of A_i.

The transitive relation can be defined in this manner: if there is a path from e_i to e_j and path from e_j to e_k then there is also a path from e_i to e_k. Mathematically, it is written as

If $e_i \to e_j$ and $e_j \to e_k$ then $e_i \to e_k$

The original digraph need not be transitive in order to have a transitive reachability relation. This is possible because the reachability relation is a transformation of the original relation. The reachability matrix, in which all entries are 1, is called the universal matrix and implies that every element is reachable from every other element. When a digraph or a subset of a digraph has a universal reachability matrix, that digraph or subset of digraph is said to be strongly connected.

6.2.1 Net

A net is a collection of connected elements or points with a given interaction (relation). A simple net is shown in Figure 6.5 consisting of only element and two nodes.

First point $P_1 = f(r_k)$ and second point $P_2 = s(r_k)$ for $r \in$ Set of relation R and $P_1, P_2 \in$ Set of points (elements) P. R and P both are finite sets. It shows that P_1 is related with P_2 and P_2 is not related with P_1, that is,

$$P_1 r_k P_2 \quad \text{and} \quad P_2 \bar{r}_k P_1$$

6.2.2 Loop

A link r_k is called a loop if the first point and second point of a line are same.
Mathematically this may be written as

$$f(r_k) = s(r_k)$$

It is shown in Figure 6.6a.

6.2.3 Cycle

It is defined as a closed sequence of lines without any parallel lines, as shown in Figure 6.6b.

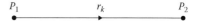

FIGURE 6.5
A simple net.

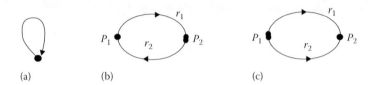

FIGURE 6.6
Components of a simple net. (a) Loop. (b) Cycle. (c) Parallel lines.

6.2.4 Parallel Lines

Two lines are said to be parallel if they have the same first and second points, that is,
First point $P_1 = f(r_1) = f(r_2)$ and second point $P_2 = s(r_1) = s(r_2)$.

To write an interaction matrix, we may consider all possible interactions between elements, but in the ISM process, we use only one contextual relation to develop a model. Contextual relations may be very general or very specific.

6.2.5 Properties of Relations

1. *Reflexive*: A relation is said to be reflexive if every point P_i is on a loop.

$$P_i R P_i \quad \text{for all } P_i \in P$$

2. *Irreflexive*: A relation is irreflexive if no point P_i is on a loop.

$$P_i \overline{R} P_i \quad \text{for all } P_i \in P$$

3. *Symmetric*: A relation is symmetric if point i is related with point j and point j is also related with point i.

$$P_i R P_j \quad \text{and} \quad P_j R P_i \quad \text{for all } P_i \text{ and } P_j \in P$$

4. *Asymmetric*: A relation is asymmetric if point i is related with point j and point j is not related with point i.

$$P_i R P_j \quad \text{and} \quad P_j \overline{R} P_i \quad \text{for all } P_i \text{ and } P_j \in P$$

5. *Transitive*: A relation is called transitive if point i is related with j and point j is related with k then point i is related with k.

$$\text{If } P_i R P_j \quad \text{and} \quad P_j R P_k \quad \text{then } P_i R P_k$$

6. *Intransitive*: A relation is intransitive if point i is related with j and point j is related with k then point i is not related with k.

$$\text{If } P_i R P_j \quad \text{and} \quad P_j R P_k \quad \text{then } P_i \overline{R} P_k$$

Interpretive Structural Modeling

7. *Complete*: A relation is complete if for every point P_i, P_j either point i is related with point j or point j is related with point i.

$$P_iRP_j \quad \text{or} \quad P_jRP_i$$

6.3 Interpretive Structural Modeling

This is a tool to use in value system design, system synthesis, and system analysis. Each stage of this methodology may be viewed as transforming a model from one format to another. This transformation has been referred to as a model exchange isomorphism (MEI). The MEIs are considered as candidate paradigms for ISM, as shown in Figure 6.7. A necessary step in this activity is to establish the preconditioning guidelines for ISM.

The following algorithm presents a convenient format for the preconditioning guidelines for ISM:

Theme	Purpose or goal or objective
Prospective	Viewed from what role position
Mode	Usually descriptive or prescriptive
Primitives	Element set (P)
	Set of relations (R)

SSIM constructed by recording participant responses as a group in a sequenced questioning session, reflecting judgment as to the existence of a specific relation R between any two elements i and j of element set S and the associated directions of the relation. This transformation of a model from one form (say mental model) to another form (say data set or SSIM) may be referred to as a MEI. Successive MEIs will transform perceived mental model to data set (SSIM)–to reachability matrix–to–canonical form–to–structural model (digraph)–to–ISM (Sage, 1977; Satsangi, 2009). Different MEIs are shown in Figure 6.7.

1. MEI-1

    At this stage modeler identifies the system elements and the relations among them. For determining the relation between the elements, we compare the elements on a pair wise basis with respect to the relation being modeled. Hence, two sets are identified at this stage:

    a. Element set P which is finite and nonempty
    b. Relational set R which is also finite

    Hence we obtain the data or information about the system in this MEI.

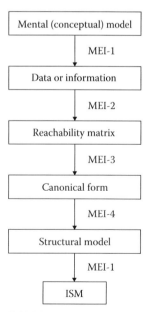

FIGURE 6.7
Model exchange isomorphism (MEI).

2. MEI-2

   The second MEI uses the above set of data or information to create the reachability matrix. The entries in the reachability matrix are 0s and 1s depending on whether elements are related or not related.

3. MEI-3

   The third MEI induces partitions on the set of element and rearranges the reachability matrix into canonical form. In this canonical form, properties of the structural model may be identified.

4. MEI-4

   At this stage the minimum edge diagram is drawn from the canonical form of matrix information. This is the first step in drawing structural model.

5. MEI-5

   The fifth MEI is the process of replacing each numbered element with the description of that element which is adequate for interpretation. This is also the final MEI for getting ISM.

Example 6.2

For the directed graph shown in Figure 6.8, write down the system subordinate (relational) matrix. Also write the reachability matrix from this and determine the maximum path length in the graph using reachability matrix.

SOLUTION

The SIM for the given directed graph is

$$A_i = \begin{matrix} & \begin{matrix} 1 & 2 & 3 & 4 & 5 & 6 & 7 \end{matrix} \\ \begin{matrix} 1 \\ 2 \\ 3 \\ 4 \\ 5 \\ 6 \\ 7 \end{matrix} & \begin{bmatrix} 0 & 0 & 0 & 0 & 0 & 0 & 0 \\ 1 & 0 & 0 & 0 & 0 & 0 & 0 \\ 1 & 0 & 0 & 0 & 0 & 0 & 0 \\ 0 & 1 & 0 & 0 & 0 & 0 & 0 \\ 0 & 0 & 0 & 0 & 0 & 1 & 0 \\ 0 & 0 & 1 & 0 & 1 & 0 & 0 \\ 0 & 0 & 1 & 0 & 0 & 0 & 0 \end{bmatrix} \end{matrix}$$

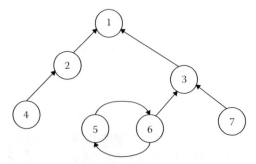

FIGURE 6.8
Directed graph for Example 6.1.

Interpretive Structural Modeling

$$
\text{Identity Matrix } I = \begin{array}{c} \\ 1 \\ 2 \\ 3 \\ 4 \\ 5 \\ 6 \\ 7 \end{array} \begin{array}{c} \begin{array}{ccccccc} 1 & 2 & 3 & 4 & 5 & 6 & 7 \end{array} \\ \left[\begin{array}{ccccccc} 1 & 0 & 0 & 0 & 0 & 0 & 0 \\ 0 & 1 & 0 & 0 & 0 & 0 & 0 \\ 0 & 0 & 1 & 0 & 0 & 0 & 0 \\ 0 & 0 & 0 & 1 & 0 & 0 & 0 \\ 0 & 0 & 0 & 0 & 1 & 0 & 0 \\ 0 & 0 & 0 & 0 & 0 & 1 & 0 \\ 0 & 0 & 0 & 0 & 0 & 0 & 1 \end{array} \right] \end{array}
$$

$$
\text{Reachability Matrix } (A_i + I) = \begin{array}{c} \\ 1 \\ 2 \\ 3 \\ 4 \\ 5 \\ 6 \\ 7 \end{array} \begin{array}{c} \begin{array}{ccccccc} 1 & 2 & 3 & 4 & 5 & 6 & 7 \end{array} \\ \left[\begin{array}{ccccccc} 1 & 0 & 0 & 0 & 0 & 0 & 0 \\ 1 & 1 & 0 & 0 & 0 & 0 & 0 \\ 1 & 0 & 1 & 0 & 0 & 0 & 0 \\ 0 & 1 & 0 & 1 & 0 & 0 & 0 \\ 0 & 0 & 0 & 0 & 1 & 1 & 0 \\ 0 & 0 & 1 & 0 & 1 & 1 & 0 \\ 0 & 0 & 1 & 0 & 0 & 0 & 1 \end{array} \right] \end{array}
$$

$$
(A_i + I)^2 = \begin{array}{c} \\ 1 \\ 2 \\ 3 \\ 4 \\ 5 \\ 6 \\ 7 \end{array} \begin{array}{c} \begin{array}{ccccccc} 1 & 2 & 3 & 4 & 5 & 6 & 7 \end{array} \\ \left[\begin{array}{ccccccc} 1 & 0 & 0 & 0 & 0 & 0 & 0 \\ 1 & 1 & 0 & 0 & 0 & 0 & 0 \\ 1 & 0 & 1 & 0 & 0 & 0 & 0 \\ 1 & 1 & 0 & 1 & 0 & 0 & 0 \\ 0 & 0 & 1 & 0 & 1 & 1 & 0 \\ 1 & 0 & 1 & 0 & 1 & 1 & 0 \\ 1 & 0 & 1 & 0 & 0 & 0 & 0 \end{array} \right] \end{array}
$$

$$
(A_i + I)^3 = \begin{array}{c} \\ 1 \\ 2 \\ 3 \\ 4 \\ 5 \\ 6 \\ 7 \end{array} \begin{array}{c} \begin{array}{ccccccc} 1 & 2 & 3 & 4 & 5 & 6 & 7 \end{array} \\ \left[\begin{array}{ccccccc} 1 & 0 & 0 & 0 & 0 & 0 & 0 \\ 1 & 1 & 0 & 0 & 0 & 0 & 0 \\ 1 & 0 & 1 & 0 & 0 & 0 & 0 \\ 1 & 1 & 0 & 1 & 0 & 0 & 0 \\ 1 & 0 & 1 & 0 & 1 & 1 & 0 \\ 1 & 0 & 1 & 0 & 1 & 1 & 0 \\ 1 & 0 & 1 & 0 & 0 & 0 & 0 \end{array} \right] \end{array}
$$

$$(A_i + I)^4 = \begin{array}{c} \\ 1 \\ 2 \\ 3 \\ 4 \\ 5 \\ 6 \\ 7 \end{array} \begin{array}{c} \begin{array}{ccccccc} 1 & 2 & 3 & 4 & 5 & 6 & 7 \end{array} \\ \begin{bmatrix} 1 & 0 & 0 & 0 & 0 & 0 & 0 \\ 1 & 1 & 0 & 0 & 0 & 0 & 0 \\ 1 & 0 & 1 & 0 & 0 & 0 & 0 \\ 1 & 1 & 0 & 1 & 0 & 0 & 0 \\ 1 & 0 & 1 & 0 & 1 & 1 & 0 \\ 1 & 0 & 1 & 0 & 1 & 1 & 0 \\ 1 & 0 & 1 & 0 & 0 & 0 & 0 \end{bmatrix} \end{array}$$

Since $(A_i + I)^3$ and $(A_i + I)^4$ matrices are same, it is clear that $(A_i + I)^3 = (A_i + I)^4\ldots$ Therefore, the maximum path length is 3 for the given digraph.

Example 6.3

For the directed graph shown in Figure 6.9, write down the system SSIM.

SOLUTION

$$\text{SSIM for given directed graph } A_i = \begin{array}{c} 1 \\ 2 \\ 3 \\ 4 \\ 5 \end{array} \begin{array}{c} \begin{array}{ccccc} 1 & 2 & 3 & 4 & 5 \end{array} \\ \begin{bmatrix} 0 & 0 & 0 & 0 & 0 \\ 1 & 0 & 0 & 1 & 0 \\ 0 & 1 & 0 & 1 & 0 \\ 0 & 0 & 0 & 0 & 0 \\ 1 & 0 & 0 & 0 & 0 \end{bmatrix} \end{array}$$

$$\text{Identity matrix } I = \begin{array}{c} 1 \\ 2 \\ 3 \\ 4 \\ 5 \end{array} \begin{array}{c} \begin{array}{ccccc} 1 & 2 & 3 & 4 & 5 \end{array} \\ \begin{bmatrix} 1 & 0 & 0 & 0 & 0 \\ 0 & 1 & 0 & 0 & 0 \\ 0 & 0 & 1 & 0 & 0 \\ 0 & 0 & 0 & 1 & 0 \\ 0 & 0 & 0 & 0 & 1 \end{bmatrix} \end{array}$$

$$\text{Reachability matrix } A_i + I = \begin{array}{c} 1 \\ 2 \\ 3 \\ 4 \\ 5 \end{array} \begin{array}{c} \begin{array}{ccccc} 1 & 2 & 3 & 4 & 5 \end{array} \\ \begin{bmatrix} 1 & 0 & 0 & 0 & 0 \\ 1 & 1 & 0 & 1 & 0 \\ 0 & 1 & 1 & 1 & 0 \\ 0 & 0 & 0 & 1 & 0 \\ 1 & 0 & 0 & 0 & 1 \end{bmatrix} \end{array}$$

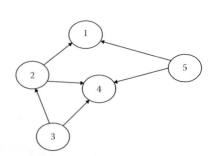

FIGURE 6.9
Directed graph for Example 6.2.

Interpretive Structural Modeling

$$(A_i + I)^2 = \begin{array}{c} \\ 1 \\ 2 \\ 3 \\ 4 \\ 5 \end{array} \begin{bmatrix} 1 & 2 & 3 & 4 & 5 \\ 1 & 0 & 0 & 0 & 0 \\ 1 & 1 & 0 & 1 & 0 \\ 1 & 1 & 1 & 1 & 0 \\ 0 & 0 & 0 & 1 & 0 \\ 1 & 0 & 0 & 1 & 1 \end{bmatrix}$$

$$(A_i + I)^3 = \begin{array}{c} \\ 1 \\ 2 \\ 3 \\ 4 \\ 5 \end{array} \begin{bmatrix} 1 & 2 & 3 & 4 & 5 \\ 1 & 0 & 0 & 0 & 0 \\ 1 & 1 & 0 & 1 & 0 \\ 1 & 1 & 1 & 1 & 0 \\ 0 & 0 & 0 & 1 & 0 \\ 1 & 0 & 0 & 1 & 1 \end{bmatrix}$$

Since $(A_i + I)^2$ and $(A_i + I)^3$ matrices are same, it is clear that $(A_i + I)^2 = (A_i + I)^3 = (A_i + I)^4...$
Therefore, the maximum path length is 2 for the given digraph.

Example 6.4

What will be the associated intent structure of a system whose SIM is given below:

$$A_i = \begin{array}{c} \\ 1 \\ 2 \\ 3 \\ 4 \\ 5 \\ 6 \\ 7 \\ 8 \end{array} \begin{bmatrix} 1 & 2 & 3 & 4 & 5 & 6 & 7 & 8 \\ 0 & 0 & 0 & 0 & 0 & 0 & 0 & 0 \\ 1 & 0 & 0 & 0 & 0 & 0 & 0 & 0 \\ 1 & 0 & 0 & 0 & 0 & 0 & 0 & 0 \\ 1 & 1 & 0 & 0 & 1 & 1 & 0 & 1 \\ 1 & 1 & 0 & 0 & 0 & 0 & 0 & 0 \\ 1 & 1 & 0 & 0 & 0 & 0 & 0 & 0 \\ 0 & 0 & 0 & 0 & 0 & 0 & 0 & 0 \\ 1 & 0 & 0 & 0 & 0 & 0 & 0 & 0 \end{bmatrix}$$

SOLUTION

Step 1 Reachability matrix can be obtained by adding identity matrix in SIM as written below:

$$A_i + I = \begin{array}{c} \\ 1 \\ 2 \\ 3 \\ 4 \\ 5 \\ 6 \\ 7 \\ 8 \end{array} \begin{bmatrix} 1 & 2 & 3 & 4 & 5 & 6 & 7 & 8 \\ 1 & 0 & 0 & 0 & 0 & 0 & 0 & 0 \\ 1 & 1 & 0 & 0 & 0 & 0 & 0 & 0 \\ 1 & 0 & 1 & 0 & 0 & 0 & 0 & 0 \\ 1 & 1 & 0 & 1 & 1 & 1 & 0 & 1 \\ 1 & 1 & 0 & 0 & 1 & 0 & 0 & 0 \\ 1 & 1 & 0 & 0 & 0 & 1 & 0 & 0 \\ 0 & 0 & 0 & 0 & 0 & 0 & 1 & 0 \\ 1 & 0 & 0 & 0 & 0 & 0 & 0 & 1 \end{bmatrix}$$

Step 2   Lower triangular matrix is obtained by rearranging the row and column of reachability matrix based on number of entries in a particular row. Row related with less number of nonzeros entries is kept first and then more number of nonzeros entries and at the bottom (i.e., the row contains maximum number of nonzero entries will be the last row).

$$L = \begin{array}{c} \\ 1 \\ 7 \\ 2 \\ 3 \\ 8 \\ 5 \\ 6 \\ 4 \end{array} \begin{array}{cccccccc} 1 & 7 & 2 & 3 & 8 & 5 & 6 & 4 \\ \left[\begin{array}{cccccccc} 0 & 0 & 0 & 0 & 0 & 0 & 0 & 0 \\ 0 & 0 & 0 & 0 & 0 & 0 & 0 & 0 \\ 1 & 0 & 0 & 0 & 0 & 0 & 0 & 0 \\ 1 & 0 & 0 & 0 & 0 & 0 & 0 & 0 \\ 1 & 0 & 0 & 0 & 0 & 0 & 0 & 0 \\ 1 & 0 & 1 & 0 & 0 & 0 & 0 & 0 \\ 1 & 0 & 1 & 0 & 0 & 0 & 0 & 0 \\ 1 & 0 & 1 & 0 & 1 & 1 & 1 & 0 \end{array}\right] \end{array}$$

Step 3   Minimum edge adjacency matrix is obtained from lower triangular matrix after deleting the entries as mentioned in Table 6.5. The following steps have been followed:

1. The diagonal entries of an adjacency matrix for a digraph are defined to be 0. $e_{ii} = 0$.
2. Search row-wise for $e_{ij} = 1$; as each e_{ij} entry of 1 is identified, the corresponding ith column is searched for all entries of $e_{ki} = 1$, where $k > i$, the corresponding e_{kj} are constraint to be 0 (i.e., $e_{kj} = 0$).

$$M = \begin{array}{c} \\ 1 \\ 7 \\ 2 \\ 3 \\ 8 \\ 5 \\ 6 \\ 4 \end{array} \begin{array}{cccccccc} 1 & 7 & 2 & 3 & 8 & 5 & 6 & 4 \\ \left[\begin{array}{cccccccc} 0 & 0 & 0 & 0 & 0 & 0 & 0 & 0 \\ 0 & 0 & 0 & 0 & 0 & 0 & 0 & 0 \\ 1 & 0 & 0 & 0 & 0 & 0 & 0 & 0 \\ 1 & 0 & 0 & 0 & 0 & 0 & 0 & 0 \\ 1 & 0 & 0 & 0 & 0 & 0 & 0 & 0 \\ 0 & 0 & 1 & 0 & 0 & 0 & 0 & 0 \\ 0 & 0 & 1 & 0 & 0 & 0 & 0 & 0 \\ 0 & 0 & 0 & 0 & 1 & 1 & 1 & 0 \end{array}\right] \end{array}$$

TABLE 6.5

Delete the Entries in Lower Triangular Matrix to get Minimum Edge Adjacency Matrix

Elements (i)	Elements (j)	k	Entries to Be Made Deleted
2	1	5	(5, 1)
		6	(6, 1)
		4	(4, 1)
5	2	4	(4, 2)

Interpretive Structural Modeling

Step 4 Minimum edge digraph

Using minimum edge adjacency matrix, minimum edge digraph may be drawn as shown in Figure 6.10. Containing following levels:

Level	Element
1	1, 7
2	2, 3
3	5, 6, 8
4	4

Example 6.5

Consider two system graphs shown in Figure 6.11, where elements 2 and 6 are common to both graphs. Determine the system structure after combining two system models given and obtain subordination (self-interaction) matrix for the final desired graph.

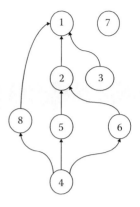

FIGURE 6.10
Minimum edge digraph for ISM.

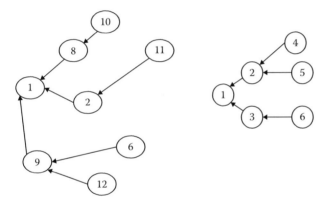

FIGURE 6.11
System directed graphs for Example 6.4.

Solution

Combined system graph shown in Figure 6.12 is obtained by combining given system graph (shown in 6.11) as element 2 and 6 are common to both graphs.

Reachability matrix

$$A_i = \begin{matrix} & \begin{matrix}1 & 2 & 3 & 4 & 5 & 6 & 7 & 8 & 9 & 10 & 11 & 12\end{matrix} \\ \begin{matrix}1\\2\\3\\4\\5\\6\\7\\8\\9\\10\\11\\12\end{matrix} & \begin{bmatrix} 0 & 0 & 0 & 0 & 0 & 0 & 0 & 0 & 0 & 0 & 0 & 0 \\ 1 & 0 & 0 & 0 & 0 & 0 & 0 & 0 & 0 & 0 & 0 & 0 \\ 1 & 0 & 0 & 0 & 0 & 0 & 0 & 0 & 0 & 0 & 0 & 0 \\ 0 & 1 & 0 & 0 & 0 & 0 & 0 & 0 & 0 & 0 & 0 & 0 \\ 0 & 1 & 0 & 0 & 0 & 0 & 0 & 0 & 0 & 0 & 0 & 0 \\ 0 & 0 & 1 & 0 & 0 & 0 & 0 & 0 & 1 & 0 & 0 & 0 \\ 0 & 0 & 0 & 0 & 0 & 0 & 0 & 0 & 0 & 0 & 0 & 0 \\ 1 & 0 & 0 & 0 & 0 & 0 & 0 & 0 & 0 & 0 & 0 & 0 \\ 1 & 0 & 0 & 0 & 0 & 0 & 0 & 0 & 0 & 0 & 0 & 0 \\ 0 & 0 & 0 & 0 & 0 & 0 & 0 & 1 & 0 & 0 & 0 & 0 \\ 0 & 1 & 0 & 0 & 0 & 0 & 0 & 0 & 0 & 0 & 0 & 0 \\ 0 & 0 & 0 & 0 & 0 & 0 & 0 & 0 & 0 & 0 & 0 & 0 \end{bmatrix} \end{matrix}$$

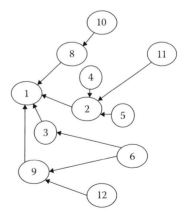

FIGURE 6.12
Combined system graph.

TABLE 6.6
SIM

0	0	0	A	A	0	0	0	0	A	A	1
0	A	0	0	0	0	0	A	A	0	2	
0	0	0	0	0	0	A	0	0	3		
0	0	0	0	0	0	0	0	4			
0	0	0	0	0	0	0	5				
0	0	0	V	0	0	6					
0	0	0	0	0	7						
0	0	A	0	8							
A	0	0	9								
0	0	10									
0	11										
12											

The SIM is written from reachability matrix A_i as shown in Table 6.6.

Example 6.6

Write SIMs for the following undirected graphs shown in Figures 6.13 through 6.15.

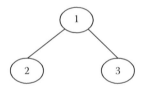

FIGURE 6.13
Undirected graph.

Solution

Self-interaction metrics are written as shown in Tables 6.8, 6.10, and 6.12. The entries made in these matrices are 0 or 1 (0 for indirect interaction and 1 for direct interaction). The corresponding SSIMs are given in Tables 6.7, 6.9, and 6.11.

Example 6.7

A directed SIM is given in Table 6.13.

1. What is the nondirected (nonoriented) SIM.
2. What is the nondirected graph corresponding to this interaction matrix.

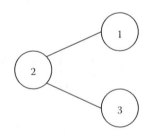

FIGURE 6.14
Undirected graph.

Solution

1. Nonoriented graph for directed SIM given in Table 6.13 is shown in Figure 6.16.
2. Nondirected SIM and corresponding SSIM is given in Table 6.14 and 6.15 respectively.

Example 6.8

For the element set from the animal world the SSIM constructed using a contextual relation is given in Table 6.12.

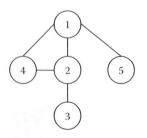

FIGURE 6.15
Undirected graph.

Interpretive Structural Modeling

TABLE 6.7

SSIM

1	1	1
0	2	
3		

TABLE 6.8

SIM

	1	2	3
1	0	1	1
2	1	0	0
3	1	1	1

TABLE 6.9

SSIM

0	1	1
1	2	
3		

TABLE 6.10

SIM

	1	2	3
1	0	1	0
2	1	0	1
3	0	1	0

TABLE 6.11

SSIM

1	1	0	1	1
0	1	1	2	
0	0	3		
0	4			
5				

TABLE 6.12

SIM

	1	2	3	4	5
1	0	1	0	1	1
2	1	0	1	1	0
3	0	1	0	0	0
4	1	1	0	0	0
5	1	0	0	0	0

TABLE 6.13

SIM

	A_1	A_2	A_3	A_4	A_5	A_6
A_1	0	1	0	0	0	0
A_2	0	0	1	0	0	0
A_3	0	1	0	0	0	0
A_4	0	0	1	0	1	0
A_5	0	0	1	1	0	0
A_6	0	0	0	0	1	0

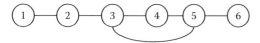

FIGURE 6.16
Nonoriented ISM.

TABLE 6.14

Nondirected SIM

	1	2	3	4	5	6
1	0	1	0	0	0	0
2	1	0	1	0	0	0
3	0	1	0	1	1	0
4	0	0	1	0	1	0
5	0	0	1	1	0	1
6	0	0	0	0	1	0

TABLE 6.15

SSIM

0	0	0	0	1	1
0	0	0	1	2	
0	1	1	3		
0	1	4			
1	5				
6					

SOLUTION

Step 1 Write reachability matrix for SSIM mentioned in Table 6.16 is

$$A_i = \begin{bmatrix} & 1 & 2 & 3 & 4 & 5 & 6 & 7 & 8 & 9 & 10 & 11 & 12 & 13 & 14 \\ 1 & 1 & 0 & 0 & 1 & 0 & 0 & 1 & 0 & 1 & 0 & 0 & 0 & 0 & 0 \\ 2 & 0 & 1 & 0 & 0 & 1 & 0 & 1 & 0 & 1 & 0 & 0 & 0 & 0 & 0 \\ 3 & 0 & 0 & 1 & 0 & 0 & 0 & 0 & 0 & 1 & 0 & 0 & 0 & 0 & 1 \\ 4 & 0 & 0 & 0 & 1 & 0 & 0 & 1 & 0 & 1 & 0 & 0 & 0 & 0 & 0 \\ 5 & 0 & 0 & 0 & 0 & 1 & 0 & 1 & 0 & 1 & 0 & 0 & 0 & 0 & 0 \\ 6 & 0 & 0 & 0 & 1 & 0 & 1 & 1 & 0 & 1 & 0 & 0 & 0 & 0 & 0 \\ 7 & 0 & 0 & 0 & 0 & 0 & 0 & 1 & 0 & 1 & 0 & 0 & 0 & 0 & 0 \\ 8 & 0 & 0 & 0 & 0 & 0 & 0 & 0 & 1 & 1 & 0 & 0 & 0 & 0 & 1 \\ 9 & 0 & 0 & 0 & 0 & 0 & 0 & 0 & 0 & 1 & 0 & 0 & 0 & 0 & 0 \\ 10 & 0 & 0 & 1 & 0 & 0 & 0 & 0 & 0 & 1 & 1 & 0 & 0 & 0 & 1 \\ 11 & 0 & 0 & 0 & 0 & 0 & 0 & 0 & 1 & 1 & 0 & 1 & 0 & 0 & 1 \\ 12 & 0 & 0 & 1 & 0 & 0 & 0 & 0 & 0 & 1 & 0 & 0 & 1 & 0 & 1 \\ 13 & 0 & 0 & 0 & 0 & 0 & 0 & 0 & 1 & 1 & 0 & 0 & 0 & 1 & 1 \\ 14 & 0 & 0 & 0 & 0 & 0 & 0 & 0 & 0 & 1 & 0 & 0 & 0 & 0 & 1 \end{bmatrix}$$

Step 2 Prepare lower triangular matrix from reachability matrix based on number of non-zeros entries in different rows. The rows of reachability matrix is re-arranged based on number of non-zero entries as given below:

Row number of reachability matrix	Number of nonzero entries
9	1
7, 14	2
3, 4, 5, 8	3
1, 2, 6, 10, 11, 12, 13	4

TABLE 6.16

SSIM

0	0	0	0	0	V	0	V	0	0	V	0	0	1
0	0	0	0	0	V	0	V	0	V	0	0	2	
V	0	A	0	A	V	0	0	0	0	0	3		
0	0	0	0	0	V	0	V	A	0	4			
0	0	0	0	0	V	0	V	0	5				
0	0	0	0	0	V	0	V	6					
0	0	0	0	0	V	0	7						
V	A	0	A	0	V	8							
A	A	A	A	A	9								
V	0	0	0	10									
V	0	0	11										
V	0	12											
V	13												
14													

Note: (1) Penguin, (2) Albatross, (3) Ungulate, (4) Nonflying Birds, (5) Flying Birds, (6) Ostrich, (7) Bird, (8) Carnivore, (9) Animal, (10) Zebra, (11) Cheetah, (12) Giraffe, (13) Tiger, (14) Mammal.

Interpretive Structural Modeling

$$L = \begin{array}{c} \\ 9 \\ 7 \\ 14 \\ 3 \\ 4 \\ 5 \\ 8 \\ 1 \\ 2 \\ 6 \\ 10 \\ 11 \\ 12 \\ 13 \end{array} \begin{array}{c} \begin{array}{cccccccccccccc} 9 & 7 & 14 & 3 & 4 & 5 & 8 & 1 & 2 & 6 & 10 & 11 & 12 & 13 \end{array} \\ \left[\begin{array}{cccccccccccccc} 1 & 0 & 0 & 0 & 0 & 0 & 0 & 0 & 0 & 0 & 0 & 0 & 0 & 0 \\ 1 & 1 & 0 & 0 & 0 & 0 & 0 & 0 & 0 & 0 & 0 & 0 & 0 & 0 \\ 1 & 0 & 1 & 0 & 0 & 0 & 0 & 0 & 0 & 0 & 0 & 0 & 0 & 0 \\ 1 & 0 & 1 & 1 & 0 & 0 & 0 & 0 & 0 & 0 & 0 & 0 & 0 & 0 \\ 1 & 1 & 0 & 0 & 1 & 0 & 0 & 0 & 0 & 0 & 0 & 0 & 0 & 0 \\ 1 & 1 & 0 & 0 & 0 & 1 & 0 & 0 & 0 & 0 & 0 & 0 & 0 & 0 \\ 1 & 0 & 1 & 0 & 0 & 0 & 1 & 0 & 0 & 0 & 0 & 0 & 0 & 0 \\ 1 & 1 & 0 & 0 & 1 & 0 & 0 & 1 & 0 & 0 & 0 & 0 & 0 & 0 \\ 1 & 1 & 0 & 0 & 0 & 1 & 0 & 0 & 1 & 0 & 0 & 0 & 0 & 0 \\ 1 & 1 & 0 & 0 & 1 & 0 & 0 & 0 & 0 & 1 & 0 & 0 & 0 & 0 \\ 1 & 0 & 1 & 1 & 0 & 0 & 0 & 0 & 0 & 0 & 1 & 0 & 0 & 0 \\ 1 & 0 & 1 & 0 & 0 & 0 & 1 & 0 & 0 & 0 & 0 & 1 & 0 & 0 \\ 1 & 0 & 1 & 1 & 0 & 0 & 0 & 0 & 0 & 0 & 0 & 0 & 1 & 0 \\ 1 & 0 & 1 & 0 & 0 & 0 & 1 & 0 & 0 & 0 & 0 & 0 & 0 & 1 \end{array} \right] \end{array}$$

Step 3 Obtain minimum edge adjacency matrix from lower triangular matrix after deleting the entries shown in Table 6.12.

$$M = \begin{array}{c} \\ 9 \\ 7 \\ 14 \\ 3 \\ 4 \\ 5 \\ 8 \\ 1 \\ 2 \\ 6 \\ 10 \\ 11 \\ 12 \\ 13 \end{array} \begin{array}{c} \begin{array}{cccccccccccccc} 9 & 7 & 14 & 3 & 4 & 5 & 8 & 1 & 2 & 6 & 10 & 11 & 12 & 13 \end{array} \\ \left[\begin{array}{cccccccccccccc} 0 & 0 & 0 & 0 & 0 & 0 & 0 & 0 & 0 & 0 & 0 & 0 & 0 & 0 \\ 1 & 0 & 0 & 0 & 0 & 0 & 0 & 0 & 0 & 0 & 0 & 0 & 0 & 0 \\ 1 & 0 & 0 & 0 & 0 & 0 & 0 & 0 & 0 & 0 & 0 & 0 & 0 & 0 \\ 0 & 0 & 1 & 0 & 0 & 0 & 0 & 0 & 0 & 0 & 0 & 0 & 0 & 0 \\ 0 & 1 & 0 & 0 & 0 & 0 & 0 & 0 & 0 & 0 & 0 & 0 & 0 & 0 \\ 0 & 1 & 0 & 0 & 0 & 0 & 0 & 0 & 0 & 0 & 0 & 0 & 0 & 0 \\ 0 & 0 & 1 & 0 & 0 & 0 & 0 & 0 & 0 & 0 & 0 & 0 & 0 & 0 \\ 0 & 0 & 0 & 0 & 1 & 0 & 0 & 0 & 0 & 0 & 0 & 0 & 0 & 0 \\ 0 & 0 & 0 & 0 & 0 & 1 & 0 & 0 & 0 & 0 & 0 & 0 & 0 & 0 \\ 0 & 0 & 0 & 0 & 1 & 0 & 0 & 0 & 0 & 0 & 0 & 0 & 0 & 0 \\ 0 & 0 & 0 & 1 & 0 & 0 & 0 & 0 & 0 & 0 & 0 & 0 & 0 & 0 \\ 0 & 0 & 0 & 0 & 0 & 0 & 1 & 0 & 0 & 0 & 0 & 0 & 0 & 0 \\ 0 & 0 & 0 & 1 & 0 & 0 & 0 & 0 & 0 & 0 & 0 & 0 & 0 & 0 \\ 0 & 0 & 0 & 0 & 0 & 0 & 1 & 0 & 0 & 0 & 0 & 0 & 0 & 0 \end{array} \right] \end{array}$$

Step 4 Draw minimum edge diagraph from edge adjacency matrix as shown in Figure 6.17.
Step 5 Draw interpretive structural model from minimum edge diagram as shown in Figure 6.18.

Example 6.9

Given the SSIM in Table 6.18 obtain an ISM using the lower triangularization method?

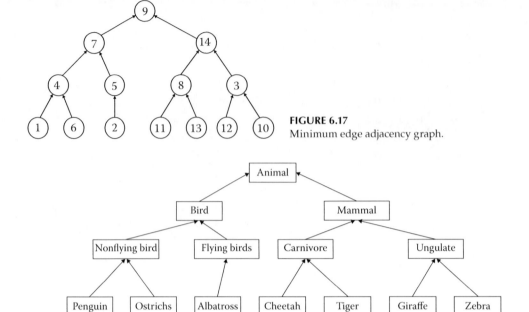

FIGURE 6.17
Minimum edge adjacency graph.

FIGURE 6.18
Interpretive structural model.

The minimum edge diagraph is drawn from minimum edge adjacency matrix obtained after removing the entries as shown in Table 6.17.

TABLE 6.17

Entries to Be Deleted from Minimum Edge Adjacency Matrix

Elements (*i*)	Elements (*j*)	*k*	Entries to Be Made Deleted
7	9	4	(4, 9)
		5	(5, 9)
		1	(1, 9)
		2	(2, 9)
		6	(6, 9)
14	9	3	(3, 9)
		8	(8, 9)
		10	(10, 9)
		11	(11, 9)
		12	(12, 9)
		13	(13, 9)
3	14	10	(10, 14)
		12	(12, 14)
4	7	1	(1, 7)
		6	(6, 7)
5	7	2	(2, 7)
8	14	11	(11, 14)
		13	(13, 14)

Interpretive Structural Modeling

TABLE 6.18

SSIM

A	0	A	A	A	A	1
A	0	0	0	A	2	
V	0	V	V	3		
0	0	0	4			
A	0	5				
0	6					
7						

TABLE 6.19

Reachability Matrix

	1	2	3	4	5	6	7
1	1	0	0	0	0	0	0
2	1	1	0	0	0	0	0
3	1	1	1	1	1	0	1
4	1	0	0	1	0	0	0
5	1	0	0	0	1	0	0
6	0	0	0	0	0	1	0
7	1	1	0	0	1	0	1

TABLE 6.20

Lower Triangular Matrix

	1	6	2	4	5	7	3
1	1	0	0	0	0	0	0
6	0	1	0	0	0	0	0
2	1	0	1	0	0	0	0
4	1	0	0	1	0	0	0
5	1	0	0	0	1	0	0
7	1	0	1	0	1	1	0
3	1	0	1	1	1	1	1

TABLE 6.21

Minimum Edge Adjacency Matrix

	1	6	2	4	5	7	3
1	**0**	0	0	0	0	0	0
6	0	**0**	0	0	0	0	0
2	1	0	**0**	0	0	0	0
4	1	0	0	**0**	0	0	0
5	1	0	0	0	**0**	0	0
7	0	0	1	0	1	**0**	0
3	0	0	0	1	0	1	**0**

TABLE 6.22

Entries to Be Deleted from Minimum Edge Adjacency Matrix

Elements (i)	Elements (j)	k	Entries to Be Made Deleted
2	1	7	(7, 1)
		3	(3, 1)
7	2	3	(3, 2)
7	5	3	(3, 5)

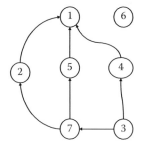

FIGURE 6.19
Minimum edge adjacency graph.

SOLUTION

Reachability matrix (Table 6.19)
Lower triangular matrix (Table 6.20)
Minimum edge adjacency matrix (Tables 6.21 and 6.22)
Minimum edge digraph (Figure 6.19)

Example 6.10

Develop ISM for paternalistic family for the relation "obey." The elements of the system could be father, mother, son, family dog, first daughter, second daughter, third daughter, and family parakeet.

Solution

Theme: To perform a simple interpretive structural modeling exercise as a learning experience.
Prospective: Student of modeling methodologies
Mode: Descriptive
Primitives:
R: will obey
S: Member of hypothetical paternalistic family
Elements in S:

1. Father
2. Mother
3. Son
4. Family dog
5. First daughter
6. Second daughter
7. Family parakeet
8. Third daughter

For developing ISM and finding contextual relationship between elements, a survey is conducted for a particular family. The participant responses reflect judgment as to the existence of a relation between any two elements and the associated direction of the relation. It could be written in the following manner:

1. If the relation holds for element i to element j and not in both directions, the modeler writes by V as symbolic of the direction from upper element to lower element.
2. If the relation holds for element j to element i and not in both directions, the modeler writes by A as symbolic of the direction from lower element to upper element.
3. If the relation is perceived by the modeler as valid in both directions, the modeler represents it by X.
4. If the relation between the elements does not appear as a valid relationship in both directions, we could write 0 to it.

Set of relations R consists of (V, A, X, 0).

The survey results could be written in the SSIM form showing the relation between the elements (Table 6.23).

The next step is to transform the SSIM form to the reachability matrix form. The reachability matrix for given system in 1s and 0s depending on whether elements are related or not is given in Table 6.24.

After obtaining the reachability matrix, it is desirable to determine the associated intent structure. The next step is to determine the lower triangular matrix by a sorting procedure, which

TABLE 6.23

SSIM

A	0	A	A	A	A	A	1
A	0	A	A	A	0	2	
0	0	0	0	A	3		
V	0	V	V	4			
0	0	0	5				
0	0	6					
0	7						
8							

Interpretive Structural Modeling

TABLE 6.24

SSIM

1	2	3	4	5	6	7	8	
1	0	0	0	0	0	0	0	1
1	1	0	0	0	0	0	0	2
1	0	1	0	0	0	0	0	3
1	1	1	1	1	1	0	1	4
1	1	0	0	1	0	0	0	5
1	1	0	0	0	1	0	0	6
0	0	0	0	0	0	1	0	7
1	1	0	0	0	0	0	1	8

identifies the highest level elements and inserts them as the first elements in a new reachability matrix, then iteratively identifies the next highest level set and transforms it until the elements are rearranged into lower triangular format.

To accomplish the lower triangularization, each row is searched in sequence, from row 1 to row 8, to identify any rows containing only one entry of 1, such as row 1 and row 7. Then search again for identifying any rows containing two entries of 1, such as rows 2 and 3, and continue this process. The last step in the process is to condense all cycles to a single representative element. The cycles are identified by entries of 1 to the right of the main diagonal. The rows are searched in sequence from row 1 to N for entries 1 to the right of the main diagonal. This identifies elements in cycle set of the row being searched. The corresponding rows and columns associated with all but the initial element of the cycle set are deleted from the matrix. This process is continued until each cycle has been identified and condensed to a single representative element. All entries to the right of the main diagonal in the triangular reachability matrix are then 0.

The lower triangular matrix obtained is shown in Table 6.25.

The next step is to transform the lower triangular reachability matrix into a minimum–edge adjacency matrix, which will represent the first approximation of the structure under development. The following steps have been followed:

The diagonal entries of an adjacency matrix for a digraph are defined to be 0 to get Table 6.26.
The minimum edge adjacency matrix is shown in Table 6.28 is obtained by deleting the entries shown in Table 6.27.
The minimum edge digraph could be drawn as shown in Figure 6.20.

Figure 6.21 shows the interpretive structural from the information contained in a minimum edge digraph or matrix.

TABLE 6.25

Lower Triangular Matrix

1	7	2	3	5	6	8	4	
1	0	0	0	0	0	0	0	1
0	1	0	0	0	0	0	0	7
1	0	1	0	0	0	0	0	2
1	0	0	1	0	0	0	0	3
1	0	1	0	1	0	0	0	5
1	0	1	0	0	1	0	0	6
1	0	1	0	0	0	1	0	8
1	0	1	1	1	1	1	1	4

TABLE 6.26
Adjacency Matrix after Making Diagonal Entries to Zero

1	7	2	3	5	6	8	4	
0	0	0	0	0	0	0	0	1
0	0	0	0	0	0	0	0	7
1	0	0	0	0	0	0	0	2
1	0	0	0	0	0	0	0	3
1	0	1	0	0	0	0	0	5
1	0	1	0	0	0	0	0	6
1	0	1	0	0	0	0	0	8
1	0	1	1	1	1	1	0	4

TABLE 6.27
Entries to Be Deleted from Minimum Edge Adjacency Matrix

Elements	Row No. (i)	j	k	Entries to Be Made Deleted
2	3	1	5	(5, 1)
			6	(6, 1)
			7	(7, 1)
			8	(8, 1)
5	5	3	8	(8, 3)

TABLE 6.28
Minimum Edge Adjacency Matrix

1	7	2	3	5	6	8	4	
0	0	0	0	0	0	0	0	1
0	0	0	0	0	0	0	0	7
1	0	0	0	0	0	0	0	2
1	0	0	0	0	0	0	0	3
0	0	1	0	0	0	0	0	5
0	0	1	0	0	0	0	0	6
0	0	1	0	0	0	0	0	8
0	0	0	1	1	1	1	0	4

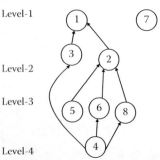

FIGURE 6.20
Minimum edge digraph.

Interpretive Structural Modeling

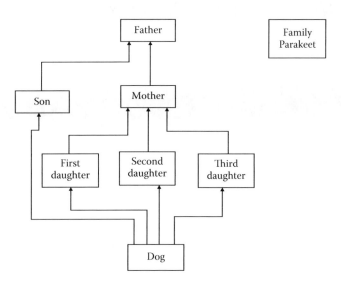

FIGURE 6.21
Interpretive structural model.

6.4 Review Questions

1. Explain the term "interpretive structural modeling."
2. Define the following terms used in interpretive structural modeling:
   (a) net, (b) cycle, (c) parallel, (d) loop, and (e) digraph.
3. For the directed graph shown in Figure 6.22, write down the system subordination matrix. Also write the reachability matrix from this and determine the maximum path length in the graph using reachability matrix.
4. Outline stepwise procedure for obtaining an interpretive structure model (ISM).
5. Explain briefly the procedure for removing transitivity and identifying and condensing cycles in a reachability matrix.
6. For the directed graph shown in Figure 6.22, write down the system SSIM.
7. What will be the associated intent structure of a system whose SIM is given in Table 6.29.
8. Develop ISM for SSIM given in Table 6.30.
9. Write self-interaction matrices for the following undirected graphs shown in Figure 6.23.
10. Explain binary and nonbinary interaction matrices in interpretive structural modeling.
11. a. What is the nondirected (nonoriented) SIM?
    b. What is the nondirected graph corresponding to this interaction matrix.

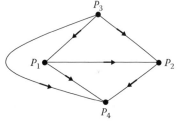

FIGURE 6.22
Digraph for Q.6.

TABLE 6.29
SSIM

A	0	A	A	A	A	1
A	0	0	0	A	2	
V	0	V	V	3		
0	0	0	4			
A	0	5				
0	6					
7						

TABLE 6.30
SSIM

N	N	N	N	N	N	N	1
A	V	A	A	N	V	2	
V	N	N	N	A	3		
N	V	N	N	4			
N	N	A	5				
N	N	6					
V	7						
8							

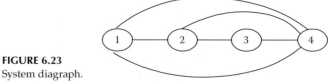

FIGURE 6.23
System diagraph.

12. Given the SSIM in Tables 6.29 and 6.30 obtain an ISM using the lower triangularization method.
13. Short answer questions
    a. What do you mean by
        i. Reflexive relation
        ii. Symmetric relation
        iii. Transitive
        iv. Complete relation
    b. Define reachability matrix in ISM.
    c. What is strongly connectedness in a digraph.
    e. Mention different MEI in ISM?
    f. Give the matrix representation of the given digraph (Figure 6.22).
    g. What is the maximum path length of the above developed matrix.
    h. Find reachability matrix for the above system.

Interpretive Structural Modeling 325

TABLE 6.31
SSIM

0	0	0	0	0	V	0	V	0	0	V	0	0	1
0	0	0	0	0	V	0	V	0	V	0	0	2	
V	0	A	0	A	V	0	0	0	0	0	3		
0	0	0	0	0	V	0	V	A	0	4			
0	0	0	0	0	V	0	V	0	5				
0	0	0	0	0	V	0	V	6					
0	0	0	0	0	V	0	7						
V	A	0	A	0	V	8							
A	A	A	A	A	9								
V	0	0	0	10									
V	0	0	11										
V	0	12											
V	13												
14													

Note: (1) Penguin, (2) Albatross, (3) Ungulate, (4) Nonflying Birds, (5) Flying Birds, (6) Ostrich, (7) Bird, (8) Carnivore, (9) Animal, (10) Zebra, (11) Cheetah, (12) Giraffe, (13) Tiger, (14) Mammal.

  i. What type of response do you expect for the following systems:
     i. $dy/dt + y(t) = U(t)$
     ii. $s^2 y(t) + sy(t) + y(t) = u(t)$
  j. A system SSIM is presumed to be

A	A	A	A	1
A	A	A	2	
A	A	3		
A	4			
5				

   i. If element 2 is also subordinate to element 4, what are the changes necessary in SSIM.
   ii. Also mention the changes to make SSIM transitive.
   iii. What is the minimum edge diagram for a given SSIM.
  k. Write reachability matrix for the given system in Q. 12.

6.5 Bibliographical Notes

Recently, the interactive management techniques (Warfiled, 2006), particularly the neutral process of Nominal Group Technique (NGT) and interpretive structural model (ISM) have been employed by many researchers (Satsangi, 2008, 2009; Dayal and Satsangi, 2008;

Dayal and Srivastava, 2008) to transform unclear, poorly articulated mental model of a system from the fictional literary/movie field, into visible well-defined hierarchical model, which assumes particular importance in managing highly complex situations. ISM and NGT as parts of interactive management have been used intensively in the field of commerce, management, engineering, and law but in the field of literature there are only a few applications.

7
System Dynamics Techniques

7.1 Introduction

The concept of how to organize scientific research is undergoing a change. Initially research was a one-man activity. But, if resources are to match adequately team research is essential. Engineering has developed the recognition of the importance of System Engineering in order to achieve the optimum goal.

System engineering is a formal awareness of the interaction between the parts of the system. The interconnection, the compatibility, the effect of one upon the other, the objectives of the whole, the relationship of the system to the users, and the economic feasibility must receive even more attention than the parts, if the final result is to be successful.

In management as in engineering, we can expect that the interconnection and interaction between the components of the system will often be more important than the separate components themselves.

7.2 System Dynamics of Managerial and Socioeconomic System

The nature of managerial and socioeconomic systems and their associated problems are examined in terms of their counterintuitive behavior, nonlinearity, and system dynamics. The need for obtaining the intermediate feedback about the efficacy policies before being implemented in the real life is highlighted, which can be fulfilled effectively by using system dynamics (SD) as a simulation methodology.

The managerial and socioeconomic systems and their associate problems are complex in nature compared to engineering and physical systems. According to the hierarchy of system given by Boulding (1956) the social system lies next to transcendental system in complexity, that is, they have complexity of highest order among all worldly systems, and the level of knowledge about their functioning is comparatively limited. This complexity is a result of many factors including a large number of components and interactions, nonlinearty in interaction, dynamics, growth, causality, endogenization of factors, and their counterintuitive nature.

7.2.1 Counterintuitive Nature of System Dynamics

The problem is not shortage of data but rather our inability to perceive the consequences of the information we already possess. The SD approach starts with the concepts and information on which people are already acting. These are usually sufficient. The available perception is then assembled in a computer model that can show the consequence of the well-known and properly perceived parts of the system. Generally the consequences are unexpected.

7.2.2 Nonlinearity

In most of the cases, there exists a high degree of nonlinearity, which is difficult to handle by either traditional or quantitative approaches in management science.

7.2.3 Dynamics

The behavior of managerial and social systems exhibits high order dynamics.

7.2.4 Causality

The dynamic behavior of these systems is to a great extent governed by their structure, which is composed of various cause–effect relationships. A methodology that can help in identifying the right casual structure and behavior will thus help a great deal in policy evaluation and design.

7.2.5 Endogenous Behavior

The managerial and social systems are governed predominantly by endogenous relationships rather than the external influences.

7.3 Traditional Management

The practice of management has been treated traditionally as an art, which is primarily being governed by the mental data base, whereas the major weakness is the absence of unifying, underlying structure of fundamental principle.

7.3.1 Strength of the Human Mind

1. It can gather a lot of information on multiple fronts, both tangible and intangible, and parallel. No amount of quantitative data can be substituted for this information.
2. While working with a system, it can easily be first-order relationship, that is, "what is affecting what" in a paired manner.

7.3.2 Limitation of the Human Mind

1. The human mind has cognitive limitations. Cognitively, in a normal sense, it can handle up to three independent dimensions (or seven possible combinations of these) at a time. If the number of dimensions goes beyond three, it becomes cognitively overloaded and the conceptualization becomes weak and incomplete.
2. Though human mind is strong in establishing first-order relationships, it is weak in generating higher order consequences.

System Dynamics Techniques

7.4 Sources of Information

The sources of information used in different approaches to manage the socioeconomic and socio-technical problems are multiple and can be broadly categorized into three types: mental, written/spoken, and numerical, as shown in Figure 7.1.

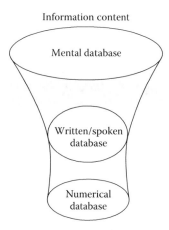

FIGURE 7.1
Source of information.

7.4.1 Mental Database

The mental database present within every human being is information rich and is the primary source of information used in the conduct of human affairs. It is generated and enriched over time with their experience in dealing with different problem situations, persons, and systems.

The mental database contains all information on conceptual and behavioral information and technical fronts. The conceptual and behavioral information available with mental database cannot be substituted with any other form of information available more formally. It is effective to put this information into words and numbers.

It can be easily and directly put into practice. The processing time is negligible.

Limitation

1. The level of information varies from individual to individual.
2. The perspective of individual governs the content of information.
3. The information available may be highly biased by the background and knowledge of the individual.
4. New information is collected and assimilated in the light of existing information.
5. Amount of information available is very high; clarity in its use and selection for a purpose is dependent upon the level of development and clarity of individual.
6. It is difficult to be freely exchanged.
7. Always there is a perceived gap, as shown in Figure 7.2.

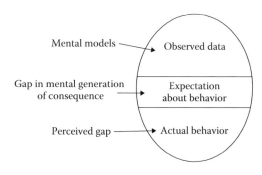

FIGURE 7.2
Categories of information in mental database and gaps between them.

7.4.2 Written/Spoken Database

The written database in terms of different reports, memos, letters, notes, etc., is being very commonly used. The written database is both system-specific and general.

The general information is available with books, periodicals, magazines, news papers, films, etc. Most of the existing literature has emphasized on the product approach rather than the process approach.

Variation of this is the spoken data base which is in the form of stored audio information, say audio cassettes, CDs, and DVDs of discussions, lecturers, seminars, presentations, etc.

Strength

1. These are more formal and clear than the mental database.
2. These are more communicable.
3. It prevents the loss of knowledge with time.
4. These can be authenticated, and thus govern the managerial action on a formal basis to avoid ambiguity.

Limitations

1. These have a lot of redundancy being mostly expressed in words.
2. All the information about the system is difficult to be expressed in written/spoken form.
3. As information content increases, storage and retrieval becomes a big problem.
4. Difficult to use for instantaneous decision.
5. Lack of precision in expressing system relationships.

7.4.3 Numerical Database

It is comparatively narrower and is available in the form of brochures, reports, etc.

Strength

1. It is more precise.
2. It can be mathematically manipulated and used for sophisticated analysis.
3. It can easily support systems modeling and simulation exercises.
4. It gives a feeling of tangibility and clarity about systems functioning.
5. Its redundancy level is low.
6. It can be easily computerized and thus makes use of computer technology possible for managerial support.
7. It can be easily stored, retrieved, and communicated.

Limitation

1. The information content is comparatively low.
2. Accuracy of the data becomes very important or else it can mislead the decisions very easily.
3. It does not reveal the cause-and-effect directions among variables.

7.5 Strength of System Dynamics

1. System dynamics (SD) methodology is able to cater a modeling framework that is casual, captures nonlinearity, and dynamics and generates endogenous behavior.
2. It uses the strength of human mind and mental models and overcomes their weaknesses by making a division of labor between the manager and the technology.
3. It uses multiple sources of information, that is, mental, written, and numerical in different phases of modeling.

There are mainly six problem-solving steps through which the SD methodology operates:

1. Problem identification and definition
2. System conceptualization
3. Model formulation
4. Simulation and validation of model
5. Policy analysis and improvement
6. Policy implementation

Industrial dynamics is the investigation of the information-feedback characters of the industrial system and the use of the models for the design of improved organizational form and guiding policy. It provides a single framework for integrating the functional areas of management marketing, production, accounting, and research and development.

An industrial dynamics approach to enterprise design progresses through several steps:

- Identify a problem.
- Isolate the factors that appear to interact.
- Trace the cause-and-effect of information-feedback loops that link decisions to action to resulting information changes and to new decisions.
- Formulate acceptable formal decision policies that describe how decisions result from the available information streams.
- Construct a mathematical model of the decision policies, information sources, and interaction of the system components.
- Generate the behavior of the system as described by the model. Digital computer is used to execute lengthy calculations.
- Compare results against all pertinent available knowledge about the actual system.
- Revise the model if required.

Information feedback control theory

System of information feedback controls is fundamental to all life and human endeavor, from the slow pace of biological evolution to the launching of the latest space satellite.

For example, consider a thermostat which receives information and decides to start the furnace; this raises the temperature, and when the temperature reaches to a desired value, the furnace is stopped. The regenerative process is continuous, and new results lead to new decision, which keep the system in continuous motion. Information feedback system, whether they are mechanical, biological, or social owe their behavior to three characteristics:

- Structure
  The structure of the system tells how the parts are related to one other.
- Delays
  Delay always exists in the availability of information, in making decisions based on the information, and in taking action on the decisions.
- Amplifications
  Amplification usually exists throughout such system. Amplification is manifested when an action is more forceful than might at first seem to be implied by the information inputs to the governing decisions.

Decision-making processes

As in military decision, it is seen that there is an orderly basis, (more or less) in present managerial decision making. Decisions are not entirely "free will" but are strongly conditioned by the environment. In engineering exercises, the environmental effects play the important role.

7.6 Experimental Approach to System Analysis

The foundation for industrial dynamics is the experimental approach to understanding system behavior.

If mathematical analysis is not powerful enough to yield general analytical solutions to situations, then alternative experimental approaches are taken.

Use of simulation methods requires extending mathematical ability. Detail setting up a simulation model needs to be monitored by experts, because special skills required for proper formulation.

7.7 System Dynamics Technique

The SD technique was developed for the first time by Forrester in the 1940s in the context of industrial applications. The technique had been used to understand such diverse problems as technical obsolesce, urban decay, drug addiction, etc. It explains the model causal mechanism and identifies the feedback for modeling.

Advantages of SD technique

1. It proposes the use of various diagrammatic tools for modeling the problem.
2. Easy to understand the model.
3. Easy to illustrate the model.

System Dynamics Techniques

4. Easy to observe the effect of variables on the response.
5. It can be used for nonlinear, ill-defined, time variant problems.
6. Delays (information, transportation, implementation, etc.) and nonlinearities can be easily incorporated in the model.
7. Easy to observe the effect of variables on system response.
8. Easy to develop Expert System if model is developed by SD approach.
9. Quick and easy insight can be gained.

Steps involved in the development of SD models

1. Identification of system variables and their causal relations.
2. Draw causal links and causal loop diagram.
3. Draw flow diagram.
4. Write dynamo equations from the causal loop diagram and flow diagrams.
5. Finally, simulate the above developed model using numerical integration techniques and also the software available for simulating dynamo equations such as Dynamo or Dynamo Plus, Dyner, and Vinsim.

The SD technique of modeling has three main parts namely,

1. Causal loop diagram
   Arrow of causal links in a causal loop diagram indicates the direction of influence and sign (+ or –) shows the type of influence. Causal loop diagrams play an important role:
   a. They serve as preliminary sketch of causal hypothesis.
   b. Simplifies illustration of the model.
   c. Encourages the modeler to conceptualize the real-world systems.
2. Flow diagram
   Flow diagram consists of rates, levels, and auxiliary variables organized into a consistent network.
3. Dynamo equations
   The dynamo model consists of first-order differential equation (ODE) to represent system states.

7.8 Structure of a System Dynamic Model

The form of a SD model should be such as to achieve several objectives. The model should have the following characteristics:

- Able to describe any statement of cause–effect relationship
- Simple in mathematical nature

- Closely synonymous in nomenclature and terminology
- Extendable to large numbers of variables without exceeding the practical limits of digital computers

7.9 Basic Structure of System Dynamics Models

The preceding requirement can be met by an alternating structure of reservoirs or levels interconnected by controlled flows, as shown in Figure 7.3. It contains

- Several level variables
- Flow-rate variables that control the flow of quantities into level variables
- Decision functions (draws as values) that control the rate of flow between levels

7.9.1 Level Variables

The level variables represent the accumulations within the system. The typical level variables may be the inventory level in a warehouse, the good in transit, bank balance, factory space, and number of employees. Levels are the present values of those variables that have resulted from the accumulated difference between inflow and outflow. Levels exist in the information network of material. It is denoted by a rectangle, as shown in Figure 7.3 and describes the condition of the system at any particular instant of time.

7.9.2 Flow-Rate Variables

The flow-rate variables define the present (instantaneous) flow between the level variables in the system. The rate variables correspond to activity, while the level variables measure the resulting state to which the system has been brought by the activity. The flow-rate variables are determined by the level variables of the system according to rules defined by the decision functions. In turn, the rates determine the levels. It tells how fast the levels are changing. The rate variable does not depend on its own past values, neither on the time interval between computations nor on the other rate variables.

No flow-rate variable can be measured instantaneously. All instruments that purport to measure rate variables actually require time for their functioning. They measure average rate over some time interval. At zero rate also, there is some value of level variable.

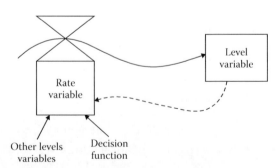

FIGURE 7.3
Schematic representation of level (state) variables and rate variables.

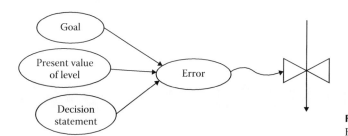

FIGURE 7.4
Four concepts in flow-rate variable.

Consider an example of the growth of a tree. The height of a tree cannot be zero although the growth rate becomes zero.

Four concepts are to be found in a rate equation, as shown in Figure 7.4:

1. Goal
2. An observed system condition
3. Discrepancy
4. A statement how action is to be taken

7.9.3 Decision Function

The decision function (also called the rate equation) is the policy statement that determines how the available information about level leads to the decision (current rate). All decisions pertain to impending action and are expressible as flow rates (generations or orders, construction of equipment, hiring of people). Decision functions therefore pertain both to managerial divisions and to those actions that are inherent results of the physical state of the system.

There are three major view points of modeling, as shown in Figure 7.5.

1. To identify the cause–effect, relations (causal links) in a system are based on disjunctive viewpoint of modeling.
2. Another viewpoint is linear, control viewpoint in which cause produces the effect, which may be the cause for other such causes and effects form chains like dominoes falling.
3. Lastly, there is a causal loop, nonlinear viewpoint, in which the causal loops are identified.

There are two types of cause–effect relationships (causal links) that can be identified in any system

1. Negative causal links
2. Positive causal links

    1. Negative causal links
       In a system, if a change in one variable affects the other variable in the opposite direction under "cetris paribus" condition (keeping other variables constant), then it called a negative causal link. For example, consider the heat transfer from a hot body to the atmosphere, as shown in Figure 7.6. The heat transfer (Q) from a hot

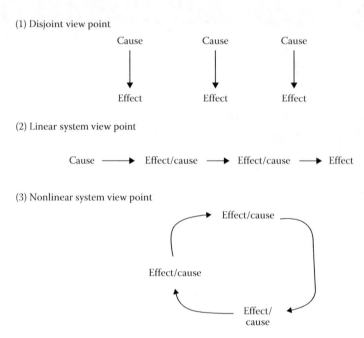

FIGURE 7.5
Different viewpoints of causal modeling.

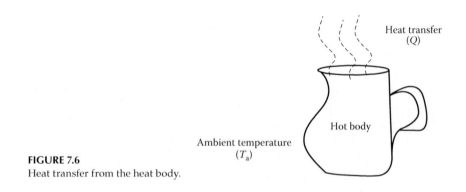

FIGURE 7.6
Heat transfer from the heat body.

body is a function of ambient temperature (T_a) and the temperature of the hot body (T_{hb}) is given by

$$\text{heat transfer } (Q) = f(T_a, T_{hb}) \tag{7.1}$$

If the temperature of the hot body is kept constant, then it is a function of ambient temperature only. Hence, the heat transfer decreases as the ambient temperature increases and vice versa. In this cause–effect relationship, the ambient temperature (cause) negatively affects the heat transfer (effect). Therefore, the causal link is called negative causal link, as shown in Figure 7.7.

System Dynamics Techniques

FIGURE 7.7
Causal link between ambient temperature and heat transfer.

2. Positive causal links

   In a system, if a change in one variable affects the other variable in the same direction under "cetris paribus" condition, for example, the relationship between the job availability (JA) and migrant (M), then it is called a positive causal link. If the job availability is more in a particular part of the region, then the migration will also be more in that part of the region and as the job availability decreases, the migration will also reduce. In this example, the job availability is a cause for migration of people from one part of a region to other part of a region. Here, the cause has positive influence on the effect. Hence, it a positive causal link, as shown in Figure 7.8.

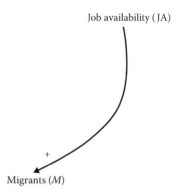

FIGURE 7.8
Positive causal link between job availability and migrants.

Negative causal loop: When feedback loop responses to a variable change opposes the original perturbation, it is called negative causal loop. A typical negative causal loop system and its responses are shown in Figures 7.9 and 7.10, respectively. The decision process sector tells us how the action process sector (Level variable) is controlled. It works on the basis of error that is the difference between the present state of the system and the reference input. The constant in this sector controls the change in level variable. This constant is a part of decision function or policy.

Positive causal loop: When a loop response reinforces the original perturbation, the loop is called positive. The typical responses of positive causal loop systems are shown in Figure 7.11. Take an example of a snow ball falling from the top of the snow mountain, as shown in Figure 7.12. Its size and weight increases as it comes down and finally, broken

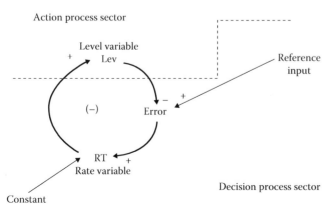

FIGURE 7.9
Negative causal loop diagram.

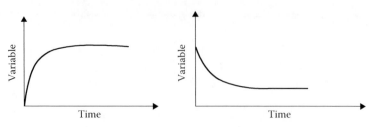

FIGURE 7.10
Responses of a negative feedback system.

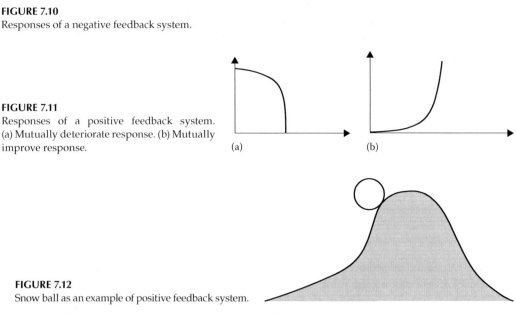

FIGURE 7.11
Responses of a positive feedback system.
(a) Mutually deteriorate response. (b) Mutually improve response.

FIGURE 7.12
Snow ball as an example of positive feedback system.

down into many small balls. This shows that the effect reinforces the cause and finally the system response becomes unstable.

S-shaped growth: Social and biological systems commonly exhibit exponential growth followed by asymptotic growth. This mode of behavior occurs in any structure where loop dominance shifts from positive to negative feedback. Finally the system response exhibits steady state when positive effect and negative effect in the system balance each other, as shown in Figure 7.13.

Examples:

1. Population trends of various plants and animals
2. Growth of biological spices
3. Propagation of a contagious disease
4. The behavior of damped pendulum
5. Diffusion of riots, rumors, news, etc.

Interconnected network

The basic model structure shows only one network with set of information from levels to rates, as illustrated in Figure 7.14. It could be noted that flow rates transport the

System Dynamics Techniques

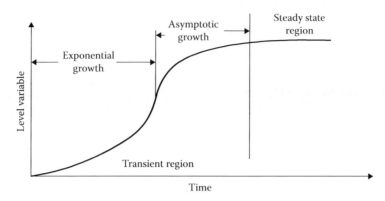

FIGURE 7.13
Sigmoid (S-shaped) growth.

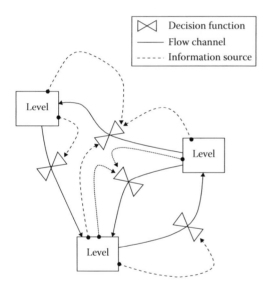

FIGURE 7.14
Basic model structure.

contents of one level to another as shown in Figure 7.15. Therefore, the levels within one network must all have the same kind of content. Inflow and outflow connecting to a level must transport the same kind of items that are stored in the level. For example, the material network deals only with material and accounts for the transport of the material from one inventory to another. Items of one type must not flow into levels that store another type.

Dynamo equations

The model structure leads to a system with an equation that suffices for representing information-feedback system. The equations tell how to generate the system conditions for a new point in time, given the conditions known from the previous point in time.

The equations should be adequate to describe the situations, concepts, interactions, and decision processes. The equations of the model are evaluated repeatedly to generate

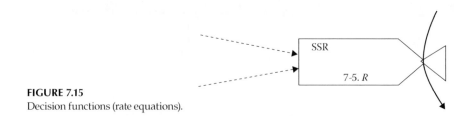

FIGURE 7.15
Decision functions (rate equations).

sequence of steps equally spaced in time. Level equations are rate equations and generate the levels and rates of the basic model structure. In addition, auxiliary supplementary and initial-value equations are used. The interval of time between solutions must be relatively short, determined by the dynamic characteristics of the real system that is being modeled.

Computing sequence

A system equation is written in the context of certain connections that state how the equations are to be evaluated. We are dealing here with a system equation that controls the changing interactions of a set of variables as time advances. Subsequently the equations will be computed periodically to yield the successive new states of the system.

The sequences of computation implied are shown in Figure 7.16. The continuous advance of time is broken into small intervals of equal length DT. It should be short enough so that we are willing to accept constant rates of flow over the interval as a satisfactory approximation to continuously varying rates in the actual system. This means that decisions made at the beginning of the interval will not be affected by any changes that occur during the interval. At the end of the interval, new values of levels are calculated and from these levels new rates (decisions) are determined for the next interval.

The time interval could be spaced closely enough so that straight line segments over the intervals will approximate any curve as closely as required. Figure 7.16 shows such a straight approximation.

As shown in Figure 7.17 J, K, and L are the designations given to successive points in time.

The instant of time K is used to develop the "present."

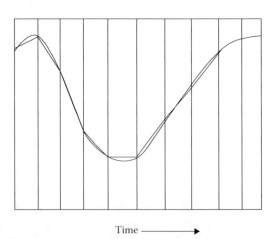

FIGURE 7.16
Straight line approximation to a variable level.   Time ———▶

System Dynamics Techniques

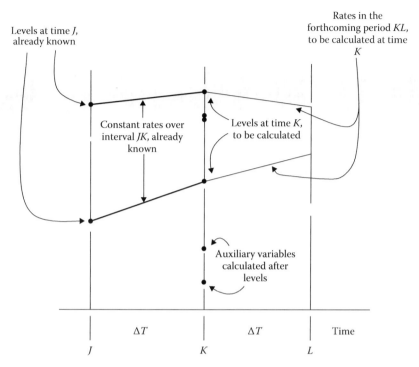

FIGURE 7.17
Calculations at time K.

- The interval JK has just passed the information about it and earlier times is, in principle, available for use.
- No information from a time later than K, like the interval KL, or time L, or beyond, can ever be available for use in an equation being evaluated at present time K.

For the purpose of numerical evaluation, the basic equations of a model are separated into two groups:

- The level equation in one group
- Rate equation in another

For each time step, the level equations are evaluated first, and the result becomes available for use in the rate equations. Auxiliary equations are an identical convenience and are evaluated between the level and rate groups.

The equations are to be evaluated at the moment of time that are separated by the solution interval ΔT. The equations are written in terms of the generalized time step J, K, and L using the arbitrary convention that K represents the "present" point in time at which the equations are being evaluated. In other words, we assume that the progress of the solution has just reached time K, but that the equations have neither been solved for levels at time K nor for rates over the interval KL.

The level equations show how to obtain levels at time K, based on level at time J, and on rates over the interval JK at time K, when the level equations are evaluated, all necessary information is available and has been carried forward from the proceeding time step.

The rate equations are evaluated at the present time K after the level equations have been evaluated. The rate equations can, therefore, be available as inputs of the present values of the level at K. The value determine the rate (decision) equations determine the rates that represent the actions that will be taken over the forthcoming interval KL constant rates imply a constant rate of change in levels during a time interval. The slopes of the straight lines in Figure 7.17 are proportional to the rates and connect the values of the levels at the J, K, and L points in time.

After evolution of the levels at time K and the rates for interval KL, time is "indexed." That is, the J, K, and L positions are moved one time interval to the right. The K levels just calculated are relabeled as J levels. The KL rates become the JK rates. Time K "the present," is thus advanced by one interval of time of ΔT length. The entire computation sequence can then be repeated to obtain a new state of the system at a time that is one ΔT later than the previous state.

7.10 Different Types of Equations Used in System Dynamics Techniques

7.10.1 Level Equation

Levels are the varying content of the reservoir of system. They would exist even though the systems were brought to rest and no flow existed. New values of levels are calculated at each of the closely spaced solution intervals. Levels are assumed to change at a constant rate between solution times, but no values are calculated between such times.

The following is an example of a typical level equation:

$$L \quad L_1.K = L_1.J + (\Delta T)*(R_1.JK - R_2.JK) \qquad (7.2)$$

The symbols represent variables as follows, given the dimension of measurement:

L_1 inventory actual at retail (units), actual being used to distinguish from "desired" and other inventory concepts.
ΔT delta time (week) interval between evaluations of the set of equations.
R_1 shipment received at retail (unit/week).
R_2 shipment sent from retail (unit/week).

The indication of L_1 on the left shows that this is a level equation.

The equation states that the present values of L_1 at time K will equal the previously computed $L_1.J$ plus the difference between the inflow rate $R_1.JK$ during the last time interval and the outflow rate $R_2.JK$, the difference rates multiplied by the length of time ΔT during which the rates persisted.

Level equations are independent of one another. Each depends only on information before time K. It does not matter in what order level equations are evaluated. At the evaluation time K, no level equation uses information from other level equations at the same time. A level at time K depends on its previous value at time J and on rates of flow during the JK interval.

7.10.2 Rate Equation (Decision Functions)

The rate equations define the rates of flow between the levels of system. The rate equations are the "decision functions." A rate equation is evaluated from the presently existing value of the level in the system, varies often including the level from which the rate comes and the one into which at goes. The rates in turn cause the changes in the level. The rate equation, being the decision equation in the broad sense controls what is to happen next in the system. A rate equation is evaluated at time K to determine the decision governing the rate of flow over the forthcoming interval KL. Rate equations, in principle, depend only on the value of levels at time K.

Rate equations are evaluated independent of one another within any particular time step, just as level equations. A rate equation determines an immediately forthcoming action.

The rate equations are independent of one another and can be evaluated in any sequence. Since they depend on the values of the levels, the group of rate equations is better to put after the level equations.

An example of a rate equation given by Equation 7.3 shows the outflow rate of a first-order exponential delay:

$$R \quad R_2.KL = L_2.K/\text{DELAY} \tag{7.3}$$

where
  R_2 is the outflow rate (unit/week)
  L_2 is the present amount stored in the delay (units)
  DELAY is a constant and average length of time to transverse the delay

The equations define the rate R_2 and give the value over the next time interval KL. The rate is to be equal to the value of a level L_2 at the present time K, divided by the constant that is called DELAY.

7.10.3 Auxiliary Equations

Often rate equations will become very complex if it is actually formulated only from levels. Furthermore, a rate may best be defined in terms of one or more concepts, which have independent meaning and, in turn, arise from the level of the system. It is often convenient to break down a rate equation into component equations called "auxiliary equations."

The auxiliary equation is a kind of great help in keeping the model formulation in close correspondence with the actual system, since it can be used to define separately, the many factors that enter decision making.

The auxiliary equations are evaluated at the time K but after evaluating the level equations for time K because, the rate for which they are the part, they make use of present values of levels. They must be evaluated before the rate equations, because their values are obtained for substitution into rate equation. Unlike the level and rate equations, the auxiliary equations cannot be evaluated in an arbitrary order.

For example, two auxiliary equations between two levels and a rate equation are given below:

$$A \quad A_1.K = C_1 * L_3.K \tag{7.4}$$

$$A \quad A_2.K = C_3 + C_2(A_1.K)/L_5.K \tag{7.5}$$

where
 L_5 is a level
 C_3 and C_2 are constant

$$R \quad R_3.KL = L_6 \cdot K / A_2.K \tag{7.6}$$

where L_6 is a level.

Equation 7.4 could be substituted into Equation 7.5 and finally substituted in Equation 7.6 to give

$$R \quad R_3.KL = (L_6.K)/C_3 + C_2(C_1^* L_3.K / L_5.K)$$

The auxiliary equations have now disappeared leaving the rate R_3 dependent only on levels and constants.

7.11 Symbol Used in Flow Diagrams

A pictorial representation of an equation encourages the modeler to understand the complex system. It is easy to visualize the interrelationships when these are shown in a flow diagram rather than merely listing the equations.

7.11.1 Levels

A level is shown by a rectangle as in Figure 7.18 at the upper corner is the symbol group (L) that denotes this particular level variable. At the lower right corner is the equation number to tie the diagram to the equation.

Decision functions (rate equations):
Decision functions determine the rate of flow. They act as values in the flow channels, as shown in Figure 7.19 to equivalent forms are displayed. The symbol shows the flow that is being controlled. The equation number that defines the rate is also given.

7.11.2 Source and Sinks

Often a rate is to be controlled whose source or destination is considered to lie outside the consideration of the model. For example, in an inventory system, an order flow must start from somewhere, but clarity of flow system terminology does not permit mere extension of information lines into the line symbols for order. Orders are properly thought of as starting from a supply a blank paper that we need not treat in the model dynamics.

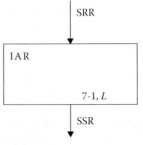

FIGURE 7.18
Levels.

System Dynamics Techniques

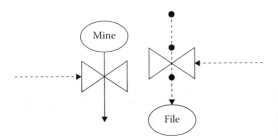

FIGURE 7.19
Source and sinks.

Likewise, orders that have been filled must be discarded from the system into a completed order file, which usually has no significant dynamic characteristics.

If it is assumed that the materials are readily available and that the characteristics of the source do not enter into system behavior, in such circumstances the controlled flow is from a source or to a sink, as shown in Figure 7.19 and given no further consideration in the system.

7.11.3 Information Takeoff

Information flows interconnect many variables in the system. The takeoff of information flow does not affect the variable from which the information is taken. An information takeoff, as shown in Figure 7.19, is indicated by a small circle at the source and by a dashed information line.

7.11.4 Auxiliary Variables

Auxiliary variables lie in the information flow channels between levels and decision functions that control rates. They have independent meanings. They can be algebraically substituted into the rate equations.

Auxiliary variables are shown by circle, as in Figure 7.20 within the circle, the name of the variable and the equation number that defines it, is mentioned. The incoming information depends on level or other auxiliary variables and the outflow is always an information takeoff. The auxiliary variable is not a one computational time step to the next. Any number of information lines can enter or leave.

7.11.5 Parameters (Constants)

Many numerical values that describe the characteristics of a system are considered constant, at least for duration of computation of a single model run. A line with circle shows the symbols of the constant, with information takeoff as shown in Figure 7.21.

7.12 Dynamo Equations

The dynamo is a special-purpose compiler for generating the executable code to simulate the system model. The dynamo compiler takes

FIGURE 7.20
Auxiliary variable.

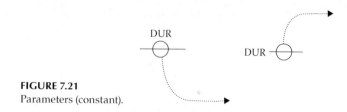

FIGURE 7.21
Parameters (constant).

equations in the form that has been discussed. It checks equations for the kinds of logical errors that represent inconsistencies within equations.

Example 7.1: Modeling of Heroin Addiction Problem

The explosive growth in the number of heroin addicts in a community might result from a positive feedback process. We assume that the addiction rate depends on the level of addicts and is not restrained by the number of potential addicts. We also assume that each addict brings one nonaddict into the addict pool every 3 years (the addicted population increases by a fractional addiction rate (*FAR*) of 0.33 per year). We ignore outflow of addicts from level through arrests, dropouts, and rehabilitation. The model, although obviously oversimplified, illustrates some interesting behavior. The causal loop diagram for heroin addiction problem is shown in Figure 7.22 and the flow diagram is shown in Figure 7.23.

Dynamo equation

From the above developed causal loop and flow diagrams, the dynamo equations may be written as given below.
   Addiction level in a given population $ADCTS.K = ADCTS.J + dt*AR.KL$
   Initial value of addiction level $ADCTS = 10$
   Addiction rate $AR.JK = ADCTS.K*FAR$
   Fractional addiction rate $FAR = 0.33$

Results

The above-mentioned dynamo equations could be easily simulated in MATLAB® environments. The results show that the addicted population and addiction rate both continuously increase with time, as shown in Figures 7.24 and 7.25.

Example 7.2: Modeling of Population Problem

The first-order positive feedback structure can explain much, but not all of the basic character of human population growth.

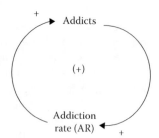

FIGURE 7.22
Causal loop diagram for addiction problem.

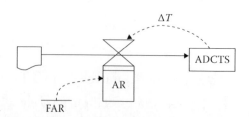

FIGURE 7.23
Flow diagram for addiction problem.

System Dynamics Techniques

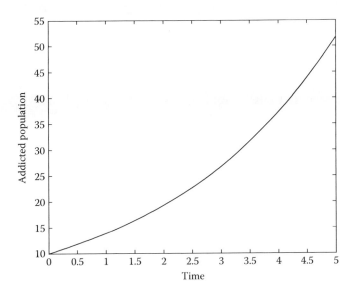

FIGURE 7.24
Addicted population trend with time.

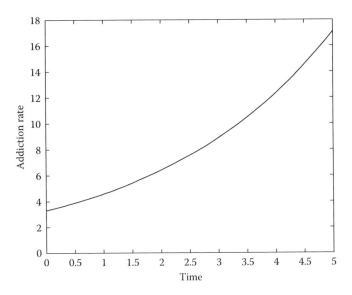

FIGURE 7.25
Addiction rate vs. time characteristic.

The causal representation of relationship between the population (POP) and net birth rate (NBR) (births per year less deaths per year) shows that as POP increases, NBR increases, as shown in Figure 7.26, and its flow diagram is shown in Figure 7.27. From these causal loop diagrams and flow diagrams, the following dynamo equations can be written and simulated in MATLAB to get the population trend, as shown in Figures 7.28 and 7.29:

```
L    POP.K = POP.J + (ΔT) (NBR.JK)
N    POP = 0.5
R    NBR.KL = NGF*POP.K
C    NGF = 0.003
```

Example 7.3: Inventory Control Problem

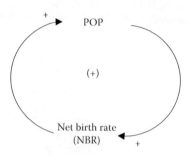

FIGURE 7.26
Positive causal loop diagram for human population growth.

A dealer likes to maintain a desired level of inventory as shown in Figure 7.30. When stock falls below the desired level, the dealer places orders to the factory to replenish the supply; orders stop when stock reaches the desired level. With too much inventory, depends on the market condition outside the system boundary assume that the dealer has no influence on the demand for his goods. The demand suddenly increases from 0 to 50 in the 6th week.

The causal relations between different variables combined are diagrammatically represented in the causal loop diagram, as shown in Figure 7.31. From the causal loop diagram, the flow diagram may be drawn as shown in Figure 7.32.

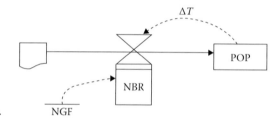

FIGURE 7.27
Flow diagram for human population growth.

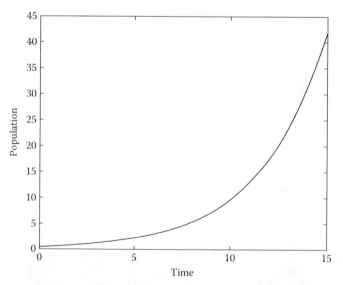

FIGURE 7.28
Population trend with time.

System Dynamics Techniques

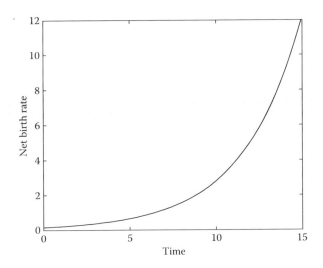

FIGURE 7.29
Net birth rate vs. time.

FIGURE 7.30
Inventory control problem.

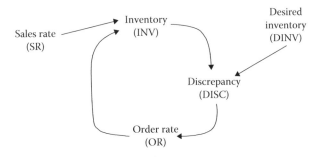

FIGURE 7.31
Causal loop diagram.

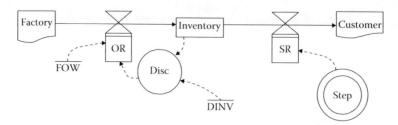

FIGURE 7.32
Flow diagram for inventory control system.

Dynamo model

L   INV.K = NV.J + (ΔT) (OR.JK − SR.JK)
N   INV = DINV
R   OR.KL = FOW*DISC.K
C   FOW = 0.2
A   DISC.K DINV − INV.K
C   DINV = 200
R   SR.KL = STEP (50, 6)

Results

The above developed dynamo equation may be simulated in MATLAB environment. The inventory level decreases when the demand is suddenly increased, as shown in Figure 7.33. As soon as the inventory level decreases, the discrepancy increases and to maintain the inventory level constant, it is necessary to order the goods. Hence the order rate increases, as shown in Figure 7.34. When the order rate and the sales rate are equal to each other, the inventory is fixed at level. The sales rate in this example is suddenly changed from 0 to 50 at the 6th week, as shown in Figure 7.35.

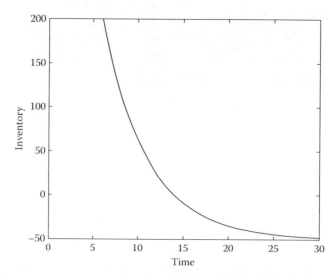

FIGURE 7.33
Change in inventory with time.

System Dynamics Techniques

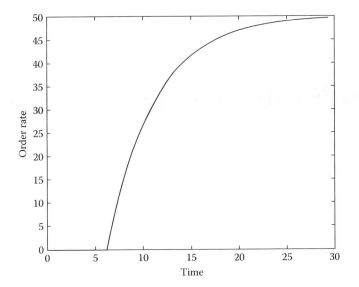

FIGURE 7.34
Change in order rate with time.

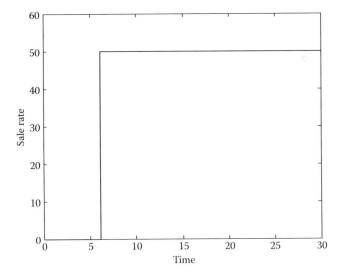

FIGURE 7.35
Demand suddenly increases in 6th week from 0 to 50.

Example 7.4: Rat Population

The rat birth rate (RBR) is defined as the number per month of infant rats that survive to adulthood the normal rat fertility NRF, the average number of infants per months produced by each adult female rat, equals 0.4 (rats/female/month) for a relatively low or normal population density. The adult rat death rate (ARDR) is a function of the number of adult rats and average rat life (ARL) time. Average rat life time ARL, defined as the number of months an average rat survives during normal conditions, equals 22 months; therefore 4.5% of the population dies every month.

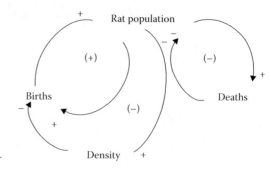

FIGURE 7.36
Casual loop diagram for rat population.

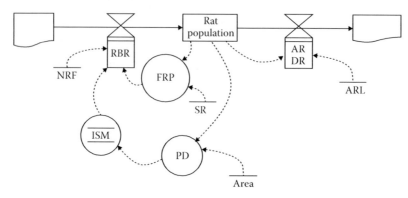

FIGURE 7.37
Flow diagram for rate population model.

The infant survival multiplier (ISM) makes infants survival dependent upon the rat population density. A low population density, approximately 100% survive. The causal loop diagram and flow diagram for rat population are shown in Figures 7.36 and 7.37, respectively.

Dynamo equation

L   RP.K = RP.J + (ΔT) (RBR.JK − ARDR.JK)
N   RP = 10
R   RBR.KL = NRF*FRP.K*ISM.K
C   NRF = 0.4
R   ARDR.KL = RP.K/ARL
C   ARL = 22
A   FRP.K = SR*RP.K
C   SR = 0.5
A   ISM.K = TABLE (ISMT, PD, K, 0, .025, .0025)
T   ISMT = 1/1/.96/.92/.82/.7/.52/.34/.20/.14/.1
A   PD.K = RP.K/A
A   FRP = RP*SR
C   A = 11000
Spec ΔT = 1, length = 200

Results

The above developed dynamo equations were simulated in MATLAB and the following results have been obtained. Figure 7.38 shows that the birth rate increases initially and then gets saturated.

System Dynamics Techniques

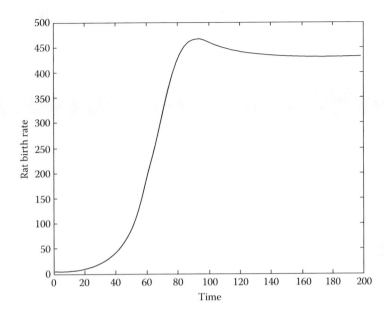

FIGURE 7.38
Rat birth rate.

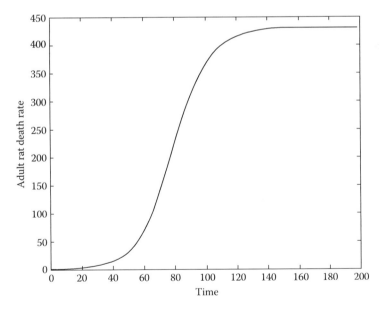

FIGURE 7.39
Death rate of adult rat.

As the rat population increases, the death rate also increases, as shown in Figure 7.39. In the initial phase of simulation, birth rate is dominating and hence the rat population increases exponentially, then in the later portion of simulation death rate increases at a faster rate and when both death rate and birth rate are equal to each other, then the rat population is constant, as shown Figure 7.40.

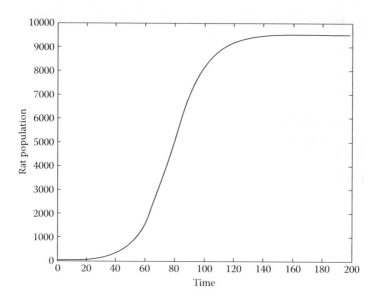

FIGURE 7.40
Rat population.

Example 7.5: Modeling of Infected Population

The infection of any disease depends on the size of susceptible population (SP), normal contacts of infected people (IPC) with healthy people, and normal contact fraction (NCF). Derive a model to predict the infected population.

Solution

The causal links for infected population model have been combined together to get a causal loop diagram as shown in Figure 7.41 and flow diagram (shown in Figure 7.42). From these diagrams, the dynamo equation can be obtained as follows:

L $IP.K = IP.J + (\Delta T)(CR.JK)$
N $IP = 10$
R $CR.KL = IPC*NCF*IP.K*SP.K$

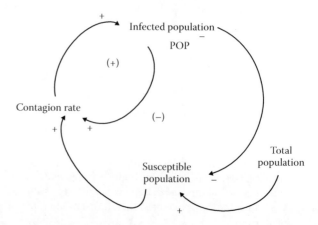

FIGURE 7.41
Causal loop diagram for infected population.

System Dynamics Techniques

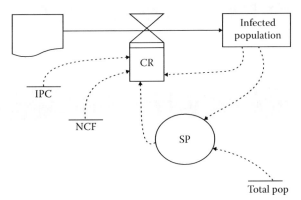

FIGURE 7.42
Flow diagram for infected population.

```
C   IPC = 0.1
C   NCF = 0.02
A   SP.K = P − IP.K
C   P = 100
```

```
%MATLAB program for simulation of addiction problem
clear all;

IP = 10
IPC = 0.1
NCF = 0.02
P = 100
SP=P-IP;
CR=0;
t=0;
dt=1;
Tsim=50;
n=round(Tsim-t)/dt;
for i=1:n
    x1(i,:)=[t,IP, SP,CR];
    SP= P-IP;
    CR = IPC* NCF*IP*SP;
    IP=IP+dt*CR;
    t=t+dt;
end
figure(1)
plot(x1(:,1),x1(:,'2),k-')
xlabel('Time')
%ylabel('Infected Population')
hold
%figure(2)
plot(x1(:,1),x1(:,3),'k-')
xlabel('Time')
%ylabel('Susceptible population')

%figure(3)
plot(x1(:,1),x1(:,4),'k:')
xlabel('Time')
%ylabel('Contagion rate ')
legend('IP', 'SP', 'CR')
```

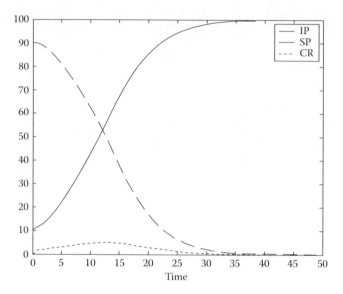

FIGURE 7.43
Simulation results of infected population model.

Results

The infected population model is simulated in MATLAB and the results shown in Figure 7.43.

Example 7.6: Market Advertising Interaction Model

Modern business management is closely related with marketing strategies and effective advertising policies.
  Aims of advertisements

- Awareness about the product
- Comprehension about the product
- Highlight its salient features
- Attract customer and ultimately increase sale

Assumptions

1. Advertising is defined as impersonal communications.
2. Only television is considered as the media, other media are excluded.
3. Different aspects of merchandising such as competition, offers, gifts, packaging, etc., were excluded.
4. Short-term effect is considered.

Postevaluation criterion of advertisements

1. Advertising cost per thousand of buyers for each media category of advertisement (cost effective)
2. Percentage of potential buyer attracted (toward the product advertised) by each media category (public response)
3. Opinion of consumers on the content, presentation, and other parameters of the advertisement for its effective (public comments)

System Dynamics Techniques

4. Awareness of potential buyers toward the product before and after going through the advertising (awareness about product)
5. Number and depth of inquiries stimulated by each category of the advertisement
6. Cost per inquiry

Variable identification

SR	sales rate
SALES	sales of product
ADV	advertising campaign
PD	consumer delay in purchasing
AAP	consumer's awareness of product after advertisement
RAAP	rate of awareness of product after advertisement
MSALES	maximum sales
MAAP	maximum awareness about product
CIA	consumer interest toward advertising
FOI	frequency of advertisement
DAPI	duration of advertisement
PMA	popular models in advertisement
AJS	attractive jingles/slogans in advertisement
TA	Timings of advertisement
CRIA	consumer's rate of interest toward advertisement
MCIA	consumer's maximum interest in advertisement

The causal loop diagram showing the cause–effect relationship between the above-mentioned variable is shown in Figure 7.44. The flow diagram differentiate between the type of variables like level variables, rate variables, auxiliary variables, and constants, as shown in Figure 7.45.

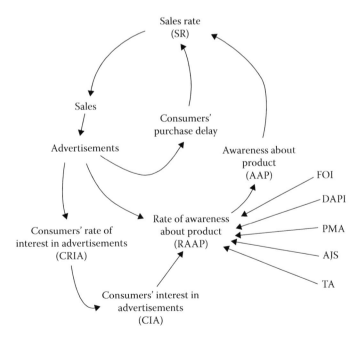

FIGURE 7.44
Causal loop diagram for advertisement interaction model.

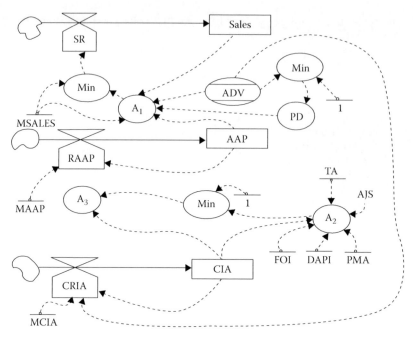

FIGURE 7.45
Flow diagram for advertisement interaction model.

Dynamo equations
Level equations

There are three level variables considered in this system such as sales, awareness about the product, and consumer's interest toward watching advertisements. The present values of level variables are calculated as the sum of previous values and the increment in them during the finite small interval of time.

$$SALES.K = SALES.J + DT*SR.JK$$

$$AAP.K = AAP.J + DT*RAAP.JK$$

$$CIA.K = CIA.J + DT*CRIA.JK$$

Rate equations

To calculate the above-mentioned level variables, three rate equations are written for sales rate, rate of awareness about the product, and consumer's rate of interest toward watching advertisements:

$$SR.KL = Min(A1.K, MSALES)$$

$$RAAP.KL = ((MAAP - AAP.K)/MAAP)*A3.K - AAP.K$$

$$CRIA.KL = (MCIA - CIA.K)/MCIA*ADV.K - CIA.K$$

System Dynamics Techniques

Auxiliary equations

When advertising is more than zero (ADV > 0), sales rate (SR) is proportional to advertisement (ADV). Although, sale is dependent on many other factors such as quality of the product, cost, and packing, it is assumed that the sale is only controlled by advertisements, just to simplify the model. This relationship is true up to saturation of the market, that is ((MSALES − SALES.K)/MSALES). The auxiliary variable A3.K represents consumer's interest toward advertisements.

$$A1.K = ((MSALES - SALES.K)/MSALES)*ADV.K*AAP.K - SALES.K*PD.K$$

$$A2.K = FOI + DAPI + PMA + AJS + TA$$

$$A3.K = CIA.K*Min (A1.K, A2.K)$$

Initialization

```
CIA = 0
SALES = 8
MSALES = 80
AAP = 0
MAAP = 80
MCIA = 100
DAPI = 0.1
PMA = 0.1
AJS = 0.1
TA = 0.1
ADV.K = 0/1, 30/1, 30/2, 30/3, 0/4;
Length = 24 h
```

Results

The simulation results shown in Figure 7.46 of the above developed model validated the results. The conclusions drawn from the above simulation are as follows:

1. The advertiser must create interesting commercials with cleverness that captures the viewer's interest.
2. He includes attractive jingles and popular models in the advertisements, which helps in remembering the advertisements and draws attentions of the viewers.
3. Advertisement should have clear messages/slogans.
4. Scheduling of advertisements is also very important. It should be such that maximum prospective consumers could watch it.
5. The awareness of the consumer does not increase immediately. Always there is some delay. As awareness increases, sales also increase, although it is not linearly proportional to each other.
6. The duration of advertisement is not directly affecting the sales, but the duration of advertisement between 15 and 30 s is sufficient.

Example 7.7: Modeling and Simulation of Production Distribution System

Every manufacturing company has three main departments, namely, finance, production, and marketing, as shown in Figure 7.47.

1. Finance department concerned with mobilizing and regulating the resources.
2. Production department concerned with production of products.
3. Marketing department related to sales of products.

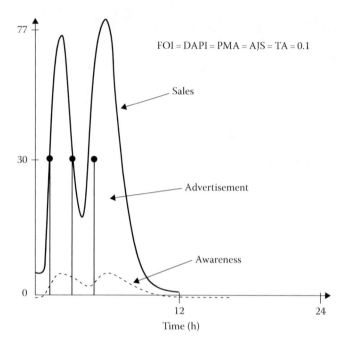

FIGURE 7.46
Simulation results of market advertisement interaction model.

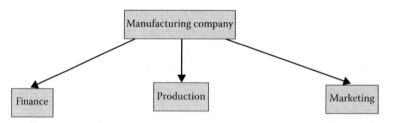

FIGURE 7.47
Main departments of a manufacturing company.

The central core of many industrial organizations is the process of production and distribution system. When the product is produced and finally sold, there is a large time delay (due to number of reasons, such as delay in quality control department, in dispatch department, transportation delay, and delay by distributor). A recurring problem is to match the production rate to the rate of final consumer sales. Production rate fluctuates more widely than the actual customer's purchase rate (due to cascade inventories of distribution system). The production distribution network is shown in Figure 7.48.

The causal loop diagram shows the relationship between pending orders which depends upon the satisfaction level of customers (distributor or industries), items available for supply and its quality, and production rate, etc., as shown in Figure 7.49. The flow diagram for production distribution system is shown in Figure 7.50 (Chaturvedi et al., 1995).

System Dynamics Techniques

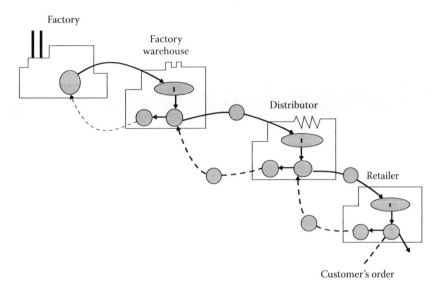

FIGURE 7.48
Production–distribution system.

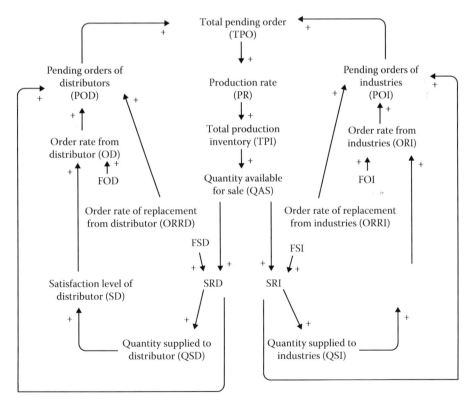

FIGURE 7.49
Causal loop diagram.

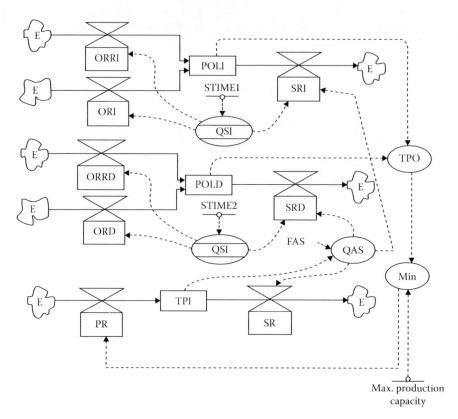

FIGURE 7.50
Flow diagram for production–distribution system.

Dynamo equations

Level equations

The pending order level of industries (POLI), pending order level of distributors (POLD), and total produced inventory (TPI) levels may be determined by adding the incremental value of these variables to the previous value. The increment in level variable may be determined from the net rate, which is the difference between incoming rate and outgoing rate:

$$POLI.K = POLI.J + dt^*(ORRI.JK - SRI.JK)$$

$$POLD.K = POLD.J + dt^*(ORD.JK - SRD.JK)$$

$$TPI.K = TPI.J + dt^*(PR.JK - SRI.JK - SRD.JK)$$

Rate equations

Rate equations for flow-rate variables such as order rate for replacement from industries (ORRI), order rate for replacement from distributor (ORRD), supply rate to industries (SRI), supply rate to distributor (SRD), production rate (PR), and supply rate (SR) may be found out with the following equations:

$$ORRI.KL = EORRI + FORI^*QSI.K$$

System Dynamics Techniques

$$SRI.KL = FSI*QAS.K$$

$$ORRD.KL = EORRD + FORD*QSD.K$$

$$SRD.KL = FSD*QAS.K$$

$$PR.KL = MIN(TPO.K, 3650)$$

$$SR.KL = QAS.K*FAS$$

Auxiliary equation

The auxiliary equations normally help in calculations of rate variables in the system.

$$ORI.K = EORI + FOI*QSI.K$$

$$ORD.K = EORD + FOD*QSD.K$$

$$QSI.K = SMOOTH(SRI.KL\ STIME1)$$

$$QSD.K = SMOOTH(SRD.KL,\ STIME1)$$

$$TPO.K = POLI.K + POLD.K$$

Constants

FORI = 0.05
FOI = 0.015
EORI = 1750
EORRI = 4
STIME1 = 10
FSI = 0.5
FORD = 0.01
FOD = 0.014
EORD = 1850
EORRD = 2
STIME2 = 10

Results

The aforesaid model is simulated on computer and the results obtained are shown in Figures 7.51 and 7.52. The order rate from industries and distributors increases rapidly initially, then saturates after sometime. The pending order level for products increases even if the company is producing to its maximum capacity. The results match satisfactorily with the real data of a computer monitor manufacturing company. The shortage of raw material, machine breakdown, labor strikes etc., is not considered in the system.

Example 7.8: Modeling of AIDS/HIV Population

AIDS is not only a health problem, but it is a societal problem with important social, cultural, and economic dimensions. It threatens the basic social institutions at the individual, family, and community levels. Its economic consequences are equally serious as it could claim up to half of national expenditure for health if the needs of AIDS patients were to be fully met.

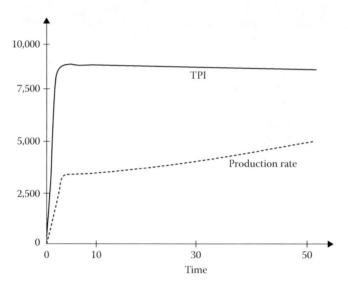

FIGURE 7.51
Simulation results of total pending order level and production rate.

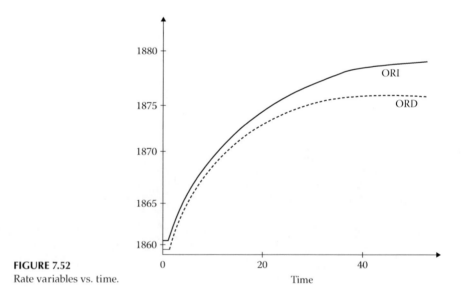

FIGURE 7.52
Rate variables vs. time.

Till today, medical research could not find a cure for AIDS or a vaccine to prevent its infection. We must rely on change in personal behavior to reduce the spread of HIV infection. Information and education play a vital role in the fight against HIV infection.

Keeping in view the complexities of HIV infection, nonrandom distribution, and occurrence of behaviors influencing HIV transmission, it is difficult to make exact estimates of HIV prevalence. It is more so in the Indian context, having its typical and varied cultural characteristics, traditions, and values with special reference to sex-related risk behaviors. The model is also required to predict HIV-infected population and its growth rate, which is very essential for planning of HIV/AIDS prevention and control programs. These predictions can also be useful for mapping of specific vulnerable groups and areas so as to plan targeted intervention in major urban areas and other areas in the states.

Applications of data from sero-surveys of HIV in a sentinel population are considered for the model development of HIV for finding approximate estimate of the magnitude of the HIV in the community. The character and number of the epidemic-affected people varies from region to region, for example, Africa and Asia show more transmission through heterosexual sex, Latin America through homosexual sex, and Eastern Europe and newly independent states through drug injections (Renee, 1988). In India and China, HIV has infected 5% of the people who are engaged in high risk behavior (Thomas, 1994). Overall, more than 6.4 million people are currently believed to be living with HIV in Asia and Pacific—just over 1%–5% of the world's total.

This is an alarming situation and it is a challenge to stop/reduce the rate of infection. Hence, there is a genuine need to develop a model for HIV-infected population.

This chapter highlights the SD methodology and presents its applications for modeling and simulation of the HIV-infected population. The SD modeling technique offers a number of advantages over the conventional modeling techniques specially for modeling such systems where exact model cannot be obtained using conventional techniques. The diagrammatic tools used in the model development encourages the modeler and give better understanding and quick insight about the system which is quite complex in nature and difficult to model due to inherent non-linearities and delays in it (Forrester, 1968). The results obtained by the model developed here is compared with the results predicted by National Aids Control Organization (NACO), India. The model is simulated up to 2003 A.D.

Variable identification

The first step in model development using SD technique is the identification of key variables. The following variables are identified for HIV-infected population model:

1. RISK
   Some of population groups considered to be at high risk of HIV infections because of their behaviors and selected as sentinel groups for HIV sentinel Surveillance (HSS) have included men who have unprotected sex with many men (MSM), injecting drug users (IDU), and heterosexuals who have unprotected sex with multiple sex partners (HET) subgroups-female sex workers (FSW), and sexually transmitted disease (STD) patients. Lower risk HET subgroups used in HSS systems include blood donors, military recruits, and antenatal clinic attendants (ANC) (Gong, 1985; Feldman and Johnson, 1986; Gupta, 1986; Panos, 1988; Contwell, 1991). The authors have conducted a survey to find the risk level of different population groups.

   The causal links between the variables and risk of HIV infection have been shown in Figure 7.53.

2. Awareness level
   The awareness has significant influence on the risk and ultimately, to control the HIV-infected population. To raise awareness about HIV in general population, an information, education and communication (IEC) campaign can be used. IEC is a process that informs, motivates, and helps people to adopt and maintain healthy practices and life styles in order to prevent them from acquiring infection and ill health (AIDS Manual 1993, 1994; Banerjee, 1999; Srinivasa, 2003; Chaturvedi, 2001).

   India is not only a vast country but also a country of numerous culture and linguistic diversities. This poses a great challenge for developing suitable IEC strategies and approaches. The survey was conducted in 1998 to find the awareness level of the population within the age group of 15–49 years in rural and urban areas and some states of India; the results are tabulated in Table 7.1.

3. Research and development
   a. For testing facility for HIV—tests and also help in rehabilitation.
   b. For new methods for IEC to create awareness the society about this disease.
   c. Lastly, it could develop some vaccine for the same.

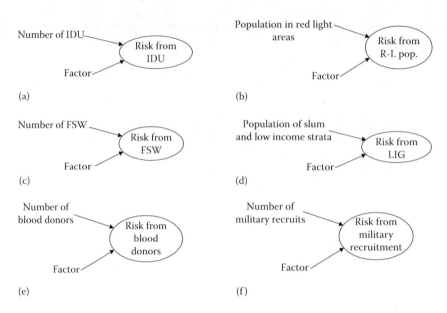

FIGURE 7.53
Causal links between risk and other factors. (a) Causal links for risk from IDU. (b) Causal links for risk from red light population. (c) Causal links for risk from FSW. (d) Causal links for risk from LIG. (e) Causal links for risk from blood donors. (f) Causal links for risk from military recruits.

TABLE 7.1
Awareness Level about HIV (%)

S. No.	State	Rural	Urban
		Awareness Level about HIV (%)	
1	Delhi/Haryana	43.6	57.2
2	West Bengal	13.4	54.4
3	Maharashtra	27.5	55.1
4	Tamil Nadu	63.8	77.9

The complete list of the variables of interest identified for the model is given in Table 7.2.

Causal loop diagram

Figure 7.54 shows the causal loop diagram with four positive loops and three negative feedback loops for HIV-infected population model. It identifies not only the direction of causality but also the strength of relationships involved on the basis of knowledge acquired through relevant literature and survey results.

Causal loop diagram for HIV/AIDS population model
The four positive causal loops are made between the following variables:

1. Risk, susceptible population (SP), rate of HIV infection (RIP), HIV-infected population (IP)
2. Risk, RIP, IP
3. Risk, IEC, rate of awareness (RA), awareness level (A)
4. SP, RIP, IP

TABLE 7.2
List of Variables

1. Risk of getting infected from HIV	16. Rate of HIV infection
2. Susceptible population	17. Information, Education, and Communication (IEC)
3. Crude death rate	18. Blood demand level/blood donors
4. Crude birth rate	19. Drug addicted population
5. Total population	20. HIV-infected population
6. Awareness level among the people	21. Funds allocated for reducing HIV infection rate
7. Man power available to educated for HIV/AIDS	22. Efforts by government and nongovernment organizations
8. Population in red-light areas	23. HIV blood testing facility
9. Population in slums and low income strata	24. Antenatal females
10. Migration level	25. Prisoners, truck, and autorickshaw drivers
11. Female sex workers	26. No. of nonlicensed blood banks
12. Meliorate recruits	
13. Efforts in R&D for HIV/AIDS	
14. Rehabilitation measure	
15. Prolonging death due to AIDS by effective rehabilitation	

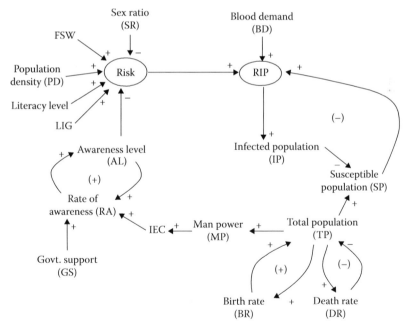

FIGURE 7.54
Causal loop diagram for HIV/AIDS population prediction problem.

Similarly, the three negative causal loops are made between the following variables:

1. Risk, IEC, awareness, blood screening facility, and RIP
2. Total population (TP), SP, RIP, IP, Risk, IEC, RA, A, rehabilitation (RH), and death rate (DR)
3. TP, manpower available for IEC, IEC, RA, A, RH, and DR

Flow diagram

The flow diagram is nothing but elaborated diagram of causal loop diagram, which clearly shows the level variables, rate variables, auxiliary variables, and exogenous variables. Figure 7.55 shows the corresponding flow diagram with the infected population (IP), blood demand (BD), total population (TP), and awareness (A) as the four-level or state variables; the rate of change of HIV infection (RIP), birth rate (BR), death rate (DR), rate of blood demand (BDR), and rate of awareness (RA) as the five flow-rate variables; the total risk due to various factors (RISK), delay in AIDS death, rehabilitation measure (RH), IEC, and susceptible population (SP) as the five auxiliary variables; and funds allocated for HIV control, R&D activities and rehabilitation measures (FUNDS), FSW, military recruit (MR), low income strata (LIG), migration (M), injecting drug users (IDU), efforts from NGOs and people in red-light areas are the exogenous (externally supplied) variables and parameters.

Dynamo model

The dynamo equation can be written from the flow diagram drawn in the earlier stage. Dynamo equations are used for characterizing higher order dynamics of the HIV-infected population, which can be directly simulated in terms of the time profile of level variables, flow-rate variables, and other illustrative characteristics for given values of exogenous variables and parameters:

L $IP.K = IP.J + DT^*RIP.JK$

L $TP.K = TP.J + DT^*(BR.JK - DR.JK)$

L $BD.K = BD.J + DT^*BDR.JK$

L $A.K = A.J + DT^*RA.JK$

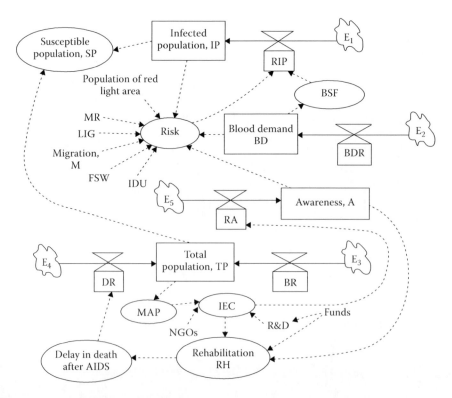

FIGURE 7.55
Flow diagram for HIV-infected population model.

The equations from 1 to 4 are the level equations representing level (state) variables of the system. The level variables (such as infected population, total population, blood demand, and awareness level) at kth time instant are equal to the level variables at Jth time plus increment:

$$R \quad RIP.KL = RISK.K*SP.K + K1*BD.K$$

$$R \quad BR.KL = \text{Table function}$$

$$R \quad DR.KL = \text{Table function}$$

$$R \quad BDR.KL = \text{Table function}$$

$$R \quad RA.KL = IEC.K*K2$$

The rate equations, which will give the rate of state or level variables over a finite time (say DT) are mentioned above. The total population is controlled by two rate variables, one is birth rate and the other is death rate. These rates are taken from the survey results conducted by the Government of India and published in the year book of 1998–1999. Rate of infection is mostly dependent on the risk (which is again controlled by various factors like awareness level, no. of injection drug users, no. of female sex workers, military recruitment, blood demand, population in red-light and slum areas and migration, etc.) and healthy population. The rate of awareness depends on the IEC programmes run by NGOs and GOs:

$$A \quad IEC.K = RISK.K*FUNDS*K3$$

$$A \quad SP.K = TP.K-IP.K$$

$$A \quad RISK.K = FSW*K4 + MR*K5 + M*K6 + IDU*K7 + LIG*K8 - IEC.K*K9 - R\&D.K*K10$$

$$A \quad MPA.K = TP.K*K11$$

$$A \quad R\&D.K = FUNDS.K*K12$$

$$A \quad RH.K = A.K*FUNDS*K13$$

$$C \quad K1 \text{ to } K13$$

The auxiliary equations are basically required for simplifying the calculations. In these equations, parameters and constants, which are initially decided based on the experience are then optimized during refining of the model.

The above developed dynamo model had been simulated using MATLAB and compared with the results in the growth in seropositive cases in India, as given by NACO from 1986 to 2003. The simulation results are shown in Figures 7.56 and 7.57 and also in Table 7.3.

From the above results, the following points are important to note:

1. The result of SD model shows that the rate of HIV seropositive is increasing and has doubled within a brief span of 6 years.
2. The awareness level is an important and crucial variable, which is affecting the behavior of HIV-infected population/rate of HIV infection at a great extent. The awareness level is dependent on the IEC and IEC is dependent on the funds and manpower available from the government and/or from NGOs. From the simulation results, it is very clear that, as the rate of infection increases, the awareness level also increases exponentially. It was found from the model that average awareness level (combined for urban, suburban, and rural areas) will be around 70%–75% in 2003 with the present rate.

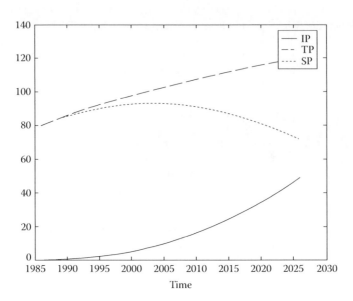

FIGURE 7.56
Prediction of infected population, total population, and susceptible population.

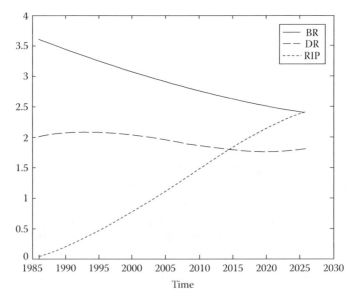

FIGURE 7.57
Time profile of birth rate, death rate, and rate of infected population.

3. The high awareness level may help in reducing the rate of infection and a sigmoid shape of infected population curve may be obtained. The model is simulated only for limited time span and therefore, the saturation part is not apparent.
4. The reasons of spreading HIV infection in Indian scenario are the gradual decline in moral values and premarital sexual experiences, increase in travel for economic reasons, flourishing sex industry, and extension beyond red-light areas, which is the main cause for the high rate of infection in India.

TABLE 7.3

Comparison of Results Obtained from the Model

S. No.	Years	Cumulative HIV Sero Positive Rate per Thousand	
		Actual	Simulated
1	1991	11.3	6.877
2	1996	16.8	25.431
3	2001	—	58.654
4	2006	—	108.820
5	2011	—	177.164
6	2016	—	263.575
7	2021	—	366.319

Example 7.9: Basic Commutating Machine

The basic commutating machine shown in Figure 7.58 has three ports namely field port, armature port, and mechanical port, which are shown in Figure 7.59 (Chaturvedi and Satsangi, 1992). It has three modes of operation:

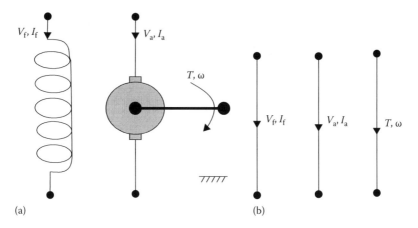

FIGURE 7.58
Constructional diagram and complete assembled diagram of basic commutating machine. (a) Schematic diagram. (b) Terminal graph.

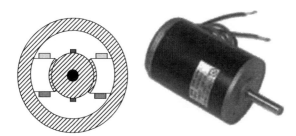

FIGURE 7.59
Basic commutating machine.

- Electromechanical transducer mode
  The field current is constant, that is, fc(t) = FC
- Electromechanical amplifier mode
  The armature current is constant, that is, ac(t) = AC
- Rotating amplifier mode
  The shaft speed is constant, that is, w(t) = W

The complete causal loop diagram for basic commutating machine is shown in Figure 7.60, and its flow diagram in Figure 7.61.

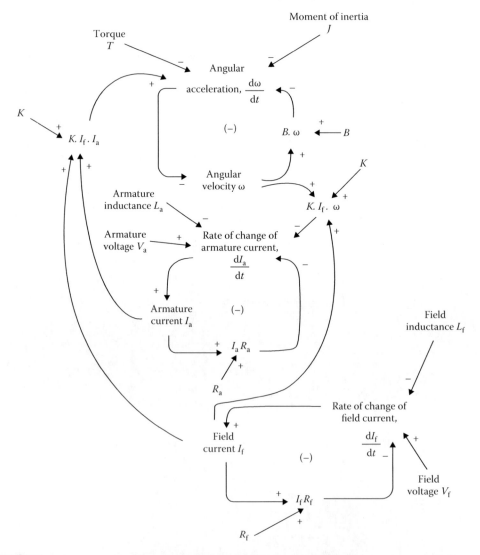

FIGURE 7.60
Causal loop diagram for basic commutating machine.

System Dynamics Techniques

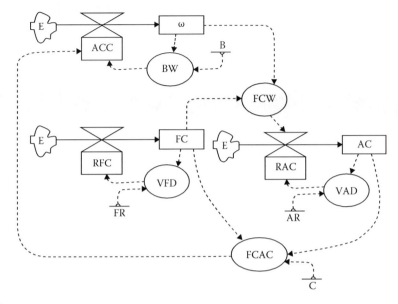

FIGURE 7.61
Flow diagram for basic commutating machine.

Dynamo equation

Field port
   Level equation for field current $FC.K = FC.J + DT*RFC.JK$
   Rate equation for rate of change of field current $RFC.KL = (VF - VFD.K)/FI$
   Auxiliary equation for armature drop in voltage $VFD.K = FR*FC.K$

Armature port
   Level equation for armature current $AC.K = AC.J + DT*RAC.JK$
   Rate equation for rate of change of field current $RAC.KL = (VA - VAD.K - C*FC.K*W.K)/AI$
   Auxiliary equation for drop in field voltage $VAD.K = AC.K*RA$

Shaft port
   Level equation for shaft speed $W.K = W.J + DT*ACC.JK$
   Rate equation for angular acceleration $ACC.K = (T + FCAC.K - BW.K)/Jx$
   Auxiliary equation for generated torque $FCAC.K = C*AC.K*FC.K$
   Auxiliary equation for torque required to overcome friction $BW.K = B*W.K$

Constants
   Field resistance FR = 100 ohm
   Armature resistance RA = 0.5 ohm
   Torque coefficient C = 150
   Moment of inertia Jx = 10

MATLAB program

```
%MATLAB program for simulation of basic commutating machine
% in electromechanical transducer mode of operation
clear all;

VA=125;  % Armature voltage, V
VF=100;  % Field voltage, V
FI=0.06; % Field inductance, H
AI=0.2;  % Armature inductance, H
```

```matlab
FR=100;     % Field resistance, ohm
RA=2;       % Armature resistance, ohm
C=0.3;      % Torque coefficient
Jx=0.010;   % Moment of inertia
B = 0.015;  % Damper
T=0;        % Applied torque

t=0;        % Initial time
DT=0.001;   % Step size
Tsim=3;     % Length of simulation
n=round(Tsim-t)/DT; % Number of iterations
% Initialization of level and rate variables
W=0; FC=0; AC=0;
ACC=0; RFC=0; RAC=0;

for i=1:n
    % Store data during simulation in x1 vector
    x1(i,:)=[t, W, FC, AC, ACC, RAC, RFC];

    % Auxialary equations
    FCAC= C*AC*FC;
    BW= B*W;
    VAD=AC*RA;
    VFD = FR*FC;
    % Rate Equations
    RFC=(VF-VFD)/FI;
    RAC=(VA-VAD-C*FC*W)/AI;
    ACC=(T+FCAC-BW)/Jx;
    % Level equation
    FC=FC+DT*RFC;
    AC=AC+DT*RAC;
    W=W+DT*ACC;

    % increament the time t
    t=t+DT;
    if t>=0.5
    %T=-5;          % Applied torque, N-m
end
end
figure(1)
plot(x1(:,1),x1(:,2))
xlabel('Time, sec.')
ylabel('Speed, r/s')

figure(2)
plot(x1(:,1),x1(:,3))
xlabel('Time, sec.')
ylabel('Field current, A')

figure(3)
plot(x1(:,1),x1(:,4))
xlabel('Time, sec.')
ylabel('Armature current, A')

figure(4)
plot(x1(:,1),x1(:,5:6))
xlabel('Time, sec.')
ylabel('Rate variables,unit/sec')
legend('Acceleration','RAC')
```

System Dynamics Techniques

Results

The above developed causal model simulated for given conditions and data and the following results have been obtained as shown in Figures 7.62 through 7.64. From the results, it is clear that when the supply is given to the machine in motoring mode the armature current increases rapidly in the beginning and then decreases and reaches steady state. The speed of machine gradually increases and settles down to some steady state value.

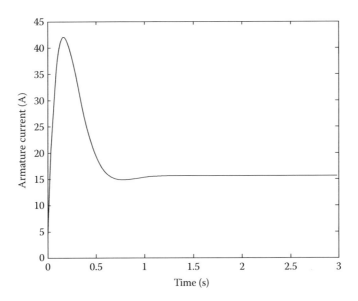

FIGURE 7.62
Armature current vs. time.

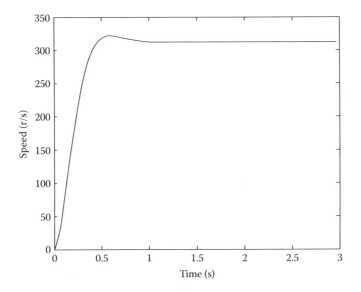

FIGURE 7.63
Speed vs. time.

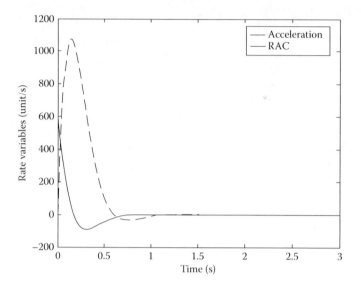

FIGURE 7.64
Rate of change of armature current (RAC) and acceleration vs. time.

7.13 Modeling and Simulation of Parachute Deceleration Device

A parachute constitutes of a round surface that matches the shape of a hemisphere. They are used for various applications, namely, paratroopers, supply dropping, aircraft landing, aircrew (for pilot of fighter aircraft), weapons, and missiles; their main purpose is to reduce the speed of descending mass to its specified terminal velocity. It depends on the diameter of parachute.

Parachute canopies are made of nylon material. Parachute canopy materials are available in different strengths and elongations. According to the requirement, the particular materials are selected. While doing the preliminary calculation for requirement of the strength of material, it is assumed that the strength requirement is uniform throughout the fabric, whereas it is not, as the load imposed on the canopy reduces during parachute inflations, an efficient design could be evolved. It will help in reducing the mass, volume, and cost of parachute.

During the operation of a typical parachute system, say personnel parachute, it goes through a sequence of the following functional phases:

- Deployment
- Inflation
- Deceleration
- Descent
- Termination

The maximum forces are imposed during inflation of the parachute. These maximum forces are the product of dynamic pressure $\left(\frac{1}{2}\rho v^2\right)$ and drag area of the parachute canopy (CdS).

In this example, the aim is to develop a simulation model to determine the variation in (1) force, (2) velocity, (3) stress, and (4) flight path angle of parachute canopy during the period of its inflation with respect to time.

7.13.1 Parachute Inflation

The shape of canopy gores in the skirt area governs the inflation process, canopy mass its porosity parameters. The stages in the inflation of a typical parachute in a finite mass system are illustrated schematically in Figure 7.65. It has been observed that when the canopy mouth opens

1. Substantial volume of air enters down the length of the simply streaming tube of fabric to the apex.
2. The crown begins to fill continuously like a balloon being inflated through a conical duct.
3. Canopy expansion is resisted by structural inertia and tension.

By definition, full inflation is completed when the canopy has first reached its normal steady state projected area.

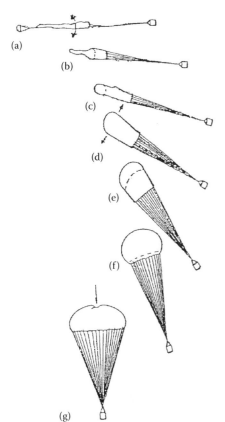

FIGURE 7.65
Different stages in parachute inflation. (a) Opening of canopy mouth. (b) Air mass moves along canopy. (c) Air mass reaches crown of canopy. (d) Influx of air expands crown. (e) Expansion of crown resisted by structural tension and inertia. (f) Enlarged inlet causes rapid filling. (g) Skirt over-expanded, crown depressed by momentum of surrounding air mass.

The inflation of circular parachutes is attended by increase in inflated diameter, or projected frontal area with the time (inflation/opening of parachute). The changing shapes during canopy inflation cause the drag coefficient to increase. During the inflation of a parachute, an aerodynamic force is developed tangential to the flight path that varies with time. The upper crown of the canopy is subjected to this maximum force and lower part of the crown is subjected to less force. The parachute could be made with a combination of two to three types of fabrics keeping in view the variation of forces/stresses on canopy surfaces.

7.13.2 Canopy Stress Distribution

The importance of knowing the stress distribution in canopy and its relation to canopy stress lies in the necessity for minimizing the weight of the canopy and its required packing volume. In lightweight designs, where maintenance of canopy strength is an important factor, stress analysis is an indispensable element.

The stresses in a canopy are caused by aerodynamic loads acting on and between the various structural components. The stresses are both dynamic and static. Dynamic stresses usually exist for comparatively short time during the transition period from the moment the suspension lines are stretched when the canopy filling begins to the moment a steady state is reached (the moment the canopy assumes its fully inflated shape). Although the stresses are not constant during the steady state, their variation is small compared to that encountered during the transition period. Therefore, the stresses in steady state can be considered as static.

The load-carrying elements are primarily fabric that has little stiffness and, therefore, can take no bending loads. Loads are resisted by tension in the members. Many types of fabric can be utilized, each having its own strength, elongation characteristics, shape, and construction, which vary with the type of canopy.

Maximum stresses occur during the opening process, which is the period of rapidly changing shape and load. Experiences with structural failures in canopies, as well as measurement of snatch and opening forces, indicate that the critical stresses occur during the opening shock. Nevertheless, the steady state problem is of interest because the maximum stress in the infinite-mass case, which approximates the conditions of opening for such applications as first-state deceleration, aircraft landing deceleration, and others seem to occur when the canopy is fully or almost fully opened. The steady state case also serves as a useful, preliminary to the more difficult and important problems of critical transient stresses during opening. The maximum force is imposed on the upper part of the parachute. When the parachute opens fully, the magnitude of the forces reduces substantially.

In order to work out the efficient design, the history of variation of forces on different part of canopy are essential so that the lower portion of the parachute fabric could be selected based on the actual reduced loads.

7.13.3 Modeling and Simulation of Parachute Trajectory

The following steps have been followed in the modeling and simulation of parachute dynamics using SD technique:

Step 1 Variable identification and drawing causal links

The first step in modeling and simulation of parachute system is to identify the key variables. Secondly, the cause–effect relationship is determined and systematically represented in the causal loop diagram shown in Figure 7.66.

System Dynamics Techniques

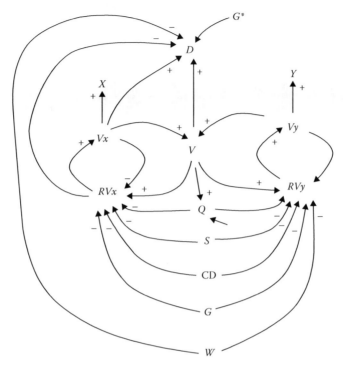

FIGURE 7.66
Causal loop diagram for modeling and simulation of parachute.

Step 2 Draw causal loop and flow diagram

After drawing the causal loop diagram, the flow diagram is drawn illustrating the different types of variables, as shown in Figures 7.66 and 7.67.

Step 3 Dynamo equations

The dynamo equation can be written from the flow diagram drawn in the earlier stage. Dynamo equations for characterizing higher order dynamics of the parachute system can be directly simulated in terms of the time profile of level variables, flow-rate variables, and other illustrative characteristics for given values of exogenous variables and parameters.

Level equations

Equations mentioned below are the level equations representing level (state) variables of the system. The level variables (such as horizontal velocity, vertical velocity, horizontal distance, and vertical distance) at k^{th} time instant is equal to the level variables at J^{th} time plus increment:

$$L \quad Vx.K = Vx.J + \Delta T^* RVx.JK$$

$$L \quad Vy.K = Vy.J + \Delta T^* RVy.JK$$

$$L \quad x.K = x.J + \Delta T^* Rx.JK$$

$$L \quad H.K = H.J + \Delta T^* Ry.JK$$

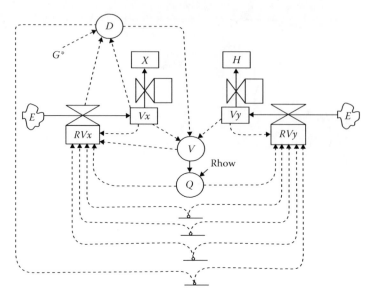

FIGURE 7.67
Flow diagram for modeling and simulation of parachute system.

Rate equations

The rate equations, which will give the rate of state or level variables over a finite time (say DT) are:

$$R \quad RVx.KL = -(Q.K*Cd*S*G*Vx.K)/(W*V.K)$$

$$R \quad RVy.KL = -(Q.K*Cd*S*G*Vy.K)/(W*V.K) - G$$

$$R \quad Rx.KL = Vx.K$$

$$R \quad Ry.KL = Vy.K$$

Auxiliary equation

The auxiliary equations are basically required for simplifying the calculations and are given below:

$$A \quad V.K = sqrt(Vx.K^2 + Vy.K^2)$$

$$A \quad Q.K = (rhow*V.K^2)/2$$

$$A \quad Drag = -(W/G)*(RVx.KL/Vx.K)*V.K$$

$$A \quad theta = 90 + atan(Vx.K/Vy.K)*180/pi$$

The above developed dynamo equations are simulated on a computer in MATLAB environment. The results are shown in Figures 7.68 and 7.69. It is very clear from the results that

System Dynamics Techniques

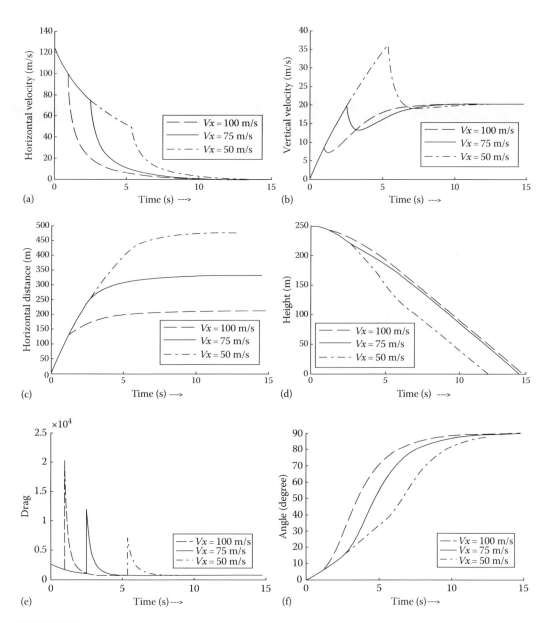

FIGURE 7.68
Simulation results with two parachutes (pilot and main) when pilot parachute is detached at different horizontal velocities.

the time required to reach the ground is different if pilot parachute is detached at different timings. The value of the drag is also dependent on the time at which the pilot parachute is detached, as shown in Table 7.4.

The results also indicate that the time required to reach the ground is different if pilot parachute is detached at different horizontal velocities. The value of snatch force and the drag is also quite different for different velocities, as shown in Table 7.5.

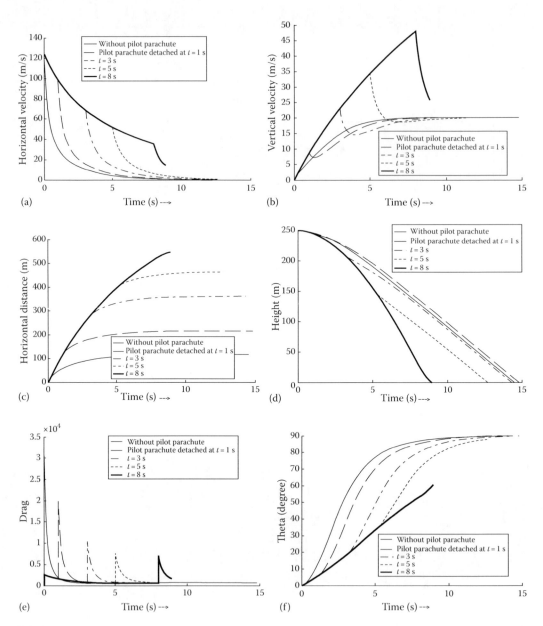

FIGURE 7.69
Simulation results with two parachutes (pilot and main) when pilot parachute is detached at different times.

7.14 Modeling of Heat Generated in a Parachute during Deployment

A large amount of heat is generated during power packing and deployment of parachute. The heat developed during packing may be controlled, but it is uncontrolled when parachute is deployed. Heat generated is due to interlayer friction and wall-fabric friction. The canopy is a poor conductor of heat and irregular in shape. A large temperature

TABLE 7.4

Simulation Results When Pilot Parachute Is Detached at Different Times

Time of Detachment (s)	Time Required to Reach to Ground (s)	Vertical Velocity When Detached (m/s)	Drag	Angle from Horizontal	Total Horizontal Distance Traveled (m)
0	14.52	0	31,535	89.9	117.8
1	14.83	9	19,866	89.8	216.53
3	14.34	22.9	10,643	89.6	362.50
5	12.71	34.45	7,815.7	88.0	464.05
8	8.9	48.03	7,140.2	60.5	546.89

TABLE 7.5

Simulation Results When Pilot Parachute Is Detached at Different Horizontal Velocities (Vx)

Vx	Time Required to Reach to Ground (s)	Vertical Velocity When Detached (m/s)	Drag	Angle from Horizontal	Total Horizontal Distance Traveled (m)
100	14.82	8.9	20,277	89.84	212.08
75	14.58	20.18	12,028	89.68	331.23
50	12.35	36.16	7,605.2	87.38	477.72

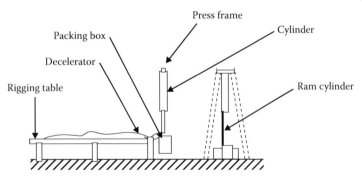

FIGURE 7.70
Schematic diagram of general utility decelerator packing press facility.

gradient is developed in the canopy surface. This may cause a major failure in the parachutes (Chaturvedi and Gupta, 1995). The schematic diagram of a general utility decelerator packing press is shown in Figure 7.70.

The key variables affecting the heat generation during deployment is identified and the causal links are drawn to get complete causal loop diagram, as shown in Figure 7.71. The different types of variables are separated and systematically shown in the flow diagram, as shown in Figure 7.72.

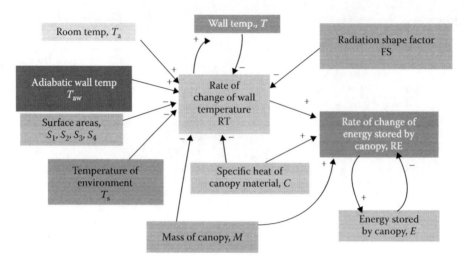

FIGURE 7.71
Causal loop diagram for heat generation model for parachute.

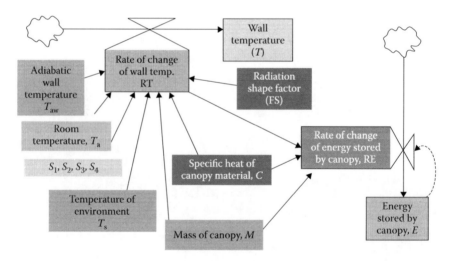

FIGURE 7.72
Flow diagram for heat generation model for parachute.

7.14.1 Dynamo Equations

From the flow diagram, the dynamic equations for heat generation during deployment are written as

1. Level equations for temperature and energy

$$T.K = T.J + dt^*RT.JK$$

$$E.K = E.J + dt^*RE.JK$$

System Dynamics Techniques

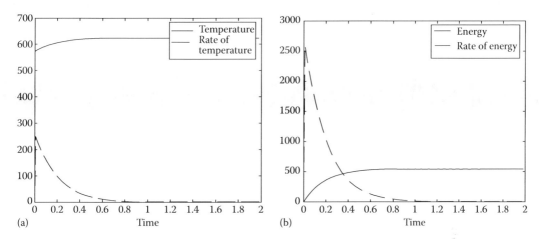

FIGURE 7.73
Trend of increase in temperature, energy, and their rate variables.

2. Rate equations

$$RT.KL = (CC.K + L^*A^*S_2 - BB.K - AA.K)/(M^*C)$$

$$RE.KL = M^*C^*RT.KL$$

3. Auxiliary equations

$$AA.K = c_1^*FS^*E^*S_4^*(T.K - Ta)$$

$$BB.K = c_1^*S_3^*E^*(T.K - Ta)$$

$$CC.K = H^*S_1^*(Taw - T)$$

These equations are simulated in MATLAB environment and the results are shown in Figure 7.73.

7.15 Modeling of Stanchion System of Aircraft Arrester Barrier System

The aircraft arrester barrier nets are installed at the end of the runways for the purpose of arresting combat aircrafts overshooting runways' length during aborted takeoff and emergency landings. Two stanchion systems one at each end of the net are required in a system to support and to provide electrical controlled movement to the net. During deployment of the net, certain forces are generated and imposed on the subsystem of the aircraft arrester barrier system. Arrestment of a landing or aborted aircraft is accomplished by the engagement of the aircraft with multiple element net assembly stretched across the runway which is lifted by two stanchion system. There are two energy absorbing systems installed at both ends of the aircraft arrester barrier system. The aircraft arrester barrier system is shown in Figure 7.74; it consists of the following subsystems:

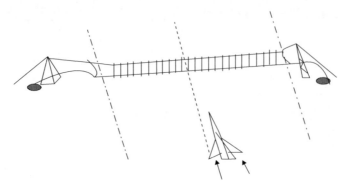

FIGURE 7.74
Aircraft energy absorbing system.

1. Stanchion system
2. Energy absorbing system
3. Engagement system
4. Tape retrieval system
5. Pressure roller system
6. Suspension system
7. Sheave assembly
8. Drive tape
9. Shear-off coupling
10. Tape connector
11. Net anchoring mechanism
12. Electrical control

These subsystems perform various functions during the operation of the aircraft arrester barrier system. After the landing of the aircraft, the aircraft arrester barrier system initiates its operation when the aircraft is stuck in the net assembly; then operation of energy absorbing systems, which are installed at both ends of the runway start working and hence the braking torque is applied through net tape so that the aircraft stops within the prescribed length of the runway. Consequent upon the stopping of aircraft, the tape retrieval is done.

The stanchion system consists of an electrically operated winch device, which helps in rising and lowering the net of aircraft arrestor barrier system, as shown in Figure 7.74. The system is fixed on a strong concrete ground base on either side across the runway through its metallic base frame.

As the up button is pressed on the panel, the current passes to the motor of the stanchion system, the motor starts rotating, and the lift cable starts winding on the winch drum lifting the stanchion system frame from horizontal to vertical position. As the frame reaches its extreme position, its base presses the lever of limit switch mounted on the base frame, switching off the current to the brake motor. As the current to the brake motor is switched off, the solenoid-operated brake of the motor is actuated, gripping to the shaft of the motor and stopping it instantaneously. To lower the stanchion, the down button is pressed, which passes current to the motor again, the brakes are released and the motor starts rotating again but in reverse direction. The winch cable thus gets unwound from the winch, lowering the stanchion frame. When frames come down to its extreme position, it presses another limit switch stopping its moment further. The maximum forces are imposed during fully deployment of net with minimum allowable sag. This means when the angle of the stanchion frame is at a lower limit in up position, that is, 79° (79°–82°) and the net height in the center of the runway is at an upper limit, that is, 3.9 m (3.7 ± 2 m). the maximum tension T is the product of three functions, that is, weight of the net (W), the horizontal distance between two extreme points of the stanchion frame (l), and the sag of the net (Y_c). This can be expressed in Equation 7.7:

$$T_1 = 0.5 * W * \sqrt{\left(l + (l/4 * Y_c)\right)^2} \qquad (7.7)$$

System Dynamics Techniques

In this work, the stanchion system is modeled and simulated under different operating conditions for the tension in winch cable and support cable under dynamic conditions when the system is erected. Also, the rate of change of tension in the suspension cable and the lift cable has been studied.

7.15.1 Modeling and Simulation of Forces Acting on Stanchion System Using System Dynamic Technique

The net deployment process is governed by the lifting of the net, thus increasing weight and center height of the net, which results in an increase in tensions in the suspension cable and the lift cable. The stages in the lifting of the net are illustrated schematically in Figure 7.75.

When the stanchion is in up position, different forces are acting on it, as shown in Figure 7.76. The total angle suspended by the 5.9 m long stanchion frame is ϕ in time t. Therefore the angular velocity ω is given by

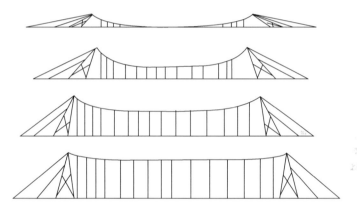

FIGURE 7.75
Different stages of deployment of aircraft arrestor barrier system.

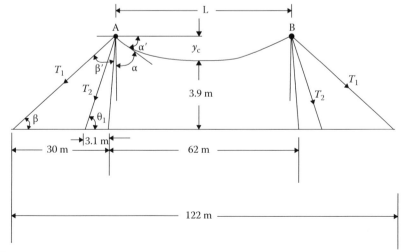

FIGURE 7.76
Forces on stanchion systems.

$$\omega = \phi * 3.14/(180 * t) \tag{7.8}$$

If H is the horizontal pull and V is the vertical reaction then

$$\tan \alpha = H/V \tag{7.9}$$

Where l is the horizontal length between points AB and Y_c is sag can be expressed as

$$l = A_1 = 62 - 2 * A_2 \tag{7.10}$$

where

$$A_2 = 5.9 \cos(\omega * t) \tag{7.11}$$

Similarly

$$Y_c = A_3 - H \tag{7.12}$$

where

$$A_3 = \sin(\omega * t) \tag{7.13}$$

and H is the height of the net.
Therefore from Equations 7.10 and 7.11, Equation 7.9 becomes

$$\tan \alpha = A_1/4 * (A_3 - H) \tag{7.14}$$

Similarly from Equations 7.10 and 7.13 we get

$$\tan \beta = A_3/(30 + A_2) \tag{7.15}$$

The look of the winch cable is attached on the stanchion frame at a distance of 5.63 m than

$$\tan(\theta_1) = A_5/3.1 + A_6 \tag{7.16}$$

where

$$A_5 = 5.63 * \sin(\omega * t) \tag{7.17}$$

and

$$A_6 = 5.63 * \cos(\omega * t) \tag{7.18}$$

Similarly RA is the reaction due to tension T_1 that can be expressed as

$$RA = T_1^* \cos \alpha + T_1 \sin \beta \tag{7.19}$$

From Equation 7.19, tension in the winch cable T_2 and compression in the stanchion system T_3 can be expressed as

$$T_2 = RA/\sin\theta_1 \qquad (7.20)$$

$$T_3 = RA*5.9/A_3$$

W is the weight of the net which can be expressed as

$$W = \rho*V \qquad (7.21)$$

where
  ρ is the density of the material which is a constant
  V is the volume of the material

Considering net as a flexible certain with B is the thickness and A is the area of the net, Equation 7.21 becomes

$$W = \rho*A*B \qquad (7.22)$$

Area A can be calculated as
  A = area of the net
    = area of the rectangle QABP – area of parabola
Finally $A = 1*(A_3 + 2H)/3$

$$\text{area} = A_1*(A_3 + 2H)/3 \qquad (7.23)$$

Put this value in Equation 7.17

$$W = V*\rho*A_1*(A_3 + 2H)/3 \qquad (7.24)$$

Considering Equations 7.8 through 7.24, the net weight W, horizontal distance l, and sag Y_c is the function of ϕ or angular velocity ω therefore change in the ϕ or angular velocity results substantial change in tensions T_1, T_2, and T_3. By simulation, prediction can be made how quickly the net is deployed with particular HP of the motor and rate of the change of tension is predicted.

In the performance characteristic evaluation of the stanchion system, it is important to determine the maximum forces or tension in the suspension cable, lift cable in order to calculate the power requirement of the electrical motor and the required strength of material to be used for above-mentioned elements. In preliminary design and trade of studies, the designer requires a simple theory, which will enable rapid and reasonably accurate estimate of peak loads in a particular deployment situation. Based on the maximum design value, the strength of shear pin is selected. HP of electrical motor is predicted by SD technique.

7.15.2 Dynamic Model

In this model, an attempt is made to simulate conditions when the net started to deploy from down position to the position when suspension strap suspended on point A to B of the stanchion system, as shown in Figure 7.77, in such a way that the center of the suspension strap made a point contact with ground. In this position, the geometrical profile of the suspension

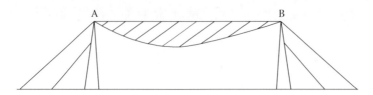

FIGURE 7.77
Calculation of area of net.

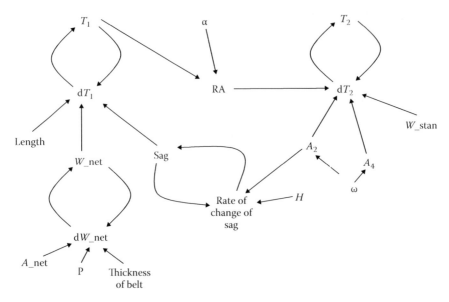

FIGURE 7.78
Causal loop diagram.

strap, on which the net is attached, is a parabola. The total weight of the net is therefore equal to the weight of the bunch of the horizontal straps plus the wait of the suspended portion of the net. It varies with time. This model is simulated up to 0.72 s. The variation in the tension T_1, T_2, and T_3 and rate of change of tension is observed at different intervals of time.

The simulation of time profile of tensions, weight of the net, and deployment of the stanchion system have been attempted using SD technique. In this technique, casual loops are identified, as shown in Figure 7.78 of casual loop diagram. From the casual loop diagram, the flow diagram is drawn, as shown in Figure 7.79, which encourages the modeler to understand the model. After drawing the flow diagram, dynamo equations are written, the DYNAMO model was simulated directly on the computer in MATLAB package without any change in the model. This is an advantage of the model, when developed by this technique.

7.15.3 Results

The models obtained in the result match with the values obtained from imperial relations. The results obtained using the simulation is quite encouraging. The maximum values of tension obtained from simulation are comparatively less than the maximum values attempted with the conventional method.

The variation of tension, weight of the net, aircraft engagement time profile, rate of the change of tension, and rate of change of weight of the net, obtained from simulation

System Dynamics Techniques

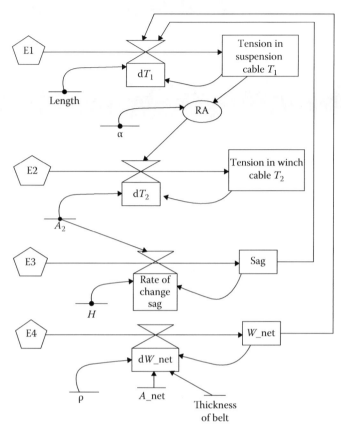

FIGURE 7.79
Flow diagram for modeling and simulation of Stanchion system.

studies are given in Figures 7.80 and 7.81. From Figure 7.81, the results are quite close to the actual results obtained from the system. Figures 7.82 and 7.83 show the simulation results of stanchion system of aircraft arrestor barrier system for 20 ton aircraft.

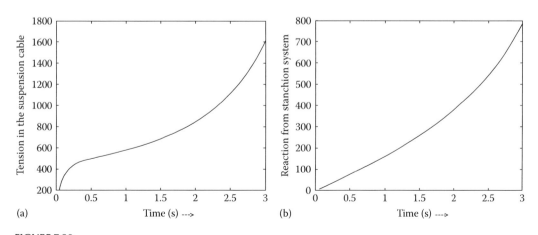

FIGURE 7.80
Simulation results of stanchion system during deployment of 20 ton aircraft arrester barrier system.

(continued)

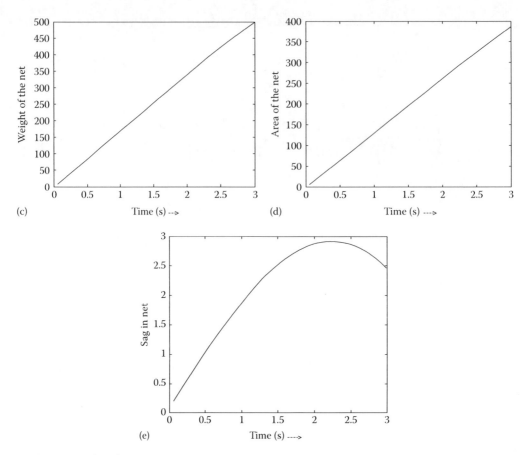

FIGURE 7.80 (continued)

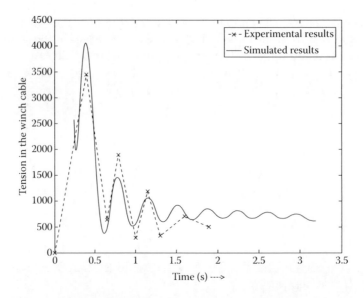

FIGURE 7.81
Time profile of tension T_1, T_2, reaction and weight of the net during deployment of 40 ton aircraft arrester barrier system.

System Dynamics Techniques

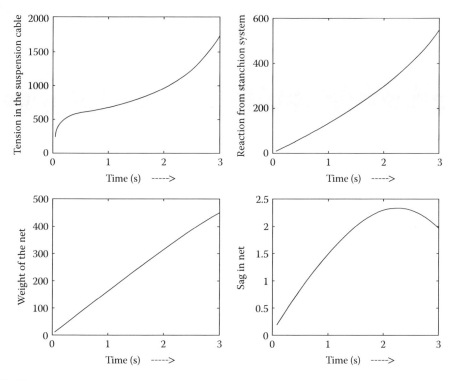

FIGURE 7.82
Time profile of tension in suspension cable, reaction from stanchion system, weight of the net and sag of the net during deployment of 20 ton ABBS.

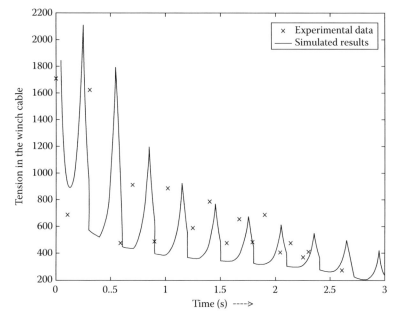

FIGURE 7.83
Comparison of experimental data and simulated data.

```matlab
%MATLAB Program for simulation of stanchion system
clear all;
% Simulation parameters
t=0.05;
tsim=3.0;
dt=0.001;tc=t;
% Constants
W_stan=500; %kg
Hmax=4.9; %m
dh=dt*Hmax/(tsim-t);
H=0;
B1=5;
w=82*pi/(180*(tsim-t));
Row=1140; % kg/m3
thick=11.37e-4; %m
c=1.02;
n=round((tsim-t)/dt); % no. of iterations
% Loop
num=.1;
den=[.5,0.9,150];
[A B C D]=tf2ss(num,den);
k=00; u=1;x=[0;0];
for i=1:n
    A1=7.4*cos(w*t);
    A2=7.4*sin(w*t);
    A3=6.9*cos(w*t);
    A4=6.9*sin(w*t);
    length=64.5-2*A1;
    sag=(A2-H);
    A5=length/(4*sag);
    alpha=atan(A5);
    beta=atan(A2/(20+A1));
    theta1=atan(A4/(A3+3.1));
    A_net=length*(A2-H)/2+length*H;
    W_net=Row*A_net*thick;
    T1=0.5*W_net*sqrt(1+A5^2);
    RA=T1*cos(alpha)+T1*sin(beta);
    dx=A*x+B*u;
    x=x+dt*dx;
    B1=45e5*(C*x+D*u);
    tc=tc+dt;
    T2=(RA*A1+W_stan*(1.25*cos(w*t))+B1*w)/(cos(theta1)*A4)-1500*exp(-
    t/0.5);%;
    H=H+dh;
    t=t+dt;
    x1(i,:)=[t,T1,T2,RA,W_net, B1, A_net sag, theta1];
end
T2exp=[10 3450 630 1905 300 1190 320 700 500
     .6 1 1.25 1.4 1.6 1.75 1.9 2.2 2.5]';
figure(3)
plot(x1(:,1),x1(:,2))
xlabel('Time (sec.) -->')
ylabel('Tension in the suspension cable')

figure(1)
plot(T2exp(:,2)-0.6, T2exp(:,1),':x')
hold
```

```
plot(x1(:,1)+0.2,x1(:,3))
xlabel('Time (sec.) --->')
ylabel('Tension in the winch cable')
legend('Experimental results', 'Simulated results')

figure(2)
plot(x1(:,1),x1(:,4))
xlabel('Time (sec.) --->')
ylabel('Reaction from stanchion system')

figure(4)
plot(x1(:,1),x1(:,5))
xlabel('Time (sec.) --->')
ylabel('weight of the Net')

figure(5)
plot(x1(:,1),x1(:,7))
xlabel('Time (sec.) --->')
ylabel('Area of the net')

figure(6)
plot(x1(:,1),x1(:,8))
xlabel('Time (sec.) --->')
ylabel('Sag in net')

figure(7)
plot(x1(:,1),x1(:,9))
xlabel('Time (sec.) --->')
ylabel('Angle theta1')

if (tc>=0.1 & tc<0.25); B=B*1.2;
   %elseif (tc>=0.25 & tc <0.26) B=B;
   elseif (tc>= 0.25 & tc <0.4) B=B/1.2;
   elseif (tc>=0.4); tc=0.1;B=.05*exp(-t/1.5);
   end
```

7.16 Review Questions

1. Briefly describe three main approaches of representing qualitative simulation models.
2. In the context of the SD simulation methodology, explain the following concepts: causal links, causal loop diagram, flow diagram, dynamo equations, and S-shaped growth.
3. Differentiate the following:
    a. Positive and negative causal link
    b. Positive and negative causal loops
    c. Level and rate variables
    d. Auxiliary variables and constants
    e. Decision sector and process sector

3. i. In a delayed ordering system, the goods on order are increased by the order rate and depleted by the receiving rate (goods on order/delay in ordering). Suppose that the goods on order are initially zero and that an order rate of 800 units per week suddenly begins at the 4th week and continues up to 20th week. Draw causal loop diagram, flow diagram, and dynamo equations to simulate for receiving rate and goods in order.

   ii. The amount of heat in thermometer determines its temperature; the greater the heat content, the higher the temperature. The heat flow rate depends on the difference in temperature between the thermometer and its surroundings. Develop a dynamo model for this thermal system and simulate it on digital computer. Assume that the initial temperature of thermometer is 60°C and the temperature of surrounding is 20°C.

   iii. Consider a water storage system the float drops, it turns on the valve to admit water in the tank. The water level rises, causes the float to rise, and this is in turn gradually shuts off the valve, because the valve controls the water level (WL) which controls the valve. Assume that the initial water level equals to zero and maximum water level may be equal to 4 gal.

   iv. Radioactive material spontaneously disintegrates at a rate that depends on the amount of material that remains. That is, a given fraction of the remaining material disintegrates everyday. Depending on the particular kind of atom, the rate of decay can be so fast that it is difficult to observe the material before it disintegrates or so slow that it takes thousands of years to lose half of the material, assume we have 1 mg of material which decays at 5% per day.

      a. Draw causal loop and flow diagram for above system.

      b. Draw decay rate and amount of material vs. time.

4. Sketch a causal loop diagram and a flow diagram. Also write DYNAMO equations for the S-shaped growth structure represented by either population model or an infection disease model.

5. Write DYNAMO equations corresponding to the flow diagram shown in Figure 7.84. The variable dimensions are given in Table 7.6.

Q. Short Answer Questions

1. Draw general trends of:
   a. Positive feedback systems.
   b. Negative feedback systems.

2. Define and give one example for each:
   a. Positive causal links.
   b. Negative causal links.

3. Arms proliferation causal loop diagram shown as:
   a. Draw flow diagram for this system.
   b. Write dynamo equations for the arms proliferation system.
   c. Draw system response with time (approximately).

System Dynamics Techniques

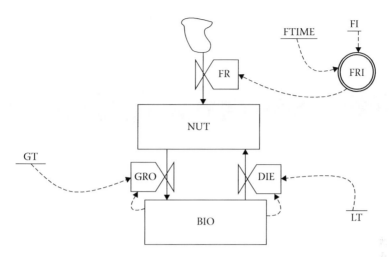

FIGURE 7.84
Flow diagram for nutrition system.

TABLE 7.6

Variables and Their Dimensions

Variable	Dimension
NUT, BIO	Food unit
FR, FOOD, GRO, DIE, FRT	Food unit/time
FTIME, LT	Time
GT	Food units-Time

4. Define:
   a. Causal links.
   b. Level variables.
   c. Rate variables.
   d. Auxiliary variables.
   e. Exogenous variables.
5. Establish the causal relationships for the following:
   a. Job availability attracts migrants to the city.
   b. More jobs affect job availability.
   c. Migration affects job availability.
   d. Rate of change of current to inductance in series R-L circuit.
   e. Weapons of nation "A" increases the threat perceived by nation "B."
   f. Addiction rate depends on fraction addiction rate.
6. Draw the level variable for the rate variable, which is shown in Figure 7.85.

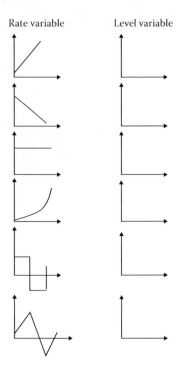

FIGURE 7.85
Rate variable and corresponding level variable.

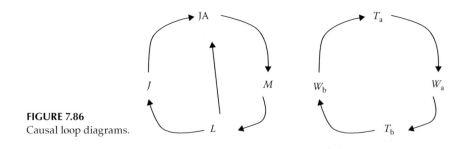

FIGURE 7.86
Causal loop diagrams.

7. A causal loop diagram is shown below; supply proper signs for each causal link and mention the type of causal loops shown in Figure 7.86.

   JA Job availability
   J Job
   L Labor
   M Migration
   T_a Threat perceived by nation A
   T_b Threat perceived by nation B
   W_a Weapons of nation A
   W_b Weapons of nation B

7.17 Bibliographical Notes

The technique of SD was developed for the first time by J.W. Forrester in 1940s. The technique had been used to understand many diverse problems. The excellent description of SD is given in the book on "Study Notes in System Dynamics" by Goodman (1983). The applications of system dynamics technique in industrial problems have only recently reemerged after a lengthy slack period. Current research on SD technique in industrial problems focuses on inventory decision and policy development, time compression, demand amplification, supply chain design and integration, and international supply chain management (Bernhard and Marios, 2000).

8
Simulation

8.1 Introduction

The word simulation is derived from the Latin word *simulare*, which means pretend. Simulation is thus an inexpensive and safe way to experiment with the system model. However, the simulation results depend entirely on the quality of the system model. It is a powerful technique for solving a wide variety of problems.

There are various types of simulations depending upon the type of model. Physical simulation deals with the experimentation of physical prototype of real system. In this case, the real input (scaled) is applied to the model and its output is compared with the real system. Sometimes, electrical analogous systems are also used in this type of simulation.

Another type of simulation is numerical simulation of mathematical models. This is very common in systems approach. After the development of digital computers, this type of simulation has gained popularity because it is easy to develop and modify. Numerical simulation is also sequential simulation because the calculations proceed in a time sequence. Computers are used to imitate or simulate the operations of various kinds of real-world facilities or processes. The facility or process of interest is usually called a system, and in order to study it scientifically, we often have to make a set of assumptions about how it works. These assumptions usually take the form of mathematical or logical relationships and constitute a model that is used to gain some understanding of how the corresponding system behaves.

If the relationships that compose the model are simple enough, it may be possible to use mathematical methods (such as algebra, calculus, or probability theory) to obtain exact information on questions of interest. This is called an analytical solution. Most real-world systems are too complex to allow realistic models to be evaluated analytically and these models must be studied numerically. However, a careful simulation study could shed some light on the question by simulating the operation of the complex system, as it currently exists, and the effect of various modifications in the existing system.

Most of the systems of engineers' concern are represented by different equations; the solutions of these equations describe the time response of the system to the given initial conditions and the input history. Electrical, mechanical, hydraulic, thermal, and other systems that contain energy storage elements can be represented by differential equations. Often the assumption is made that the system is lumped rather than distributed parameter, so that it can be described by ordinary differential equations (ODEs) rather than partial differential equations (PDE). Digital simulation refers to the simulation on a digital computer of the equations that describes a particular system. The solution is obtained either by numerical integration of the system ODEs or from the solution of difference equations.

Simulation can also be viewed as the manipulation of a model in such a way that it operates on time or space to compress it, thus enabling one to perceive the interactions that would not otherwise be apparent because of their separation in time or space.

Simulation generally refers to a computerization of a developed model, which is run over time to study the implication of the defined interactions of the parts of the system. It is generally iterative in their development. One develops a model, simulates it, learns from the simulation, revises the model, and continues the iteration unit until an adequate level of understanding/accuracy is obtained. Simulation points out modeler's own ignorance. Simulation is an operational research tool which can be used to get top performance from the given system. Simulations are good at helping people to understand complex problems where mistakes are expensive. Simulation helps to improve an existing process/system or plan a new facility and fills the gap between exact analysis and physical intuition.

8.2 Advantages of Simulation

Simulation could play a more vital role in biology, sociology, economics, medicine, psychology, engineering, etc., where experimentation is very expensive, time consuming, and risky.

1. Get deeper understanding about system
   Important design issues that may have been overlooked frequently come to light. Quick and useful insight about the systems can be gained easily.

2. Improve system efficiency
   Simulation gives a good look at how the resources are performing. It makes the whole system more productive by spotting problems like excess capacity, bottlenecks, and space shortages.

3. Try out many alternatives
   With simulation model, one can avoid the risk and cost of trial and error experimentation on the real system. It tells us how key design changes can affect the entire system.

4. Save time and money
   The cost of carrying out a simulation is often small compared to the potential reward of making a correct management decision or the cost of making a poor decision. Simulations allow rapid prototyping of an idea, strategy, or decision.

5. Create virtual environments
   With the simulation models one can generate a virtual environment, which is very useful for many applications such as training of army personnel and training to fire fighters and players.

6. Improves the quality of analysis
   System analysis can be effectively done if the system model is developed and simulated for different operating conditions and/or with different system parameters.

8.3 When to Use Simulations

The performance of a complex system (such as weather, traffic jam, monthly production in an industry) is difficult to predict, either because the system itself is too complex or the theory is not sufficiently developed. The difficulties in handling such problems using conventional mathematical modeling also arise due to the effect of uncertainties or due to dynamic interaction between system components or due to complex interdependencies among variables in the system or due to some combinations of these.

- Changing one part of your strategy results in cascading changes throughout the organization. For example, Ford company realizes that when they reduced delivery times, it was possible for dealers to offer better service with smaller inventories. The consequence was that dealer orders to Ford became more reliable and Ford could reduce delivery times further still. Once the system began to improve, these cascading effects allowed it to improve even more.
- You don't see the results of the decision you have made for several years, so you would not know that you have gone down the wrong path until it is too late. Simulation can help you in understanding when that short- and long-term trade-offs are worthwhile and when they are costly. They help you in gaining the confidence to make tough choices.
- Many of the opportunities you are evaluating have not been tried before (new idea). When there is limited historical information, people sometimes like to rely on hunches alone, but a good simulation is just a simple version of the real world. It provides you with experiences but the future is more relevant than past experience.
- Your mere involvement in the problem changes the environment. Many of the consequences of your decision might only partially be under your control. This makes decision messy, but simulation is good at handling these messes.
- Making mistakes would be painful. Making a mistake in a simulation is inexpensive and has no effect on people's lives. In a simulation, it is a good idea to experiment with making intentionally bad decisions to try to crash the organization just to learn where the pitfalls exist and help in creating brilliant strategies.
- The decision you are making is complicated and difficult to think.

8.4 Simulation Provides

1. Efficiency

   Cost quantification in terms of savings or avoidance
   - Higher mission availability
   - Increased operational system availability
   - Transportation avoidance
   - Reduced/eliminated expendable costs
   - Less procurement and operational costs

2. Effectiveness
   Positive contribution that are seldom quantified as
   - Improved proficiency/performance
   - Provides activities otherwise impossible short of combat
   - Provides neutral and opposition forces
   - Greater observation/assessment/analysis capability
3. Risk reduction
   - Safety
   - Environment
   - Equipment

What is the risk?
Risk is the possibility of loss, damage, or any other undesirable event.

Where is the risk?
Whenever we make any changes in the system that may be good or bad poses some risk.

How significant is the risk?
Once the risk is identified, a model helps you in quantifying it (putting a price on it). It also helps you in deciding whether risk is worth taking or not.

8.5 How Simulations Improve Analysis and Decision Making?

Simulation can improve analysis and decision making by

- Improving your ability to understand the consequences of choices and the effect on your organization or the world.
- Identifying both strengths and weaknesses of specific options you are considering.
- Challenging your organization's assumptions about its internal or external environment.
- Providing a nonthreatening forum for discussion about what might happen if certain policies are pursued.
- Helping to identify comprehensive strategies and policies that meet your objectives in a variety of future scenarios.
- Developing and communicating analyses that include multiple perspectives, showing consistencies and inconsistencies among a variety of viewpoints.
- Improving the speed and efficiency of understanding and solving complex strategy and policy problems.

8.6 Application of Simulation

Application areas for simulation are numerous and diverse. A list of some particular kinds of problems for which simulation has been found to be a useful and powerful tool is as follows:

- Designing and analyzing of systems
- Evaluating hardware and software requirements for a computer system
- Evaluating a new military weapons system or tactic
- Determining ordering policies for an inventory system
- Designing communications systems and message protocols for them
- Designing and operating transportation facilities such as freeways, airports, subways, or ports
- Evaluating designs for service organizations such as hospitals, post offices, or fast-food restaurants
- Analyzing financial or economic systems

8.7 Numerical Methods for Simulation

The numerical methods can be classified into the following two classes:

1. One-step or single-step method
2. Multistep method

One-step methods: In one-step method, to find out the value of a dependent variable Y_{i+1} at a future point X_{i+1}, we require the information at a single point X_i, for example, Euler's method, Runge–Kutta method, etc.

Multistep function: In case of multistep methods, the information at more than one point is needed to find out the value of Y_{i+1} at X_{i+1}.

The numerical methods yield solution either

i. As power series in t from which the values of y can be found by direct substitution (Picard's and Taylor's methods)
ii. As a set of values of t and y (Euler, Runge–Kutta, Milne, Adams–Bashforth, etc.)

Note:

a. In (ii), the values of y are calculated in short steps for equal intervals of t and are, therefore, termed as step-by-step methods.
b. Euler and Runge–Kutta methods are used for computing y over a limited range of t values whereas Milne, Adams–Bashforth methods may be applied for finding y over a wide range of t values. These later methods require starting values, which are found by Picard and Taylor's methods or Runge–Kutta methods.

Relatively simple functions cannot be integrated in terms of elementary functions and it is futile to make this unattainable goal the aim of the integral calculus. On the other hand, the definite integral of a continuous function does exist and this fact raises the problem of finding methods for calculating it numerically. Here, we shall discuss the simplest and most obvious of these methods with the aid of geometrical intuition and then consider error estimation.

Our objective is to calculate the integral

$$I = \int_a^b f(x)dx \qquad (8.1)$$

where $a < b$. We imagine the interval of integration to have been subdivided into n equal parts of length $h = (b-a)/n$ and denote the points of subdivision by $x_1 = a$, $x_2 = a + h,\ldots$, $x_n = b$, the values of the function at the points of subdivision by $f_0, f_1, \ldots, f_n$ and, similarly, the values of the function at the midpoints of the intervals by $f_{1/2}, f_{3/2}, \ldots, f_{(2n-1)/2}$. We interpret our integral as an area and cut up the region under the curve into strips of width h in the usual manner. We must now obtain an approximation for each such strip of surface, that is, for the integral

$$I_y = \int_{x_y}^{x_y+h} f(x)dx \qquad (8.2)$$

8.7.1 The Rectangle Rule

The crudest and most obvious method of approximating the integral I is directly linked to the definition of the integral; we replace the area of the strip I_v by the rectangle of area $f_v h$ and obtain for the integral the approximate expression

$$I \approx h(f_0 + f_1 + \cdots + f_{n-1}) \qquad (8.3)$$

From here on, the symbol $\approx$ means is approximately equal to

$$x_y, x_{y+1} = x_y + h \quad \text{and} \quad x_{y+2} = x_y + 2h \qquad (8.4)$$

8.7.2 The Trapezoid and Tangent Formulae

We obtain a closer approximation with no greater trouble if we do not replace the area of the strip I_v, as above, by a rectangular area, but by the trapezoid of area $(f_v + f_{v+1})h/2$ (Figure 8.1). For the whole integral, this process yields the approximate expression

$$I \approx h(f_1 + f_2 + \cdots + f_{n-1}) + \frac{h}{2}(f_0 + f_n) \qquad (8.5)$$

(Trapezoid formula), because, when the areas of the trapezoids are added, each value of the function except the first and the last occurs twice.

As a rule, the approximation becomes even better if, instead of choosing the trapezoid under the chord AB as an approximation to the area of I, we select the trapezoid under the tangent to the curve at the point with the abscissa $x = x_v + h/2$. The area of this trapezoid is simply $hf_{v+1/2}$, so that the approximation for the entire integral is

$$I \approx h(f_{1/2} + f_{3/2} + \cdots + f_{(2n-1)/2}) \qquad (8.6)$$

which is called the tangent formula.

Simulation

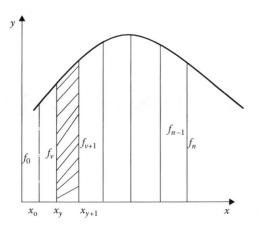

FIGURE 8.1
Trapezoid formula.

8.7.3 Simpson's Rule

By means of Simpson's rule, we arrive with very little trouble at a numerical result, which is generally much more exact. This rule depends on estimating the area $I_v + I_{v+1}$ of the double strip between the abscissa $x = x_v$ and $x = x_v + 2h = x_{v+2}$ by considering the upper boundary to be no longer a straight line but a parabola—in order to be specific, that parabola which passes through the three points of the curve with the abscissa x_v, $x_{v+1} = x_v + h$, and $x_{v+2} = x_v + 2h$ (Figure 8.2). The equation of this parabola is

$$y = f_v + (x - x_v)\frac{f_{v+1} - f_v}{h} + \frac{(x - x_v)(x - x_v - h)}{2} \cdot \frac{f_{v+2} - 2f_{v+1} + f_v}{h^2} \tag{8.7}$$

When we substitute the three values of x in question, this equation gives the proper values of y, that is, f_v, f_{v+1} and f_{v+2}, respectively. If we integrate this second-degree polynomial between the x_v and $x_v + 2h$, we obtain, after a brief calculation, for the area under the parabola

$$\int_{x_v}^{x_v+2h} y \, dx = 2hf_v + 2h(f_{v+1} - f_v) + \frac{1}{2}\left(\frac{8}{3}h - 2h\right)(f_{v+2} - 2f_{v+1} + f_v) \tag{8.8}$$

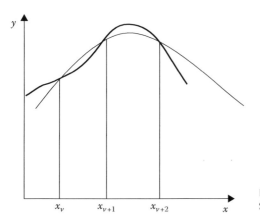

FIGURE 8.2
Simpson's formula.

This represents the required approximation to the area of our strip $I_v + I_{v+1}$.

If we now assume that $n = 2m$, that is, that n is an even number, we obtain, by addition of the areas of such strips, Simpson's rule

$$I \approx \frac{4h}{3}(f_1 + f_3 + \cdots + f_{2m-1}) + \frac{2h}{3}(f_2 + f_4 + \cdots + f_{2m-2}) + \frac{h}{3}(f_0 + f_{2m}) \tag{8.9}$$

Example 8.1

Solve $\log_e 2 = \int_1^2 \frac{dx}{x}$ using

a. Trapezoidal rule
b. Tangent rule and
c. Simpson's rule

If we divide this integral from 1 to 2 into 10 equal parts, h will be equal to 0.1 and results obtain by the Trapezoidal rule as given in Table 8.1.

MATLAB® codes

```
% Codes for Trapezoidal rule
clear all
h=0.1;    % step size
y=0;
x=1;      % Initial value of x
n=9;      % number of iterations
sum_f=0;
for i=1:n
    x=x+h;
    f=1/x;
    sum_f=sum_f+f;
    x1(:,i)=[x f];
end
x1'
sum_f=0.1*(sum_f+(1/2)*(1/1+1/2))
```

The value, as was to be expected, is too large, since the curve has its convex side turned towards the x-axis.

By the tangent rule as shown in Table 8.2, we find owing to the convexity of the curve, this value is too small.

For the same set of subdivisions, we obtain the most accurate result by means of Simpson's rule given in Table 8.3.

As a matter of fact, $\log_e 2 = 0.693147...$

Estimation of the error: It is easy to obtain an estimate of the error for each of our methods of integration, if the derivatives of the function $f(x)$ are known throughout the interval of the integration. We take $M_1, M_2, \ldots$ as the upper bounds of the absolute values of the first, second, ... derivative, respectively, that is, we assume that throughout the interval $|f^{(v)}(x)| < M_v$. Then the estimates are as follows:

TABLE 8.1
Results of Trapezoidal Rule

$x_1 = 1.1$	$f_1 = 0.90909$
$x_2 = 1.2$	$f_2 = 0.83333$
$x_3 = 1.3$	$f_3 = 0.76923$
$x_4 = 1.4$	$f_4 = 0.71429$
$x_5 = 1.5$	$f_5 = 0.66667$
$x_6 = 1.6$	$f_6 = 0.62500$
$x_7 = 1.7$	$f_7 = 0.58824$
$x_8 = 1.8$	$f_8 = 0.55556$
$x_9 = 1.9$	$f_9 = 0.52632$
	Sum 6.18773
$x_0 = 1.0$	$\frac{1}{2}f_0 = 0.5$
$x_{10} = 2.0$	$\frac{1}{2}f_{10} = 0.25$
	$6.93773 \times \frac{1}{10}$
	$\log_e 2 \approx 0.69377$

TABLE 8.2
Results of Tangent Rule

$x_0 + 1/2h = 1.05$	$f_{1/2} = 0.95238$
$x_1 + 1/2h = 1.15$	$f_{3/2} = 0.86957$
$x_2 + 1/2h = 1.25$	$f_{5/2} = 0.80000$
$x_3 + 1/2h = 1.35$	$f_{7/2} = 0.74074$
$x_4 + 1/2h = 1.45$	$f_{9/2} = 0.68966$
$x_5 + 1/2h = 1.55$	$f_{11/2} = 0.64516$
$x_6 + 1/2h = 1.65$	$f_{13/2} = 0.60606$
$x_7 + 1/2h = 1.75$	$f_{15/2} = 0.57143$
$x_8 + 1/2h = 1.85$	$f_{17/2} = 0.54054$
$x_9 + 1/2h = 1.95$	$f_{19/2} = 0.51282$
	$6.92836 \times \frac{1}{10}$
	$\log_e 2 \approx 0.69284$

TABLE 8.3

Results of Simpson's Rule

$x_1 = 1.1$	$f_1 = 0.90909$	$x_2 = 1.2$	$f_2 = 0.83333$
$x_3 = 1.3$	$f_3 = 0.76923$	$x_4 = 1.4$	$f_4 = 0.71429$
$x_5 = 1.5$	$f_5 = 0.66667$	$x_6 = 1.6$	$f_6 = 0.62500$
$x_7 = 1.7$	$f_7 = 0.58824$	$x_8 = 1.8$	$f_8 = 0.55556$
$x_9 = 1.9$	$f_9 = 0.52632$		
	Sum 3.45955×4		Sum 2.72818×2
			5.45636
	13.83820		13.83820
		$x_0 = 1.0$	$f_0 = 1.0$
		$x_{10} = 2.0$	$f_{10} = 0.5$
		$20.79456 \times 1/30$	
		$\log_e 2 = 0.693147$	

For the rectangle rule:

$$\left|I_v - hf_v\right| < \frac{1}{2}M_1 h^2 \quad \text{or} \quad \left|I_v - h\sum_{v=0}^{n-1} f_v\right| < \frac{1}{2}M_1 n h^2 = \frac{1}{2}M_1(b-a)h \tag{8.10}$$

For the tangent rule:

$$\left|I_v - hf_{v+\frac{1}{2}}\right| < \frac{1}{24}M_2 h^3 \quad \text{or} \quad \left|I - h\sum_{v=0}^{n-1} f_{v+\frac{1}{2}}\right| < \frac{1}{24}M_2(b-a)h^2 \tag{8.11}$$

For the trapezoid rule:

$$\left|I_v - \frac{h}{2}(f_v + f_{v+1})\right| < \frac{M_2}{12}h^3 \tag{8.12}$$

For Simpson's rule:

$$\left|I_v + I_{v+1} - \frac{h}{3}(f_v + 4f_{v+1} + f_{v+2})\right| < \frac{M_4}{90}h^5 \tag{8.13}$$

The last two estimates also lead to estimates for the entire integral I. We see that Simpson's rule has an error of much higher order in the small quantity h than the other rules, so that when M_4 is not too large, it is very advantageous for practical calculations. In order to avoid tiring the reader with the details of the proofs of these estimates, which are fundamentally quite simple, we shall restrict ourselves to the proof for the tangent formula. For this purpose, we expand the function $f(x)$ in the $(v+1)$th strip by Taylor's theorem:

$$f(x) = f_{v+\frac{1}{2}} + \left(x - x_v - \frac{h}{2}\right)f'\left(x_v + \frac{h}{2}\right) + \frac{1}{2}\left(x - x_v - \frac{h}{2}\right)^2 \tag{8.14}$$

where ξ is a certain intermediate value in the strip. If we integrate the right-hand side over the interval $x_v \leq x \leq x_v + h$, the integral of the middle term is zero. Since, as is easily verified,

$$\frac{1}{2}\int_{x_v}^{x_v+h} \left(x - x_v - \frac{h}{2}\right)^2 dx = \frac{h^3}{24} \tag{8.15}$$

it follows immediately that

$$\left| \int_{x_v}^{x_v+h} f(x)dx - hf_{v+1} \right| < M_2 \frac{h^3}{24} \tag{8.16}$$

which proves the above assertion.

8.7.4 One-Step Euler's Method

The linear or nonlinear dynamic system may be represented by first-order differential equations (state space model).

$$\frac{dx(t)}{dt} = f(x,t) \tag{8.17}$$

where
$x(t)$ is state vector of size $n \times 1$
$x(t_0) = \alpha$ is initial condition

We are interested in estimating $x(t_i)$ for $t_0 < t_i < T$, where t_i are equally spaced points between the interval t_0 and T. $t_i = t_0 + kh$, where h is the step size and $i = 1$ to m.

$$h = \frac{T - t_0}{m} \quad m\text{—number of points in simulation}$$

$$k_1 = hf(x_n, y_n)$$

$$x_{n+1} = x_n + k_1 + O(h^2)$$

However, the Euler method has limited value in practical usage, because solution diverges if the norm of the difference between the computed solution x_{n+1} and the exact solution $x(t)$ grows without bound with increasing time. Also, the numerical solution diverges when the step size is large.

8.7.5 Runge–Kutta Methods of Integration

Euler's method for solving ODE can be regarded as the first member of family of higher order methods called the Runge–Kutta methods. Each of the methods is one-step method in the sense that only values of $f(t, x)$ in the intergration interval (t_0, t_1) is used. Each Runge–Kutta method is derived from appropriate Taylor method in such a way that the final global error is of the order $O(h^N)$. A trade-off is made to perform several function evaluations at each step and eliminate the necessity to compute the higher order derivatives. These methods can be constructed for any order N.

Simulation

The Runge–Kutta method of order N = 4 is the most popular and called fourth-order Runge–Kutta method. It is a good choice for common purposes because it is quite accurate, stable, and easy to program. Most authorities proclaim that it is not necessary to go to a higher order method because the increased accuracy is offset by additional computation efforts. If more accuracy is required, then either a smaller size or an adaptive method should be used.

8.7.5.1 Physical Interpretation

Consider the solution curve $x = x(t)$ over the interval $[t_0, t_1]$ and the functional values $k_1, k_2, k_3,$ and k_4 which are approximations for the slopes to this curve. k_1 is the slope at the left, k_2 and k_3 are two estimates for the slope in the middle, k_4 is the slope at the right. The next point (t_1, x_1) is obtained by integrating the slope function.

$$x(t_1) - x(t_0) == \int f(t, x(t))dt$$

The Euler's method of integration may be generalized using the average of the slope of $x(t)$ at $t = t_i$ and the slope at $t = t_{i+1}$. The slope at $t = t_{i+1}$ is not known because $x(t_{i+1})$ is not known. However, the slope at t_{i+1} can be approximated by using $f(t_{i+1}, x_{i+1})$, where x_{i+1} is computed by $x_{i+1} = x_i + h^*dx$. This yields the following system of three equations called a second-order Runge–Kutta method.

$$k_1 = hf(t_n, x_n)$$

$$k_2 = hf(t_n + h, x_n + k_1)$$

$$x_{n+1} = x_n + \frac{k_1}{2} + \frac{k_2}{2} + O(h^3) \qquad (8.18)$$

Equation 8.18 has a local truncation error of order $O(h^3)$ and global truncation error of order $O(h^2)$. Hence, this is a second-order method. This is also known as midpoint method, which improves the Euler method by adding a midpoint in the step which increases the accuracy.

8.7.6 Runge–Kutta Fourth-Order Method

The fourth-order Runge–Kutta method is used to calculate the value x_{i+1} from past value of x_i with the help of the following four equations.

$$k_1 = hf(t_n, x_n)$$

$$k_2 = hf\left(t_n + \frac{h}{2}, x_n + \frac{k_1}{2}\right)$$

$$k_3 = hf\left(t_n + \frac{h}{2}, x_n + \frac{k_1}{2}\right)$$

$$k_4 = hf(t_n + h, x_n + k_3)$$

$$x_{n+1} = x_n + \frac{k_1}{6} + \frac{k_2}{3} + \frac{k_3}{3} + \frac{k_4}{6} + O(h^5) \qquad (8.19)$$

The coefficient of fourth-order Runge–Kutta method is chosen to ensure that its local truncation error of order $O(h^5)$, and its global truncation error is order $O(h^4)$.

The second- and fourth-order Runge–Kutta method is more accurate than the Euler's method; they still can become unstable, if the step size h is too large. For example, for the one-dimensional linear system, the fourth-order Runge–Kutta method gives a solution which diverges from exact solution for step sizes in the range $h > 2.785/c$ as opposed to $h > 2/c$ for Euler's method. The computational efficiency of fourth-order Runge–Kutta method can be significantly improved by allowing the step size to vary as the solution progresses. For example, for certain values of x, the function $f(t, x)$ may change very rapidly, in which case small steps are needed. However, in the other regions the function $f(t, x)$ might be quite smooth, even flat, in which case much larger steps are permitted. The key to implementing adaptive step size control is to develop an estimate of local truncation error. The step size can then be decreased or increased dynamically during execution in order to maintain a desired level for error. There are many methods presented by Schilling and Harris (2000) for making adaptive step size algorithms such as interval halving, Runge–Kutta–Fehlberg, and step size adjustment using local error criterion.

8.7.7 Adams–Bashforth Predictor Method

The Euler and Runge–Kutta methods are examples of single-step method because they use only the recent value x_n, to compute the solution. In multistep methods, we consider the many past solution points $\{x_n, x_{n-1}, x_{n-2}, \ldots X_{n-m}\}$ to predict x_{n+1}. The general formula for calculating the next point is as follows:

$$x_{n+1} = x_n + h \sum_{j=-1}^{m} b_j f(t_{n-j}, x_{n-j}) \qquad (8.20)$$

To generate a Adams–Bashforth predictor multistep method, we start approximating using Newton backward difference interpolation of degree 3. In third-order Adams–Bashforth predictor the following expression is used.

$$x_{n+1} = x_n + h\left(f_n + \frac{1}{2}f_{n-1} + \frac{5}{12}\Delta^2 f_{n-2} + \frac{3}{8}\Delta^3 f_{n-3}\right) \qquad (8.21)$$

where
$\Delta f_{n-1} = f_n - f_{n-1}$
$\Delta^2 f_{n-2} = f_{n-2} f_{n-1} + f_{n-3}$

Similarly fourth-order Adams–Bashforth formula may be written as

$$x_{n+1} = x_n + \frac{h}{24}(55f_n - 59f_{n-1} + 37f_{n-2} - 9f_{n-3}) \qquad (8.22)$$

The four-point estimate of $x(t_{n+1})$ has a local truncation error of order $O(h^5)$ and global truncation error of order $O(h^4)$. In this method, four previous solution points must be known to start. Hence, it is not a self-starting method. A single-step method, such as Euler's method or Runge–Kutta method is used to compute the first four points and then this method could be used.

8.7.8 Adams–Moulton Corrector Method

The accuracy of the results obtained from the Adams–Bashforth method can be improved by using a second application of multistep formula. The basic idea is to use predictor formula of Equation 8.22 first and then use implicit formula given in Equation 8.23 for refinement or correction to the solution obtained in earlier step.

$$x_{n+1} = x_n + h\left(f_{n+1} - \frac{1}{2}\Delta f_n - \frac{1}{12}\Delta^2 f_{n-1} - \frac{1}{24}\Delta^3 f_{n-2}\right) \quad (8.23)$$

The equations pair (8.22) and (8.23) together forms the predictor–corrector pair. Equation 8.23 can be made more direct by expanding the forward differences and collecting terms. This yields the Adams–Moulton formula given in equation.

$$x_{n+1} = x_n + \frac{h}{24}(9f_{n+1} + 19f_n - 5f_{n-1} + f_{n-2}) \quad (8.24)$$

This method has a local truncation error of order $O(h^5)$ and a global truncation error of order $O(h^4)$.

8.8 The Characteristics of Numerical Methods

1. Number of starting points
2. Rate of convergence stability
3. Accuracy
4. Breadth of the application
5. Propagation efforts required
6. Ease of application
7. Presence of error

8.9 Comparison of Different Numerical Methods

An ODE is an equation which consists of one independent variable and one dependent variable. Such an equation plays an important role in engineering because many physical phenomena are one of the best formulated mathematically n term of their rate of change.

TABLE 8.4
Comparison of Different Numerical Methods for Solving Differential Equations

Methods	Starting Values	Iteration Required	Global Error	Programming Efforts
One-step Euler's method	1	No	$O(H)$	Easy
Improved Euler's method	1	Yes	$O(H^2)$	Moderate
Modified Euler's method	1	No	$O(H^2)$	Moderate
Runge–Kutta method	1	No	$O(H^2)$	Moderate
Fourth-order Runge–Kutta method	1	No	$O(H^4)$	Moderate
Milne's method	4	Yes	$O(H^5)$	Moderate to difficult
Fourth-order Adam's method	4	Yes	$O(H^5)$	Moderate to difficult

To discuss various numerical methods, the solution can be found in two forms as follows:

1. A series of y in terms of power of x from which the value of y can be obtained by direct substitution, for example, Picard's method and Taylor's method.
2. A set of tabulated values of x and y, for example, Euler's method, Runge–Kutta method, etc.

The performance of different numerical methods for solving ODEs has been compared and shown in Table 8.4.

8.10 Errors during Simulation with Numerical Methods

The errors are present in numerical methods because we take a number of approximations and by numerical methods we get approximate result so the relationship between result and approximate result is

$$\text{True value} = \text{Approximation} + \text{Error}$$

or

$$\text{Error} = \text{True value} - \text{Approximation}$$

There are mainly two types of errors in numerical methods:

1. Truncation error
2. Round off error

8.10.1 Truncation Error

The truncation errors are those which are produced due to using approximation an m place of an exact mathematical procedure. For example, we have a derivative term $\dfrac{dx}{dt}$.

$$\frac{dx}{dt} = \frac{\Delta x}{\Delta t} = \frac{x_i(t_n + h) - x(t_n)}{t_{n+1} - t_n} \tag{8.25}$$

Simulation

The truncation errors are composed of two parts: local truncation error and propagated truncation error. The sum of these two is called global truncation error.

1. Local truncation error: It results from an application of method in question over a single step.
2. Propagated truncation error: It results from the approximation produced during the previous step.

The truncation error arises in Euler's method because the curve $y(x)$ is generally not a straight line between the neighboring grid points x_n and x_{n+1}, as assumed above. The error associated with this approximation can easily be assessed by Taylor's expansion.
Initial condition $x = x_n$

$$x(x_n+h) = x(t_n)+hx'(t_n)+(h^2/2)x''(t_n)+\cdots \qquad (8.26)$$

In other words, every time we take a step using Euler's method we incur a truncation error of $O(h^2)$ where h is the step length. Suppose that we use Euler's method to integrate ODE over an x interval of order unity. This requires $O(h^{-1})$ steps. If each step incurs an error of $O(h^2)$ and the errors are simply cumulative (a fairly conservative assumption), then the net truncation error is $O(h)$. In other words, the error associated with integrating an ODE over a finite interval using Euler's method is directly proportional to the step length. Thus, if we want to keep the relative error in the integration below about 10^{-6}, then we would need to take about 1 million steps per unit interval in x. Incidentally, Euler's method is termed a first-order integration method because the truncation error associated with integrating over a finite interval scales like h^1. More generally, an integration method is conventionally called nth order if its truncation error per step is $O(h^{n+1})$.

Note that truncation error would incur even if computers perform floating-point arithmetic operations to infinite accuracy. Unfortunately, computers do not perform such operations to infinite accuracy. In fact, a computer is only capable of storing a floating-point number to a fixed number of decimal places.

8.10.2 Round Off Error

The round off error is an error which is present in every type of computer due to the fact that the computer can represent the quantities with a finite number of digits. For example, we know that the value of π is 3.141592653897285... and if we are using a computer that can retain only seven significant figures so this computer might store or use π as $\pi = 3.141592$ which omitted the term resulting in an error called round off error. This error equals to 0.00000065. Every floating-point operation incurs a round off error of $O(p)$ which arises from the finite accuracy to which floating-point numbers are stored by the computer. Suppose that we use Euler's method to integrate our ODE over a given interval of time. This entails $O(h^{-1})$ integration steps, and, therefore, $O(h^{-1})$ floating-point operations. If each floating-point operation incurs an error of $O(p)$ and the errors are simply cumulative, then the net round off error is $O(p(h))$.

The total error, E associated with integrating ODE over an x-interval of order unity is (approximately) the sum of the truncation and round off errors. Thus, for Euler's method

$$E = p/h + h$$

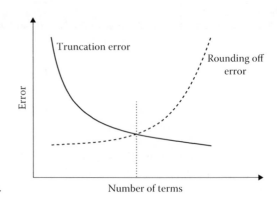

FIGURE 8.3
Error curves.

Clearly, at large step size (h), the error is dominated by the truncation error, whereas the round-off error dominates at small step size, as shown in Figure 8.3. The net error attains its minimum value when $h = h_o \sim p^{1/2}$. There is clearly no point in making the step size, h, any smaller than h_o, since this increases the number of floating-point operations but does not lead to an increase in the overall accuracy. It is also clear that the ultimate accuracy of Euler's method (or any other integration method) is determined by the accuracy, p, to which floating-point numbers are stored on the computer performing the calculation.

The value of p depends on how many bytes the computer hardware uses to store floating-point numbers. For IBM-PC clones, the appropriate value for *double precision* floating-point numbers is $p = 2.12 \times 10^{-15}$. It follows that the minimum practical step size for Euler's method on such a computer is $h_o = 10^{-8}$, yielding a minimum relative integration error of $e_o = 10^{-8}$. This level of accuracy is perfectly adequate for most scientific calculations. Note, however, that the corresponding p value for *single-precision* floating-point numbers is only $p = 1.19 \times 10^{-7}$, yielding a minimum practical step length and a minimum relative error for Euler's method of $h_o = 3 \times 10^{-4}$ and $e_o = 3 \times 10^{-4}$, respectively. This level of accuracy is generally not adequate for scientific calculations, which explains why such calculations are invariably performed using double rather than single-precision floating-point numbers.

The errors in the trapezoid rule and Simpson's rule

If $f''(x)$ is continuous in $[t_{n+1}, t_n]$, then the error in the trapezoid rule is no larger than

$$\frac{(t_{n+1} - t_n)^3}{12n^2} |f''(M)| \tag{8.27}$$

where $|f''(M)|$ is the largest value of $|f''(x)|$ in $[t_{n+1}, t_n]$.

If $f^{(4)}(x)$ is continuous in $[t_{n+1}, t_n]$, then the error in Simpson's rule is no larger than

$$\frac{(t_{n+1} - t_n)^5}{180n^4} |f^{(4)}(M)| \tag{8.28}$$

where $|f^{(4)}(M)|$ is the largest value of $f^{(4)}(x)$ in $[t_{n+1}, t_n]$.

Example 8.2

Solve the following integration using trapezoidal integration rule with $n = 4$.

$$\int_1^2 e^{-x^2} dx \qquad (8.29)$$

In order to estimate the error, we need to find the largest value of $|f''(x)|$ in the interval $[1, 2]$ for

$$f(x) = e^{-x^2}$$

Calculating, $f'(x) = -2x\, e^{-x^2}$ and $f''(x) = 2(2x^2 - 1)e^{-x^2}$.
Since, we want to find the extreme values of f'', we calculate its derivative

$$f^{(3)}(x) = 4x(3 - 2x^2)e^{-x^2}$$

Now $f^{(3)}(x) = 0$ only when $x = 0$ or $3 - 2x^2 = 0$, so $x = 0$ or $\pm(3/2)^{1/2} = \pm 1.225$. Checking values in the interval $[1, 2]$, we get the following (where we have rounded up the values of $f''(x)$ rather than simply rounding to two decimal places).

X	1	1.225	2
f''(x)	0.74	0.90	0.26

The largest value of $|f''(x)|$ is therefore 0.90. This tells us that the error is no larger than

$$\frac{(2-1)^3}{12 * 4^2} \approx 0.0047$$

Estimating the error in the Simpson rule

How large would n have to be to obtain a Simpson rule approximation of $\int_1^2 (x^3 + e^{-x}) dx$ accurate to five decimal places?

Solution

"Accurate to five decimal places" means an error of less than 0.000 005. In this problem, we do not know the value of n, but we know an upper bound for the error.

Our formula for the error in Simpson's rule says that

$$|\text{Error}| \leq \frac{(t_{n+1} - t_n)^5}{180 n^4} |f^{(4)}(M)| \qquad (8.30)$$

A quick calculation shows that the fourth derivative of f is

$$f^{(4)}(x) = e^{-x}$$

so that it is positive, and its largest value in the interval $[-1, 2]$ occurs when $x = -1$:

$$|f^{(4)}(M)| = e \approx 3$$

As before, we have overestimated rather than underestimated the quantity e. This gives

$$|\text{Error}| \leq \frac{(t_{n+1}-t_n)^5}{180n^4}\left|f^{(4)}(M)\right|$$

$$\leq \frac{3^5}{180n^4}\cdot 3 = \frac{81}{20n^4} \quad (8.31)$$

8.10.3 Step Size vs. Error

The methods we introduce for approximating the solution of the initial value problem (IVP) are called difference methods or discrete variable methods (DVM). The solution approximated at a set of discrete points is called a grid (or mesh) of points. An elementary single-step method has the form: $x_{n+1} = x_n + h\Phi(t_k, x_n)$ for some function called an increment function.

In case of DVM for IVP, there are two types of errors:

- Discretization error
- Round off error

8.10.4 Discretization Error

Let $\{(t_n, x_n)\}$ be a set of approximation and $x = x(t)$ the unique solution to IVP then

1. Global discretization error

$$e_k = x(t_n) - x_n \quad (8.32)$$

for $k = 0, 1, 2, 3, \ldots, M$
This the difference between unique solution and solution obtained by DVM.

2. Local discretization error

$$\epsilon_{k+1} = x(t_{n+1}) - x_n - h\Phi(t_k, x_n) \quad (8.33)$$

It is the error committed in the single step from t_k to t_{k+1}.

Final global error

After M steps within $[t_{n+1}, t_n]$, error accumulated $= \sum \epsilon_{k+1} = O(h)$

Note that the Taylor method of order N has the property that the final global error is of the order of $O(h^{N+1})$.

The numerical methods for solving ODEs are methods of integrating a system of first-order differential equations, since higher order ODEs can be reduced to a set of first-order ODEs. For example

$$p(t)\frac{d^2x}{dt^2} + q(t)\frac{dx}{dt} = r(t) \quad (8.34)$$

Let $x(t) = x_1(t)$ and $\dfrac{dx}{dt} = x_2(t)$, then the above equation may be written as

$$\Rightarrow \begin{cases} \dfrac{dx_1}{dt} = x_2(t) \\ \dfrac{dx_2}{dt} = [r(t) - q(t)x_2(t)]/p(t) \end{cases}$$

An nth order ordinary differential can be similarly reduced to

$$\dfrac{dx_n(t)}{dt} = f_n(t, x_1, x_2, \ldots, x_n)$$

where $k = 1, 2, \ldots, n$.

Example 8.3: Pendulum Problem

Consider a simple pendulum having length L, mass m, and instantaneous angular displacement θ (theta), as shown below in Figure 8.4.

The force equation for pendulum is

$$mg\sin\theta + mL\dfrac{d^2\theta}{dt^2} = 0$$

Thus $\dfrac{d^2\theta}{dt^2} = -\dfrac{g}{L}\sin\theta$.

For all angles θ, we make the assumption that $\sin\theta = \theta$ leading to an analytical solution. However, for large angles θ, we resort to a numerical method to solve the differential equation above. We first reduce the second-order differential equation to a set of two first-order differential equations by introducing ω (angular velocity), thus

$$\dfrac{d\theta}{dt} = \omega$$

$$\dfrac{d\theta}{dt} = -\dfrac{g}{L}\sin(\theta)$$

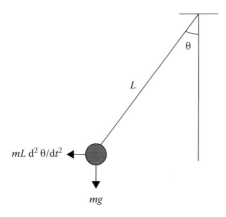

FIGURE 8.4
Pendulum swing.

MATLAB codes

```
function [x,dx] = pend(t,x)
% Pendulum differential equations
% G [m/s^2] acceleration due to gravity
% L pendulum length in meter
global G L
THETA = 1;              % index of angle in rad
OMEGA = 2;              % index of angular velocity in rad/sec
dx(THETA) = x(OMEGA);
dx(OMEGA) = -(G/L)*sin(x(THETA));
dx = [dx(THETA);dx(OMEGA)]; % Column vector
```

Notice the use of upper case characters for both the indices and the global variables, as is the convention in MATLAB. Notice also that returning both x and dx (instead of only dx as is normally done). Consider now a typical main function for this case

```
% Solve the large angle pendulum problem

THETA = 1;              % index of angle in rad
OMEGA = 2;              % index of angular velocity in rad/sec
global G L
G = 9.807;              % [m/s^2} acceleration due to gravity
L = input('Enter pendulum length in meter');
angle0 = input('Enter initial pendulum angle degrees');
period = 2*pi*sqrt(L/G);
fprintf('small angle period is %.3f seconds\n', period);
tfinal = 2*period;
n = input('Enter the number of solution points >');
dt = tfinal/(n - 1);
t = 0;
x = [angle0*pi/180;0];   % Initialize y as a Column vector
time(1) = t;
angle(1) = angle0;
for i = 2:n
[t, x, dx] = rk4('pend',2,t,dt,x);
            time(i) = t;
            angle(i) = x(THETA)*180/pi;
            fprintf('%10.3f%10.3fn',time(i),angle(i));
end
plot(time, angle)
```

The MATLAB functions mentioned above are very short and mostly self-documenting. Notice that after specifying the system parameters, we use the *for* loop to build the 'time' and 'angle' arrays in order to plot them. The most interesting aspect of this program is the 'rk4' function call to integrate one-step 'dt' of the differential equation set given in function 'pend' as follows:
[t, x, dx] = rk4('pend',2,t,dt,x);

Thus, the input arguments are the string constant representing the derivative function 'dpend,' the number of differential equations in the set (2), the independent variable and increment (t,dt), and the dependent variable vector (y). The output arguments are the solution variables and derivatives (t,x,dx) integrated over one time step dt. MATLAB provides five separate routines for solving ODEs of the form $dx = f(t,x)$ where t is the independent variable and x and dx are the solution and derivative vectors. One of the most widely used of all the single-step Runge–Kutta methods (which include the Euler and Modified Euler methods) is the so-called classical fourth-order method with Runge's coefficients. Being a fourth-order method (equivalent in accuracy to including the first five terms of the Taylor series expansion of the solution), it requires four evaluations of the derivatives over each increment dx, denoted by dx1, dx2, dx3, dx4, respectively. These are then weighted and summed in a very specific manner to obtain the final derivative dx.

Simulation

```
function [t, x, dx] = rk4(deriv,n,t,dt,x)
% Classical fourth order Runge - Kutta method
% integrates n first order differential equations
% dx(t,x) over interval x to t+dt
t0 = t;
x0 = x;
[x,dx1] = feval(deriv,t0,x);
for i = 1:n
        x(i) = x0(i) + 0.5*dt*dx1(i);
end
tm = t0 + 0.5*dt;
[x,dx2] = feval(deriv,tm,x);
for i = 1:n
        x(i) = x0(i) + 0.5*dt*dx2(i);
end
[x,dx3] = feval(deriv,tm,x);
for i = 1:n
        x(i) = x0(i) + dt*dx3(i);
end
t = t0 + dt;
[x,dx] = feval(deriv,t,x);
for i = 1:n
        dx(i) = (dx1(i) + 2*(dx2(i) + dx3(i)) + dx(i))/6;
        x(i) = x0(i) + dt*dx(i);
end
```

The most interesting aspect of this function is the use of the MATLAB function 'feval' to evaluate the function argument which is passed as a string. This is a fundamental aspect of MATLAB programming and understanding this is essential to developing MATLAB programs. The plotting capabilities of MATLAB are extremely simple to use and very versatile. A typical output plot for the pendulum case study is shown in Figure 8.5.

We see that 20° is about the limit of the small angle period (2.006 s), however at 80°, the period is seen to be much larger (2.293 s). This could only have been determined by numerically solving the differential equation set. The model may also be simulated using Simulink, as shown in Figure 8.6 and the results are shown in Figure 8.7.

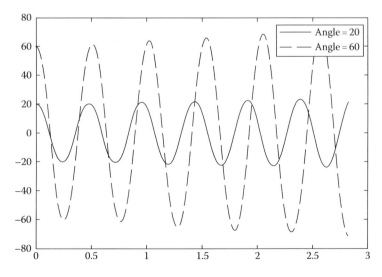

FIGURE 8.5
Simulation results.

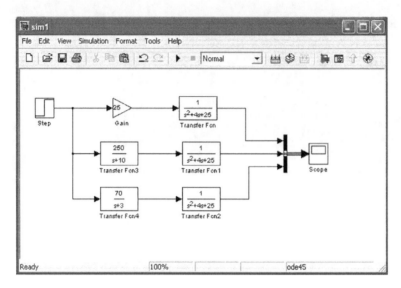

FIGURE 8.6
Simulink model.

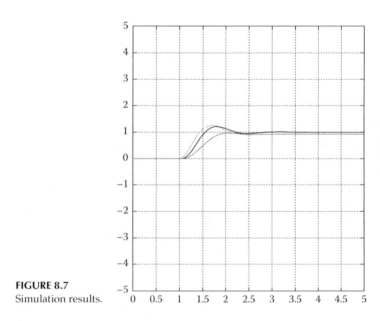

FIGURE 8.7
Simulation results.

Example 8.4: Continuous System Simulation (Water Reservoir System)

Consider the water reservoir system in which two inputs are river inflow and direct rainfall and one output is demand of water for irrigation and for other purposes, as shown in Figure 8.8. Some of the water is lost in the form of seepage and evaporation. It is known that the evaporation loss depends upon the surface area and the atmospheric temperature and the seepage loss depends upon the volume water stored in the reservoir. Develop a model to determine the volume of water stored in the reservoir so that it does not spill over and there is no shortage in summer.

Simulation

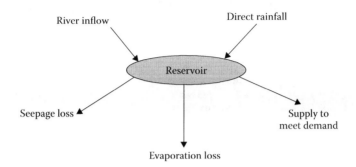

FIGURE 8.8
Water reservoir system.

Solution

As given in the problem, seepage loss is function of volume.

$$\text{Seepage loss} = f(\text{volume})$$

Evaporation loss is function of area of exposed surface and coefficient of evaporation.

$$\text{Evaporation loss} = f(\text{surface area}, C_{evap})$$

The state variable is the volume which changes with time.

Model development

Total input V_{in} = Rain + river flow
Gross volume = V_{in} + earlier volume
Total loss = Seepage loss + evaporation loss
Net volume = Gross volume − total loss − Demand
If Demand >= Net volume *then* reservoir is dry and shortage of water (=Dem-V_{net})
Otherwise Diff = Demand − V_{net}
If Diff > Cap *then* Spill over

MATLAB program

```
% Matlab program for simulation of Water reservoir
clear all;
Cevap=0.1; % Coefficient of evaporation
River_flow=[5000 4500 4000 3000 2500 2000 2000 3000 5000 5500 5000
5000]; % input from river
V=500;          % initial volume of water
Dem=[100, 400, 400, 200, 200, 100, 50 50 50 100 150 200];;    % Water
from rain
Rain=[0 0 0 0 0 50 300 500 0 0 0];         % Initial Demand
t=0;            % STart time
DT=1;           % Step size - 1-month
Tsim=120;       % Simulation time (100 Yrs.)
n=round(Tsim-t)/DT;
Cap=20000;i1=1;
for i=1:n
    x1(i,:)=[t,V, Dem(i1)];
    if i1==12;
        i1=1;
    end
    Demand=Dem(i1)*exp(0.003*t);
```

```
        Vin= Rain(i1) + River_flow(i1);
        Asurface=0.01*V;
        Evaporation=Asurface*Cevap;
        Seepage = 0.2*V;
        Tloss = Seepage + Evaporation;
        V = V + Vin -Tloss-Demand;
        if Demand >= V;
            % disp('shortage of water')
            Vshortage=Demand-V;
    else
        % disp('Excess of water')
        Diff1= V-Demand ;
        if Diff1 > Cap
        % disp('Spill over')
        Vspil=Diff1-Cap;end

    end
        i1=i1+1;
        t=t+DT;
    end

time1=x1(:,1);
figure(1)
plot(time1,x1(:,2),'k-')
xlabel('Time(months)')
ylabel('Water in Rservior')
figure(2)
plot(time1,x1(:,3),'k--')
xlabel('Time(months)')
ylabel('Demand');
axis([0 120 0 400]);
```

The above developed model is simulated in MATLAB for 10 years and the results are shown in Figures 8.9 and 8.10.

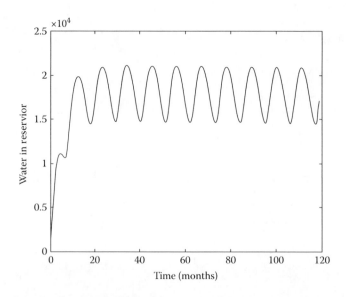

FIGURE 8.9
Net volume of water in reservoir.

Simulation

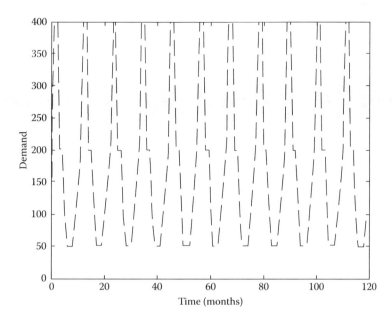

FIGURE 8.10
Change in demand of water.

Example 8.5: Chemical Reactor

In a certain chemical reaction, substances A and B produce a third chemical substance C.
A (1 g) + B (1 g) ↔ C (2 g)
The rate of formation of C is proportional to the amounts of A and B (Forward reaction). The decomposition of C is proportional to amount of C present in the mixer (backward reaction).
Develop a model for the above-mentioned reaction.

Solution

The differential equations for the system are

$$\frac{da}{dt} = k_2 c - k_1 ab$$

$$\frac{db}{dt} = k_2 c - k_1 ab$$

$$\frac{dc}{dt} = 2k_1 ab - 2k_2 c$$

and the initial conditions are
K_1, K_2, a, b are given
$c = 0$, $t = 0$, choose suitable value of dt
Use some integration technique to get the time profile of a, b, and c.

MATLAB program

```
% Matlab program for simulation of Water reservoir
clear all;
K2 = 0.1;
K1=0.01;
c = 0.02;
a= 0.01;
b = 0.02;

t=0;                    % Start time
DT=.01;                 % Step size - 1-month
Tsim=10;                % Simulation time (100 Yrs.)
n=round(Tsim-t)/DT;
Cap=20000;i1=1;
for i=1:n
    x1(i,:)=[t,a, b, c];
    da = K2 *c - K1* a* b;
    db = K2 *c - K1 *a* b;
    dc =2 *K1 *a *b - 2 * K2* c;
    a=a+DT*da;
    b=b+DT*db;
    c=c+DT*dc;
    t=t+DT;
end

time1=x1(:,1);
figure(1)
plot(time1,x1(:,2:4))
xlabel('Time(months)')
legend('Concentration of a', 'concentration of b', 'concentration of c')
```

The simulation results obtained are shown in Figure 8.11.

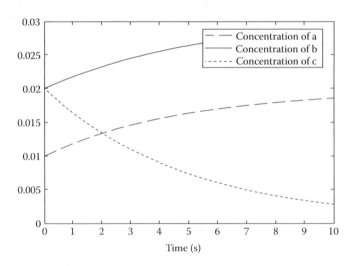

FIGURE 8.11
Simulation results of chemical reaction.

Simulation

Example 8.6: Rocket Dynamics

Consider the dynamics of a rocket constraint to travel in a vertical line over the launch site. Ignore or linearize the nonlinearity due to changing distance from the center of the earth and aerodynamic drag effects. The linearized time-varying equation of motion for the system is (Figure 8.12)

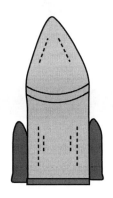

FIGURE 8.12
Dynamics of a rocket when launched from a site.

$$m(t)\frac{d^2x(t)}{dt^2} + k(t)\frac{dx(t)}{dt} = u - m(t)g$$

where
  $m(t)$—Mass of rocket (varying due to fuel consumption)
  $k(t)$—Aerodynamic drag
  g—Acceleration due to gravity
  u—Thrust

The mass of rocket varies as
  $m(t) = (120,000 - 2,000t - 20,000g(t-40))$ kg   for $t <= 40$ s.
  $m(t) = 20,000$ kg   for $t > 40$ s.
  $k = 1000$
  $u = 10^6$ for $t <= 40$ s.
  $u = 0$ for $t > 40$ s.
Write a MATLAB program to simulate it.

MATLAB program

```
% Simulation program of rocket dynamics
k=1000;
dt=0.01;
n=300;
time=0.0;
x=[0;0];
for i=1:n
        if time<=40
                g=10;
                u=10E6;
                m=(120000-2000*time-20000*g*(time-40));
        else
                g=0;
                u=0;
                m=20000;
        end
        a=[0 1;0 -k/m];
        b=[0;u/m-g];
        dx=a*x+b;
        x=x+dx*dt;
        x1(i,:)=[x'];
        time=time+dt;
        t1(i,:)=time;
end
plot(t1,x1)
legend('Displacement', 'Velocity')
```

Results (Figure 8.13)

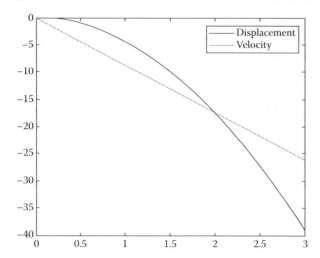

FIGURE 8.13
Simulation of rocket dynamics.

Example 8.7: Butterworth Filter

Simulate digitally the operation of a third-order Butterworth filter governed by

$$\frac{d^2x(t)}{dt^2} = -2\frac{dx}{dt} - x + u$$

using Adam–Bashforth method. Compare the results with Euler's method.

MATLAB program

```
% Matlab simulation of Butterworth filter using AB-3.
dt=0.1;
xe=0;
dx=0;
x=[0 0 0]';
f=x;
fs=f;
fss=f;
u=1;
for n=1:150
        f(1)=x(2);
        f(2)=x(3);
        f(3)=-x(1)-2*x(2)-2*x(3)+u;
        x=x+(dt/12)*(23*f-16*fs+5*fss);
        fss=fs;
        fs=f;
    ddx=-2*dx-xe +u;
    dx=dx+dt*ddx;
    xe=xe+dt*dx;
        y1(n,:)=[x(1), xe];
end
plot(0.1:0.1:15,y1)
xlabel('Time')
ylabel('Filter output')
legend('AB-3 with dt=0.1', 'Euler Method')
```

Results (Figure 8.14)

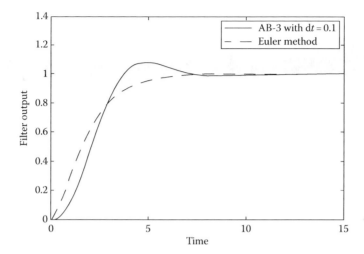

FIGURE 8.14
Simulation results of Butterworth filter.

Example 8.8: Modeling of Water Pollution Problem

Let us consider the water pollution problem and the associated dynamics. The river water can decompose significant amount of untreated industrial waste products without upsetting the living processes that occur within and around them. However, pollution level increases when more untreated waste water is mixed in the river. The decomposition processes may use considerable oxygen, which is dissolved in the water. If the amount of oxygen in water gets too low, the water is not usable and also unfit for fish, drinking, and recreational activities. Rivers replenish their oxygen level by drawing oxygen from atmosphere (Figure 8.15).

Develop a model to predict the oxygen content in a river when the waste water is dumped in upstream.

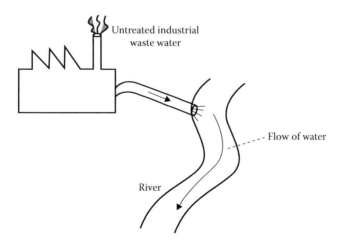

FIGURE 8.15
Untreated waste industrial water modeling for river pollution.

Solution

First, we develop a model for pollution level of river water. It is very clear from the above discussion that the pollution level in the river water is proportional to the rate at which the amount of waste water ($w(t)$) is dumped into the river by the industry and the chemical concentration. It also depends on the decomposition of wastage. The decomposition reduces the pollution level.

Hence, the rate of pollution level

$$\frac{d}{dt}P(t) = C_{chem}W(t) - D_{chem} * P(t)$$

where

C_{chem}—coefficient of chemical concentration of waste water
$W(t)$—amount of waste water dumped into river
D_{chem}—decomposition of waste water
$P(t)$—pollution level

Similarly, the oxygen content in the river water is proportional to the rate of oxygen used by wastes in decomposition and some oxygen is replenished in the water by air.

The rate of oxygen content in river water is given by

$$\frac{d}{dt}O(t) = C_o * (O_{Max} - O(t)) - C_d * Dchem * P(t)$$

where

$O(t)$—oxygen content at a particular time t
C_o—coefficient representing the time delay in taking oxygen from air
C_d—coefficient to represent oxygen requirement for particular wastes

8.11 Review Questions

1. a. Differentiate between analog and digital simulations.
   b. Discuss three major simulation characteristics for digital simulation.
   c. Explain the errors in this simulation.
2. Describe the essential features of PC-based simulation packages for analog and discrete simulations.
3. Use Euler's method to develop a difference equation to solve the differential equation: $dy/dt = 2x(t) = 1.0$ with $x(0) = 0$.
   a. Calculate the first 20 values for $T = 0.1$ s and also $T = 0.05$ s.
   b. Determine the exact solution and compare the values at corresponding points in time.
   c. Comment on the impact of integration time interval on the magnitude of the error.
   d. Write a MATLAB program for solving the above equation using Euler's method.

Simulation

4. Write a computer program for simulating a linear time invariant system whose state space model is given below:

$$\frac{d}{dt}\begin{bmatrix} x_1(t) \\ x_2(t) \\ x_3(t) \end{bmatrix} = \begin{bmatrix} 1 & 2 & 0 \\ -1 & -1 & 3 \\ 1 & 4 & 5 \end{bmatrix} \begin{bmatrix} x_1(t) \\ x_2(t) \\ x_3(t) \end{bmatrix} + \begin{bmatrix} 1 \\ 0 \\ 0 \end{bmatrix} u(t)$$

5. A double-pole LC circuit shown in Figure 8.16. Develop the system model and simulate it in Matlab for the frequency analysis.
6. Write a MATLAB program for determining the transfer function for the network shown in Figure 8.17.
7. Develop a four-segment lumped model for coaxial power cable shown in Figure 8.18. The parameter values per segment are as follows: $R = 0.0193\,\Omega$, $L = 0.163\,H$, $C = 68.4\,pf$, $V_s = 10\,V$ dc, $R_s = 50\,\Omega$, $R_l = 50\,\Omega$. Also write a MATLAB program for simulating it.

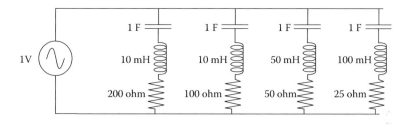

FIGURE 8.16
LC circuit.

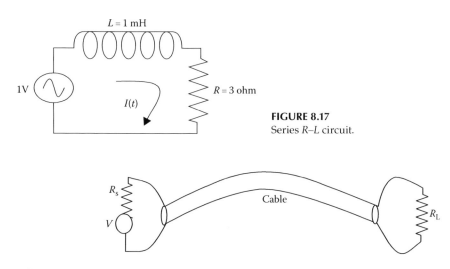

FIGURE 8.17
Series R–L circuit.

FIGURE 8.18
Coaxial electric cable.

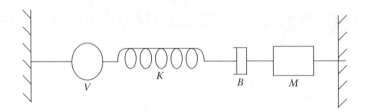

FIGURE 8.19
Schematic diagram of a nonlinear mechanical system.

8. Simulate the following nonlinear system (shown in Figure 8.19).

   $M = 10\,\text{kg}$, $K = 360\,\text{N/m}$, and damper B is a quadratic damper governed by the equation $Fb = BVb2$.

9. Explain local and global truncation errors.

10. What do you mean by adaptive step size and how is it advantageous in simulating ODE?

12. Compare truncation errors for different methods.

9

Nonlinear and Chaotic System

Too little liberty brings stagnation and too much brings chaos.

Bertrand Russell (1872–1970)

Chaos is the score upon which reality is written.

Henry Miller

Everybody's a mad scientist, and life is their lab. We're all trying to experiment to find a way to live, to solve problems, to fend off madness and chaos.

David Cronenberg

I have a great belief in the fact that whenever there is chaos, it creates wonderful thinking. I consider chaos a gift.

Septima Poinsette Clark

9.1 Introduction

Life is a nonlinear entity. Almost everything around us is associated with some fundamentally nonlinear system. Commonly the nonlinearities are undesirable in mathematical modeling of systems. The mathematics associated with nonlinear dynamic systems is not very accessible or useful. The mathematics for linear systems however, is relatively simple. Also, there are powerful techniques available in linear systems theories. Validity of linear system stability can be evaluated by Root locus techniques, Nyquist plots, Bode plots, and other techniques. Thus, to begin with, engineers initially consider most systems to be linear in their behavior. However, it is not always true as we are aware what nonlinear systems can do. From the point of view of control theory, if we do not know what the system is going to do we cannot control it. From a simulation point of view, if we do not know what the system is capable of doing, we will not know if our simulation is reasonable. Thus, it is important to be aware of the types of behavior demonstrable by nonlinear system so that the strange responses that sometimes occur are not totally unexpected.

We may categorize nonlinearities in dynamic system into continuous and discontinuous nonlinearity. Continuous nonlinearities usually arise out of the physics of the system. These are more mathematical in nature. A good example is the gravitational force acting on the pendulum, which is proportional to $\sin(x)$. Discontinuous nonlinearities are more often engineered into a system. Good examples of discontinuous nonlinearities are relays, saturation, hysteresis, and gear backlash, etc. None of these nonlinearities are particularly easy to deal with; however, they each have developed specific tools.

9.2 Linear versus Nonlinear System

The most fundamental property of a linear system is the validity of the principle of superposition. A nonlinear system does not hold the superposition principle. Nonlinearity is an undesirable phenomenon in a system. Nonlinearity is to be as less as possible in a system. Let us take any linear time invariant system in this world, we can find some nonlinearities in that system up to some extent. Unfortunately there is a difficulty in evaluating the stability of a nonlinear system.

Perfect linear systems do not exist in practice, as all practical systems are nonlinear to some extent. Linear feedback control systems are idealized models, which are made by analyst purely for the simplicity of analysis and design (Kuo, 1990).

When the input signal to the control system has magnitude that is limited to a range in which the system components possess linear characteristics, the system is essentially linear. But when the input signal magnitude extends outside the range of linear operation, the system is no longer considered linear. For example, amplifiers used in control system often exhibit saturation effect; when their input signal becomes large, the magnetic field of motor usually has saturation properties. Other common nonlinear effects found in system are backlash or dead play between coupled gear members, nonlinear characteristics in springs, nonlinear frictional force or torque between moving members.

Some times nonlinear characteristics are intentionally introduced in a control system to improve its performance or provide more effective control. For example, to achieve minimum time control, an on–off (bang-bang or relay) type of controller is used. This type of controller is used in many missiles and spacecrafts. Linear systems have a wealth of analytical and graphical techniques for design and analyses purposes. Nonlinear systems are very difficult to treat mathematically and there are no general methods that may be used to solve a wide class of nonlinear systems.

9.3 Types of Nonlinearities

Following are the types of nonlinearities (Nagrath and Gopal, 2001; Fielding and Flux, 2003) in a system:

1. Saturation
2. Friction
3. Backlash/hysteresis
4. Dead zone
5. Relay
6. Delay
7. Jump resonance

Friction: In all the moving parts of a system, friction exists between two rubbing surfaces.

Saturation: All system components operate within a permissible range. Beyond this range, it may refuse to operate at all. This situation may be called as saturation.

Dead zone: In this region, the output is zero for a particular range of input. Such a situation occurs due to initial conditions of inertia, friction, etc.

Hysteresis: This is a common type of phenomena that occurs when the system is subjected to an alternative input. This phenomenon causes an energy waste or hysteresis loss in a system.

Relay: Relay is a nonlinear amplifier.

Delay: The information delay and transformation delay in the system also produce non linearity in the system response.

Jump resonance: It is usually associated with nonlinearity in spring due to change in spring stiffness with frequency.

9.4 Nonlinearities in Flight Control of Aircraft

The flight control system (autopilot) is designed for automatic stabilization of the aircraft attitude about its center of gravity. As every system possesses some nonlinearities, the flight control system (autopilot) also has some nonlinearities, that is, backlash in servo system compliances, dead zone in pilot's control column, and saturation in servo amplifier.

This study investigates the effects of backlash in servo system compliances, dead zone in pilot's control column, and saturation in servo amplifier of flight control system (FCS). Starting from pilot's column, a realistic nonlinear Simulink® model of FCS is evolved including the applicable nonlinearities.

Different pilot's commands have been given to the Simulink model of nonlinear FCS in the form of step input, ramp input, constant input, and sinusoidal input considering different applicable nonlinearities separately and combined.

Proportional, proportional-integral, proportional-integral-derivative (PID) (based on root locus method), and fuzzy controllers (Mamdani type) are designed and devolved. Performance of all these controllers has been compared. Fuzzy controller is able to handle the system nonlinearities and give the desired output.

9.4.1 Basic Control Surfaces Used in Aircraft Maneuvers

In order to steer an aircraft, a system of flaps called control surfaces is used. Control surfaces deflect the airflow around an aircraft and turn or twist the aircraft so that it rotates about the center of gravity. This movement is made by a control stick and pedals.

Basic control surfaces used in aircraft maneuvers, as shown in Figure 9.1, are as follows:

1. Aileron—For roll
2. Elevator—For pitch
3. Rudder—For yaw

9.4.2 Principle of Flight Controls

In addition to the engines and control surfaces that can be used to change the present state of motion of the aircraft, every aircraft contains some motion sensors which provide

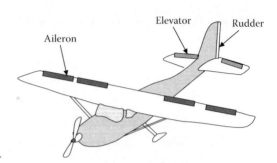

FIGURE 9.1
Basic control surfaces used in aircraft maneuvers.

measure of change which have occurred in the measured motion variables as the aircraft responses to the pilot's commands or it encounters a disturbance (McLean, 1999). In modern aircraft, hydraulic systems or electric motors called actuators move control surfaces by responding to control signals sent from a flight computer connected to the control stick. The actuator command signal travels over a wide area from the control computer to the actuator, as shown in Figure 9.2.

The signals from these sensors can be used for display in cockpit or can be used for feedback signal to FCS (Figure 9.3). These sensors are gyroscopes; and aircraft sensors such as orientation, altitude, velocity sensors, etc. Control effectors include throttle control, rudder control, elevator control, aileron control, etc. Pilots issue control commands for flight path control, velocity command, altitude command, etc.

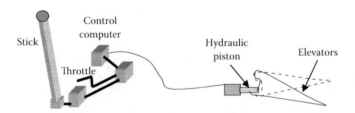

FIGURE 9.2
Movement of control surfaces.

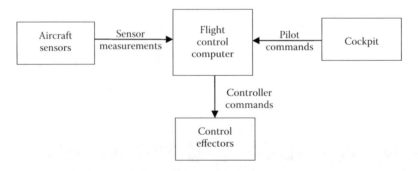

FIGURE 9.3
Arrangement of FCS.

9.4.3 Components Used in Pitch Control

Elevators are used to pitch the aircraft up and down causing it to climb or dive. To climb, the pilot pulls the control stick back causing the elevators to be deflected up. This in turn causes the airflow to force the tail down and the nose up thereby increasing the pitch angle, as shown in Figure 9.4. Similarly, to dive, the pilot pushes the control stick forward causing the elevator to deflect down. This in turn causes the airflow to lift the tail up and the nose down thereby decreasing the pitch angle.

Components, which are used in FCS (pitch control), are shown in Figures 9.5 and 9.6.

1. *Vertical gyro*: The vertical gyro generates electrical signals proportional to the aircraft's pitch attitude.
2. *Rate gyro*: The rate gyro generates electrical signals proportional to the aircraft's pitch rate.
3. *Altitude controller*: Altitude controller is designed to give altitude hold signal when altitude hold mode is selected as shown in Figure 9.5.
4. *Servo unit*: When error signal is given in case of stabilization mode and control signal is given from pilot in case of control mode, servo unit is designed to move the elevators.

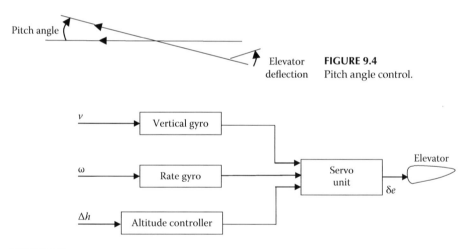

FIGURE 9.4
Pitch angle control.

FIGURE 9.5
Pitch channel operation in altitude hold mode and synchronization mode block diagram.

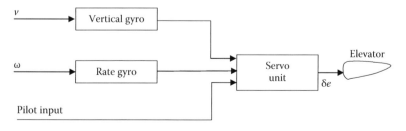

FIGURE 9.6
Pitch channel operation in control mode block diagram.

9.4.4 Modeling of Various Components of Pitch Control System

Though our mathematics—given certain axioms—can be shown to be true in an absolute sense, any attempt to model a physical system, say, a car suspension or an aeroplane in flight will result in something that is "lesser" than the original system. Even if our estimates of all relevant physical quantities are correct, it is likely that we will have taken a simplified view of the system dynamics and ignored the potential for interaction with the rest of the world.

Transfer functions simplify working with linear differential equations, which describe the transition behavior of linear dynamical systems. The servo unit diagram is shown in Figure 9.7.

Transfer function of pitch control system components

Transfer function of servo amplifier = $G_a(s) = K$

Transfer function of gears = $\dfrac{1}{15} = 0.66$

Transfer function of pitch gyro = K_{g1}

Transfer function of pitch rate gyro = K_{g2}

To get the transfer function of servomotor, we have to model its components.

$$\text{Rotor circuit } e_a = i_a R_a + L_a \frac{di_a}{dt} + V_b \tag{9.1}$$

$$\text{Rotor EMF } V_b = K_b \frac{d\theta_m}{dt} \tag{9.2}$$

$$\text{Mechanical torque } T_m = K_t i_a \tag{9.3}$$

$$\text{Equation governing mechanical port } J_m \frac{d^2\theta_m}{dt^2} = T_m(t) - D_m \frac{d\theta_m}{dt} \tag{9.4}$$

$$e_a = \frac{L_a}{K_t}\frac{dT_m}{dt} + \frac{R_a}{K_t}T_m + K_b \frac{d\theta_m}{dt} \tag{9.5}$$

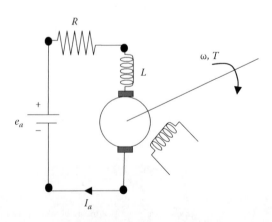

FIGURE 9.7
Servomotor.

From Equations 9.4 and 9.5

$$e_a = \frac{L_a}{K_t} J_m \frac{d^3\theta_m}{dt^3} + \left(\frac{L_a}{K_t} D_m + \frac{R_a}{K_t} J_m\right)\frac{d^2\theta_m}{dt^2} + \left(\frac{R_a}{K_t} D_m + K_b\right)\frac{d\theta_m}{dt} \quad (9.6)$$

Taking Laplace transform of both sides

$$E_a(s) = \frac{L_a}{K_t} J_m s^3 \theta_m(s) + \left(\frac{L_a}{K_t} D_m + \frac{R_a}{K_t} J_m\right)s^2\theta_m(s) + \left(\frac{R_a}{K_t} D_m + K_b\right)s\theta_m(s)$$

$$E_a(s) = \left[s\left(\frac{L_a}{K_t} J_m s^2 + \left(\frac{L_a}{K_t} D_m + \frac{R_a}{K_t} J_m\right)s + \left(\frac{R_a}{K_t} D_m + K_b\right)\right)\right]\theta_m$$

$$\frac{\theta_m}{E_a(s)} = \frac{1}{\left[s\left(\frac{L_a}{K_t} J_m s^2 + \left(\frac{L_a}{K_t} D_m + \frac{R_a}{K_t} J_m\right)s + \left(\frac{R_a}{K_t} D_m + K_b\right)\right)\right]} \quad (9.7)$$

Let us introduce

Motor time constant $T_m = \dfrac{J_m R_a}{K_t}$

Armature time constant $T_a = \dfrac{L_a}{R_a}$

Damping factor $\gamma = \dfrac{D_m R_a}{K_t}$

$$\text{Transfer function of servomotor} = \frac{\theta_m}{E_a(s)} = \frac{1}{s[T_a T_m s^2 + (T_m + \gamma T_a)s + \gamma + K_b]} \quad (9.8)$$

Designing the values

$$K_t = 0.5\,\text{N m/A}, \quad K_b = 0.5\,\text{V s/rad}$$

$$J_m = 0.03\,\text{kg-m}^2,$$

$$D_m = 0.02\,\text{N m s/rad}, \quad R_a = 8\,\Omega$$

$$L_a = 8\,\text{H},$$

$$T_m = \frac{J_m R_a}{K_t} = 0.68s, \quad \gamma = \frac{D_m R_a}{K_t} = 0.32, \quad T_a = 1$$

$$\text{Transfer function of servomotor} = \frac{1}{s(0.68\,s^2 + 1s + 1.82)}$$

9.4.5 Simulink Model of Pitch Control in Flight

Simulink model of pitch control in flight is shown in Figure 9.8 after obtaining the transfer functions of different components by connecting them as their applicable order (James, 1993).

9.4.5.1 Simulink Model of Pitch Control in Flight Using Nonlinearities

Simulink model of pitch control in flight is made after obtaining the transfer functions of different components by connecting them as their applicable order and considering some nonlinearities, that is, backlash in servo system compliances, dead zone in pilot's control column, and saturation in servo amplifier (Wills, 1999) of FCS (Figures 9.9 and 9.11).

9.4.6 Study of Effects of Different Nonlinearities on Behavior of the Pitch Control Model

Study of effects of different nonlinearities on the behavior of the pitch control model with different input signals is carried out and the results are discussed below:

9.4.6.1 Effects of Dead-Zone Nonlinearities

Dead-zone nonlinearity is considered in pilots control column. The system performance is observed at different dead-zone nonlinearities, that is, at (−0.05 to 0.05), (−0.25 to 0.25), and at (−0.50 to 0.50). The system response is shown in Figure 9.10 with different nonlinearities and with step input. The system performance (with above-said nonlinearities) is compared

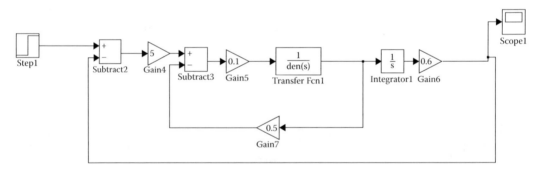

FIGURE 9.8
Simulink model of pitch control in flight.

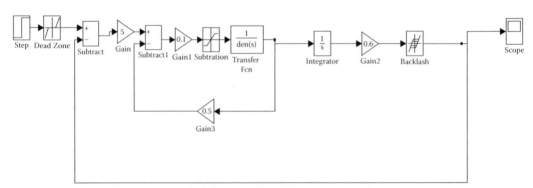

FIGURE 9.9
Simulink model of pitch control in flight using some nonlinearities.

Nonlinear and Chaotic System

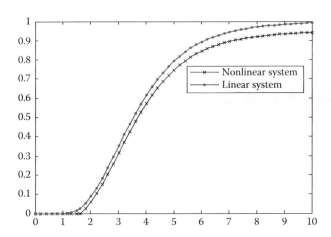

FIGURE 9.10
Comparison of performances of linear and nonlinear models.

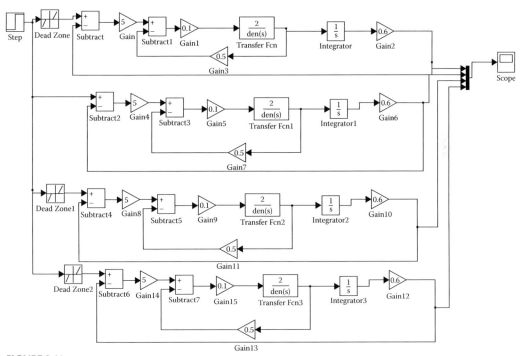

FIGURE 9.11
Simulink model of FCS (pitch control) with dead-zone nonlinearities.

with the system performance (without considering nonlinearities) at step input, as shown in Figure 9.10. As shown in Figures 9.12 through 9.15, when the dead zone increases the system responses get more distorted.

9.4.6.2 Effects of Saturation Nonlinearities

Saturation nonlinearity is considered in servo amplifier. The system performance is observed at different saturation nonlinearities, that is, at (−0.05 to 0.05), (−0.10 to 0.10), and

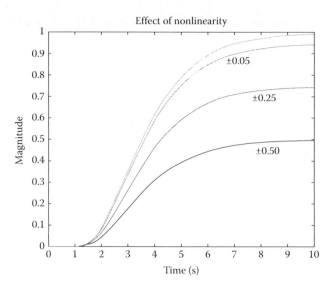

FIGURE 9.12
System performance at different dead-zone nonlinearities with step input.

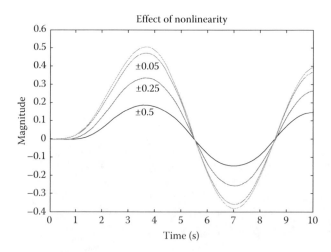

FIGURE 9.13
System performance at different dead-zone nonlinearities with sinusoidal input.

at (−0.25 to 0.25). The system is shown in Figure 9.16 with different nonlinearities and with step input. The system performance (with above-said nonlinearities) is compared with the system performance (without considering nonlinearities) at step input, as shown in Figure 9.17. As shown in Figures 9.17 through 9.21 when saturation nonlinearities decrease, the system responses gets more distorted.

9.4.6.3 Effects of Backlash Nonlinearities

The backlash nonlinearity is considered in pilots control column. The system performance is observed at different backlash nonlinearities, that is, at 1.0, 2.0, and at 3. The system is

Nonlinear and Chaotic System

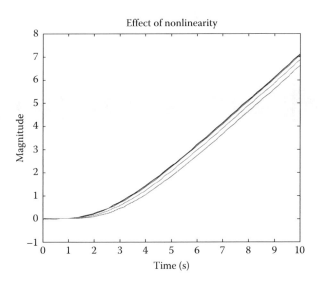

FIGURE 9.14
System performance at different dead-zone nonlinearities with ramp input.

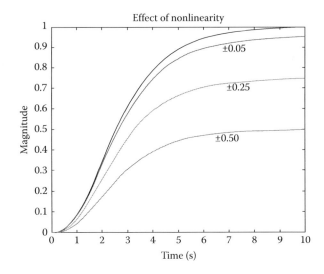

FIGURE 9.15
System performance at different dead-zone nonlinearities with constant input.

shown in Figure 9.22 with different nonlinearities and with step input. The system performance (with above-said nonlinearities) is compared with the system performance (without considering nonlinearities) at step input. As shown in Figures 9.23 through 9.27 when the dead zone increases, the system responses get more distorted.

9.4.6.4 Cumulative Effects of Backlash, Saturation, Dead-Zone Nonlinearities

Effects of backlash at 2, saturation is at (0.15 and −0.15), dead zone at (−0.25 and 0.25) nonlinearities collectively is analyzed in Figures 9.28 through 9.30 at different inputs.

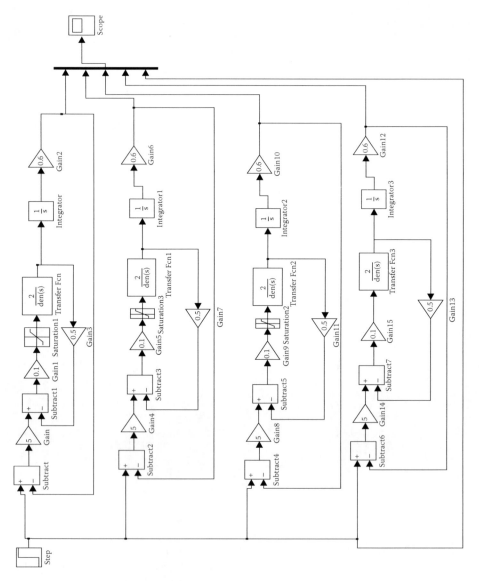

FIGURE 9.16
Simulink model of FCS at different saturation nonlinearities with step input.

Nonlinear and Chaotic System

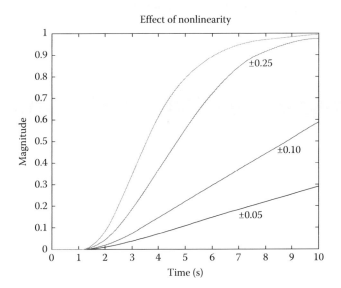

FIGURE 9.17
System performance at different saturation nonlinearities with step input.

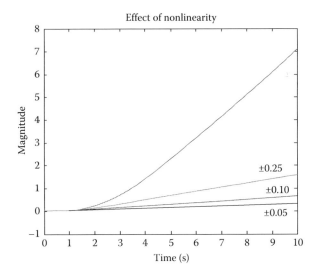

FIGURE 9.18
System performance at different saturation nonlinearities with ramp input.

9.4.7 Designing a PID Controller for Pitch Control in Flight

9.4.7.1 Designing a PID Controller for Pitch Control in Flight with the Help of Root Locus Method (Feedback Compensation)

We design a feedback system to meet the specifications which are given below:

1. Velocity error as small as possible
2. Damping ratio $\zeta = 0.9$
3. Settling time $\leq 10\,\text{s}$

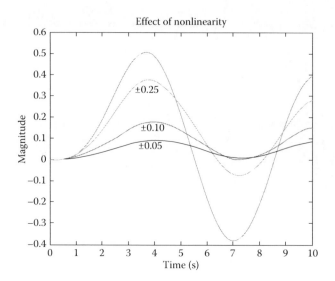

FIGURE 9.19
System performance at different saturation nonlinearities with sinusoidal input.

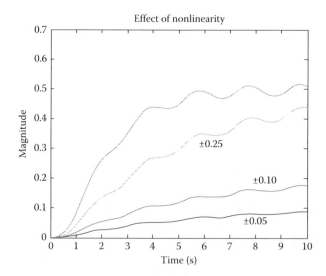

FIGURE 9.20
System performance at different saturation nonlinearities with waveform generator input.

First, we can introduce a proportional controller (a gain, say A).
Open loop transfer function of uncompensated system is as follows:

$$G(s) = \frac{0.24A}{s(0.9s^2 + 1.6s + 1.64)} \tag{9.9}$$

Characteristic equation of the uncompensated system is

$$1 + \frac{0.24A}{s(0.9s^2 + 1.6s + 1.64)} = 0 \tag{9.10}$$

Nonlinear and Chaotic System

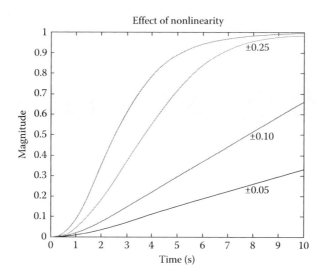

FIGURE 9.21
System performance at different saturation nonlinearities with constant input.

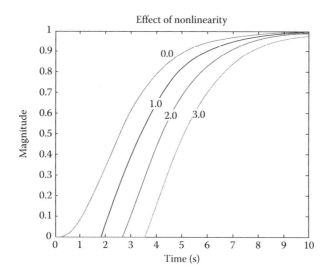

FIGURE 9.22
System performance at different backlash nonlinearities with constant input.

The root locus of Equation 9.2 can be found as follows:
With the help of MATLAB®,

```
num=[1];
den=[0.96 1.6 1.64 0];
rlocus(num,den),
```

In order for the feedback system to have a settling time ≤10 s, all the closed loop poles in root locus must lie on the left-hand side of the vertical line passing through the point $(-4/t_s = -4/10 = -0.4\,\text{s})$.

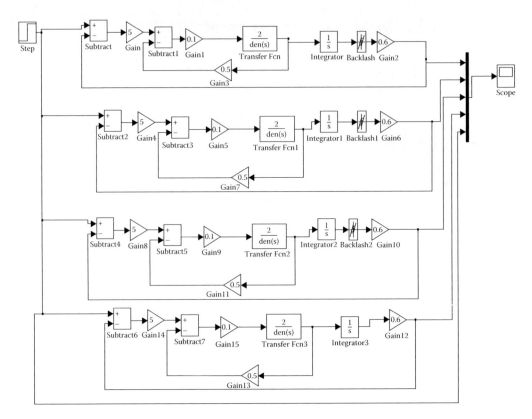

FIGURE 9.23
System performance at different backlash nonlinearities with step input.

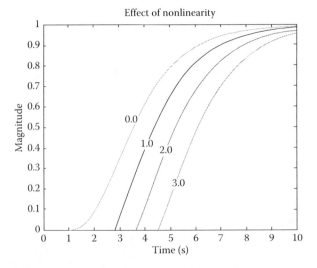

FIGURE 9.24
System performance at different backlash nonlinearities with step input.

Nonlinear and Chaotic System

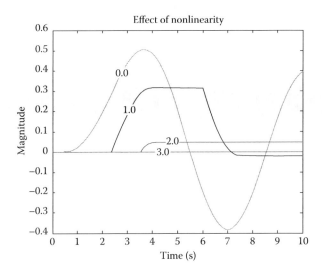

FIGURE 9.25
System performance at different backlash nonlinearities with sinusoidal input.

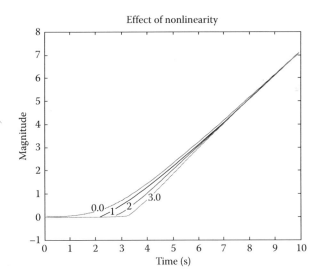

FIGURE 9.26
System performance at different backlash nonlinearities with ramp input.

From the root locus of the system without compensator, as shown in Figure 9.31, we see this is not possible for any $A > 0$.

The characteristic equation (Equation 9.11) of the compensated system becomes

$$1 + A(0.24)/0.96s^3 + 1.6s^2 + 1.64s + 0.24sK_t = 0 \tag{9.11}$$

Partitioning the characteristic equation and selecting the value of $A = 5$, it gives

$$0.96s^3 + 1.6s^2 + 1.64s + 1.20 = -0.24sK_t$$

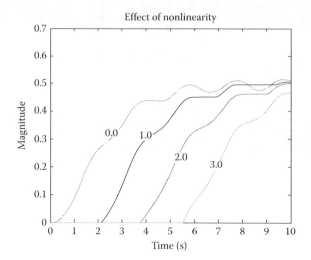

FIGURE 9.27
System performance at different backlash nonlinearities with waveform generator input.

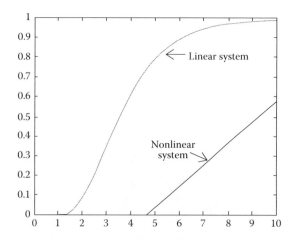

FIGURE 9.28
System performance at different nonlinearities with step input.

$$\text{or} \quad 1 + (0.24sK_t)/(0.96s^3 + 1.6s^2 + 1.64s + 1.20) = 0 \tag{9.12}$$

Now we plot root locus for Equation 9.12.

The root locus of Equation 9.12 is shown in Figure 9.32:

```
num = [1 0];
den = [0.96 1.6 1.64 1.20];
rlocus(num,den),
```

The root locus of the system with compensator (Figure 9.32) is analyzed now. The damping ratio ζ 0.9 line intersects the root locus at two points. Both the points lie on the left-hand side of the vertical line passing through (−0.4).

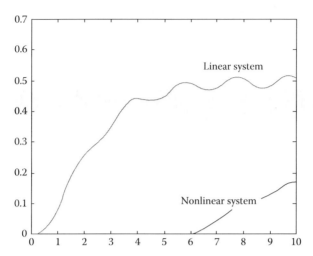

FIGURE 9.29
System performance at different nonlinearities with waveform generator input.

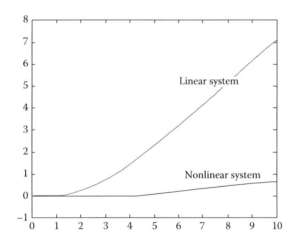

FIGURE 9.30
System performance at different nonlinearities with ramp input.

One of these two points is $(-0.41 + j1.1)$.

Now putting the value of $s = -0.41 + j1.1$ in Equation 9.12 gives $K_t = 1.5670$.

So, now selecting the value for $K_t = 1.5670$, $A = 5.0$ (already taken), a new compensated Simulink model in combination with uncompensated Simulink model is developed, as shown in Figure 9.33.

9.4.7.1.1 Study of Steady State Behavior of the Compensated System

Let us now investigate the study of steady state behavior of the compensated system:

$$K_v = \lim_{s \to 0} \left[\frac{AG(s)}{1 + sK_t G(s)} \right]$$

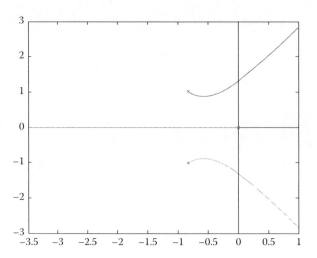

FIGURE 9.31
Root locus of the system without compensator.

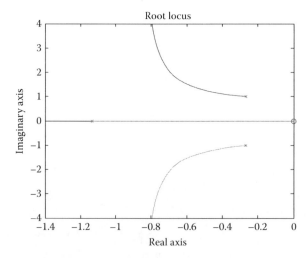

FIGURE 9.32
Root locus of the system with compensator.

where
$A = [5.0]$
$K_t = [1.5670]$
$$G(s) = \frac{0.24}{0.96s^3 + 1.6s^2 + 1.64s}$$

It gives

$K_v = 0.5970$

Figure 9.34 shows a comparison between compensated and uncompensated systems outputs. The graph shown by yellow line denotes uncompensated system and graph shown by other line denotes compensated system.

Nonlinear and Chaotic System

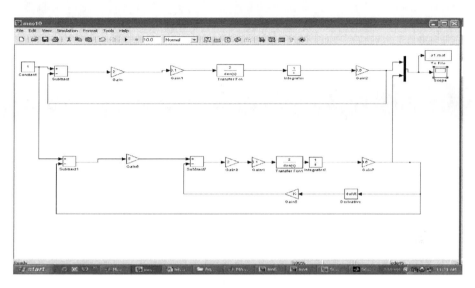

FIGURE 9.33
Comparison of Simulink models (between compensated and uncompensated).

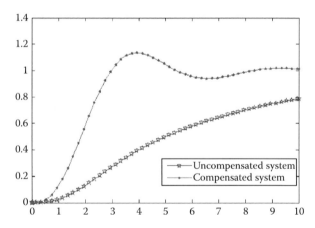

FIGURE 9.34
Results of comparison of Simulink models (between compensated and uncompensated systems).

9.4.7.1.2 Stability Check (Using Routh Table)

Stability check of the obtained compensated system (using Routh table) is carried out. The characteristic equation of the obtained compensated system can be written as follows:

$$0.96s^3 + 1.6s^2 + 1.64s + 0.24sK_t + 1.20 = 0$$

Putting $K = 0.24K_t$

$$0.96s^3 + 1.6s^2 + (1.64 + K)s + 1.20 = 0$$

Routh table is as under

s^3	0.96	$(1.64 + K)$
s^2	1.6	1.20
s^1	$[1.6(1.64 + K) - 1.152]/1.6$	0
s^0	1.20	0

For stable operation

$$[1.6(1.64 + K) - 1.152]/1.6 > 0$$

it gives $K > -0.92$
or, $K_t > -3.83$

This condition is satisfied by our obtained compensated system as $K_t = 1.5670$. Now we can say that our obtained compensated system is stable.

Comments

1. The above-mentioned PID controller is developed using feedback compensation.
2. The feedback compensation provides greater stiffness against load disturbances.
3. Suitable rate gyroscope is available for feed back compensation.
4. Its performance indices are as given below:
   - Percentage overshoot = 14%
   - Rise time = 2.9 s
   - Settling time = 8.3 s

9.4.7.2 Designing a PID Controller (Connected in Cascade with the System) for Pitch Control in Flight

$$G(s) = [0.24/0.96s^3 + 1.6s^2 + 1.64s]$$

$$C(s)/R(s) = Kp[0.24/(0.96s^3 + 1.6s^2 + 1.64s)]/(1 + Kp[0.24/0.96s^3 + 1.6s^2 + 1.64s])$$

$$= Kp(0.24)/(0.96s^3 + 1.6s^2 + 1.64s + Kp0.24)$$

The critical gain is determined by Routh array for the characteristic equation:

$$0.96s^3 + 1.6s^2 + 1.64s + Kp0.24 = 0 \text{ is as given below.}$$

Routh table is as follows:

s^3	0.96	1.64
s^2	1.6	Kp0.24
s^1	$[0.96Kp(0.24) - (1.6)(1.64)]/1.6$	0
s^0	Kp0.24	0

For stable operation

$$[0.96Kp(0.24) - (1.6)(1.64)]/1.6 > 0$$

it gives Kp > 11.39

Hence, the critical gain K_{er} = 11.39,

9.4.7.2.1 To Find the Frequency of Oscillations (T_{er})

$1.6s^2$ + Kp0.24 = 0 (Auxiliary equation found from above Routh table)

putting Kp = 11.9

$s = \pm j1.3$ rad/s

$T_{er} = 2\pi/1.3 = 4.83$ s

Substituting these values in the following standard equations:

Kp = 0.6 K_{er}

$\tau_i = 0.5\ T_{er}$

$\tau_d = 0.125\ T_{er}$

We get

Kp = 6.834

$\tau_i = 2.46$

$\tau_d = 0.60$

The transfer function of developed PID controller may be in the following form:

$$G_c(s) = Kp\ (1 + 1/\tau_i s + \tau_d s)$$

This PID controller is connected in cascade to the system's transfer function.

The PID controller is prepared on the above discussion basis and then Simulink model is developed, as given in Figure 9.35, and compared with the results of the predeveloped PID controller (feedback compensation), as shown in Figure 9.36.

Comments: Comparisons between both PID controllers for Design 1 and 2

1. We can easily see from Figure 9.36 that there is an appreciable difference in percentage overshoots of the outputs for the given step input.
2. Rise time in feedback compensation controller is better.
3. There is no remarkable difference in settling times of both the designs.

Parameters	Design 1	Design 2
Percentage over shoot	14%	62%
Rise time	2.9 s	2.1 s
Settling time	8.3 s	8.4 s

From the above discussion, it is concluded that design 1 is the better design option.

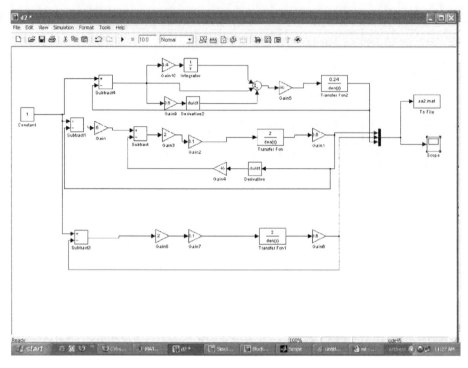

FIGURE 9.35
Arrangement of cascade proportional controller and predeveloped PID controller (feedback compensation) at a place.

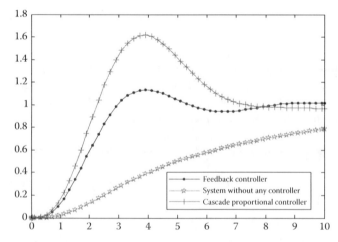

FIGURE 9.36
Comparing the results with the predeveloped PID controller (feedback compensation).

9.4.7.3 Design of P, I, D, PD, PI, PID, and Fuzzy Controllers

Design of P, I, D, PD, PI, PID, and fuzzy controllers and their use on linear as well as nonlinear system is carried out. The compared results of application of these controllers (for linear as well as nonlinear system) are shown after each model (Figure 9.37).

Nonlinear and Chaotic System

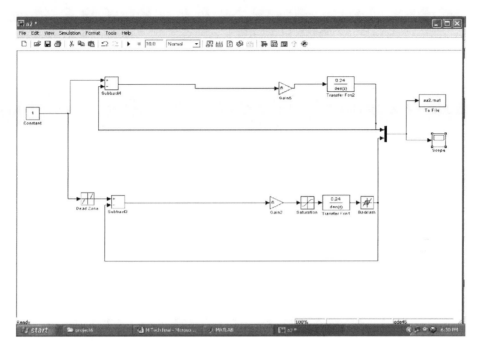

FIGURE 9.37
Proportional controller for linear and nonlinear systems.

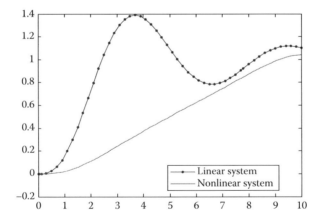

FIGURE 9.38
Results of application of proportional controller.

From Figures 9.38 through 9.46, it is clear that the performance of linear system can be controlled and improved by different controllers, that is, P, I, PD, PI, PID, and fuzzy controller. The blue line curves show linear system and green line curves show nonlinear system. But the performance of nonlinear system cannot be controlled and improved by different controllers, that is, P, I, PD, PI, and PID controllers. Only fuzzy controller is able to control the nonlinear system.

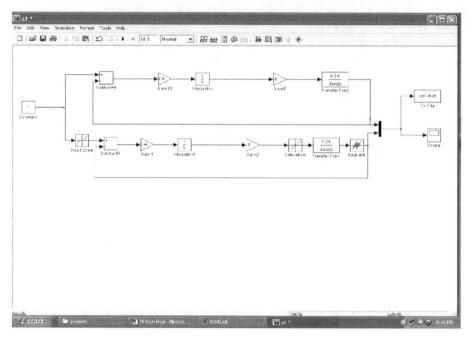

FIGURE 9.39
Arrangement of integral controller.

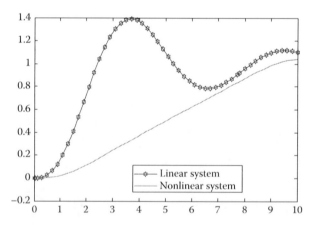

FIGURE 9.40
Results of application of integral controller.

The results of application of fuzzy controller on both the systems are satisfactory.

1. Proportional controller
   The proportional mode adjusts the output signal in direct proportion to the controller input (which is the error signal e). The adjustable parameter is the controller gain.
   A proportional controller reduces error but does not eliminate it (unless the process has naturally integrating properties), that is, an offset between actual and the desired value will normally exist.

$$V(s)/e(s) = K_c$$

Nonlinear and Chaotic System

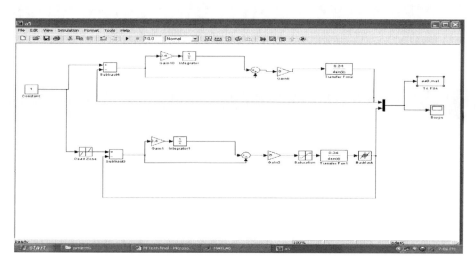

FIGURE 9.41
Arrangement of proportional-integral controller.

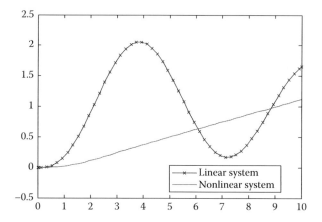

FIGURE 9.42
Results of application of proportional-integral controller.

2. Integral controller
3. Proportional-integral controller
   The additional integral mode (often referred to as reset) corrects for any reset (error) that may occur between the desired value (set point) and the process output automatically. The adjustable parameter to be specified is the integral time of the controller:

$$V(s)/e(s) = K_c[1 + 1/T_i(s)]$$

4. Proportional-derivative controller

$$V(s)/e(s) = K_c[1 + T_d(s)]$$

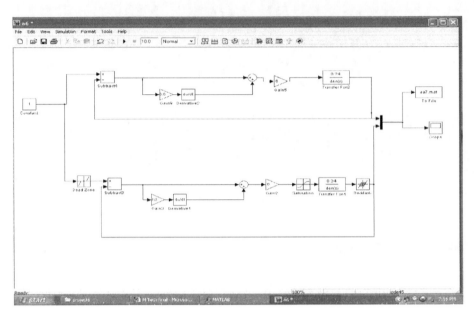

FIGURE 9.43
Arrangement of proportional-derivative controller.

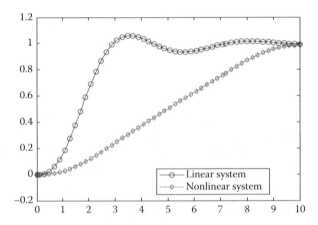

FIGURE 9.44
Results of application of proportional-derivative controller.

5. PID controller
   The PID control constitutes the heuristic approach to controller design that has found wide acceptance in industrial applications.

   The PID controller is the most popular feedback controller used within the process industries. It has been successfully used over many years. It is a robust easily understood algorithm that can provide excellent control performance despite the varied dynamic characteristics of process plant. The mathematical representation of the PID controller is as shown in Figure 9.45:

$$V(s)/(e) = K_c[1 + 1/T_i(s) + T_d(s)]$$

Nonlinear and Chaotic System

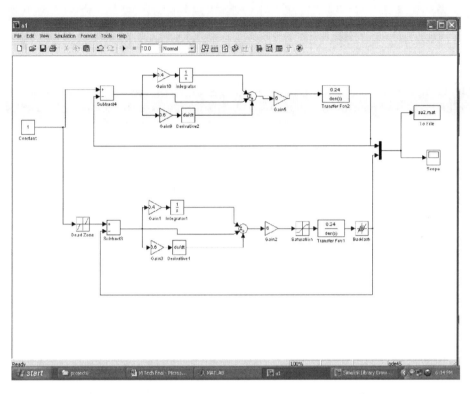

FIGURE 9.45
Arrangement of PID controller.

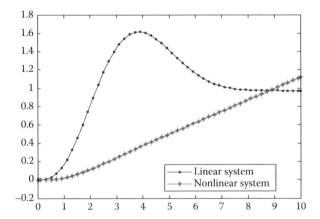

FIGURE 9.46
Results of application of PID controller.

9.4.8 Design of Fuzzy Controller

There is no design procedure in fuzzy control such as root-locus design, frequency response design, pole placement design, or stability margins, because the rules are often nonlinear.

9.4.8.1 Basic Structure of a Fuzzy Controller

A fuzzy controller can be handled as a system that transmits information like a conventional controller with inputs containing information about the plant to be controlled and an output, that is, the manipulated variable. From outside, there is no vague information visible, both, the input and output values as crisp values. The input values of a fuzzy controller consist of measured values from the plant that are either plant output values or plant states, or control errors derived from the set-point values and the controlled variables (Mamdani and Assilian, 1975; Chaturvedi, 2008). A control law represented in the form of a fuzzy system is a static control law. This means that the fuzzy rule-based representation of a fuzzy controller does not include any dynamics, which makes a fuzzy controller a static transfer element, like the standard state-feedback controller. In addition to this, a fuzzy controller is, in general, a fixed nonlinear static transfer element, which is due to those computational steps of its computational structure that have nonlinear properties. The computational structure of a fuzzy controller consists of three main steps, as illustrated by the three blocks in below Figure 9.47:

1. Signal conditioning and filtering at the input (input filter)
2. Fuzzy system
3. Signal conditioning and filtering at the output (output filter)

9.4.8.2 The Components of a Fuzzy System

Figure 9.48 shows the contents of a fuzzy system.

The input signals are crisp values, which are transformed into fuzzy sets in the fuzzification block. The output u comes out directly from the defuzzification block, which transforms an output fuzzy set back to a crisp value. The set of membership functions responsible for the transforming part and the rule base as the relational part contain as

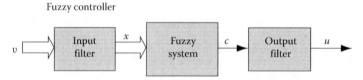

FIGURE 9.47
Basic structure of a fuzzy controller.

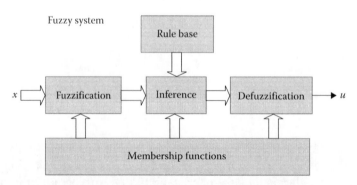

FIGURE 9.48
Basic components of fuzzy system.

Nonlinear and Chaotic System

a whole the modeling information about the system, which is processed by the inference machine (discussed later). This rule-based fuzzy system is the basis of a fuzzy controller, which is described in the following section.

9.4.8.2.1 Representation Using 3D Characteristics

For the case of two inputs and one output, a 3D representation is possible. The fuzzy controller with a fuzzy system having two inputs e_1 and e_2 and one output u, is shown in Figures 9.49 and 9.50.

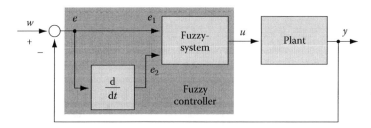

FIGURE 9.49
Block diagram of fuzzy controller.

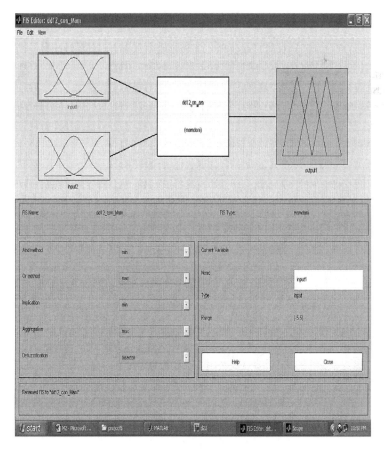

FIGURE 9.50
Fuzzy controller.

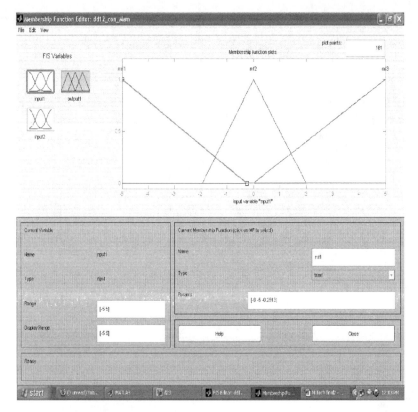

FIGURE 9.51
Membership function plot of input 1 of fuzzy controller.

Fuzzification:

The first block inside the fuzzy system is fuzzification, which converts each piece of input data to degrees of membership by a lookup in one or several membership functions (Figures 9.51 and 9.52).

Rule base:

The rules may use several variables both in the condition and the conclusion of the rules. The controllers can therefore be applied to both multi-input-multi-output (MIMO) problems and single-input-single-output (SISO) problems. The typical SISO problem is to regulate a control signal based on an error signal. The controller may actually need both the *error*, the *change of error*, and the *accumulated error* as inputs, but we will call it single-loop control, because in principle all three are formed from the error measurement. To simplify, this section assumes that the control objective is to regulate some process output around a prescribed set point or reference.

Rule format:

Basically a linguistic controller contains rules in the *if-then* format, but they can be presented in different formats. In many systems, the rules are presented to the end user in a format similar to the following (Figures 9.54 and 9.55):

Nonlinear and Chaotic System

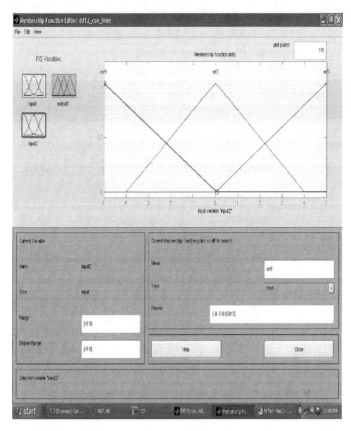

FIGURE 9.52
Membership function plot of input 2 of fuzzy controller.

1. If error is low and change in error is low then output is low.
2. If error is med and change in error is med then output is med.
3. If error is high and change in error is high then output is high.
4. If error is low and change in error is med then output is med.
5. If error is high and change in error is med then output is high.
6. If error is med and change in error is high then output is med.
7. If error is high and change in error is med then output is high.

The names *low*, *med*, and *high* are labels of fuzzy sets.
A more compact representation is as follows:

Error	Change in Error	Output
Low	Low	Low
Med	Med	Med
High	High	High
Low	Med	Med
High	Med	High
Med	High	Med
High	Med	High

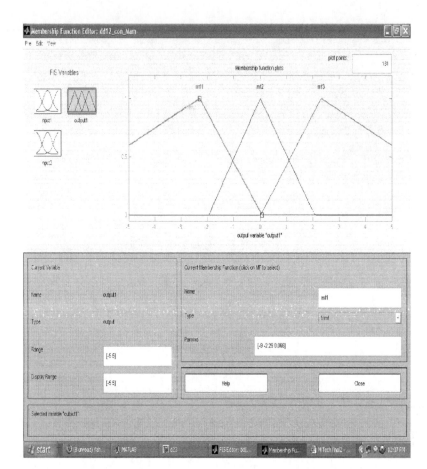

FIGURE 9.53
Membership function plot of output of fuzzy controller.

Defuzzification:

The resulting fuzzy set must be converted to a number that can be sent to the process as a control signal. This operation is called defuzzification. The output surface is shown in Figure 9.56 for both inputs.

The Simulink model of Fuzzy controller is shown in Figure 9.57 and the results are shown in Figure 9.58.

In the graph in Figure 9.58, dotted line shows linear system and simple line shows nonlinear system.

Cumulative arrangement of different controllers (P, I, D, PD, PI, PID, and fuzzy controllers) for linear system (Figures 9.59 through 9.64):

```
load xyy1;
plot(ans(1,:),ans(2,:),'-',ans(1,:),ans(3,:),'-',ans(1,:),ans(4,:),'-',
ans(1,:),ans(5,:),'-',ans(1,:),ans(6,:),'-'),
```

In the above comparison, nonlinearities included are the following:

Dead-zone parameters: ±0.01

Saturation parameters: ±0.85

Backlash parameters: ±0.002

Nonlinear and Chaotic System

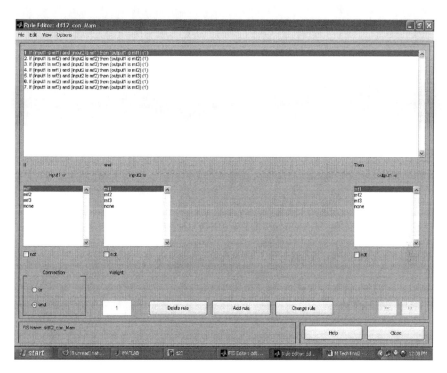

FIGURE 9.54
Rule editor of fuzzy controller.

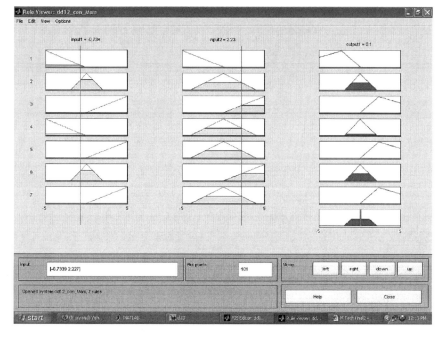

FIGURE 9.55
Rule viewer of fuzzy controller.

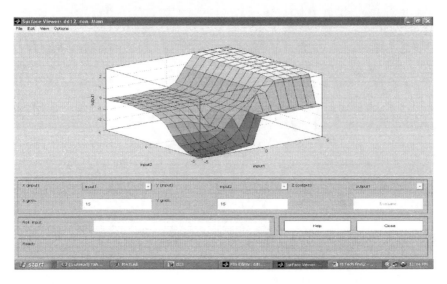

FIGURE 9.56
Surface view of fuzzy controller.

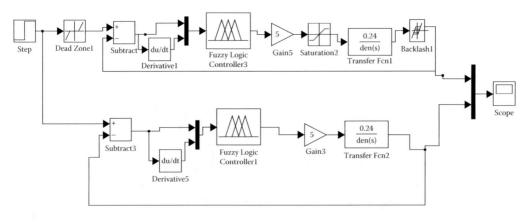

FIGURE 9.57
Arrangement of fuzzy controller in linear system as well as in nonlinear system.

Cumulative arrangement of different controllers (P, PD, PI, PID, and fuzzy controllers) for increased dead-zone nonlinearities:
Now, the considered nonlinearities are following:

Dead-zone parameters: (−0.15 to 0.15)
Saturation parameters: (−0.85 and 0.85)
Backlash parameters: (−0.002 and 0.002)

Nonlinear and Chaotic System

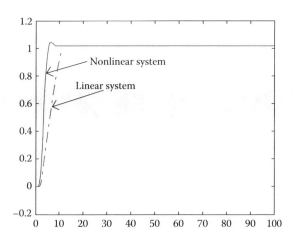

FIGURE 9.58
Results of application of fuzzy controller in linear as well as nonlinear system.

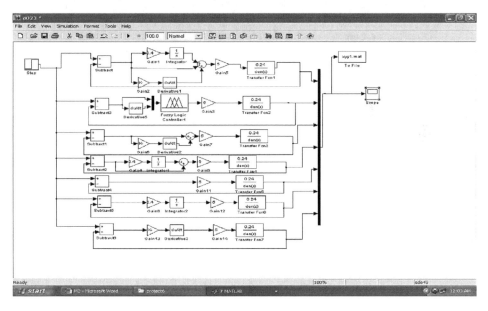

FIGURE 9.59
Arrangement of different controllers (P, I, D, PD, PI, and fuzzy controllers) connected at a place for linear system.

9.4.9 Tuning Fuzzy Controller

The tunned fuzzy system is shown in Figure 9.65 and the fuzzy membership function for inputs and outputs are shown in Figure 9.66 through 9.68.

Rule format

Basically, a linguistic controller contains rules in the *if-then* format, but they can be presented in different formats. In many systems, the rules are presented to the end user in a format shown in Figure 9.69.

470 *Modeling and Simulation of Systems Using MATLAB and Simulink*

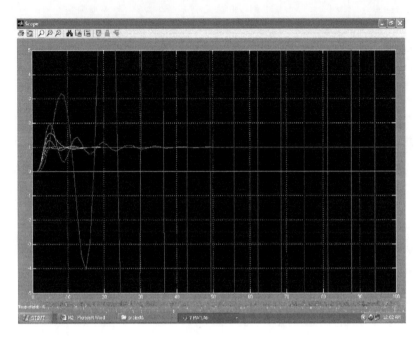

FIGURE 9.60
Results of different controllers (P, I, D, PD, PI, PID, and fuzzy controllers) connected at a place for linear system.

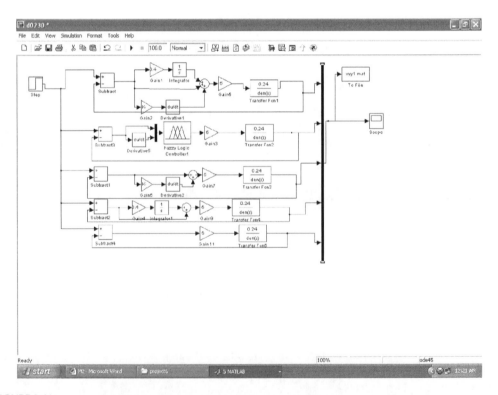

FIGURE 9.61
Arrangement of different controllers (P, PD, PI, PID, and fuzzy controllers) connected at a place for linear system.

Nonlinear and Chaotic System

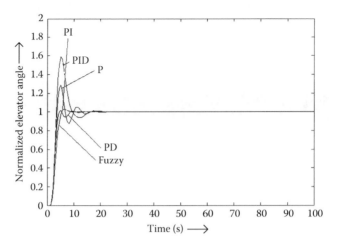

FIGURE 9.62
Results of different controllers (P, PD, PI, PID, and fuzzy controllers).

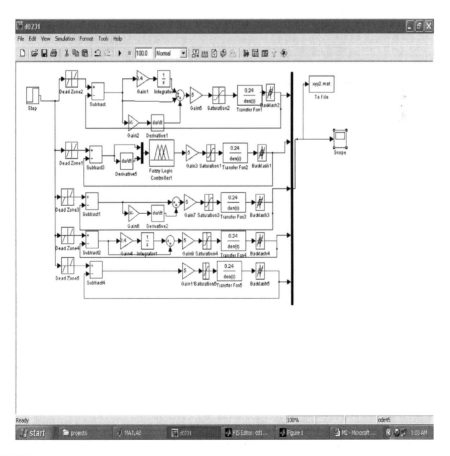

FIGURE 9.63
Arrangement of different controllers (P, PD, PI, and fuzzy controllers) connected at a place for nonlinear system.

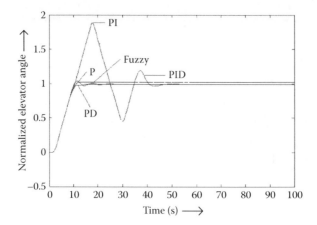

FIGURE 9.64
Results of different controllers (P, PD, PI, PID, and fuzzy controllers) with some nonlinearities connected at a place.

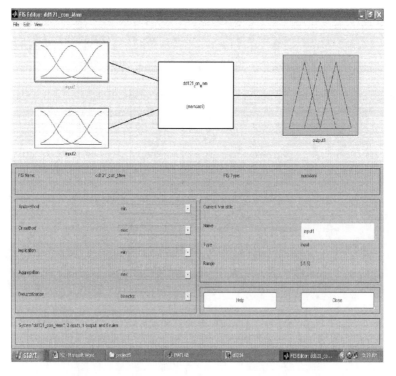

FIGURE 9.65
Tuned fuzzy controller.

1. If error is low and change in error is low then output is low.
2. If error is med and change in error is med then output is med.
3. If error is high and change in error is high then output is high.
4. If error is low and change in error is med then output is med.

Nonlinear and Chaotic System

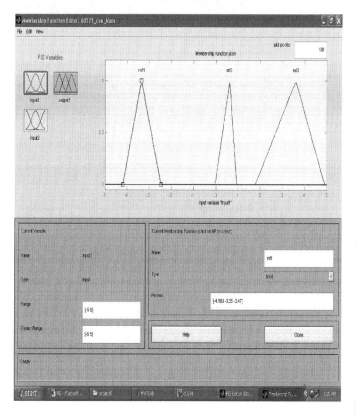

FIGURE 9.66
Input-1 membership function of tuned fuzzy controller.

5. If error is high and change in error is med then output is high.
6. If error is med and change in error is high then output is med.

The names *low, med,* and *high* are labels of fuzzy sets.

A more compact representation of rules is shown in Table 9.1. The 3-D surface of tuned fuzzy controller is shown in Figure 9.70.

9.5 Conclusions

It is quite clear from the results that the nonlinearities are affecting the performance of the flight control system. Flight control system (FCS) Simulink model has been evolved using some applicable nonlinearities. Proportional controller (P), Integral controller (I), Derivative controller (D), Proportional integral controller (PI), Proportional-derivative controller (PD), Proportional-integral-derivative controller (PID), evolved and tuned. Proportional controller also evolved with the help of root locus techniques. The performance of P, PI, PD, and PID controllers are analyzed separately. Their responses for step input, ramp input, and constant input have been studied. Fuzzy controller has also been developed and results have been compared with the aforesaid controllers; the results show that fuzzy controller is able to handle the applicable nonlinearities more efficiently.

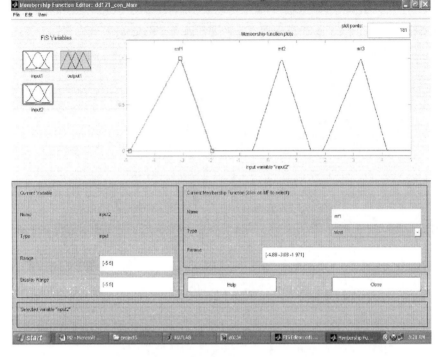

FIGURE 9.67
Input-2 membership function of tuned fuzzy controller.

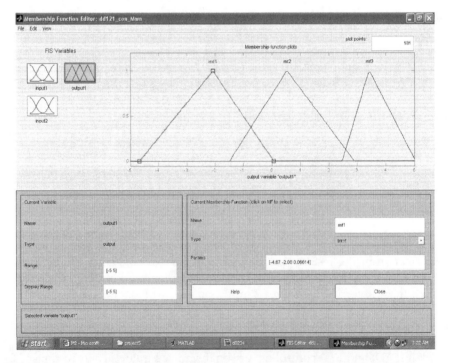

FIGURE 9.68
Output membership function of tuned fuzzy controller.

Nonlinear and Chaotic System

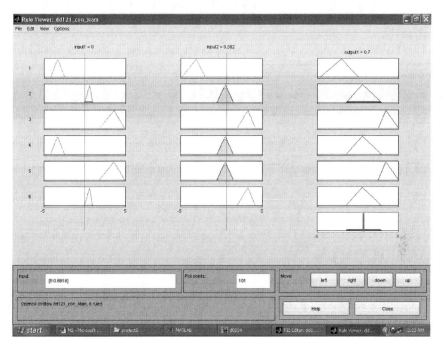

FIGURE 9.69
Rule view of tuned fuzzy controller.

This work may be further extended as follows:

- Adaptive/neuro-fuzzy controller may be developed.
- Online tuning of fuzzy logic controller (FLC) may be done using genetic algorithm.
- Yow and Roll control can be done with fuzzy/neuro-fuzzy controllers.

Example

Consider a simple mass–dashpot and spring system.

TABLE 9.1
Fuzzy Rules Table

Error	Change in Error	Output
Low	Low	Low
Med	Med	Med
High	High	High
Low	Med	Med
High	Med	High
Med	High	Med

$K = 360\,\text{N/m}$
$M = 10\,\text{kg}$
$B = 24–60\,\text{N s/m}$

In this system, the damping (friction) is not linear but is nonlinear; the damping equations for linear damping and nonlinear damping are as follows:
For linear damping, $f = bv_{23}$
For nonlinear damping, $f = bv_{23}^2$

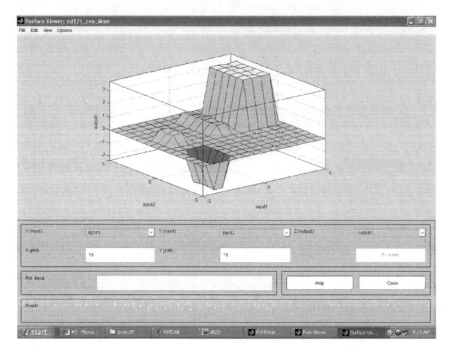

FIGURE 9.70
Surface view of tuned fuzzy controller.

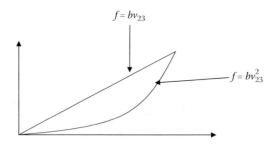

System equations

Linear system

$$\frac{d^3v}{dt^3} + \frac{k}{mb}\frac{dv}{dt} + \frac{k}{m}v = \frac{k}{m}Vin$$

Nonlinear system

$$\frac{d^3v}{dt^3} + \sqrt{\frac{k}{mb}\frac{dv}{dt}} + \frac{k}{m}v = \frac{k}{m}Vin$$

MATLAB program

```
% Non-linear Mass-spring dashpot system Simulation
clear all
clc
K = 360; % N/m
M = 10;  % Kg
```

Nonlinear and Chaotic System

```
B = 24;      % N-s/m - 60N-s/m
dt = 0.001;
t = 0;
tsim = 2;
n = round((tsim-t)/dt);
x1 = 0;dx1 = 0;x2 = 0;dx2 = 0;
vin = 0;
for i = 1:n
    x11(i,:) = [t,dx1,dx2,x1,x2,vin];
    vin = sin(2*pi*5*t);
    ddx1 = (K/M)*(vin-x1-dx1/B);
    ddx2 = (K/M)*(vin-x2)-sqrt(K*dx2/M*B);
dx1 = dx1 + dt*ddx1;dx2 = dx2 + dt*ddx2;
    x1 = x1 + dt*dx1;x2 = x2 + dt*dx2;
    t = t + dt;
end
figure(1)
plot(x11(:,1),x11(:,2:3))
title('Simulation of non-linear')
legend('Linear parameters', 'non-linear parameter')
xlabel('Time (Sec.)')
ylabel('Velocity (m/s)')
figure(2)
plot(x11(:,1),x11(:,4:5))
title('Simulation of non-linear')
legend('Linear parameters', 'non-linear parameter')
xlabel('Time (sec.)')
ylabel('Displacement (m)')
```

Results

From a simulation point of view, if we do not know what the system is capable of doing, we will not know if our simulation is reasonable, thus it is important to be aware of the types of behavior demonstrable by nonlinear system so that the strange responses that sometimes occur are not totally unexpected. The response of given mechanical system for linear parameters and nonlinear parameters are in figures below.

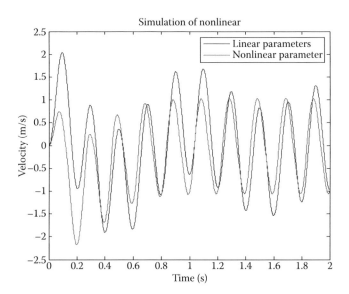

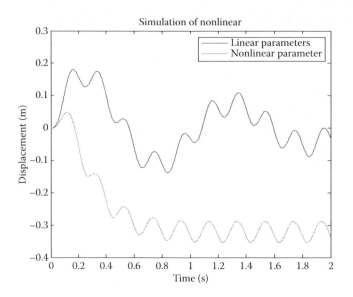

9.6 Introduction to Chaotic System

The word chaos (derived from the Ancient Greek Χάος, Chaos) typically refers to unpredictability. Chaos is the complexity of causality or the relationship between events. It has both a general meaning and a scientific meaning. As is usually the case, the general meaning tends to convey little of the strict definition that scientists and mathematicians apply to the chaos.

9.6.1 General Meaning

1. A condition or place of total disorder or utter confusion: "emotions in complete chaos."
2. Chaos is the disordered state of unformed matter and infinite space supposed by some religious cosmological views to have existed prior to the ordered universe.
3. The confused, unorganized condition or mass of matter before the creation of distinct and orderly forms.
4. Any confused or disordered collection or state of things or a confused mixture.

9.6.2 Scientific Meaning

What scientists and mathematicians mean by chaos is very much related to the spirit of the definitions given above.

The systems are chaotic if they are deterministic through description by mathematical rules and these mathematical descriptions are nonlinear in some way. Unpredictable behavior of deterministic systems has been called *chaos*.

However, we can talk about chaos in regard to its typical features:

1. Nonlinearity
2. Determinism
3. Sensitivity to initial conditions

Nonlinear and Chaotic System

4. Sustained irregularity
5. Long-term prediction impossible

Chaos is related to the irregular behavior of the system. The system contains a "hidden order" which itself consists of a large or infinite number of periodic patterns (or motions). Effectively this is "order in disorder." Chaotic processes are not random; they follow rules but even simple rules can produce extreme complex outputs. This blend of simplicity and unpredictability also occurs in *music* and *art*. Music consisting of random notes or of an endless repetition of the same sequence of notes would be either be disastrously discordant or unbearably boring. Chaos is all around us. For example, consider the human brain function and heart beat. It is suggested that pathological "order" may lead to epilepsy and too much periodicity in heart rate may indicate disease!

9.6.3 Definition

No definition of the term chaos is universally accepted yet, but almost everyone would agree on the three ingredients used in the following working definition:

Chaos is a periodic long-term behavior in a deterministic system that exhibits sensitive dependence on initial conditions.

1. A periodic long-term behavior
   It means that the system behaviors do not settle down to fixed points, period orbits, or quasi periodic orbits as $t \to \infty$.
2. Deterministicism
   It means that the system has no random or noisy inputs or parameters. The irregular behavior arises from the system's nonlinearity, rather than from noisy driving forces.

   A system which is deterministic has an attribute that its behavior is specified without probabilities and the system behavior is predictable without uncertainty once the relevant conditions are known.
3. Sensitive dependence on initial conditions
   If the nearby trajectories separate exponentially fast, it means the system has a positive Lyapunov exponent. Small changes in the initial state can lead to radically different behavior in its final state. This gives rise to "butterfly effect" in the system behavior.

   An *iterated function system* (IFS) consists of one or more functions that are repeatedly applied to their own outputs.

   An attractor is a set to which all neighboring trajectories converge. Stable fixed points and stable limit cycles are examples of attractors. More precisely, it is a closed set A with the following properties:

1. A is an invariant set
   Any trajectory $x(t)$ that starts in A stays in A for all time.
2. A attracts an open set of initial conditions
   A attracts all trajectories that start sufficiently close to it. The largest open set containing As is called the basin of attraction of A.
3. A is minimal
   There is no proper subset of A that satisfies conditions 1 and 2.

If an attractor is a *fractal*, then it is said to be a *strange attractor*. A strange attractor is an attractor that exhibits sensitive dependence on initial conditions. Chaotic systems generally have *strange attractors*.

When we say a chaotic system is *unpredictable*, we do not mean that it is *nondeterministic*.

- A deterministic system is one that always gives the same results for the same input/initial values.
- *Real-world chaotic systems* are unpredictable in practice because we can never determine our input values *exactly*. Thus, sensitive dependence on initial conditions will always mess up our results, *eventually*.

A chaotic system does not need to be very complex. For example, $f(x) = ax(1 - x)$ is pretty simple, but it does need to be nonlinear. So methods that approximate functions by linear/affine functions (e.g., Newton's method) can sometimes give qualitatively different behavior than that which they purport to approximate.

Chaos refers to the *complex, difficult-to-predict behavior* found in nonlinear systems. One of the important properties of chaos is *sensitive dependence on initial conditions*, informally known as *the Butterfly effect*.

- *Sensitive dependence on initial conditions* means that a *very* small change in the initial state of a system can have a large effect on its later state, as shown in Figure 9.71.
- In particular, you can eventually get the large effect, no matter how small the initial change was. This effect is very common in time series prediction such as weather forecasting and electrical load forecasting, as shown in Figure 9.72.

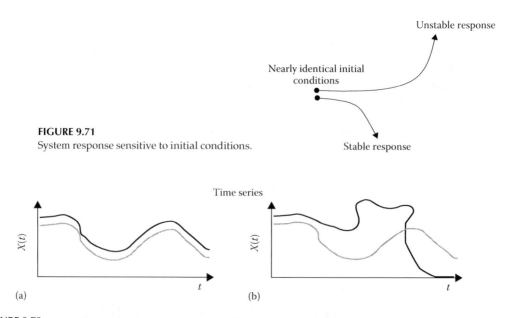

FIGURE 9.71
System response sensitive to initial conditions.

FIGURE 9.72
Chaotic behavior of time series predictions. (a) Non chaotic. (b) Chaotic.

9.7 Historical Prospective

The first discoverer of chaos was Henri Poincaré in 1890 while studying the three-body problem. Henri Poincaré found that there can be orbits which are non-periodic, and yet not forever increasing nor approaching a fixed point (Poincaré, 1890). In 1898, Jacques Hadamard published an influential study of the chaotic motion of a free particle gliding frictionless on a surface of constant negative curvature (Hadamard, 1898). He had shown that all trajectories are unstable in that all particle trajectories diverge exponentially from one another with a positive Lyapunov exponent.

Chaos was observed by a number of experimenters before it was recognized; for example, in 1927 by van der Pol (van der Pol and van der Mark, 1927) and Ives in 1958 (Ives, 1958). Edward Lorenz in 1963 recorded chaotic behavior while studying mathematical models of weather (Lorenz, 1963). Lorenz had discovered that small changes in initial conditions produced large changes in the long-term outcome. Lorenz's discovery, which gave its name to Lorenz attractors, proved that meteorology could not reasonably predict weather beyond a certain period. Lorenz explained the basic concepts of chaos in his book (Lorenz, 1996). The history of dynamics and chaos is given in Table 9.2. The chaotic and systematic knowledge handling tools are compared in Table 9.3.

TABLE 9.2

Dynamics and Chaos: A History

1666	Newton	Invention of calculus, explanation of planetary motion
1700		Flowering of calculus and classical mechanics
1800		Analytical studies of planetary motion
1890	Poincare	Geometric approach, nightmare of chaos
1898	Jacques Hadamard	Particle trajectories diverge exponentially from one another
1920–1950		Nonlinear oscillators in physics and engineering, invention of radio, radar, laser
1950–1960	Birkhoff Kolmogorov Arnol'd Moser	Complex behavior in Hamiltonian mechanics
1963	Edward Lorenz	Chaos in weather predictions and found strange attractor in simple model of convection
1970	Ruelle and Takens	Turbulence and chaos
	May	Chaos in logistic map
	Feigenbaum	Universality and renormalization, connection between chaos and phase transitions
	Winfree	Experimental studies of chaos
	Mandelbrot	Nonlinear oscillators in biology Fractals
December 1977		First symposium on chaos organized by New York Academy of Sciences
1979	Albert J. Libchaber	Experimental observation of the bifurcation cascade leads to chaos
1980		Widespread interest in chaos, fractal, oscillators, and their applications

TABLE 9.3

Chaotic versus Systematic Knowledge Handling

Chaotic	Systematic
Heuristics	Rules
Unsound reasoning methods	Formal logic
Inconsistent knowledge	Consistency
Jumping to conclusions	Proofs
Ill-defined problems	Well-defined problems
Unclear boundaries of knowledge	Domain-specific
Informal, continuous meta-reasoning	Knowledge
	Expensive, distinct meta-reasoning

TABLE 9.4

Chaos in Cryptography

Chaotic Property	Cryptographic Property	Description
Ergodicity	Confusion	The output has the same distribution for any input
Sensitivity to initial conditions/control parameter	Diffusion with a small change in the plaintext/secret key	A small deviation in the input can cause a large change at the output
Mixing property	Diffusion with a small change in one plain block of the whole plaintext	A small deviation in the local area can cause a large change in the whole space
Deterministic dynamics	Deterministic pseudorandomness	A deterministic process can cause a random-like (pseudorandom) behavior
Structure complexity	Algorithm (attack) complexity	A simple process has a very high complexity

The main catalyst for the development of chaos theory was the digital computer. Much of the mathematics of chaos theory involves the repeated iteration of simple mathematical formulas, which would be impractical to do by hand. Digital computers made these repeated calculations practical and powerful software made figures and images made it possible to visualize these systems. The chaotic properties are explained in Table 9.4 to use chaos in cryptography.

Orbits and Periodicity

- Suppose we have an IFS with exactly one function: $f:S \to S$.
- Recall: Given a point x in S, the *orbit* of x is the collection of all points that the IFS takes x to that is, the orbit of x contains $x, f(x), f(f(x)), f(f(f(x)))$, etc.
- x is a *periodic point* if repeatedly applying f eventually takes x back to itself.
  - Put another way: A periodic point is a point whose orbit is finite.

TABLE 9.5

Linear and Nonlinear Systems

		$n = 1$	$n = 2$	$n \geq 3$	$n \gg 1$	Continuum
Linear		Growth decay or equilibrium	Oscillation		Collective phenomena	Waves and patterns
		Exponential growth	Linear oscillator mass and spring	Civil engineering, structure	Coupled harmonic oscillators solid state physics	Elasticity wave equations
		RC circuit	RLC circuit	Electrical engineering	Molecular dynamics	Electromagnetism (Maxwell)
		Radioactive decay	2-body problem (Kepler, Newton)		Equilibrium statistical mechanics	Quantum mechanics (Schrodinger, Heisenberg, Dirac) Heat and Diffusion Acoustics Viscous fluids
				Chaos		Spatiotemporal complexity
Nonlinear		Fixed points	Pendulum	Strange attractor (Lorenz)	Coupled nonlinear oscillators	Nonlinear waves (shocks, solitons)
		Bifurcations	Harmonic oscillator	3-body problem (Poincare)	Laser, nonlinear optics	Plasmas
		Overdamped systems	Limit cycles	Chemical kinetics	Nonequilibrium statistical mechanics	General relativity(Einstein)
		Relaxational dynamics	Biological oscillators (neurons, heart cells)	Iterated maps (Feigenbaum)	Nonlinear solid state physics (semiconductors)	Quantum field theory
		Logistic equation for single species	Predator–prey cycles	Fractals (Mandelbrot)	Josephson arrays	Reaction–diffusion Biological and chemical waves
			Nonlinear electronics (van der Pol, Josephson)	Forced nonlinear oscillators (Levison, Smale)	Heart cell synchronization Neural networks Immune system Ecosystems	Fibrillation Epilepsy Turbulent fluid (Navier–Stokes) Life
				Practical use of chaos, quantum chaos	Economics	

Three main categories where chaos may be applied are as follows.

1. Stabilization and control
   The idea is that the extreme sensitivity of chaotic systems to tiny perturbations can be manipulated to control the system artificially. Artificially generated systems incorporate perturbations to
   a. Keep a large system stable (stabilization)
   b. Direct a large system into a desired state (control)

2. Synthesis
   The idea is that "regular is not always best." So, take artificially generated chaotic systems and apply them to certain problems to make certain systems (whether chaotic or not) work better, for example, artificially stimulated chaotic brain waves may help inhibit epileptic seizures. Similarly, chaos may be used in encryption for communications via "synchronization." Two identical sequences of chaotic signals may be used via the sender superimposing a message on one sequence and the receiver simply strips of the chaotic signal. It may also be used in solving optimization problems (such as in neural networks).
3. Analysis and prediction
   It is possible to predict for the future trend of a chaotic system based on sufficient chaotic data points of past.

9.8 First-Order Continuous-Time System

First-order autonomous time invariant system can be written as

$$\overset{\circ}{x} = f(x) \tag{9.13}$$

Several important things can be determined from this equation. The values of x where $f(x) = 0$ are the steady state points of the system. These are also called fixed points. Contrary to linear thinking more than one steady state can exist simultaneously. The slope of $f(x)$ at these points in the system Jacobian, which is df/dx, determines the local stability of the points. If df/dx is negative at a fixed point, the point is stable and is called an attractor. If df/dx is positive at a fixed point, the point is unstable and is called a repeller. If $df/dx = 0$, the points are saddled, which means that they are attracted from one side and repelled from the other. The geometry of this equation indicates that the attractor and repeller alternate on the x-axis. It should be noted that no oscillation or undamped responses are possible. Basically first-order systems are damped.

Consider the system

$$\overset{\circ}{x} = x - x^3 + u \tag{9.14}$$

with the input $u = 0$. The plot of the derivative versus x is given in Figure 9.73. The fixed points can be found either from the intersection with the x-axis or from

$$\overset{\circ}{x} = 0 = x - x^3 \tag{9.15}$$

which gives $x = 0, +1,$ and -1. The local stability of these points can be determined graphically from the slopes at these points. Mathematically, this is shown by Equation 9.16:

$$J - \left[\frac{\partial \dot{x}}{\partial x}\right]_{x_{ss}} = 1 - 3x_{ss}^2 \tag{9.16}$$

Nonlinear and Chaotic System

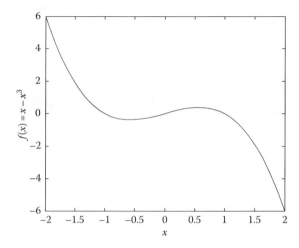

FIGURE 9.73
Function trajectory versus x.

Thus the slope at $x = 0$ is +1, and the slope at $x = \pm 1$ is 2. The fixed points at the origin are unstable and are called repellers. The fixed points at $x = \pm 1$ are stable and are called attractors, as shown in Figure 9.73. It should be observed that these slopes near fixed points are effectively equivalent to the system poles of a linearized version of the system.

Another thing to observe is that each attractor possesses a basin of attraction that is a region in space where any initial condition in the space is attracted to that particular attractor. For this example, the basin of attraction for the $x = +1$ points is all x. The basin of attraction for the $x = -1$ points is all x.

Linearization

Before going on to second-order system, it is important to talk about the linearization procedure as it is used by almost all control engineers at some time. The general vector form is given first, and then it is applied to the scalar system of the last section.

Consider the vector system

$$\dot{x} = f(x, u) \tag{9.17}$$

Assume that u reaches some fixed value u. It is then possible to find the steady state value of x where the derivative vector is identically zero by solving

$$f(x_{ss}, u_{ss}) = 0 \quad \text{for } x_{ss} \tag{9.18}$$

The purpose of linearization is to find the behavior of the system near one of the fixed points. Thus we set

$$x = x_{ss} + \delta x \quad \text{and} \quad u = u_{ss} + \delta u$$

Substituting this back into the original vector system, we get

$$\dot{x}_{ss} + \delta \dot{x} = f(x_{ss} + \delta x, u_{ss} + \delta u) \tag{9.19}$$

Notice that the derivative of x is zero. Now expanding the right side we obtain

$$\delta \dot{x} = f(x_{ss}, u_{ss}) + \left[\frac{\partial f}{\partial x}\right]_{x_{ss},u_{ss}} \delta x + \left[\frac{\partial f}{\partial u}\right]_{x_{ss},u_{ss}} \delta u \tag{9.20}$$

Notice that the first term in this series is zero by definition. We then have

$$\delta \dot{x} = J\delta x + G\delta u \tag{9.21}$$

Equation 9.21 should be recognized as a useful state space representation used by most control engineers. Thus the eigenvalues of the system Jacobian matrices are the poles of the linearized system and determine the stability of the resulting system. Improved techniques for computing the input Jacobian G are given in Danielli and Krosel.

Example

Find the stability of system $\overset{o}{x} = x^2 - 1$ after finding all fixed points.

SOLUTION

Here $f(x) = x^2 - 1$. To find the fixed points, we set $f(x^*) = 0$ and solve for x^*. Thus $x^* = \pm 1$. To determine the stability, we plot the function $f(x)$. Figure 9.74 shows that the flow is to the right where $x^2 - 1 > 0$ and to the left where $x^2 - 1 < 0$. Thus $x^* = -1$ is stable and $x^* = 1$ is unstable.

MATLAB codes

```
x = -2:.1:2;
fx = x.^2-1;
plot(x,fx)
```

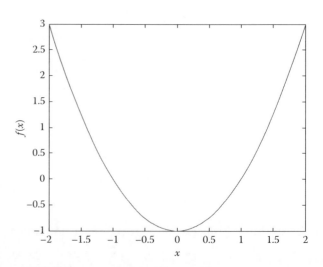

FIGURE 9.74
Function $f(x) = x^2 - 1$ versus x.

9.9 Bifurcations

The dynamics of vector fields on the line is very limited: all solutions either settle down to equilibrium or head out to ±∞. In one-dimensional system, the dynamics depends upon parameters. The qualitative structure of the flow can change as parameters are varied. In particular, fixed points can be created or destroyed, or their stability can change. These qualitative changes in the dynamics are called bifurcations, as shown in Figure 9.75, and the parameter values at which they occur are called bifurcation points. Bifurcation is important scientifically; they provide models of transitions and instabilities as some control parameters are varied. Bifurcations occur in both continuous systems (described by ODEs, DDEs, or PDEs) and discrete systems.

Types of bifurcations

1. Local bifurcation
2. Global bifurcation

1. Local bifurcation
   It can be analyzed entirely through changes in the local stability properties of equilibria, periodic orbits, or other invariant sets as parameters cross through critical thresholds. For example:

   Saddle-node (fold) bifurcation
   Transcritical bifurcation
   Pitchfork bifurcation
   Period-doubling (flip) bifurcation
   Hopf bifurcation
   Neimark (secondary Hopf) bifurcation

2. Global bifurcation
   It occurs when larger invariant sets of the system "collide" with each other or with equilibria of the system. They cannot be detected purely by a stability analysis of the equilibria (fixed points). For example:

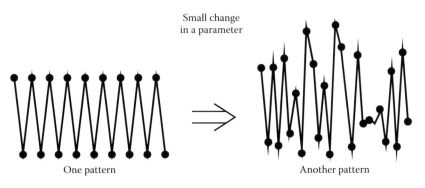

FIGURE 9.75
Bifurcation.

Homoclinic bifurcation in which a limit cycle collides with a saddle point.
Heteroclinic bifurcation in which a limit cycle collides with two or more saddle points.
Infinite-period bifurcation in which a stable node and saddle point simultaneously occur on a limit cycle.
Blue sky catastrophe in which a limit cycle collides with a non-hyperbolic cycle.

9.9.1 Saddle Node Bifurcation

The saddle node bifurcation is the basic mechanism by which fixed points are created and destroyed. As a parameter is varied, two fixed points move toward each other, collide, and mutually annihilate.

The prototype example of a saddle node bifurcation is given by the first-order system:

$$\overset{o}{x} = r + x^2 \tag{9.22}$$

where r is a parameter which may be positive, negative, or zero. When r is negative there are two fixed points, one stable and one unstable, as shown in Figure 9.76.

As r approaches 0, the parabola moves up and the two fixed points move toward each other. When $r = 0$, the fixed points coalesce into a half-stable fixed point at $x^* = 0$. This type of fixed point is extremely delicate—it vanishes as soon as $r > 0$, and now there are no fixed points at all.

Hence, in this example the bifurcation occurred at $r = 0$, since the vector fields for $r < 0$ and $r > 0$ are qualitatively different, as shown in Figure 9.77.

9.9.2 Transcritical Bifurcation

There are certain scientific situations where a fixed point must exist for all values of a parameter and can never be destroyed. For example, in the logistic equation and other simple models for the growth of a single species, there is a fixed point at zero population, regardless of the value of the growth rate. However, such a fixed point may change its stability as the parameter is varied, as shown in Figure 9.78. The transcritical bifurcation is the standard mechanism for such changes in stability.

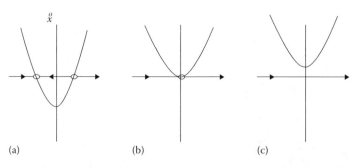

FIGURE 9.76
Effect of r on fixed points. (a) $r < 0$. (b) $r = 0$. (c) $r > 0$.

Nonlinear and Chaotic System

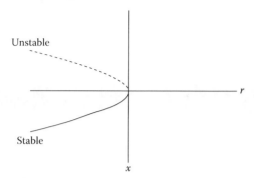

FIGURE 9.77
The bifurcation diagram for the saddle node bifurcation.

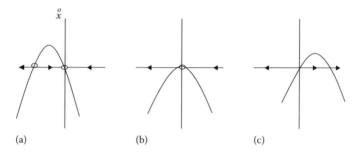

FIGURE 9.78
Effect of r on fixed points. (a) $r < 0$. (b) $r = 0$. (c) $r > 0$.

The normal form for a transcritical bifurcation is

$$\overset{o}{x} = rx - x^2 \tag{9.23}$$

Note that there is a fixed point at $x^* = 0$ for all values of r. The important difference between the saddle node and transcritical bifurcations is in transcritical case, the two fixed points do not disappear after the bifurcation—instead they just switch their stability, as shown in Figure 9.79.

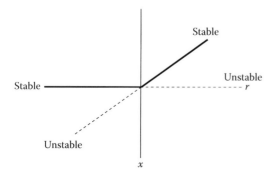

FIGURE 9.79
The bifurcation diagram for the transcritical bifurcation.

9.9.3 Pitchfork Bifurcation

This bifurcation is common in physical problems that have symmetry. There are two types of pitch fork bifurcation. The simpler type is called supercritical pitchfork bifurcation.

9.9.3.1 Supercritical Pitchfork Bifurcation

The normal form of the supercritical pitchfork bifurcation is expressed with Equation 9.24:

$$\overset{o}{x} = rx - x^3 \qquad (9.24)$$

This equation is invariant under the change of variable $x \to -x$.

In real physical systems, such as explosive instability is usually opposed by the stabilizing influence due to higher order terms. In the supercritical case, $\overset{o}{x} = rx - x^3$, the cubic term is stabilizing, which acts as a restoring force that pulls $x(t)$ back toward $x = 0$, as shown in Figures 9.80 and 9.81. On the other hand, if the cubic term is positive as $\overset{o}{x} = rx + x^3$, then this term was destabilizing, which is called subcritical pitchfork bifurcation, as shown in Figure 9.82.

$x^* = \pm\sqrt{(-r)}$ are unstable and exist only below the bifurcation ($r < 0$), which motivate the term "subcritical." More importantly, the origin is stable for $r < 0$ and unstable for $r > 0$ as in the supercritical case, but now the instability for $r > 0$ is not opposed by the cubic term—in fact the cubic term lends a helping hand in driving the trajectories out to infinity. This shows that $x(t) \to \pm\infty$ in finite time starting from any initial condition $x_o \neq 0$ (Figure 9.83).

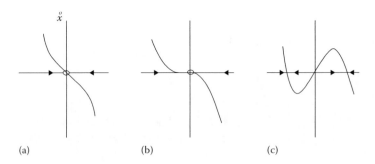

(a) (b) (c)

FIGURE 9.80
Effect of r on fixed points. (a) $r < 0$. (b) $r = 0$. (c) $r > 0$.

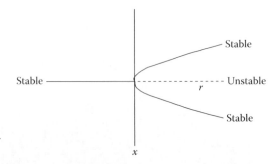

FIGURE 9.81
The bifurcation diagram for the supercritical pitchfork bifurcation.

Nonlinear and Chaotic System

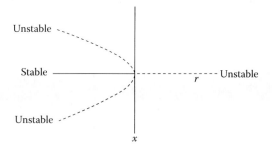

FIGURE 9.82
The bifurcation diagram for the subcritical pitchfork bifurcation.

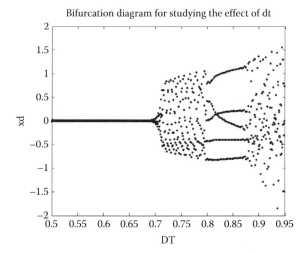

FIGURE 9.83
Bifurcation diagram for studying the effect of ΔT.

```
%Matlab program for studying the effect of simulation time step
u = -1;
b = 1.8;
x = -0.99;
xd = 0;
dt = 0.5;
axis([0.5 0.95 -2 2]);
plot(0.5,0,'.')
hold
for n = 1:900
        f1 = xd;
        f2 = -b*xd-x^3 + u;
        xd = xd + dt*f2;
        x = x + dt*f1;
        plot(dt,xd,'.')
        dt = dt + 0.0005;
end
hold
title('Biffurcation diagram for studying the effect of dt')
xlabel('DT'), ylabel('xd')
```

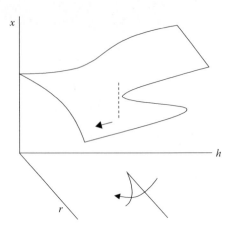

FIGURE 9.84
Catastrophe due to parameter change.

9.9.4 Catastrophes

The term catastrophe is motivated by the fact that as parameters change, the state of the system can be carried over the edge of the upper surface, after which it drops discontinuously to the lower surface, as shown in Figure 9.84. This jump could be truly catastrophic for the equilibrium of a bridge or a building.

9.9.4.1 Globally Attracting Point for Stability

Consider an uncoupled system state; equate $\overset{\circ}{x} = ax$ and $\overset{\circ}{y} = -yx$. The solution to these equations are $x(t) = x_0 e^{at}$ and $y(t) = y_0 e^{-t}$.

The phase portraits for different values of "a" are shown in Figure 9.85. When $a < 0$, $x(t)$ also decays exponentially and so all trajectories approach the origin as $t \to \infty$. However, the direction of approach depends on the size of "a" compared to -1.

In nonlinear systems, if $x^* = 0$, then it results in attracting a fixed point as shown in Figure 9.85. It means all trajectories that start near x^* approach to it as $t \to \infty$, that is,

$$x(t) \to x^* \text{ as } t \to \infty$$

In fact x^* attracts all trajectories in the phase plane, so that it could be called globally attracting fixed point.

There is a completely different notion of stability, which is related to the behavior of trajectories for all time, not just as $t \to \infty$. A fixed point $x^* = 0$ is Lyapunov stable if all trajectories that start sufficiently close to x^* remain close to it for all time.

Figure 9.85(a) shows that a fixed point can be Lyapunov stable but not attracting, which is called neutrally stable. This type of stability is commonly encountered in mechanical systems in the absence of friction. Conversely, it is possible for a fixed point to be attracting but not Lyapunov stable; thus neither notion of stability implies the other.

However, in practice, the two types of stability often occur together. If a fixed point is both Lyapunov stable and attracting, we will call it stable or sometimes asymptotically stable.

Figure 9.85(e) is unstable, because it is neither attracting nor Lyapunov stable.

Nonlinear and Chaotic System

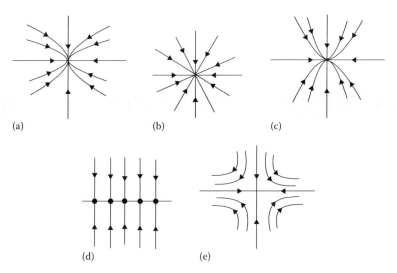

FIGURE 9.85
Phase portraits for different values of "a." (a) $a < -1$. (b) $a = -1$. (c) $-1 < a < 0$. (d) $a = 0$. (e) $a > 0$.

9.10 Second-Order System

Second-order systems are considerably more complicated than first-order system. Although multiple attractor and repeller are still possible oscillation in addition to the damped responses of a first-order system. Furthermore a new form of attractor or repeller is available. This is usually called a limit cycle. A limit cycle is an isolated closed trajectory. Isolated means that neighboring trajectories are not closed; they spiral either toward or away from the limit as shown in Figure 9.86.

In the phase plane, this closed curve that shows continuous oscillation, whereas a fixed point is zero. It should be noted that limit cycle should not be confused with conservative oscillation. A conservative system such as

$$\ddot{x} + x = 0$$

or

$$\ddot{x} + \sin(x) = 0$$

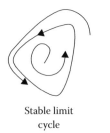

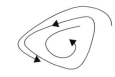

Stable limit cycle   Unstable limit cycle   Half stable limit cycle

FIGURE 9.86
Different types of limit cycles.

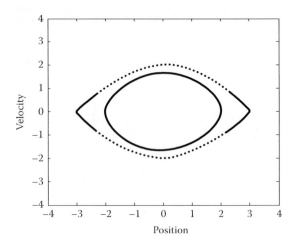

FIGURE 9.87
Simulation of the pendulum using Halijak double integrator, $T = 0.08$, $x0 = 2$ and 3.

will oscillate forever with an amplitude determined by its initial condition. Any slight perturbation will cause it to oscillate with different amplitude. These oscillations are called conservative since there is no damping and energy is conserved. Volumes in phase space are constant. A MATLAB program for simulating the pendulum is given below and results are shown in Figure 9.87.

```
% Matlab program for simulation of the pendulum using Halijak double
% integrator, T = 0.08, x0 = 2 & 3.

clear all;
xss = 2; % or 3
T = 0.08;
xd = 0;
xs = xss + T*xd + 0.5*T*T*(-sin(xss));
plot(xs, (xs-xss)/T,'.k')
hold on
for n = 1:300
    f = -sin(xs);
    x = 2*xs-xss + T*T*f;
    xss = xs;
    xs = x;
    xd = (xs-xss)/T;
    plot(xs,xd,'.k')
end
xlabel('position')
ylabel('velocity')
axis([-4 4 -4 4])
```

Unlike conservative oscillations, a limit cycle attracts trajectories. Any slight perturbation will still be attracted back to the cycle, equivalently, volumes in phase space contract. Good example of a limit cycling system is van der Pol's oscillator shown in Equation 9.25:

$$\ddot{x} + (x^2 - 1)\dot{x} + x = 0 \tag{9.25}$$

Nonlinear and Chaotic System

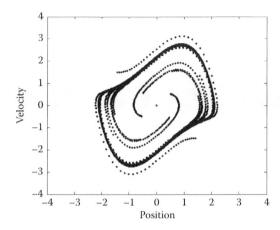

FIGURE 9.88
Simulation of van der Pol equation using Euler, $T = 0.05$, several ICs.

This equation looks like a simple harmonic oscillator, but with a nonlinear damping term $(x^2 - 1)\dot{x}$. Here, this damping is negative when $|x| < 1$, indicating instability. When $|x| > 1$, the damping is positive, indicating stability. In other words, it causes large amplitude oscillations to decay, but it pumps them back up if they become too small. Thus the trajectory is basically stuck in the middle and a continuous limit cycle occurs as shown in Figure 9.88. A MATLAB program for producing the simulation is written below.

```
%Matlab program for simulation of van der Pol equation using Euler, T = 0.05,
% % several initial conditions

Clear all
plot(0, 0,'.r')
hold on
T = 0.05;

for x10 = -1.5:1:1.5
    for x20 = -1.5:1:1.5
        x1 = x10;
        x2 = x20;
for n = 1:200
    f1 = x2;
    f2 = -x1-(x1*x1-1)*x2;
    x1 = x1 + T*f1;
    x2 = x2 + T*f2;

    plot(x1,x2,'.r')
end
end
end

xlabel('position')
ylabel('velocity')
title('simulation of van der Pol equation using Euler, T = 0.05, several ICs')
axis([-4 4 -4 4])
```

When second-order systems are forced by a periodic forcing function, several frequency dependent phenomena are possible. These include subharmonics super-harmonics, jump, and entrainment, among others. Since a time dependent input can simply be considered to be the function of another state of a system, for example $x = 1$, these are also subsets of the next section.

9.11 Third-Order System

Third-order systems display all the above behavior in addition to some interesting feature. It should be noted that as the dimension of the phase space increases the possible dimension of the attractor and repeller also increases. The limiting factor is that the dimension of the attracting and repelling object must be strictly less than the dimension of the phase space. Thus, the first object that logically follows is a two dimensional attractor. A two dimensional object basically looks like a donut and can be thought of as a limit cycle that is being forced at a different frequency. Trajectories are than attracted to or repelled from this object, which is usually referred to as a tours.

An interesting and unexpected phenomenon that can also exist in third and higher order continuous-time system is called chaos. Although difficult to explain briefly, chaos usually is caused by the tangled intersection of several unstable repellers. The resulting trajectory, which basically is being repelled from everywhere, is called a chaotic, or strange attractor, the dimension of this object is usually not integer, for example, dim = 2.6 and is called fractal, at any given cross section, the flow is converging in one direction and diverging in the other, so that volume in phase space is still getting smaller. However, since the flow is diverging in one direction, it is usually folded back over on to itself, and the attractor can resemble taffy being stretched and folded on a machine. Some authors have referred this behavior as sensitive dependence on initial conditions. We preferred to describe chaos as the apparent random motion demonstrated by a completely deterministic system. It should also be remembered that the chaotic behavior just like a limit cycle is a steady-state behavior. That is although, the system is oscillating with no apparent period and the trajectories have converged to their attractor and the steady state chaotic behavior is observed.

A simple system that behaves in this way is the chaotic oscillator of Cook:

$$x''' + x'' + x' - a(x - x^3) = 0 \qquad (9.26)$$

As the parameter a is increased, the system goes through a series or period doubling limit cycle bifurcation and eventually becomes chaotic as shown in Figures 9.89 through 9.92. It is truly remarkable that the interesting behavior of such simple system went unnoticed until relatively recently. The MATLAB program for this simulation is given below:

```
%Matlab program for simulation of Cook system, a = 0.55, ic = [.1.1.1]
using Euler, T = 0.1
clear all;
plot(0, 0,'.r')
hold on
T = 0.1;
a = 0.55;
x1 = 0.1;
x2 = 0.1;
x3 = 0.1;
```

Nonlinear and Chaotic System

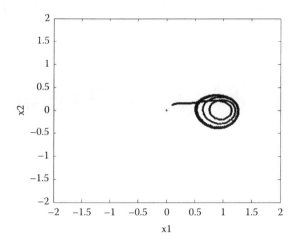

FIGURE 9.89
Simulation of Cook system, $a = 0.55$, ic = [.1.1.1] using Euler, $T = 0.1$.

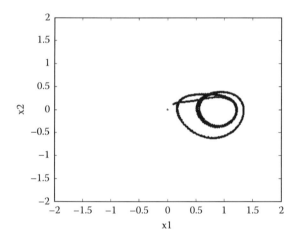

FIGURE 9.90
Simulation of Cook system, $a = 0.7$, ic = [.1.1.1] using Euler, $T = 0.1$.

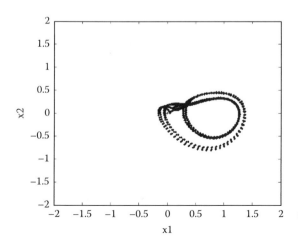

FIGURE 9.91
Simulation of Cook system, $a = 0.85$, ic = [.1.1.1] using Euler, $T = 0.1$.

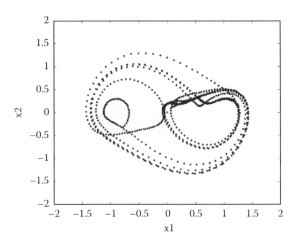

FIGURE 9.92
Simulation of Cook system, $a = 1.4$, ic = [.1.1.1] using Euler, $T = 0.1$.

```
    for n = 1:800
        f1 = x2;
        f2 = x3;
        f3 = -x3-x2 + a*(x1-x1^3);
        x1 = x1 + T*f1;
        x2 = x2 + T*f2;
        x3 = x3 + T*f3;
    plot(x1,x2,'.k')
end

xlabel('x1')
ylabel('x2')
axis([-2 2 -2 2])
```

With respect to chaos, this complicated deterministic phenomenon is sometimes confused with the fact of stochastic or noisy signals. Although it can be analyzed as if it were a noisy signals, chaos is a completely deterministic feature of a nonlinear system and is not known to be related to the unpredictable behavior resulting from physically occurring randomness. Alternatively if we wished to simulate a system being forced by a stochastic signal that is represented in the computer by a random number generator. A continuous–time stochastic signal is usually represented by its probabilistic properties such as mean and variance; although chaotic signals can use these properties is one form of measure, they do not characterize the system. It should be noted in simulating a continuous-time noisy input signal that the sampling of these noise required its variance to be divided by the simulation time step to preserve the same level of excitement in the system as the continuous-time noise. This is not surprising when it is remembered that the frequency spectrum of white noise is identical to that of an impulse.

Nonetheless, *how do you know nonlinear behavior when you see it?* Let's start by looking at an example.

9.11.1 Lorenz Equation: A Chaotic Water Wheel

In 1963 Edward Lorenz derived three dimensional system from a drastically simplified model of convection rolls in the atmosphere. The mechanical model of Lorenz equations

Nonlinear and Chaotic System

(Equation 9.27) was invented by Willem Malkus and Lou Howard at MIT in the 1970s. The simplest version is a toy waterwheel with leaky paper cups suspended from its rim.

$$\dot{x} = \sigma(y - x)$$
$$\dot{y} = rx - y - xz \tag{9.27}$$
$$\dot{z} = xy - bz$$

where $\sigma, r, b > 0$ are parameters.

In this system, the nonlinearities are due to product terms such as xy and xz. Another important property of this equation is symmetry. If (x, y) replaced by $(-x, -y)$ in Equation 9.27, then the equation stays the same.

The numerical solution to these equations at $\sigma = 10$, $r = 8/3$, $b = 28$ and the results are shown in Figures 9.93 through 9.95 for different initial conditions.

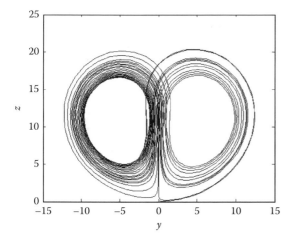

FIGURE 9.93
Simulation of Lorenz equations at $r = 12$.

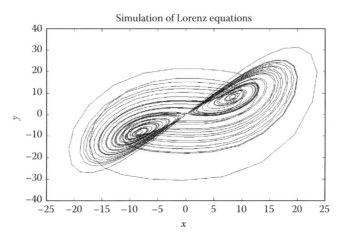

FIGURE 9.94
Response of Lorenz equations for initial conditions $r = 15$.

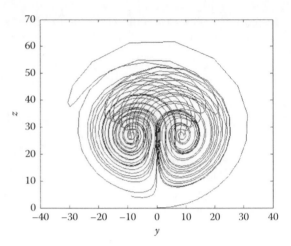

FIGURE 9.95
Response of Lorenz equations for initial conditions $r = 28$.

```
%Matlab Code for Simulation of Lorenz equations
clear all;
T = 0.02;
a = 10;
b = 8/3;
r = 28;
x = 0.1;
y = 0.1;
z = 0.1;
    for n = 1:2000
        dx = a*(y-x);
        dy = (r*x-y-x*z);
        dz = (x*y-b*z);
        x = x + dx*T;
        y = y + dy*T;
        z = z + dz*T;
        var(n,:) = [x,y,z];
end
plot(var(:,2),var(:,3),'-k')
xlabel('y')
ylabel('z')
title('simulation of Lorenz Equations')
```

Consider the Rössler system whose system equations (Equation 9.28) are

$$\begin{aligned} \overset{o}{x} &= -y - z \\ \overset{o}{y} &= x + ay \\ \overset{o}{z} &= b + z(x - c) \end{aligned} \qquad (9.28)$$

where a, b, and c are parameters.

When the system is simulated for different values of c the following results are obtained as shown in Figure 9.96.

Nonlinear and Chaotic System

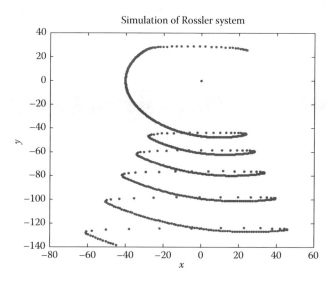

FIGURE 9.96
Results of Rossler system.

9.12 Review Questions

1. What do you mean by nonlinearity of a system?
2. Explain different types of nonlinearities in a system?
3. How could nonlinearities be modeled mathematically?
4. Define chaos?

9.13 Bibliographical Notes

General concepts of chaos theory as well as its history is described in many books (Gutzwiller, 1990; Moon, 1990; Tufillaro et al. 1992; Baker, 1996; Gollub and Baker, 1996; Alligood et al. 1997; Badii and Politi, 1997; Douglas and Euel, 1997; Smith, 1998; Hoover, 1999; Strogatz, 2000; Ott, 2002; Devaney, 2003; Sprott, 2003; Zaslavsky, 2005; Tél and Gruiz, 2006). The mathematics of chaos is explained by Stewart (1990). Currently, chaos theory continues to be a very active area of research, involving many different disciplines (mathematics, topology, physics, population biology, biology (Richard, 2003), meteorology, astrophysics, information theory, economic systems (Hristu-Varsakelis and Kyrtsou, 2008), social sciences (Douglas and Euel, 1997), etc.).

10

Modeling with Artificial Neural Network

> Look deep into nature, and then you will understand everything better.
> Logic will get you from A to B. Imagination will take you everywhere.
>
> **Albert Einstein**
>
> Art is the imposing of pattern on experience, and our aesthetic enjoyment is in the recognition of the pattern.
>
> **Alfred North Whitehead**

10.1 Introduction

Neural networks are very sophisticated modeling techniques capable of modeling extremely complex functions. In particular, neural networks are *nonlinear*. Linear modeling has been the commonly used technique in most modeling domains since linear models have well-known optimization strategies. Where the linear approximation was not valid, the models suffered accordingly. Neural networks also keep in check the *curse of dimensionality* problem that bedevils attempts to model nonlinear functions with large numbers of variables.

Neural networks *learn by example*. The neural network user gathers representative data, and then invokes *training algorithms* to automatically learn the structure of the data. Although the user does need to have some heuristic knowledge of how to select and prepare data, how to select an appropriate neural network, and how to interpret the results, the level of user knowledge needed to successfully apply neural networks is much lower than would be the case using (for example) some more traditional nonlinear statistical methods.

We can explain neural networks as a broad class of models that mimic functioning inside the human brain and are known as biological neurons.

10.1.1 Biological Neuron

The biological neuron is the most important functional unit in human brain and is a class of cells called as NEURON. A human brain consists of approximately 10^{11} computing elements called neurons, which communicate through a connection network of axons and synapses having a density of approximately 10^4 synapses per neuron. A typical neuron has three major regions: cell body, axon, dendrites. Dendrite receives information from neurons through axons—long fibers that serve as transmission lines. The axon dendrite contact organ is called a synapse, as shown in Figure 10.1. The signal reaching a synapse and

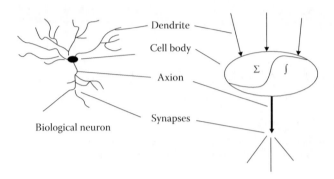

FIGURE 10.1
Biological neuron and its schematic.

received by dendrites are electrical impulses. Apart from the electrical signaling, there are other forms of signaling that arise from neurotransmitter diffusion, which have an effect on electrical signaling. As such, neural networks are extremely complex.

Neuron elements and their functions

Dendrites   Receives information
Cell body   Process information
Axon   Carries information to other neurons
Synapse   Junction between axon end and dendrites of other neuron

10.1.2 Artificial Neuron

When creating a functional model of the biological neuron, there are three basic components of importance. First, the dendrites of the neuron are modeled as weights. The strength of the connection between an input and a neuron is noted by the value of the weight. Negative weight values reflect inhibitory connections, while positive values designate excitatory connections. The next two components model the actual activity within the neuron cell. An adder sums up all the inputs modified by their respective weights. This activity is referred to as a linear combination. Finally, an activation function controls the amplitude of the output of the neuron. An acceptable range of output is usually between 0 and 1, or –1 and 1. The output is passed through axon to give outputs.

Mathematically, this process is described in Figure 10.2.

The artificial neuron receives the information $(X_1, X_2, X_3, X_4, X_5, \ldots, X_p)$ from other neurons or environments. The inputs are fed in connection with weights, where the total input is the weighted sum of inputs from all the sources represented as

$$I = w_1 \times X_1 + w_2 \times X_2 + w_3 \times X_3 + \cdots w_p \times X_p$$

$$= \sum w_i x_i$$

The input I is fed to the transfer function or activation function which converts input I into output Y.

$$Y = f(I)$$

The output Y goes to other neuron or environment for processing.

Modeling with Artificial Neural Network 505

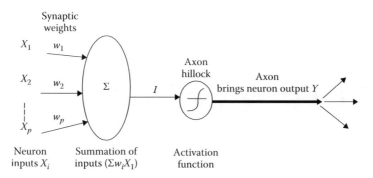

FIGURE 10.2
Artificial neuron.

10.2 Artificial Neural Networks

ANN consists of many simple elements called neurons. The neurons interact with each other using weighted connection similar to biological neurons. Inputs to artificial neural net are multiplied by corresponding weights. All the weighted inputs are then segregated and then subjected to nonlinear filtering to determine the state or active level of the neurons.

Neurons are generally configured in regular and highly interconnected topology in ANN. The networks consist of one or more layers between input and output layers. There is no clear-cut methodology to decide parameters, topologies, and method of training of ANN. Hence, to build the ANN is time consuming and computer intensive. However these can be used in real time because of inherent parallelisms and noise immunity characteristics. The development of ANN is performed in two phases:

10.2.1 Training Phase

In this phase, the ANN tries to memorize the pattern of learning data set. This phase consists of the following modules.

10.2.1.1 Selection of Neuron Characteristics

Neurons can be characterized by two operations: aggregation and activation function. Nonlinear filtering (also called activation function) can be characterized by several functions: sigmoidal and tangent hyperbolic functions being the most common. Selection of the function is problem dependent. For example, the logarithmic activation is used when the upper limit is unbounded while the radial basis function (RBF) is used for problems having complex boundaries.

10.2.1.2 Selection of Topology

Topology of ANN deals with the number of neurons in each layer and their interconnections. Too few hidden neurons hinder the learning process, while too many occasionally degrade the ANN generalization capability. There are no clear-cut or absolute guidelines available in the literature for deciding the topology of ANN. One rough guideline for choosing the number of hidden neurons is the geometric pyramid rule.

10.2.1.3 Error Minimization Process

When ANNs are trained, the weights of the links are changed/adjusted so as to achieve minimum error. During this process, a part of the entire data set is used as training set, and the error is minimized on this set. Calculations for error may be done using various functions. The most commonly used being the root mean square (RMS) function and is given by

$$\text{Error} = (1/2) * \sqrt{\left(\sum [(\text{actual} - \text{predicted})^k]\right)}$$

where k is the order of norm.

10.2.1.4 Selection of Training Pattern and Preprocessing

Selection of the training data is very critical for building ANN. Although preprocessing is not mandatory, it definitely improves the performance of the network and reduces the learning time.

10.2.1.5 Stopping Criteria of Training

The process of adjusting weights of an ANN is repeated until the termination condition is met. The training process may be terminated if any one of these conditions is met:

1. The error goes below a specific value.
2. The magnitude of gradient reaches below a certain value.
3. Specific number of iterations is complete.

10.2.2 Testing Phase

Here ANN tries to predict and test data sets. The performance of ANN on the testing data set represents its generalization ability. In actual practice, the data set is divided into two. One set is used for training and the other for testing. In fact, a generalized neural network will perform well for both training and testing data.

10.2.2.1 ANN Model

An ANN is a powerful data modeling tool that is able to capture and represent complex input–output relationships. The motivation for the development of neural network technology stemmed from the desire to develop an artificial system that could perform "intelligent" tasks similar to those performed by the human brain. Neural networks resemble the human brain in the following two ways:

1. Acquires knowledge through learning.
2. A neural network's knowledge is stored within interneuron connection strengths known as synaptic weights.

The true power and advantage of neural networks lies in their ability to represent both linear and nonlinear relationships and in their ability to learn these relationships directly from the data being modeled. Traditional linear models are simply inadequate when it comes to modeling data that contains nonlinear characteristics.

Modeling with Artificial Neural Network

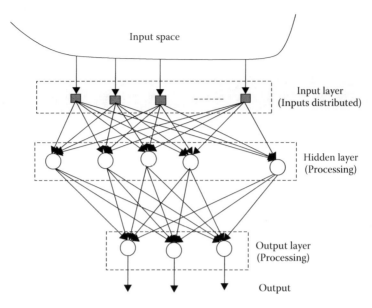

FIGURE 10.3
Artificial neural network.

The ANN consists of three groups, or layers, of units: a layer of "input" units is connected to a layer of "hidden" units, which is connected to a layer of "output" units, as represented in Figure 10.3.

- The activity of the input layer represents the raw information that is fed into the network.
- The activity of each hidden unit is determined by the activities of the input units and the weights on the connections between the input and the hidden units.
- The behavior of the output units depends on the activity of the hidden units and the weights between the hidden and output units.

This simple type of network is interesting because the hidden units are free to construct their own representations of the input. The weights between the input and hidden units determine when each hidden unit is active, and so by modifying these weights, a hidden unit can choose what it represents.

The neural network models are designed considering the following characteristics:

Input: $X_1, X_2, X_3, \ldots$
Output: $Y_1, Y_2, Y_3, \ldots$
ANN model: $f(X_1, X_2, X_3, \ldots, w_1, w_2, w_3, \ldots)$

The model is explained as an example considering the inputs and weight assignment. The weights between the input and hidden units determine when each hidden unit is active, and so by modifying these weights, a hidden unit can choose what it represents.

When the inputs ($X_1 = -1$, $X_2 = 1$, $X_3 = 2$) are given to the neural network, the output (Y) is predicted according to the weights assigned. This is explained with the help of Figure 10.4.

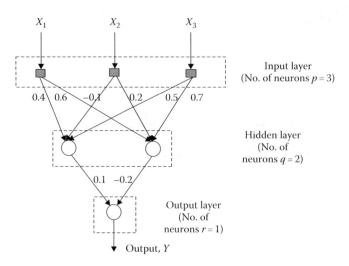

FIGURE 10.4
Typical ANN architecture.

10.2.2.2 Building ANN Model

Step 1: At the input layer, the number of inputs = number of input neurons = 3.

Step 2: At the output layer, the number of outputs = number of input neurons = 1.

Step 3: At the hidden layer, the number of neurons and layers are not fixed and may take any number more than zero to map any complex functions. Let us assume single hidden layer with two neurons.

Step 4: Assign weights.
According to the above architecture, number of weights = $p * q + q * r = 3 * 2 + 2 * 1 = 8$. Assign these weights randomly in the range ±1.0.

Step 5: Decide activation function
Here, logistic function is taken for all neurons.

$$f(x) = \frac{e^x}{1+e^x}$$

Step 6: Select appropriate training pattern, that is, input–output pairs.

	Inputs		Output
X_1	X_2	X_3	Y

Step 7: Training of ANN model
For the above selected training patterns, calculate the ANN output and compare with the desired output to determine the error E (i.e., desired output − actual output). Finally, minimize this error using some optimization technique. The sum-squared error may be written as

$$E = \sum_{i=1}^{n}(Y_i - V_i)^2$$

Modeling with Artificial Neural Network

For the above model, the output was calculated as $V = 0.531$ and let us assume that the desired output is $Y = 1$. The prediction error in the given network is given by $1 - 0.531 = 0.469$.

Hence, we have to update the weights of the network so that the error is minimized. This process is called the training of neural network. In the training of ANN, the error is fed back to the network and updates the weights. This will complete one cycle. The complete cycle is performed many times till the predicted error is reduced. The complete stage is known as an *epoch*.

10.2.2.3 Backpropagation

The adjustment of the weight is done by the method of backpropagation. The procedure of adjustment of the weights by backpropagation is done as follows:

Let V_i be the prediction for ith observation in a network and is a function of the network weights vector $\underline{W} = (W_1, W_2, \ldots)$

Hence, E, the total prediction error is also a function of W.

$$E(\underline{W}) = \sum [Y_i - V_i(\underline{W})]^2$$

10.2.2.3.1 Gradient Descent Method

For every individual weight W_I, the updating formula is given as

$$W_{new} = W_{old} + \alpha * (\partial E / \partial W)\big|_{W_{old}}$$

where α is the learning parameter and is generally taken between 0 and 1.

10.2.2.3.2 Learning Parameter

Too big a learning parameter leads to large leaps in weight space that result in the risk of missing global minima. Too small a learning parameter takes a long time to converge to the global minima and once stuck in the local minima, finds it difficult to get out of it.

In gradient descent method, the following procedure is used:

- Start with a random point (w_1, w_2)
- Move to a "better" point (w'_1, w'_2) where the height of error surface is lower
- Keep moving till you reach (w_1^*, w_2^*), where the error is minimum

10.2.2.4 Training Algorithm

The training algorithm can be explained as having two process of information flow given as back propagation and feed forward.

Decide the network architecture
(Hidden layers, neurons in each hidden layer)
    Decide the learning parameter and momentum
        Initialize the network with random weights
            Do till convergence criterion is not met
                For i = 1 to # training data points

Feed forward the i-th observation through the net
Compute the prediction error on i-th observation
Back propagate the error and adjust weights
   Next i
 Check for Convergence
End Do

How do we know, when to stop the network training?
Ideally, when we reach the global minima of the error surface

Practically:

1. Stop if the decrease in total prediction error (since last cycle) is *small*.
2. Stop if the overall changes in the weights (since last cycle) are *small*.

Partition the training data into **training set** and **validation set.** Training set to build the model, and validation set to test the performance of the model on unseen data.
   Typically as we have more and more training cycles

   Error on training set keeps on decreasing (training)
   Error on validation set keeps first decreases and then increases (over train)

Stop training when the error on validation set starts increasing. The block diagram for ANN model development is shown in Figure 10.5.

10.2.2.5 Applications of Neural Network Modeling

1. Data mining applications
   a. *Prediction*: For prediction application of ANN, past history is taken as input and based on that the future trend may be predicted, as shown in Figure 10.6.
   b. *Classification*: Classify data into different predefined classes based on the inputs, for example, if the color, shape, and dimensions of different fruits are given, we may classify then as lemon (class 1), orange (class 2), banana (class 3), etc., as shown in Figure 10.7.

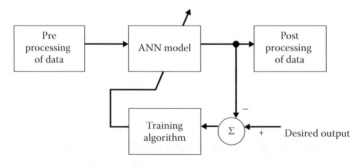

FIGURE 10.5
Block diagram for model development using ANN.

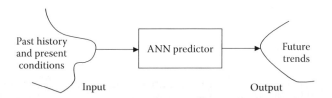

FIGURE 10.6
MISO ANN predictor.

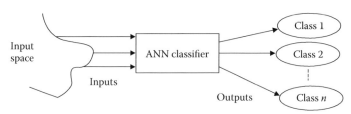

FIGURE 10.7
MIMO ANN classifier.

    c. *Change and deviation detection (interpolation)*: To find certain data that are missing from the records or to detect and determine suspicious records/data. It is quite similar to the ANN predictor. The ANN predictor normally extrapolates while this interpolates the values.

    d. *Knowledge discovery*: The ANN is a very good tool for knowledge discovery. We may apply unforeseen inputs and determine the output for study purpose. This will help in finding new relationships and nonobvious trends in the data. With the help of ANN, we may explore our own ignorance.

    e. *Response modeling*: Train the ANN model to find the systems response.

2. Industrial applications

    a. *Process/plant control*: The ANN is quite suitable for complex and ill-defined dynamical process/plant control applications. It is difficult or sometimes impossible to model these processes/plant using mathematical equations. For such systems, conventional linear control system is difficult or not possible.

    b. *Quality control*: The routine jobs with certain fuzziness like predicting the quality of products and other raw materials, product defect diagnosis, tire testing, load testing, etc. may be done with ANN.

    c. *Parameter estimation*: Unknown or complex system parameter estimation could be done with ANN estimators. This is a very useful tool for complex and ill-defined systems, where mathematical modeling is either not possible or quite difficult. ANN helps in estimating the system parameters for control applications.

3. Financial applications

    a. *Stock market prediction*: Predict the future trend of share market using the historical data and present economical conditions of countries.

    b. *Financial risk analysis*: Decide whether an applicant for a loan is a good or bad credit risk. From present financial conditions and previous performance of a company or an individual, we may find the credit rating of a company or an individual.

c. *Property appraisal*: Evaluate real estate, automobiles, machinery, and other properties. Also forecast prices of raw materials, commodities, and products for future.
4. Medical
   a. *Medical diagnosis*: Assist doctors in diagnosing a particular disease after analyzing the medical test reports a patient's personal information, symptoms, heart rate, blood pressure, temperature, etc.
   b. Prescribe the medicine based on medical diagnosis, personal information such as age and sex, heart rate, blood pressure, temperature, etc.
5. Marketing
   a. *Sales forecasting*: Predict future sales based on historical data, advertisement budget, mode of advertisements, special offers, and other factors affecting sales.
   b. *Profit forecasting*: Forecast the behavior of profit margins in the future to determine the effects of price changes at one level to other. It also helps in deciding distributors' and retailers' commissions.
6. HR management
   a. *Employee selection and hiring*: Predict on which job an applicant will achieve the best job performance based on personal information, previous experience and performance, educational levels, etc. In the hiring process of an employee, it is necessary to determine their retentivity of an employee, which could be predicted using ANN models.
   b. Predict the psychology of a person.
7. Operational analysis
   a. *Inventory control and optimization*: Forecast optimal stock level that can meet customer needs, reduce waste, and lessen storage; predict the demand based on previous buyers' activity, operating parameters, season, stock, and budgets.
   b. *Smart scheduling*: Smart scheduling of buses, airplanes, and trains based on seasons, special events, weather, etc.
   c. *Decision making in conflicting situation*: Select the best decision option using the classification capabilities of neural network.
8. Energy

   Electrical load forecasting, short- and long-term load shading, predicting natural resources and their consumption, monitoring and control applications, etc.
9. Science

   In the area of science, ANN may be used on a number of problems such as physical system modeling, image recognition, identification and optimization, ecosystem evaluation, classification, signal processing, systems analysis, etc.

Example 10.1: Artificial Neural Network Applications in Physical System's Modeling

Predict the speed for a vehicle at a given accelerating force using ANN (Figure 10.8).
  Let us define
  F = force generated by the car = $80t$ N assumed.
  V = velocity (m/s)
  M = mass of the vehicle = 1000 kg
  b = frictional coefficient = 40 N-s/m

Modeling with Artificial Neural Network

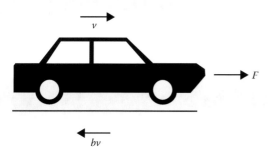

FIGURE 10.8
Automobile modeling for predicting speed.

Applying Newton third law

$$F = M\frac{dv}{dt} + bv$$

$$80t = 1000\frac{dv}{dt} + 40v$$

Taking Laplace transform

$$\frac{80}{s^2} = 1000V(s) + 40V(s)$$

$$V(s) = \frac{0.08}{s^2(s+0.04)}$$

Applying partial fraction and after calculating the residues we have

$$V(s) = \frac{2}{s^2} - \frac{50}{s} + \frac{50}{s+0.04}$$

Taking inverse Laplace transform we obtain

$$v(t) = 2t - 50 + 50e^{-0.04t}$$

Using neural network we can train the velocity equation and obtain the response; we can also predict the future value (Figures 10.9 through 10.11).

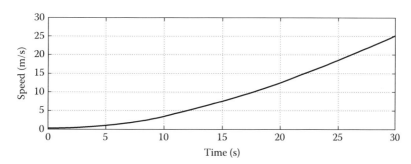

FIGURE 10.9
Speed profile of an automobile.

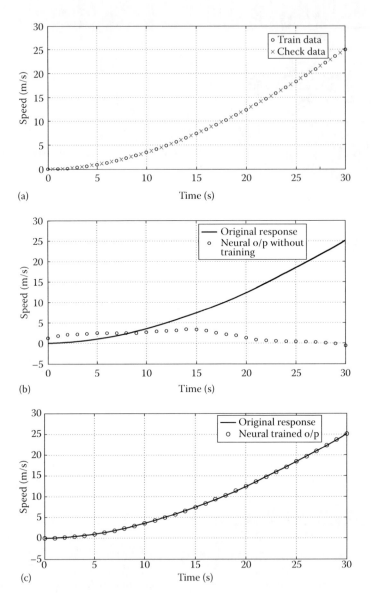

FIGURE 10.10
Comparison of ANN response and system response.

```
% MATLAB CODE FOR PREDICTION OF SPEED FOR A VEHICLE AT AN
% ACCELERATING FORCE:

clear;
% Define and plot time vs velocity for the model
t=[0:.5:30]';
v=2*t-50+50*exp(-0.04*t);
grid
figure(1);
plot(t,v);
xlabel('time in seconds');ylabel('speed in m/s');
```

Modeling with Artificial Neural Network

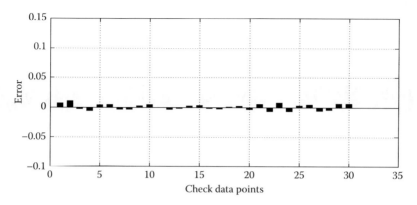

FIGURE 10.11
Error during testing.

```
%Define and plot training as well as checking data
data=[t v];
trndata=data(1:2:size(t),:);
chkdata=data(2:2:size(t),:);
grid
figure(2);
plot(trndata(:,1),trndata(:,2),'o',chkdata(:,1),chkdata(:,2),'x');
xlabel('time in seconds');ylabel('speed in m/s');

%initialize ANN
trainpoint=trndata(:,1)';
trainoutput=trndata(:,2)';
net=newff(minmax(trainpoint),[10 1],{'tansig' 'purelin'});
%Simulate and plot the network ouput without training
Y=sim(net,trainpoint);
grid
figure(3);
plot(trainpoint,trainoutput,trainpoint,Y,'o');

%Initialize parameter
net.trainParam.epochs=100;
net.trainParam.goal=0.0001;

%Train the network and plot the output
net=train(net,trainpoint,trainoutput);
Y=Sim(net,trainpoint);
figure(4);
plot(trainpoint,trainoutput,trainpoint,Y,'o');

%test the network and plot the error
checkpoint=chkdata(:,1)';
checkoutput=chkdata(:,2)';
W=Sim(net,checkpoint);
e=W-checkoutput;
grid
figure(5);
plot(e);
```

Example 10.2: Weighing Machine Model (Figure 10.12)

Modeling the given diagram, we have taken following parameters
Mass (M) = 1 kg
Damping coefficient = 20 N-s/m
Spring constant = 10 N/m
Force = F
Displacement = x m
Modeling the system

$$F = M\frac{d^2x}{dt^2} + B\frac{dx}{dt} + kx;$$

$$F = 1.\frac{d^2x}{dt^2} + 20\frac{dx}{dt} + 10x;$$

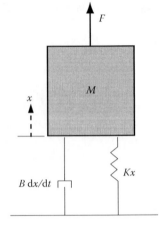

FIGURE 10.12
Schematic diagram of weighing machine.

Taking the Laplace transform

$$F(s) = s^2X(s) + 20sX(s) + 10X(s)$$

$$X(s) = \frac{F(s)}{s^2 + 20s + 10}$$

In the given model, the random force (F) was applied and given to the neural network for training the system with input "force" and output "displacement." Using the trained neural network, the output displacement (x) is predicted using the check data. In the previous example, we have given constant input force but in this example the input is a random force as shown in Figure 10.13.

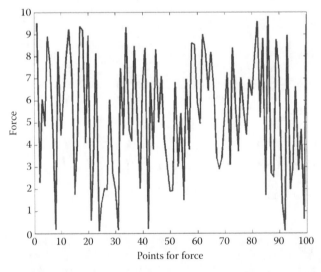

FIGURE 10.13
Random force as system input.

```
%MATLAB CODE FOR PREDICTING DISPLACEMENT USING RANDOM INPUT FORCE

clear all;
% created random force applied on system
forceupper=10;
forcelower=0;
force=(forceupper-forcelower)*rand(100,1)+forcelower;
figure(1);
clf;
%plot for random force
plot(force,'-');
%determination of displacement
for i=1:100
num=[force(i)];
den=[1 20 10];
displace(i)=max(step(num,den));
end
%plot for force vs displacement
figure(2);
clf;
plot(force,displace,'*');
displace=displace';
data=horzcat(force, displace);
%Initialize data for ANN
trndata=data(1:2:100,:);
chkdata=data(2:2:100,:);
forcepoints=trndata(:,1)';
disppoints=trndata(:,2)';
net=newff(minmax(forcepoints),[5 1],{'tansig' 'purelin'});
net.trainParam.epochs=50;
net.trainParam.goal=.00001;
net=init(net);
%Train the system
net=train(net,forcepoints,disppoints);
Y=sim(net,forcepoints);
figure(3);
clf;
%plot the real output and trained output
plot(forcepoints,disppoints,'-',forcepoints,Y,'o');
%testing of ANN with checkdata
checkpoint=chkdata(:,1)';
checkoutput=chkdata(:,2)';
W=Sim(net,checkpoint);
e=W-checkoutput;
grid
figure(4);
bar(e);
```

The results of the program are shown in Figures 10.14 through 10.18.

Example 10.3: Approximation and Prediction for Teacher Evaluation System Using Neural Network

In the current education system, to improve education quality, teacher evaluation is must. The teacher can be evaluated on the basis of student feedback, results, peer feedback, and educational activities as shown in Figure 10.19. The evaluation is based on fixed mathematical formula and hence there are many problems.

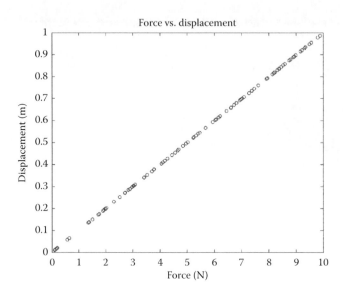

FIGURE 10.14
Results of displacement vs. force.

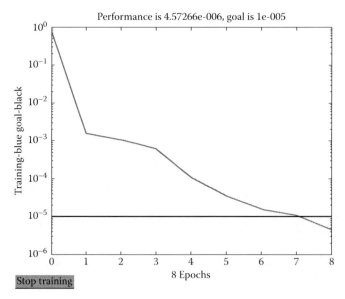

FIGURE 10.15
Error reduction during training.

- Each expert has his/her own intuition for judgment; often it is not taken into account.
- Different weights are assigned to different parameters by each expert, but it is not easy due to subjectivity in weights.
- Mathematical computation is not fast.
- Many experts get biased with candidate teacher and it influences the results.

Due to the above difficulties, the conventional methods are not so effective for this problem. The soft computing techniques such as the ANN model may be used to overcome some of the difficulties mentioned above. The block diagram for ANN approach is shown in Figure 10.20.

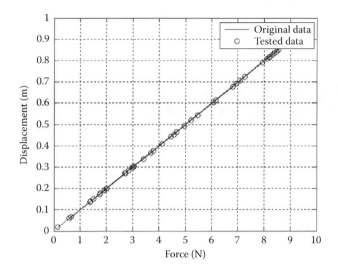

FIGURE 10.16
Data fitting during training.

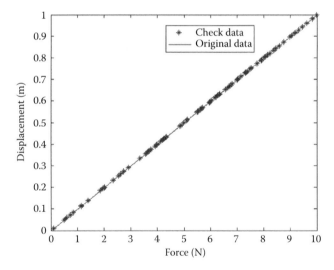

FIGURE 10.17
Data fitting during testing.

For this example, the system model has a teacher evaluation database, which has three experts who have judged the teachers on the basis of their feelings from the mind. The average of the experts' judgments were taken as the overall judgment.

The judgment of the teacher is done on a 10-point scale. The evaluated data as shown in Table 10.1 is given to the neural network for training and testing. MATLAB® code is written for simulating the ANN model with the following data set. The ANN performance during training and testing is shown in Figures 10.21 and 10.22, respectively.

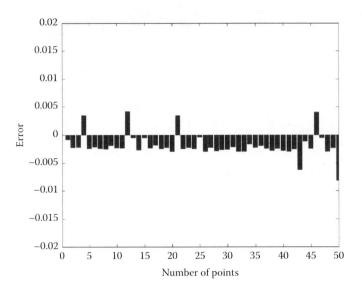

FIGURE 10.18
Error during prediction.

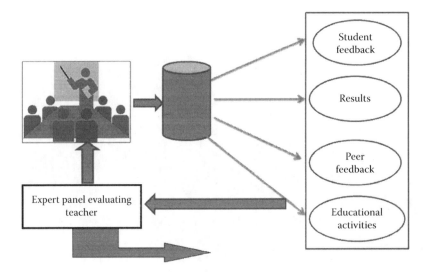

FIGURE 10.19
Teacher evaluation system.

```
% Matalb Codes of ANN model for Teachers Evaluation system

Trndata=[Student feedback, Examination results, Peer feedback, Educational
activities]
Target = [Expert overall results]

% Matlab code for approximation and prediction of
% teacher evaluating system

%Initialization of training and target data
```

```
trndata;
target;
x=1:1:58;
%Initialization of neural network
net=newff (minmax (trndata),[20 1],{'logsig' 'purelin'});
net.trainParam.epochs=500;
net.trainParam.goal=.0001;
net=init(net);
%Train the system
net=train(net,trndata,target);
Y=sim(net,trndata);
%Plot for the output
figure(1);
plot(x,target,'-',x,Y,'o');
```

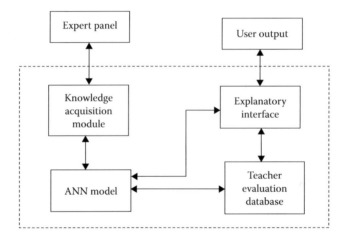

FIGURE 10.20
Block diagram for teacher's evaluation system.

TABLE 10.1

Training Data for ANN Model

Student Results	Examination Result	Peer Feedback	Educational Activities	Expert 1	Expert 2	Expert 3	Total
8.4	10	7.53	8.2	8.7	9	8.5	8.733333
8.4	5.8	7.85	8.4	7.9	8	7.8	7.9
8.84	5.8	7	9.2	8	7.8	8.1	7.966667
7.8	10	7.4	9.2	8.8	9.5	9.1	9.133333
8.33	10	9.5	9.6	8.8	10	9.2	9.333333
8.66	10	9	9.5	8.9	10	9.2	9.366667
9	10	9	9	9.1	10	9.2	9.433333
8.66	10	8.35	7.5	9	9	8.5	8.833333
8.66	10	7	7	8.8	9	8.2	8.666667
7.33	10	6.5	6.5	8.6	8	7.2	7.933333
9	7.33	8.5	9.6	7.9	9	8.5	8.466667
9.33	7	7	6.5	7.9	7	8.2	7.7
7.33	7.33	7	6	7.7	7	7.2	7.3

(*continued*)

TABLE 10.1 (continued)
Training Data for ANN Model

Student Results	Examination Result	Peer Feedback	Educational Activities	Expert 1	Expert 2	Expert 3	Total
8	7.66	7	7.5	7.8	7.5	7.9	7.733333
8.66	9.33	8	6.5	8.4	8.2	7.5	8.033333
8.66	9.33	8	8.5	8.6	8.5	8.5	8.533333
7.33	9.33	7	6	8.9	8	7.1	8
8.33	7.66	5.5	8	7.7	6.5	8.1	7.433333
8.66	7.66	5.5	6.5	7.6	6.5	8.2	7.433333
7.66	9	6.5	8	8.2	7.5	7.9	7.866667
7.66	9.33	6	8	8.3	7.5	7.9	7.9
8.33	9.33	6	5	8.4	8	7.2	7.866667
9.38	2	8	9	6	6	9	7
7.38	2	7.2	7.2	6	5.5	7.6	6.366667
6.07	2	6.4	6.6	6	5	6.9	5.966667
9.18	4	7	7	6.5	6	7.7	6.733333
8.07	2	7	6.6	6	6.5	7.5	6.666667
7.9	2	6.2	6.4	6	6.3	7.2	6.5
5.2	2	6	5.4	6	6	5.1	5.7
8.2	2	6	6.4	6	6	7.5	6.5
6.78	2	5.4	6	6	6	6.5	6.166667
9.2	2	6	7.2	6	7	8.2	7.066667
8.87	2	6.4	6.8	6	6.8	8.4	7.066667
7.16	2	5.4	5.4	6	6	7.9	6.633333
8	4.9	8.65	8.2	6.6	6	8.1	6.9
8.07	4.95	8.6	8.4	6.7	7	8.5	7.4
8.28	2.58	8.45	8.6	6.1	8	6.2	6.766667
8.3	6.44	7.35	6.4	6.9	7	7.5	7.133333
8.13	3.32	7.6	7	6.2	6	7	6.4
9.03	4.79	8.45	8.6	6.6	8	8.2	7.6
7.08	4.06	6.35	6.2	6.1	7	7.5	6.866667
7.96	4.16	6	5.2	6.1	6.5	7.5	6.7
7.5	3.55	7.35	6.5	5.9	6.5	7.4	6.6
8.15	6.16	6.8	6.6	7.2	6.8	7.2	7.066667
6.8	3.6	7.35	6.6	6.1	6	6.5	6.2
8.03	5.54	6.8	6.6	6	6.5	7.5	6.666667
9.4	6	9.55	7.8	7	8	8.2	7.733333
7.37	2	8.51	8.6	5.4	8.5	7.9	7.266667
7.87	5	9.11	7.8	6.5	7.4	7.8	7.233333
6.86	3	6.68	6.2	5.9	6	6.5	6.133333
9.08	4	6.96	8.2	6.2	7.5	9.2	7.633333
8.3	6	7.08	7.4	7.1	7	7.2	7.1
6.8	2	6.51	7	5.4	7	7.2	6.533333
7.49	4	6	7.6	6.2	6.5	7.6	6.766667
7.7	4	8.21	8.4	6.3	7.5	8.5	7.433333
8.03	2	5.86	8.2	5.6	7.8	8.2	7.2
4.23	4	2.81	2.6	5.7	4.2	4.5	4.8
3	3	3.63	2.8	4.6	4	3.5	4.033333

Modeling with Artificial Neural Network

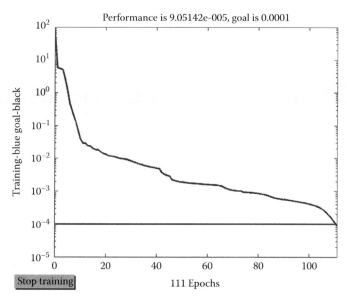

FIGURE 10.21
Training error of ANN for teacher's evaluation system.

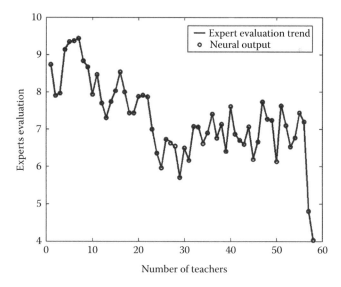

FIGURE 10.22
Comparison of results.

Example 10.4: Rainfall Prediction Using Neural Network

Rainfall forecasting has been one of the most challenging problems around the world for more than half a century. The traditional rainfall forecasters are generally based on available information to establish the mathematical models. Thus, it involves the question of pattern recognition. We introduce the situation of ANN for rainfall forecasting. To develop the ANN model, we have used the data of rainfall comprising of 194 years. Neural network technology can be applied to such a nonlinear system. Neural network has a built-in capability to adapt their synaptic weights to changes in the surrounding environment. In particular, ANN is trained to operate in a specific situation and it can easily retain the past experiences. ANN completes the whole network of information processing through the interaction between the neurons. It has self-learning and adaptive capabilities, which motivated us to use in rainfall forecasting. The historical data of rainfall for 194 years are graphically shown in Figure 10.23.

The rainfall data shown above is random in nature. Hence, it is difficult to develop a mathematical model for it except time series prediction. The time series prediction method has its own drawbacks. Here, we have used neural network as nonlinear predictor. The MATLAB code is written to simulate the problem. The training, validation, and testing performance of ANN rainfall predictor is shown in Figure 10.24.

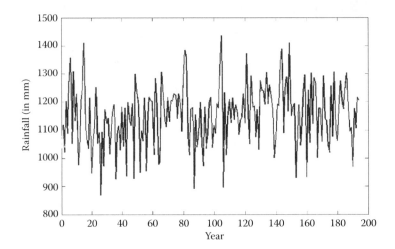

FIGURE 10.23
Historical data plot for rainfall.

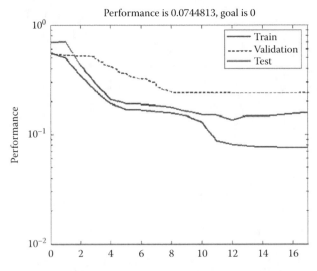

FIGURE 10.24
ANN performance during rainfall prediction.

Modeling with Artificial Neural Network

```
% Load data for training
p=year;
t=rainfall;
figure(1);
% plot the data
plot(year,rainfall);
%Normalize the data
[p2,ps] = mapminmax(p);
[t2,ts] = mapminmax(t);
%divide data for training,validation and testing
[trainV,val,test] = dividevec(p2,t2,0.2,0.15);
%Initialize network
net = newff(minmax(p2),[40 1]);
init(net);
%train the network
[net,tr]=train(net,trainV.P,trainV.T,[],[],val,test);
%simulate the network
a2 = sim(net,p2);
% reverse normalize the data for original plot
a= mapminmax('reverse',a2,ts);
[m,b,r] = postreg(a,t);
```

The best fit is represented by the red line; we can simulate the network using *sim* command for the future prediction (Figure 10.25).

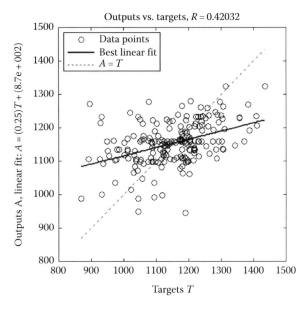

FIGURE 10.25
Best fit curve for rainfall problem.

10.3 Review Questions

1. Explain the working of biological neuron.
2. Explain different parts of artificial neuron and mention how it replicates the biological neuron.
3. What is artificial neural network?
4. How does ANN model be developed?
5. Explain different types of training used in ANN.
6. Develop a feed forward ANN model for character recognition.
7. What is gradient descent learning algorithm for ANN models?
8. Develop ANN model for electrical load forecasting problem. (Use `randn` function of MATLAB to generate data.)

11

Modeling Using Fuzzy Systems

So far as the laws of mathematics refer to reality, they are not certain and so far as the laws they are certain, they do not refer to reality.

Albert Einstein

If you do not have a good plant model or if the system is changing, then fuzzy will produce a better solution than conventional control technique.

Bob Varley

Good judgment comes from experience; experience comes from bad judgment.

Frederick Brooks

11.1 Introduction

A paradigm shift from classical thinking to fuzzy thinking was initiated by the concept of fuzzy sets and the idea of mathematics based on fuzzy sets. It emerged from the need to bridge the gap between mathematical models and their interpretations. This gap has become increasingly disturbing, especially in the areas of biology, cognitive sciences, and social and applied sciences, such as modern technology and medicine. Hence, there is a need to bridge this gap between a mathematical model and the experience, as expressed by Zadeh (1962):

> ...There is a fairly wide gap between what might be regarded as 'animate' system theorists and 'inanimate' system theorists at the present time, and it is not all certain that this gap will be narrowed, much less closed, in the near future. There are some who feel this gap reflects the fundamental inadequate of the conventional mathematics—the mathematics of precisely-defined points, functions, sets, probability measures, etc.—for coping with the analysis of biological systems, and that to deal effectively with such systems, which are generally orders of magnitude more complex than man-made systems, we need a radically different kind of mathematics, the mathematics of fuzzy or cloudy quantities which are not describable in terms of probability distribution. Indeed, the need for such mathematics is become increasingly apparent even in the realm of inanimate systems, for in most practical cases the a priori data as well as the criteria by which the performance of a man-made system is judged are far from being precisely specified or having accurately known probability distributions.

Prof. Lotfi A. Zadeh, University of California, Berkley, first coined the term "fuzzy" in 1962. This was the time for a paradigm shift from the crisp set to the fuzzy set. The fuzzy logic provides a tool for modeling human-centered systems. Fuzziness seems to pervade most human perceptions and thinking processes. One of the most important facets of human thinking is the ability to summarize information "into labels of fuzzy sets which bear an approximate relation to primary data." Linguistic descriptions, which are usually summary descriptions of complex situations, are fuzzy in essence. Fuzziness occurs when there are no sharp boundaries.

In fuzzy logic, the transition is gradual and not abrupt, from member to nonmember. It can deal with uncertainty in terms of imprecision, nonspecificity, vagueness, inconsistency, etc. Uncertainty is undesirable in science and should be avoided by all possible means; science strives for certainty in all manifestations (precision, specificity, sharpness, and consistency).

An alternative view, which is tolerant of uncertainty, insists that science cannot avoid it. Warren Weaver (1948) mentioned the problems of organized simplicity and disorganized complexity (randomness). He called some of the problems lying in these categories and most of the problems lying in between, organized complexity (nonlinear systems with large numbers of components and rich interactions), which may be nondeterministic but not as a result of randomness.

Fuzzy sets were specifically designed to mathematically represent uncertainty and vagueness, and to provide formalized tools for dealing with imprecision intrinsic to many problems. However, the detailed history of fuzzy logic was given in Chaturvedi (2008).

The goal of this chapter is to demonstrate how fuzzy logic can be used for system modeling. To demonstrate fuzzy logic applications to model real-world systems, MATLAB® with the fuzzy logic toolbox is used.

11.2 Fuzzy Sets

A fuzzy set is any set that allows its member to have different grades of the membership function in the interval of [0, 1]. It is a generalized subset of a classical set.

Degree of Truth

To what degree is person x tall?

To answer this question, it is necessary to assign a degree of membership to each person in the universe of discourse, in the fuzzy subset "tall." The easiest way to do this is to define the membership function based on a person's height.

Person	Height	Degree of Tallness
Raman	3'2"	0.0
Madan	5'5"	0.31
John	5'9"	0.67
Hari	6'1"	0.76
Kareem	7'2"	1.00

$$\text{Membership value} = \begin{cases} 0 & \text{If height }(x) < 5' \\ (\text{height}(x)-5')/2' & \text{If height }(x) \leq 7' \\ 1 & \text{If height }(x) > 7' \end{cases}$$

Modeling Using Fuzzy Systems

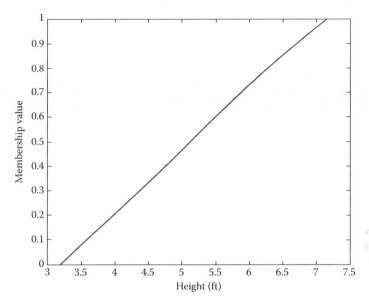

FIGURE 11.1
Fuzzy membership function for tall.

Each fuzzy set is completely and uniquely defined by one particular membership function, for example, "tall" is defined by a fuzzy set shown in Figure 11.1. A fuzzy set represents vague concepts and their contexts, for example, the concept of high temperature in context of weather may be between 25°C and 50°C. Again, in different seasons, high temperature may be different. The concept of high temperature in context of a nuclear reactor may be between 100°C and 300°C. The usefulness of fuzzy sets depends critically on our ability to construct appropriate membership functions for various concepts in various contexts. The different shapes generally used for fuzzy membership functions are shown in Figure 11.2 and mathematically expressed as in Table 11.1. In some situations, it is not possible to identify membership functions with an exact degree. Then, it is necessary to choose reasonable values between lower and upper bounds. Such fuzzy sets are called internal-valued fuzzy sets, as shown in Figure 11.3.

FIGURE 11.2
Different shapes for fuzzy membership functions. (a) Piecewise linear functions and (b) nonlinear functions.

TABLE 11.1
Different Membership Functions

Triangular membership function	$trimf(x;a,b,c) = \max\left(\min\left(\dfrac{x-a}{b-a}, \dfrac{c-x}{c-b}\right), 0\right)$
Trapezoidal membership function	$trapmf(x;a,b,c,d) = \max\left(\min\left(\dfrac{x-a}{b-a}, 1, \dfrac{d-x}{d-c}\right), 0\right)$
Gaussian membership function	$gaussmf(x;a,b,c) = e^{-\frac{1}{2}\left(\frac{x-c}{\sigma}\right)^2}$
Bell-shaped membership function	$gbellmf(x;a,b,c) = \dfrac{1}{1+\left|\dfrac{x-c}{b}\right|^{2b}}$
Sigmoidal membership function	$sigmf(x;a,b,c) = \dfrac{1}{1+e^{-a(x-c)}}$

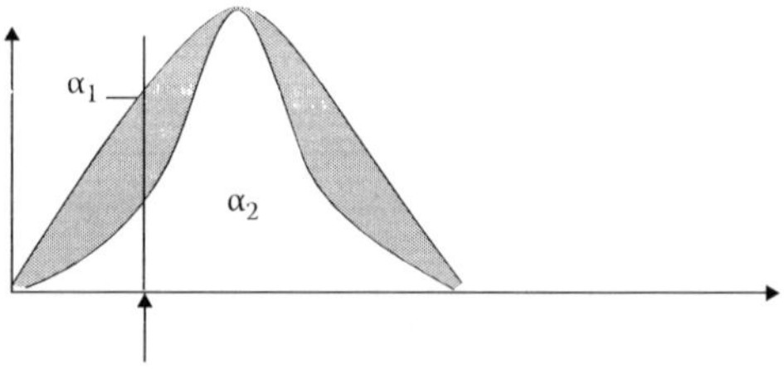

FIGURE 11.3
Internal-valued fuzzy sets.

Let X be a collection of objects or a universe of discourse. Then a fuzzy set, A, in X is a set of ordered pairs, $A = \{\mu_A(x)/x\}$, where $\mu_A(x)$ is the characteristic function (or the membership function) of x in A. If the membership function of A is discrete, then A is written as in Equation 11.1:

$$A = [\mu_1(x)/x_1 + \mu_2(x)/x_2 + \mu_3(x)/x_3 + \cdots + \mu_n(x)/x_n]$$
$$= \sum [\mu_i(x)/x_i] \qquad (11.1)$$

where

"+" or "$\sum$" sign denotes union
$\mu_i(x)$ denotes the membership value/grade
x_i denotes the variable value

When the membership function is continuous, then the fuzzy set, A, is written as

$$A = \int \mu_A(x)/x$$

where the integral denotes the fuzzy singletons.
There are three basic methods by which a set can be defined.

1. List of elements
   A set may be defined by all the elements along with their membership values or membership grades associated with it, as mentioned in Equation 11.2:

$$A = [\mu_1(x)/x_1 + \mu_2(x)/x_2 + \mu_3(x)/x_3 + \cdots + \mu_n(x)/x_n] \quad \text{or}$$
$$= [\mu_1(x)/x_1, \mu_2(x)/x_2, \mu_3(x)/x_3, \ldots, \mu_n(x)/x_n] \qquad (11.2)$$

Modeling Using Fuzzy Systems

2. **Rule method**
   A set may be represented by some rule that all the elements follow. A is defined by the following notation as the set of all elements of x for which the proposition, $P(x)$, is true:

   $$A = \{x | p(x)\}$$

   where
   | denotes the phrase "such that"
   $p(x)$ denotes that x has the property p

3. **Characteristic (membership) function**
   A set may also be defined by a characteristic (membership) function that will give the membership value of each element:

   $$\mu_A(x) = \begin{cases} f(x) & \text{for } x \in A \\ 0 & \text{for } x \notin A \end{cases}$$

   where $f(x)$ could be any function like $f(x) = 0.5x + 1$ or $f(x) = 1.5 * e^{-1.2x}$.

11.3 Features of Fuzzy Sets

1. Fuzzy logic provides a systematic basis for quantifying uncertainty due to vagueness and incompleteness of the information.
2. Classes with no sharp boundaries can be easily modeled using fuzzy sets.
3. Fuzzy reasoning is an informalism that allows the use of expert knowledge and is able to process this expertise in a structured and consistent way.
4. There is no broad assumption of a complete independence of the evidence to be combined using fuzzy logic, as required for other subjective probabilistic approaches.
5. When the information is inadequate to support a random definition, the use of probabilistic methods may be difficult. In such cases, the use of fuzzy sets is promising.

Subset

If every member of set A is also a member of set B (i.e., $x \, \varepsilon \, A$ implies $x \, \varepsilon \, B$), then A is called a subset of B and is written as $A \subseteq B$. A subset has the following properties:

1. Every set is a subset of itself and every set is a subset of the universal set.
2. If $A \subseteq B$ and $B \subseteq A$ then $A = B$ (equal set).
3. If $A \subseteq B$ and $B \neq A$ then $A \subset B$ (A is a proper subset of B).

Power set

The family of all subsets of a given set, A, is called the power set of A and is denoted by $P(A)$. The family of all subsets of $P(A)$ is called the second-order power set of A, denoted by $P^2(A)$.

$$P^2(A) = P(P(A))$$

11.4 Operations on Fuzzy Sets

11.4.1 Fuzzy Intersection

Fuzzy intersection of two sets, A and B, is interpreted as "A and B", which takes the minimum value of two membership functions, as given in Equation 11.3:

$$\mu_{A \cap B}(x) = \sum \{\mu_A(x) \wedge \mu_B(x)\}$$
$$= \min(\mu_A(x), \mu_B(x)) \quad \forall x \in X \tag{11.3}$$

11.4.2 Fuzzy Union

Fuzzy union is interpreted as "A or B," which takes the maximum value of two membership functions, as given in Equation 11.4:

$$\mu_{A \vee B}(x) = \sum \mu_A(x) \vee \mu_B(x)$$
$$= \max(\mu_A(x), \mu_B(x)) \quad \forall x \in X \tag{11.4}$$

A justification of the choice of max and min operators for fuzzy union and fuzzy intersection was given by Bellman and Giertz (1973). These are the only functions that meet the following requirements:

1. The results of these operators depend on the element membership values.
2. These are commutative, associative, and mutually distributive operators.
3. These are continuous and nondecreasing with respect to their arguments.
4. The boundary conditions imply as operator(1, 1) = 1 and operator(0, 0) = 0.

The above requirements are true for all union and intersection operators suggested by different researchers.

11.4.3 Fuzzy Complement

The complement of a fuzzy set, A, which is understood as "$NOT(A)$," is defined by Equation 11.5:

$$\mu_{\bar{A}}(x) = \sum (1 - \mu_A(x)) \quad \forall x \in X \tag{11.5}$$

where $\bar{A}$ stands for the complement of A.

The relative complement of set A with respect to set B is the set containing all members of B that are not members of A, denoted by B − A:

$$(B - A) = \{x | x \in B \text{ and } x \notin A\}$$

If set B is a universal set, then A's complement is absolute. An absolute complement is always involutive. The absolute complement of an empty set is always equal to a universal set and vice versa:

$$\bar{\phi} = X \text{ and } \bar{X} = \phi$$

11.4.4 Fuzzy Concentration

Concentration of fuzzy sets produces a reduction in the membership value, $\mu_i(x)$, by taking a power more than 1 of the membership value of that fuzzy set. If a fuzzy set, A, is written as in Equation 11.6:

$$A = \{\mu_1/x_1 + \mu_2/x_2 + \cdots + \mu_n/x_n\} \tag{11.6}$$

then a fuzzy concentrator applied to the fuzzy set, A, is defined as given in Equation 11.7 and graphically shown in Figure 11.4:

$$\text{CON}(A) = A^m = \{\mu_1^m/x_1 + \mu_2^m/x_2 + \cdots + \mu_n^m/x_n\} \tag{11.7}$$

where CON(A) represents the concentrator applied to A:

$$m > 1$$

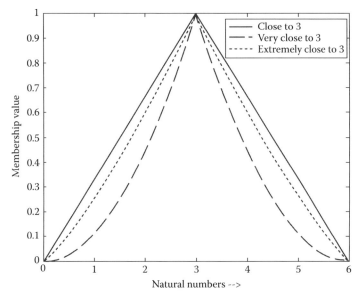

FIGURE 11.4
Strong fuzzy modifier.

```
%Matlab Program for Fuzzy concentrator
% Close_to_3=A
A =[0:.1:1, .9:-0.1:0];
x=[0:6/20:6];
Very_Close_to_3= A.^2;
Extermey_Close_to_3 = A.^1.25;
plot(x,A)
hold
plot(x,Very_Close_to_3, '--')
plot(x,Extermey_Close_to_3, ':')
legend('Close to 3', 'Very close to 3', 'Extremely close to 3')
xlabel('Natural Numbers -->')
ylabel('Membership value')
```

11.4.5 Fuzzy Dilation

Fuzzy dilation is an operation that increases the degree of belief in each object of a fuzzy set by taking a power less than 1 of the membership value, as given in Equation 11.8 and graphically shown in Figure 11.5. Dilation has an opposite effect to that of concentration:

$$\text{DIL}(A) = A^m = \{\mu_1^m/x_1 + \mu_2^m/x_2 + \cdots + \mu_n^m/x_n\} \tag{11.8}$$

where $m < 1$.

The fuzzy operations of *'plus and minus'* applied to a fuzzy set give intermediate effects of CON(A) and DIL(A) for which the values of m are 1.25 and 0.75, respectively. If a set, A, is fuzzy, then the plus and minus operations may be defined as

Plus(A) = $A^{1.25}$ and Minus(A) = $A^{0.75}$

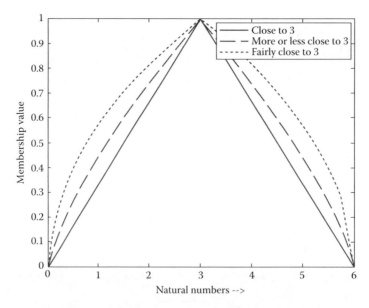

FIGURE 11.5
Weak fuzzy modifier.

Modeling Using Fuzzy Systems

```
%Matlab Program for Weak Fuzzy Modifiers
clear all
% Close_to_3=A
A=[0:.1:1, .9:-0.1:0];
x=[0:6/20:6];
More_or_less_Close_to_3=A.^0.75;
fairly_Close_to_3=A.^0.5;
plot(x,A)
hold
plot(x,More_or_less_Close_to_3, '--')
plot(x,fairly_Close_to_3, ':')
legend('Close to 3', 'More or Less close to 3', 'fairly close to 3')
xlabel('Natural Numbers -->')
ylabel('Membership value')
```

11.4.6 Fuzzy Intensification

Intensification is an operation that increases the membership function of a set above the maximum fuzzy point (at $\alpha = 0.5$) and decreases it below the maximum fuzzy point.

If A is a fuzzy set, $x \in A$, then the intensification applied to A is defined as given in Equation 11.9:

$$I(A) = \begin{cases} \mu_A^{n_1} & n_1 \geq 1 \quad \text{if } 0 \leq \mu_A \leq 0.5 \\ \mu_A^{n_2} & n_1 \leq 1 \quad \text{if } 0.5 \leq \mu_A \leq 1.0 \end{cases} \tag{11.9}$$

$$\text{For example intensified close to 3} = \begin{matrix} 2\mu_{A^2} & \text{If } 0 \leq \mu_A \leq 0.5 \\ 1-2(1-\mu_A)^2 & \text{If } 0.5 \leq \mu_A \leq 1.0 \end{matrix}$$

```
%Matlab Program for intensified fuzzy modifiers
clear all
% Close_to_3=A
A=[0:.1:1, .9:-0.1:0];
x=[0:6/20:6];
[l n] = size(A);
for i=1:n
        if A(i)<=0.5
            A1(i) =2*A(i).^2
  else
     A1(i) =1-2*(1-A(i)).^2
     end
end
plot(x,A); hold
plot(x,A1, ':')
legend('Close to 3', 'intensified close to 3')
xlabel('Natural Numbers -->')
ylabel('Membership value')
```

The intensified fuzzy modifier is shown in Figure 11.6.

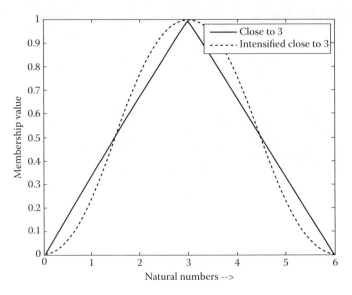

FIGURE 11.6
Intensified fuzzy modifier.

11.4.7 Bounded Sum

The bounded sum of two fuzzy sets, *A* and *B*, in the universes, *X* and *Y*, with the membership functions, $\mu_A(x)$ and $\mu_B(y)$, respectively, is defined by Equation 11.10 and also shown in Figure 11.7:

$$A \oplus B = \mu_{A \oplus B}(x) = 1 \wedge (\mu_A(x) + \mu_B(y))$$

$$= \min(1, (\mu_A(x) + \mu_B(y))) \quad (11.10)$$

where the "+" sign is an arithmetic operator.

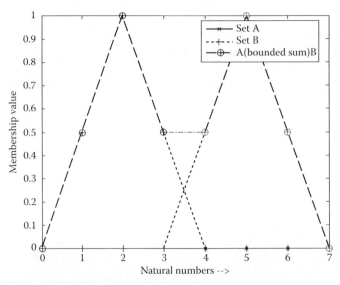

FIGURE 11.7
Results of MATLAB program for bounded sum.

Modeling Using Fuzzy Systems

TABLE 11.2

Summary of Operations on Fuzzy Sets

S. No.	Operation	Symbol	Formula
1.	Intersection	$A \cap B$	$\min(\mu_A(x), \mu_B(x))$
2.	Union	$A \cup B$	$\max(\mu_A(x), \mu_B(x))$
3.	Absolute complement	$\bar{A}$	$1 - \mu_A(x)$
4.	Relative complement of A with respect to B	$B - A$	$\mu_B(x) - \mu_A(x)$
5.	Concentration	CON(A)	μ_i^m if $m \geq 1$
6.	Dilution	DIL(A)	μ_i^m if $m \leq 1$
7.	Intensification	INT(A)	μ_i^m if $m \geq 1$ for $\mu_i < 0.5$
			μ_i^m if $m \leq 1$ for $\mu_i > 0.5$
8.	Bounded sum	$A \oplus B$	$\min(1, (\mu_A(x) + \mu_B(y))$
9.	Bounded difference	$A \ominus B$	$\max(0, (\mu_A(x) - \mu_B(x)))$
10.	Bounded product	$A \otimes B$	$\text{Max}(0, (\mu_A(x) + \mu_B(x) - 1))$
11.	Algebraic sum	$A + B$	$\mu_A(x) + \mu_B(x) - \mu_A(x)\mu_B(x)$
12.	Algebraic product	AB	$\mu_A(x)\mu_B(x)$
13.	Equality	$A = B$	$\mu_A(x) = \mu_B(x)$

These operators are summarized in Table 11.2.

```
%Matlab program for finding the bounded sum of two fuzzy numbers
clear all
% Close_to_3=A on the universe
X=[0, 1, 2, 3, 4, 5, 6]
% and Close_to_4=B on the universe X=Y
A=[0, 0, 0.5, 1, 0.5, 0, 0];
B=[0, 0, 0, 0.5, 1, 0.5, 0];
A_bounded_B= min(1, (A+B))
plot(X,A, 'b'); hold
plot(X,B,'k')
plot(X, A_bounded_B, 'r-+')
legend('Set A', 'Set B', 'Set A(bounded sum)B')
xlabel('Natural Numbers -->')
ylabel('Membership value')
```

11.4.8 Strong α-Cut

A Strong α-cut of a fuzzy set, A, is also a crisp set, $A^{\alpha+}$, that contains all the elements of the universal set, X, that have a membership grade in A greater than the specified values of α between 0 and 1, as given in Equation 11.11:

$$A^{\alpha+} = \{x | A(x) \; A > \alpha\} \tag{11.11}$$

For example, Young$^{0.8+}$ = {1/5, 1/10}

The α-cuts of a given fuzzy set, A, for two distinct values of α, say α_1 and α_2 such that $\alpha_1 < \alpha_2$, are A^{α_1} and A^{α_2} such that $A^{\alpha_1} \supseteq A^{\alpha_2}$. Similarly, for a strong α-cut, $A^{\alpha_1+} \supseteq A^{\alpha_2+}$.

Properties of an α-cut

1. $A^{\alpha+} \supseteq A^{\alpha}$.
2. $(A \cap B)^{\alpha} = (A^{\alpha} \cap B^{\alpha})$ and $(A \cup B)^{\alpha} = (A^{\alpha} \cup B^{\alpha})$.
3. $(A \cap B)^{\alpha+} = (A^{\alpha+} \cap B^{\alpha+})$ and $(A \cup B)^{\alpha} = (A^{\alpha} \cup B^{+\alpha})$.
4. $\overline{A}^{\alpha} = \overline{A}^{(1-\alpha)+}$.

11.4.9 Linguistic Hedges

Linguistic hedges, such as very, much, more, and less, modify the meaning of atomic as well as composite terms, and thus serve to increase the range of linguistic variables from a small collection of primary terms.

A hedge, h, may be regarded as an operator that transforms the fuzzy set, $M(u)$, representing the meaning of u, into the fuzzy set, $M(hu)$. For example, by using the hedge "very" in conjunction with "not," and the primary term "tall," we can generate the fuzzy sets "very tall," "very very tall," "not very tall," "tall," etc.:

Very: $x = x^2$

Very very: $x = (\text{very }(x))^2$

Not very: $x = \overline{(\text{very}(x))}$

Plus: $x = x^{1.25}$

Minus: $x = x^{0.75}$

Slightly: $A = \text{int [norm (plus } A \text{ and not (very } A))]$

From the last two sets, we have the approximate identity:

$$\text{plus plus } x = \text{minus (very } x)$$

Example

If the universe of discourse, $U = [1, 2, 3, 4, 5]$, and small $= \left[\dfrac{1}{1}, \dfrac{0.8}{2}, \dfrac{0.6}{3}, \dfrac{0.4}{4}, \dfrac{0.2}{5}\right]$, find very small, very very small, not very small, plus small, and minus small.

Solution

$$\text{Then, very small} = \left[\dfrac{1}{1}, \dfrac{0.64}{2}, \dfrac{0.36}{3}, \dfrac{0.16}{4}, \dfrac{0.04}{5}\right] = \mu_A^2/A$$

Modeling Using Fuzzy Systems

$$\text{very very small} = \left[\frac{1}{1}, \frac{0.4}{2}, \frac{0.1}{3}\right] \text{ (neglecting small terms)}$$

$$\text{not very small} = \left[\frac{0}{1}, \frac{0.36}{2}, \frac{0.64}{3}, \frac{0.84}{4}, \frac{0.96}{5}\right]$$

$$\text{plus small} = \left[\frac{1}{1}, \frac{0.76}{2}, \frac{0.53}{3}, \frac{0.32}{4}, \frac{0.13}{5}\right]$$

$$\text{minus small} = \left[\frac{1}{1}, \frac{0.85}{2}, \frac{0.68}{3}, \frac{0.5}{4}, \frac{0.3}{5}\right]$$

The modified set may be plotted using MATLAB, as shown in Figure 11.8.

```
small=[1 0.8 0.6 0.4 0.2];
very_small = small.^2
very_very_small = very_small.^2
not_very_small = (1-very_small)
plus_small=small.^1.25
minus_small=small.^0.57
plot(small, 'k'); hold on
plot(very_small, 'k--'); plot(very_very_small, 'k:')
plot(not_very_small, 'k-.'); plot(plus_small, 'k-*')
plot(minus_small, 'k-+')
legend('small','very small', 'very very small', 'not very small', ...
   'plus small', 'minus small')
xlabel('Variable value'); ylabel('membership value');
```

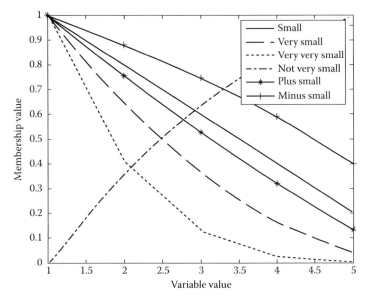

FIGURE 11.8
Fuzzy quantifiers/modifiers.

11.5 Characteristics of Fuzzy Sets

11.5.1 Normal Fuzzy Set

A fuzzy set is said to be normal if the greatest value of its membership function is unity: $Vx\ \mu(x) = 1$, where Vx stands for the supremum of $\mu(x)$ (least upper bound), otherwise the set is subnormal.

11.5.2 Convex Fuzzy Set

A convex fuzzy set is described by a membership function whose membership values are strictly monotonically increasing, or strictly monotonically decreasing, or strictly monotonically increasing and then strictly monotonically decreasing with the increasing value for elements in the universe of discourse, X, that is, if $x, y, z \in A$ and $x < y < z$ then $\mu_A(y) \geq \min[\mu_A(x), \mu_A(z)]$ (see Figure 11.9).

11.5.3 Fuzzy Singleton

A fuzzy singleton is a fuzzy set that only has a membership grade for a single value. Let A be a fuzzy singleton of a universe of discourse, X, $x \in X$, then A is written as $A = \mu/x$. With this definition, a fuzzy set can be considered the union of the fuzzy singleton. In Figure 11.10, the range from a to c is the support and element b only has a membership value greater than zero. Fuzzy sets of this type are called fuzzy singletons.

11.5.4 Cardinality

The number of members of a finite discrete fuzzy set, A, is called the cardinality of A, and is denoted by $|A|$. The number of possible subsets of a set, A, is equal to $2^{|A|}$, which is also called a power set.

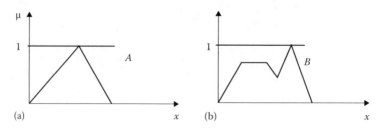

FIGURE 11.9
Convexity of fuzzy sets: (a) A is a convex fuzzy set and (b) B is a nonconvex fuzzy set.

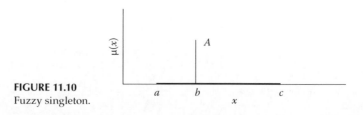

FIGURE 11.10
Fuzzy singleton.

11.6 Properties of Fuzzy Sets

The properties of fuzzy sets are summarized in Table 11.3.

11.7 Fuzzy Cartesian Product

Let A be a fuzzy set on universe X and B be a fuzzy set on universe Y; then the Cartesian product between fuzzy sets, A and B, will result in a fuzzy relation, R, which is contained within the full Cartesian product space, or

$$A \times B = R \subset X \times Y$$

where the fuzzy relation, R, has a membership function:

$$\mu_R(x,y) = \mu_{A \times B}(x,y) = \min(\mu_A(x), \mu_B(y)) \tag{11.12}$$

The Cartesian product defined by $A \times B = R$, is implemented in the same fashion as the cross product of two vectors, as given in Equation 11.12. The Cartesian product is not

TABLE 11.3

Properties of Fuzzy Sets

S. No.	Property	Expression
1.	Commutative property	$\max[\mu_A(x), \mu_B(x)] = \max[\mu_B(x), \mu_A(x)]$ $\min[\mu_A(x), \mu_B(x)] = \min[\mu_B(x), \mu_A(x)]$
2.	Associative property	$\max[\mu_A(x), \max(\mu_B(x), \mu_c(x)]$ $= \max[\mu_A(x), \max\{\mu_B(x), \mu_c(x)\}]$ $\min[\mu_A(x), \min\{\mu_B(x), \mu_C(x)\}]$ $= \min[\min\{\mu_A(x), \mu_B(x)\}, \mu_C(x)]$
3.	Distributive property	$\max[\mu_A(x), \min(\mu_B(x), \mu_c(x)]$ $= \min[\max\{\mu_A(x), \mu_B(x)\},$ $\max\{\mu_A(x), \mu_c(x)\}]$
4.	Idempotency	$\max\{\mu_A(x), \mu_A(x)\} = \mu_A(x)$ $\min\{\mu_A(x), \mu_A(x)\} = \mu_A(x)$
5.	Identity	$\max\{\mu_A(x), 0\} = \mu_A(x)$ $\min\{\mu_A(x), 1\} = \mu_A(x)$ $\min\{\mu_A(x), 0\} = 0$ and $\max\{\mu_A(x), 1\} = 1$
6.	Involution	$1 - (1 - \mu_A(x)) = \mu_{\overline{\overline{A}}}(x)$
7.	Excluded middle law	$\max\{\mu_A(x), \mu_{\bar{A}}(x)\} \neq 1$
8.	Law of contradiction	$\min\{\mu_A(x), \mu_{\bar{A}}(x)\} \neq 0$
9.	Absorption	$\max[\mu_A(x), \min[\mu_A(x), \mu_B(x)] = \mu_A(x)$ $\min[\mu_A(x), \max[\mu_A(x), \mu_B(x)] = \mu_A(x)$
10.	Demorgan's law	$\overline{(A \cup B)} = \overline{A} \cap \overline{B}$ and $\overline{(A \cap B)} = \overline{A} \cup \overline{B}$
11.	Transitive	If $A \subseteq B$ and $B \subseteq C$ Then $\subseteq C$

the same operation as the arithmetic product. Each of the fuzzy sets could be thought of as a vector of membership values; each value is associated with a particular element in each set.

For example, for a fuzzy set (vector), A, that has four elements, hence a column vector of size 4×1, and for a fuzzy set (vector), B, that has five elements, hence a row vector of size 1×5, the resulting fuzzy relation, R, will be represented by a matrix of size 4×5, that is, R will have four rows and five columns.

Example

Suppose that we have two fuzzy sets—A, defined on a universe of four discrete temperatures, $X = \{x_1, x_2, x_3, x_4\}$, and B, defined on a universe of three discrete pressures, $Y = (y_1, y_2, y_3)$—and we want to find the fuzzy Cartesian product between them. The fuzzy set, A, could represent the "ambient" temperature and the fuzzy set, B, the "near optimum" pressure for a certain heat exchanger, and the Cartesian product could represent the conditions(temperature–pressure pairs) of the exchanger that are associated with "efficient" operations.

$$\text{Let } A = 0.2/x_1 + 0.5/x_2 + 0.8/x_3 + 1/x_4$$

$$\text{and } B = 0.3/y_1 + 0.5/y_2 + 0.9/y_3.$$

Here, A can be represented as a column vector of size 4×1 and B can be represented as a row vector of size 1×3. Then the fuzzy Cartesian product results in a fuzzy relation, R, (of size 4×3) representing "efficient" conditions:

$$A \times B = R = \begin{matrix} & \begin{matrix} y_1 & y_2 & y_3 \end{matrix} \\ \begin{matrix} x_1 \\ x_2 \\ x_3 \\ x_4 \end{matrix} & \begin{bmatrix} 0.2 & 0.2 & 0.2 \\ 0.3 & 0.5 & 0.5 \\ 0.3 & 0.5 & 0.8 \\ 0.3 & 0.5 & 0.9 \end{bmatrix} \end{matrix} \qquad (11.13)$$

11.8 Fuzzy Relation

If A and B are two sets and there is a specific property between elements x of A and y of B, this property can be described using the ordered pair (x, y). A set of such (x, y) pairs, $x \in A$ and $y \in B$, is called a relation, R:

$$R = \{(x, y) | x \in A, y \in B\} \qquad (11.14)$$

R is a binary relation and a subset of $A \times B$.

The term "x is in relation, R, with y" is denoted as

$(x, y) \in R$ or xRy, with $R \subseteq A \times B$.

If $(x, y) \notin R$, x is not in relation, $\bar{R}$ with y.

If $A = B$ or R is a relation from A to A, it is written $(x, x) \in R$ or xRx, for $R \subseteq A \times A$.

Modeling Using Fuzzy Systems

Example

Consider two binary relations, R and S, on X × Y defined as

R: x is considerably smaller than y
S: x is very close to y

The fuzzy relation matrices, M_R and M_S, are given as

M_R	y_1	y_2	y_3
x_1	0.3	0.2	1.0
x_2	0.8	1.0	1.0
x_3	0.0	1.0	0.0

M_S	y_1	y_2	y_3
x_1	0.3	0.0	0.1
x_2	0.1	0.8	1.0
x_3	0.6	0.9	0.3

Fuzzy relation matrices, $M_{R \cup S}$ and $M_{R \cap S}$, corresponding to $R \cup S$ and $R \cap S$, yield the following:

$M_{R \cup S}$	y_1	y_2	y_3
x_1	0.3	0.2	1.0
x_2	0.8	1.0	1.0
x_3	0.6	1.0	0.3

$M_{R \cap S}$	y_1	y_2	y_3
x_1	0.3	0.0	0.1
x_2	0.1	0.8	1.0
x_3	0.0	0.9	0.0

Here, the Union of R and S means that "x is considerably smaller than y" OR "x is very close to y." And the Intersection of R and S means that "x is considerably smaller than y" AND "x is very close to y." Also, the complement relation of the fuzzy relation, R, shall be

$M_{\bar{R}}$	a	b	c
1	0.7	0.8	0.0
2	0.2	0.0	0.0
3	1.0	0.0	1.0

Inverse relation

When a fuzzy relation, $R \subseteq A \times B$, is given, its inverse relation, R^{-1}, is defined by the following membership function:
For all $(x, y) \subseteq A \times B$,

$$\mu_R^{-1}(y, x) = \mu_R(x, y)$$

Composition of fuzzy relation

Two fuzzy relations, R and S, are defined on sets A, B, and C. That is, $R \subseteq A \times B$, and $S \subseteq B \times C$. The composition, $S \bullet R = S \bullet R$, of two relations, R and S, is expressed by the relation from A to C, and this composition is defined by the following:

For $(x, y) \in A \times B$, and $(y, z) \in B \times C$,

$$\mu_{S \bullet R}(x, z) = \max_y [\min(\mu_R(x, y), \mu_S(y, z))]$$

$$= \vee_y [\mu_R(x, y) \wedge \mu_S(y, z)]$$

From this elaboration, $S \bullet R$ is a subset of $A \times C$. That is, $S \bullet R \subseteq A \times C$.

If the relations, R and S, are represented by the matrices, M_R and M_S, the matrix, $M_{S \bullet R}$, corresponding to $S \bullet R$, is obtained from the product of M_R and M_S:

$$M_{S \bullet R} = M_R \bullet M_S$$

Example

Consider fuzzy relations $R \subseteq A \times B$, and $S \subseteq B \times C$. The sets A, B, and C shall be the sets of events. By the relation, R, we can see the possibility of occurrence of B after A, and by the relation, S, that of C after B. For example, by M_R, the possibility of $a \in B$ after $1 \in A$ is 0.1. By M_S, the possibility of occurrence of 7 after a is 0.9.

R	a	b	c	d
1	0.1	0.2	0.0	1.0
2	0.3	0.3	0.0	0.2
3	0.8	0.9	0.1	0.4

S	α	β	γ
a	0.9	0.0	0.3
b	0.2	1.0	0.8
c	0.8	0.0	0.7
d	0.4	0.2	0.3

Here, we cannot guess the possibility of C when A has occurred. So, our main job now will be to obtain the composition, $S \bullet R \subseteq A \times C$. The following matrix, $M_{S \bullet R}$, represents this composition as given in Figure 11.11:

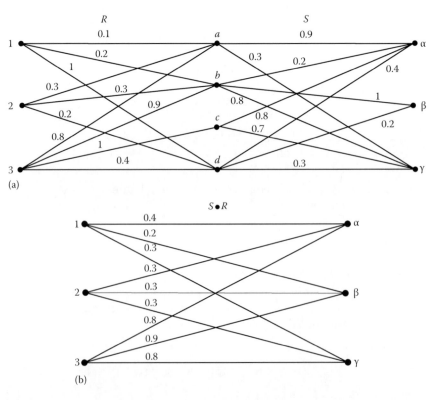

FIGURE 11.11
Composition of fuzzy relation.

Modeling Using Fuzzy Systems

$S \bullet R$	α	β	γ
1	0.4	0.2	0.3
2	0.3	0.3	0.3
3	0.9	0.9	0.8

We see the possibility of occurrence of $\alpha \in C$ after event $1 \in A$ is 0.4, that of $\beta \in C$ after event $2 \in A$ is 0.3, etc.

Presuming that the relations, R and S, are the expressions of rules that guide the occurrences of events or facts, the possibility of occurrence of event B when event A has happened is guided by the rule, R. And rule S indicates the possibility of C when B is existing. For further cases, the possibility of C when A has occurred can be induced from the composition rule, $S \bullet R$. This technique is named an "inference," which is a process producing new information.

11.9 Approximate Reasoning

In 1979, *Zadeh* introduced the theory of approximate reasoning. This theory provides a powerful framework for reasoning in the face of imprecise and uncertain information.

Central to this theory is the representation of propositions as statements assigning fuzzy sets as values to variables.

Suppose that we have two interactive variables, $x \in X$ and $y \in Y$, and the causal relationship between x and y is completely known, that is, we know that y is a function of x:

$$y = f(x)$$

Then we can make inferences easily:

premise $y = f(x)$
fact $x = x'$
consequence $y = f(x')$

This inference rule says that if we have $y = f(x), \forall_x \in X$, and we observe that $x = x'$, then y takes the value, $f(x') = y'$, as shown in Figure 11.12.

Suppose that we are given an $x' \in X$ and want to find $y' \in Y$ that corresponds to x' under the rule base.

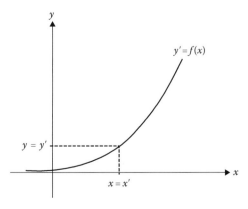

FIGURE 11.12
Simple crisp inference.

Rule 1: If $x = x_1$ Then $y = y_1$
And
Rule 2: If $x = x_2$ Then $y = y_2$
And
Rule 3: If $x = x_3$ Then $y = y_3$
And
... ...
Rule n: If $x = x_n$ Then $y = y_n$
Fact: $x = x'$
Consequence: $y = y'$

Let x and y be linguistic variables, for example, "x is big" and "y is medium." The basic problem of approximate reasoning is to find the membership function of the consequence, C, from the rule base and the fact, A.

Rule 1: If $x = A_1$ Then $y = C_1$
And
Rule 2: If $x = A_2$ Then $y = C_2$
And
... ...
Rule n: If $x = A_n$ Then $y = C_n$
Fact: $x = A$
Consequence: $y = C$

Zadeh introduced a number of translation rules that allow us to represent some common linguistic statements in terms of propositions in our language.

Entailment rule:

x is A Anjali is very beautiful
$A \subseteq B$ very beautiful $\subset$ beautiful
_____

x is B Anjali is beautiful

Conjunction rule:

x is A Anjali is beautiful
and
x is B Anjali is intelligent
_____

x is $A \cap B$ Anjali is beautiful and intelligent

Disjunction rule:

x is A Anjali is married
or
x is B Anjali is a bachelor
_____

x is $A \cup B$ Anjali is married or a bachelor

Projection rule:

(x, y) have a relation, R: x is $\Pi_X(R)$
(x, y) have a relation, R: y is $\Pi_Y(R)$
For example, (x, y) is close to $(3, 2)$: x is close to 3
(x, y) is close to $(3, 2)$: y is close to 2

Negation rule:

not (x is A): x is $\bar{A}$

For example, not (x is high): x is not high

The classical implication:

Let P = "x is in A" and Q = "y is in B" be crisp propositions, where A and B are crisp sets for the moment. The implication, $P \rightarrow Q$, is interpreted as $\neg(p \cap \neg q)$.

"P entails Q" means that it can never happen that P is true and Q is not true.

It is easily seen that

$$P \rightarrow Q = \neg P \cup Q \tag{11.15}$$

In the 1930s, Lukasiewicz, a Polish mathematician, explored for the first time logics other than the Aristotelian (classical or binary) logic (Rescher, 1969; Ross 1995). In this implication, the proposition, P, is the hypothesis or the antecedent, and the proposition, Q, is also referred to as the conclusion or the consequent. The compound proposition, $P \rightarrow Q$, is true in all cases except when a true antecedent, P, appears with a false consequent, Q, that is, a true hypothesis cannot imply a false conclusion, as given in the truth table (Table 11.4).

Generalized modus ponens

Classical logic elaborated many reasoning methods called tautologies. One of the best known is modus ponens. The reasoning process in modus ponens is given as follows:

Rule	IF X is A THEN Y is B
Fact	X is A
Conclusion	Y is B

In classical modus ponens tautology, the truth value of the premise "X is A" and the conclusion "Y is B" is allowed to assume only two discrete values, 0 and 1, and the fact considering "X is ...," must fully agree with the implication premise:

IF X is A THEN Y is B

Only then the implication be used in the reasoning process. Both the premise and the rule conclusion must be formulated in a deterministic way. Statements with nonprecise formulations such as follows are not accepted:

X is about A.
X is more than A.
Y is more or less B.

In fuzzy logic, an approximate reasoning has been applied. It enables the use of fuzzy formulations in premises and conclusions. The approximate reasoning based on the generalized modus ponens (GMS) tautology is

Rule	IF X is A THEN Y is B
Fact	X is A'
Conclusion	Y is B'

TABLE 11.4
Truth Table for Classical Implication

P	Q	$P \rightarrow Q$
T	T	T
T	F	F
F	T	T
F	F	T

where A' and B' can mean, for example, A' = more than A and B' = more or less B. A reasoning example according to the GMP tautology can be as follows:

Rule	IF (route of the trip is long) THEN (traveling time is long)
Fact	route of the trip is very long
Conclusion	traveling time is very long

Here, fuzzy implication is expressed in the following way:

$$\forall u \in U, \quad \forall v \in V$$

$$\mu_R(u,v) = \begin{cases} 1 & \text{if } \mu_A(u) \leq \mu_B(v) \\ \mu_B(v) & \text{otherwise} \end{cases}$$

$$\mu_{B'}(v) = \sup_{u \in U} \min(\mu_{A'}(u), \mu_R(u,v))$$

In fuzzy logic and approximate reasoning, the most important fuzzy implication inference rule is the GMP. The classical modus ponens inference rule says

premise   if p then q
fact p
_____
consequence q

This inference rule can be interpreted thus: If p is true and $p \to q$ is true, then q is true.

The fuzzy implication inference is based on the compositional rule of inference for approximate reasoning suggested by Zadeh:

Compositional rule of inference:

premise   if x is A then y is B
fact x is A'
_____
consequence y is B'

where the consequence, B', is determined as a composition of the fact and the fuzzy implication operator:

$$B' = A' \circ (A \to B)$$

that is

$$B'(v) = \sup_{u \in U} \min\{A'(u), (A \to B)(u,v)\}, \quad v \in V \qquad (11.16)$$

The consequence, B', is nothing else but the shadow of $A \to B$ on A'.

The GMP, which reduces to classical modus ponens when $A' = A$ and $B' = B$, is closely related to the forward data-driven inference, is particularly useful in the fuzzy logic control. The truth table for the GMP is given in Table 11.5.

The classical modus tollens inference rule is stated thus: If $p \to q$ is true and q is false, then p is false. The generalized modus tollens:

TABLE 11.5
Truth Table for Modus Ponens

P	Q	P → Q	[P ∩ (P → Q)]	[P ∩ (P → Q)] → Q
0	0	1	0	1
0	0.5	1	0	1
0	1	1	0	1
0.5	0	0	0	1
0.5	0.5	1	0.5	1
0.5	1	1	0.5	1
1	0	0	0	1
1	0.5	0.5	0.5	1
1	1	1	1	1

premise   if x is A then y is B
fact y is B'
_____
consequence x is A'

which reduces to modus tollens when $B = \bar{B}$ and $A' = \bar{A}$, is closely related to the backward goal-driven inference. The consequence, A', is determined as a composition of the fact and the fuzzy implication operator:

$$\overline{A'} = \overline{B'} \circ (A \rightarrow B)$$

Fuzzy relation schemes:

Rule 1: If $x = A_1$ and $y = B_1$ Then $z = C_1$
Rule 2: If $x = A_2$ and $y = B_2$ Then $z = C_2$
...   ...

Rule n: If $x = A_n$ and $y = B_n$ Then $z = C_n$
Fact: x is x_0 and y is y_0
_____
consequence: $z = C$

The ith fuzzy rule from this rule base:

Rule i: If $x = A_i$ and $y = B_i$ Then $z = C_i$
is implemented by a fuzzy relation, Ri, and is defined as

$$R_i(u, v, w) = (A_i x B_i \rightarrow C_i)(u, w)$$

$$= [A_i(u) \cap B_i(v)] \rightarrow C_i(w) \qquad (11.17)$$

for $i = 1, 2, \ldots, n$.

From the input, x_0, and from the rule base, $R = \{R_1, R_2, \ldots, R_n\}$, find the interpretation "C" with

- Logical connective "and"
- Sentence connective "also"
- Implication operator "then"
- Compositional operator "∘".

We first compose $x_0 \times y_0$ with each R_i producing intermediate results:

$$C_i' = x_0 \cap y_0 \circ R_i$$

for $i = 1, \ldots, n$. Here, C_i' is called the output of the ith rule:

$$C_i'(w) = A_i(x_0) \cap B_i(y_0) \rightarrow C_i(w)$$

for each w.

Then combine C_i' component-wise into C' by some aggregation operator:

$$C = \cup C_i'$$

$$C_i(w) = (A_i(x_0) \cap B_i(y_0) \rightarrow C_i(w)) \cup \ldots \cup (A_n(x_0) \cap B_n(y_0) \rightarrow C_n(w))$$

Steps involved in approximate reasoning:

- Input to the system as (x_0, y_0).
- Fuzzify the input as (x_0, y_0).
- Find the firing strength of the ith rule after aggregating the inputs as

$$A_i(x_0) \cap B_i(y_0)$$

- Calculate the ith individual rule output as

$$C_i'(w) = A_i(x_0) \cap B_i(y_0) \rightarrow C_i(w)$$

- Determine the overall system output (fuzzy) as

$$C = C_1' \cup C_2' \ldots \cup C_n'$$

- Finally, calculate the defuzzified output.

We present five well-known inference mechanisms in fuzzy rule-based systems. For simplicity, we assume that we have fuzzy IF-THEN rules of the form:

Rule 1: If $x = A_1$ and $y = B_1$ Then $z = C_1$
Rule 2: If $x = A_2$ and $y = B_2$ Then $z = C_2$
...   ...
Rule n: If $x = A_n$ and $y = B_n$ Then $z = C_n$
Fact: x is x_0 and y is y_0
_____

consequence $z = c$

Mamdani: The fuzzy implication is modeled by Mamdani's minimum operator and defined by the max operator.

Modeling Using Fuzzy Systems

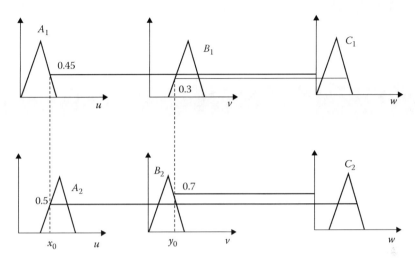

FIGURE 11.13
Fuzzy inference.

The firing levels of the rules, denoted by α_i, $i = 1, 2$, are computed, as shown in Figure 11.13, by

$$\alpha_1 = A_1(x_0) \cap B_1(y_0) \quad \text{and} \quad \alpha_2 = A_2(x_0) \cap B_2(y_0)$$

The individual rule outputs are obtained by

$$C_1'(w) = (\alpha_1 \cap C_1(w)) \quad \text{and} \quad C_2'(w) = (\alpha_2 \cap C_2(w))$$

Then the overall system output is computed by taking the union of the individual rule outputs:

$$C(w) = C_1'(w) \cup C_2'(w)$$
$$= (\alpha_1 \cap C_1(w)) \cup (\alpha_2 \cap C_2(w))$$

Finally, to obtain a deterministic control action, we employ any defuzzification strategy.

Tsukamoto: All linguistic terms are supposed to have monotonic membership functions.

The firing levels of the rules, denoted by α_i, $i = 1, 2$, are computed, as shown in Figure 11.14, by

$$\alpha_1 = A_1(x_0) \cap B_1(y_0) \quad \text{and} \quad \alpha_2 = A_2(x_0) \cap B_2(y_0)$$

In this mode of reasoning, the individual crisp control actions, z_1 and z_2, are computed from the equations:

$$\alpha_1 = C_1(z_1), \quad \alpha_2 = C_2(z_2)$$

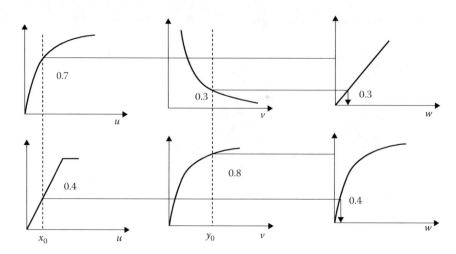

FIGURE 11.14
Tsukamoto inference.

and the overall crisp control action is expressed as

$$z_0 = \frac{\alpha_1 z_1^* + \alpha_2 z_1^*}{\alpha_1 + \alpha_2} \quad (11.18)$$

that is, z_0 is computed by the discrete center-of-gravity method.

If we have n rules in our rule base, then the crisp control action is computed as

$$z_0 = \frac{\sum_{i=1}^{n} \alpha_i z_i}{\sum_{i=1}^{n} \alpha_i} \quad (11.19)$$

where α_i is the firing level and z_i is the (crisp) output of the ith rule, $i = 1, \ldots, n$.

Sugeno: Sugeno and Takagi use the following architecture:

Rule 1: If $x = A_1$ and $y = B_1$ Then $z_1 = a_1 x + b_1 y$
Rule 2: If $x = A_2$ and $y = B_2$ Then $z_2 = a_2 x + b_2 y$
...   ...

Rule n: If $x = A_n$ and $y = B_n$ Then $z_n = a_n x + b_n y$
Fact: x is x_0 and y is y_0

---

consequence: Z_0

The firing levels of the rules are computed, as shown in Figure 11.15, by Equation 11.20:

$$\alpha_1 = A_1(x_0) \cap B_1(y_0) \quad \text{and} \quad \alpha_2 = A_2(x_0) \cap B_2(y_0) \quad (11.20)$$

Then the individual rule outputs are derived from the relationships:

$$z_1^* = a_1 x_0 + b_1 y_0 \quad \text{and} \quad z_2^* = a_2 x_0 + b_2 y_0 \quad (11.21)$$

Modeling Using Fuzzy Systems

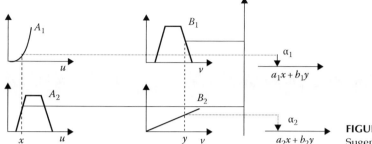

FIGURE 11.15
Sugeno inference mechanism.

and the crisp control action is expressed as

$$z_0 = \frac{\alpha_1 z_1^* + \alpha_2 z_2^*}{\alpha_1 + \alpha_2} \tag{11.22}$$

If we have n rules in our rule base, then the crisp control action is computed as

$$z_0 = \frac{\sum_{i=1}^{n} \alpha_i z_i^*}{\sum_{i=1}^{n} \alpha_i} \tag{11.23}$$

where α_i denotes the firing level of the ith rule, $i = 1, \ldots, n$.

Larsen: The fuzzy implication is modeled by Larsen's product operator, and the sentence connective "also" is interpreted as oring the propositions and defined by the max operator.

Let us denote α_i, the firing level of the ith rule, as shown in Figure 11.16, $i = 1, 2$:

$$\alpha_1 = A_1(x_0) \cap B_1(y_0) \quad \text{and} \quad \alpha_2 = A_2(x_0) \cap B_2(y_0) \tag{11.24}$$

Then the membership function of the inferred consequence, C, is pointwise, given by

$$C(w) = (\alpha_1 C_1(w)) \cup (\alpha_2 C_2(w)) \tag{11.25}$$

To obtain a deterministic control action, we employ any defuzzification strategy.

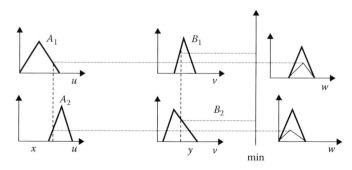

FIGURE 11.16
Inference with Larsen's product operation rule.

If we have n rules in our rule base, then the consequence, C, is computed as

$$C(w) = \bigvee_{i=1}^{n}(\alpha_i C_i(w)) \tag{11.26}$$

where α_i denotes the firing level of the ith rule, $i = 1, \ldots, n$.

11.10 Defuzzification Methods

The output of the inference process so far is a fuzzy set, specifying a possibility of the distribution of control action. In the online control, a nonfuzzy (crisp) control action is usually required. Consequently, one must defuzzify the fuzzy control action (output) inferred from the fuzzy control algorithm, namely:

$$z_0 = defuzzifier(C)$$

where
z_0 is the nonfuzzy control output
defuzzifier is the defuzzification operator

Defuzzification is a process to select a representative element from the fuzzy output, C, inferred from the fuzzy control algorithm. The most often used defuzzification methods are as follows.

1. Center-of-area/gravity
   The defuzzified value of a fuzzy set, C, is defined as its fuzzy centroid:

$$z_0 = \frac{\int_w zC(z)dz}{\int_w C(z)dz} \tag{11.27}$$

   The calculation of the defuzzified value of the center of area is simplified if we consider a finite universe of discourse, W, and thus a discrete membership function, $C(w)$:

$$z_0 = \frac{\int_w z_j C(z_j)dz}{\int_w C(z_j)dz} \tag{11.28}$$

2. First-of-maxima
   The defuzzified value of a fuzzy set, C, is its smallest maximizing element, as shown in Figure 11.17, that is:

$$z_0 = \min\{z \,|\, C(z) = \max_w C(w)\} \tag{11.29}$$

Modeling Using Fuzzy Systems

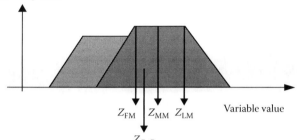

FIGURE 11.17
Fuzzy system outputs for different defuzzification methods.

3. Middle-of-maxima
   The defuzzified value of a discrete fuzzy set, C, is defined as a mean of all values of the universe of discourse, having maximal membership grades:

$$z_0 = \frac{1}{N}\sum_{j=1}^{N} z_j \qquad (11.30)$$

where $\{z_1, \ldots, z_N\}$ is the set of elements of the universe, W, that attain the maximum value of C.

If C is not discrete, then the defuzzified value of a fuzzy set, C, is defined as

$$z_0 = \frac{\int_G z\,dz}{\int_G dz} \qquad (11.31)$$

where G denotes the set of maximizing elements of C.

4. Max-criterion
   The max-criterion method chooses an arbitrary value, from the set of maximizing elements of C, that is:

$$z_0 \in \{z \mid C(z) = \max_w C(w)\} \qquad (11.32)$$

5. Height defuzzification
   The elements of the universe of discourse, W, that have membership grades lower than a certain level, α, are completely discounted, and the defuzzified value, z_0, is calculated by the application of the center-of-area method on those elements of W that have membership grades not less than α:

$$z_0 = \frac{\int_{[C]^\alpha} zC(z)\,dz}{\int_{[C]^\alpha} C(z)\,dz} \qquad (11.33)$$

where $[C]^\alpha$ usually denotes the α-level set of C.

11.11 Introduction to Fuzzy Rule-Based Systems

The inputs of fuzzy rule-based systems should be given in a fuzzy form, and, therefore, we have to fuzzify the crisp inputs, that is, preprocess sensor's outputs. Furthermore, the output of a fuzzy system is always a fuzzy output, and, therefore, to get an appropriate crisp value, we have to defuzzify and postprocess it. Fuzzy logic control systems usually consist of three major parts: fuzzification, approximate reasoning, and defuzzification, as shown in Figure 11.18.

A fuzzification operator has the effect of transforming crisp data into fuzzy sets. In most cases, we use fuzzy singletons as fuzzifiers:

$$\text{fuzzifier}(x_0) = \bar{x}_0 \tag{11.34}$$

where x_0 is a crisp input value from a process.

1. *Preprocessing* module does input signal conditioning and also performs a scale transformation (i.e., an input normalization), which maps the physical values of the current system state variables into a normalized universe of discourse (normalized domain).

2. *Fuzzification* block performs a conversion from crisp (point-wise) input values to fuzzy input values, in order to make it compatible with the fuzzy set representation of the system state variables in the rule "antecedent." The process of fuzzification of $\frac{\mu_i}{x_i}$ gives rise to two special cases:

    a. Support fuzzification (s-fuzzification)
    In this fuzzification, μ_i is kept constant and x_i is fuzzified. It is denoted by $SF(A, k)$:

$$SF(A,k) = \mu_1 k(x_1) + \mu_2 k(x_2) + \cdots + \mu_n k(x_n) \tag{11.35}$$

$\mu_i k(x_i)$ is a fuzzy set that is the product of a scalar constant and a fuzzy set, $k(x_i)$.

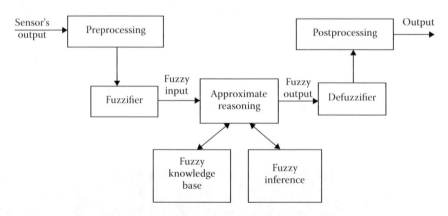

FIGURE 11.18
Fuzzy knowledge-based system (FKBS).

Modeling Using Fuzzy Systems

Example

Let the universe of discourse be $U = [1, 2, 3, 4]$ and the fuzzy set be $A = \left[\frac{0.8}{1}, \frac{0.5}{2}, \frac{0}{3}, \frac{0}{4}\right]$.

Assume that $k(1) = \left[\frac{1}{1}, \frac{0.3}{2}, \frac{0}{3}, \frac{0}{4}\right]$ and $k(2) = \left[\frac{0.3}{1}, \frac{1.0}{2}, \frac{0.2}{3}, \frac{0}{4}\right]$.

$SF(A, k) = \mu_1 k(x_1) + \mu_2 k(x_2)$

$$= \left[0.8\left[\frac{1}{1}, \frac{0.3}{2}, \frac{0}{3}, \frac{0}{4}\right]\right] + 0.5\left[\left[\frac{0.3}{1}, \frac{1.0}{2}, \frac{0.2}{3}, \frac{0}{4}\right]\right]$$

$$= \left[\frac{\min(0.8,1)}{1}, \frac{\min(0.8,0.3)}{2}, \frac{0}{3}, \frac{0}{4}\right] + \left[\frac{\min(0.5,0.3)}{1}, \frac{\min(0.5,1.0)}{2}, \frac{0}{3}, \frac{0}{4}\right]$$

$$= \left[\frac{\max(0.8,0.3)}{1}, \frac{\max(0.3,0.5)}{2}, \frac{0}{3}, \frac{0}{4}\right]$$

$$= \left[\frac{0.8}{1}, \frac{0.5}{2}, \frac{0}{3}, \frac{0}{4}\right]$$

b. Grade fuzzification (g-fuzzification)
   It is denoted by $GF(A,k)$:

$$GF(A,k) = \frac{k(\mu_1)}{x_1} + \frac{k(\mu_2)}{x_2} + \cdots + \frac{k(\mu_n)}{x_n}$$

where $k(\mu_i)$ denotes point fuzzification of μ_i.

Consider the above example to illustrate grade fuzzification. If we define

$$k(0.8) = \left[\frac{1}{0.8}, \frac{0.7}{0.6}, \frac{0.3}{0.5}\right] \quad \text{and} \quad k(0.5) = \left[\frac{1}{0.5}, \frac{0.6}{0.4}, \frac{0.5}{0.3}\right]$$

then $GF(A_i, k) = \left[\frac{1}{0.8}, \frac{0.7}{0.6}, \frac{0.3}{0.5}\right]\bigg/1 + \left[\frac{1}{0.5}, \frac{0.6}{0.4}, \frac{0.5}{0.3}\right]\bigg/2$

3. *Fuzzy knowledge base (FKB)* comprises of a knowledge of the application domain and the attendant control goals. It includes the following:

   a. Fuzzy sets (membership functions) representing the meaning of the linguistic values of the system state and control output variables.

   b. Fuzzy rules of the form: If x is A THEN y is B. x is the state variable and y is the output variable. A and B are linguistic variables. The basic function of the fuzzy rule base is to represent the control policy of an experienced process operator in a structured way and/or a control engineer in the form of set of production rules. Rules have two parts: antecedent and consequent.

4. *Fuzzy inference*: There are two basic types of approaches employed in the design of the inference engine of an FKBC: (1) composition-based inference (firing) and (2) individual rule-based inference (firing). Mostly, we use individual rule-based inference.

5. *Defuzzification* converts fuzzy output values into a single point value. Six most often used defuzzification methods are center of area (COA), center of sum (COS), center of

largest area (CL), first maxima (FM), mean of maxima (MOM), and height (weighted average) defuzzification.
6. *Postprocessing module* involves de-normalization of the defuzzified output of a fuzzy system onto its physical domain.

11.12 Applications of Fuzzy Systems to System Modeling

In the previous chapters of this book, methods have been discussed to develop mathematical models for physical as well as conceptual systems. In these models, the parameters and variables have crisp (numerical) values. Another class of models, which are called logical models, use logical-type connectives, such as "and," "or," and "if-then." These models have linguistic terms for variables and parameters. These models use qualitative information, which may be dealt with fuzzy logic. Here, it is the information gathered from domain experts to develop fuzzy rules for such models. These models are generally based on a partitioning of the domain of the problem, which reduces the difficulty of the modeling problem by allowing the use of simpler models in each of the components or subsystems.

Why fuzzy systems for modeling?

- They support the generation of fast prototype and incremental optimization.
- They are plain and simple to understand.
- The intelligence of a system is not buried in differential equations or source codes.
- The fuzzy system models are simple, and fast in development.
- System design is more transparent.
- They have greater expressive powers than classical mathematical models.
- They are easily implemented on 8 bit microcontrollers.

Fuzzy systems theory is the starting point for developing models of ambiguous thinking and judgment processes; the following fields of application are conceivable:

1. Human models for management and societal problems
2. Use of high-level human abilities for use in automation and information systems
3. Reducing the difficulties of man–machine interface
4. Other AI applications, like risk analysis and prediction, and development of functional devices

The fuzzy rule-based model, as shown in Figure 11.19, may be classified as

1. Mamdani model
2. Takagi–Sugeno–Kang (TSK) model

In Mamdani models, the input side (antecedent part or if part) as well as the output side (consequence part or then part) of the system are fuzzy, as shown in Figure 11.20. Consider a simple resistive electrical network; the voltage and current are related with the following rule:

Modeling Using Fuzzy Systems

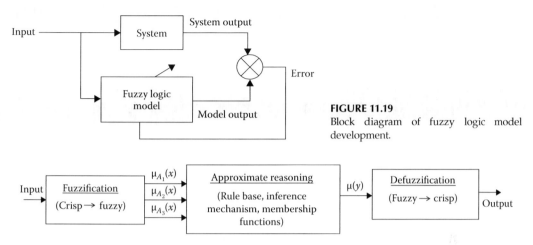

FIGURE 11.19
Block diagram of fuzzy logic model development.

FIGURE 11.20
Detailed diagram of fuzzy logic model for a SISO system.

If *voltage* is high **then** *current* flow is high.

The fuzzy input in the above system (Figure 11.20) has linguistic terms A_1, A_2, and A_3.

On the other hand, TSK models have a fuzzy antecedent part and a functional consequent; essentially, they are a combination of fuzzy and nonfuzzy models. These types of models permit a relatively easy application of powerful learning techniques for their identification of data. In this chapter, we shall discuss both kinds of models and how they apply to dynamics systems.

11.12.1 Single Input Single Output Systems

In the following section, we will discuss about single input single output system (SISO) models.

Example

Consider that the hysteresis (*B–H*) characteristic of a magnetic material for all electrical devices is nonlinear in nature, as shown in Figure 11.21.

The input universe (i.e., magnetic field intensity, *H*) and the output universe (i.e., magnetic flux density, *B*) may be divided in three linguistic terms: small, medium, and high.

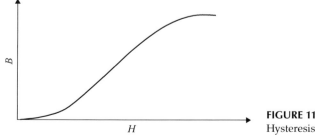

FIGURE 11.21
Hysteresis curve for magnetic material.

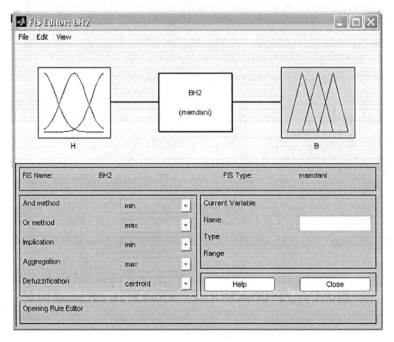

FIGURE 11.22
Fuzzy model for *B–H* characteristic in MATLAB fuzzy toolbox.

The following set of rules may be used:

If *H* is *small* **THEN** *B* is *small*
Also
If *H* is *medium* **THEN** *B* is *medium*
Also
If *H* is *high* **THEN** *B* is *high*

This model is developed for a SISO system in MATLAB fuzzy toolbox (see Figure 11.22) using the Mamdani method of reasoning, and the results are shown in Figure 11.23 with the membership

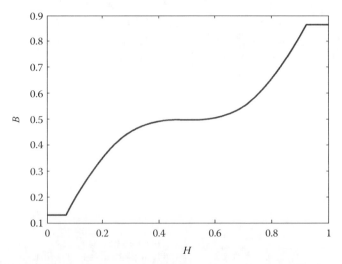

FIGURE 11.23
Result of the Mamdani fuzzy model for *B–H* characteristic before tuning.

Modeling Using Fuzzy Systems

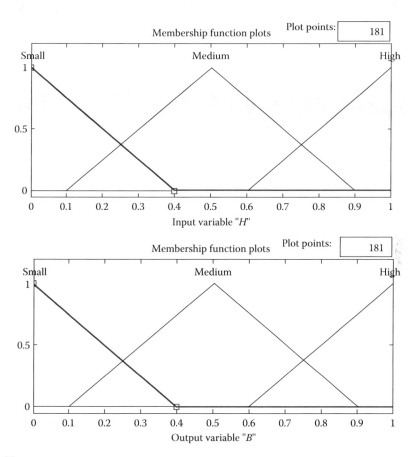

FIGURE 11.24
Membership functions for input (H) and output (B) variables before tuning.

functions shown in Figure 11.24. When the fuzzy membership functions are tuned, as shown in Figure 11.25, the results improve, as shown in Figure 11.26.

Algorithm:

1. For linguistic variables, H and B, select suitable number of linguistic terms.
2. Draw membership functions for these linguistic terms.
3. Write fuzzy rules between the input and the output.
4. For each rule of fuzzy logic model

Calculate the degree of firing (DOF_i), if the input is $A(x)$:

$$DOF_i = \vee_x[H_i(x) \wedge A(x)]$$

Find the fuzzy output, O_i, from the ith rule:

$$O_i = DOF_i \wedge B_i(y)$$

Aggregate the fuzzy output, O_i, of each rule using the max operation to get the final output:

$$O = \bigcup_{i=1}^{m} O_i(y)$$

Find the crisp output, B, from the aggregated fuzzy output using the centroid defuzzification method.

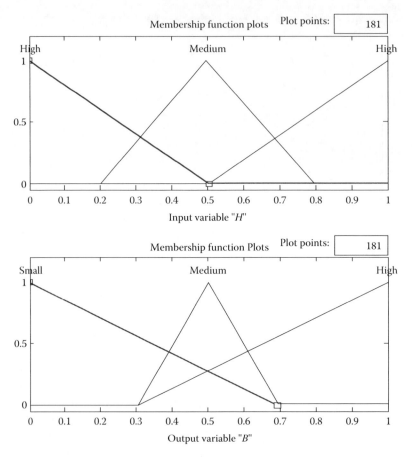

FIGURE 11.25
Membership functions for input (H) and output (B) variables after tuning.

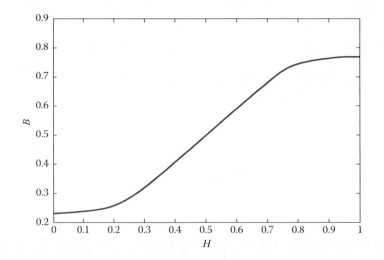

FIGURE 11.26
Result of the Mamdani fuzzy model for B–H characteristic after tuning.

Modeling Using Fuzzy Systems

Example

Model a theoretical relationship between a set of input and a set of output given below:

input = [1 2 3 4 5 6 7 8 9 10 11];
output = [20 12 9 6 5 4 5 6 9 12 20].

The plot between these sets of input and output and the best-fit curve are shown in Figure 11.27.
 To model a relationship between a set of input and a set of output, the following steps have to be adhered to:

Step 1 Create fuzzy sets for input variable

The first thing to consider while designing a system model is the range for the input variables. The input ranges for this particular problem from 1 to 11, which is the universe of discourse for the input variable. Once we have a reasonable range, then we have to divide that range into descriptive linguistic terms. Here, we have divided the input universe of discourse into six, even, triangular fuzzy sets, as shown in Figure 11.28.

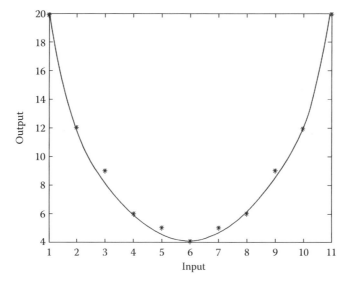

FIGURE 11.27
Original data plot with best-fit curve.

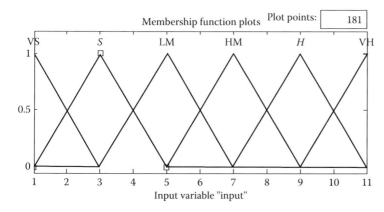

FIGURE 11.28
Fuzzy sets for input variables.

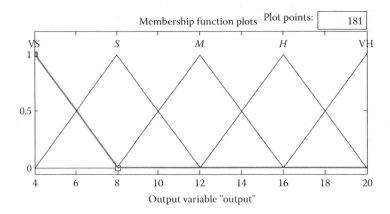

FIGURE 11.29
Fuzzy sets for output variables.

Step 2 Create fuzzy sets for output variable

Like the input variable, the range of the output variable must be divided into various membership functions. The output ranges from 4 to 20 over the input range, called the universe of discourse for the output variable, which is divided into five fuzzy sets, as shown in Figure 11.29.

Step 3 Develop fuzzy rules

Fuzzy rules represent a pair-wise relationship between an input and an output. The first part in the fuzzy rules is the input condition or the *if* part of the implication, and the second part represents the output condition or the *then* part. In this example, the input–output pairs are {(1, 20), (2, 12), ..., (10, 12), (11, 20)}, which are written in linguistic terms as (VS, VH), and would be equivalent to the verbal statement:

If the *input* is VS **then** the *output* is VH.

Fuzzy rules are based on test data or simply on observations. Here is the complete list of rules that were used in the paper presented by Stachowicz and Kochanska (1987).
   Fuzzy rule base:

If the *input* is VS **then** the *output* is VH
If the *input* is S **then** the *output* is H
If the *input* is H **then** the *output* is H
If the *input* is VH **then** the *output* is VH
If the *input* is LM **then** the *output* is VS
If the *input* is HM **then** the *output* is VS

To get a better performance from the above-developed Mamdani model, tuning of the membership function is needed.

11.12.2 Multiple Input Single Output Systems

The above discussed logical model for SISO systems may be extended for multiple input single output (MISO) systems. In MISO systems, the input variables are more than one and the output is one.

The rules for the MISO system are as follows:

If I_1 is x_{11} **and** I_2 is x_{12} **and** ... **and** I_{1n} is x_{1n} **then** O_1 is y_1
If I_1 is x_{21} **and** I_2 is x_{22} **and** ... **and** I_{2n} is x_{2n} **then** O_2 is y_2

.
.
.

If I_1 is x_{n1} **and** I_2 is x_{n2} **and** ... **and** I_{nn} is x_{1n} **then** O_1 is y_n

where
$[I_1\ I_2\ ...\ I_n]$ is the input variable vector
$[O_1\ O_2\ ...\ O_n]$ is the output variable vector
x_{ij} and y_i are the fuzzy subsets of the universe of discourse of input and output spaces

For a SISO system, the fuzzy model is straightforward, but in a MISO system, the antecedent part of the fuzzy rule base contains many inputs, which are aggregated by "and" and "or" operators. The "and" aggregation is interpreted as a fuzzy intersection (i.e., min) operator and the "or" aggregation as a fuzzy union (i.e., max) operator.

Considering the rule: **If** I_1 is x_{11} **and** I_2 is x_{12} **and** ... **and** I_{1n} is x_{1n} **then** O_1 is y_1.

The firing strength for this rule can be determined with the aggregated output:

$$\text{Agg} = x_{11} \wedge x_{12} \wedge ... x_{1n}$$

The output from this rule may be calculated as

$$O_i = \text{Agg} \wedge y_i$$

The total output is the union of the outputs from all the rules:

$$O = \bigcup_{i=1}^{m} O_i(y)$$

Example—Dynamic System Modeling

Fuzzy logic models can be used for representing dynamic knowledge-based systems. It is difficult to imagine the applicability of the concept of fuzzy modeling in the framework of physical systems, like electrical or mechanical systems. However, this strategy is attractive for soft (conceptual) systems, like economical, social, emotional, or spiritual systems. About conceptual systems, the knowledge we have is incomplete. The concept of fuzziness is to represent this incomplete knowledge in a form that is an alternative to a nonlinear state space model. A fuzzy logic model is very useful when the system information is qualitative and not quantitative. If quantitative information is available for a given system, then an artificial neural network (ANN) is the best tool for modeling.

In fuzzy logic models, the fuzzy input–output data may be partitioned in a suitable number of linguistic terms (fuzzy sets), say B_{i0}, B_{i1}, B_{i2}, ..., B_{in} for the input space, and A_{i0}, A_{i1}, A_{i2}, ..., A_{in} for the output space. Fuzzy logic models for dynamic systems may be derived from the MISO rules, such as:

If $u(k)$ is B_{i0} **and** $u(k-1)$ is B_{i1} **and** $u(k-2)$ is B_{i2} ... **and** $u(k-n)$ is B_{in} **and** $y(k-1)$ is A_{i1} **and** $y(k-2)$ is A_{i2} **and** $y(k-3)$ is A_{i3} ... **and** $y(k-n)$ is A_{in} **Then** $y(k)$ is A_{i0}
where $I = 1$ to m.

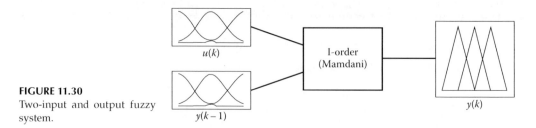

FIGURE 11.30
Two-input and output fuzzy system.

It is easy to see that this model is the fuzzy counterpart of an nth-order nonlinear model. The output of the system is a nonlinear function, which depends upon the input and the past outputs of this system, and is represented as mentioned in Equation 11.36:

$$y(k) = f(u(k), u(k-1), u(k-2), \ldots, u(k-n),$$
$$y(k-1), y(k-2), y(k-3), \ldots, y(k-n)) \qquad (11.36)$$

This model can be viewed as a fuzzified extension of the auto regression moving average (ARMA) model, widely used in digital control.

Modeling of first-order systems

A first-order system may be written as $y(k) = f(u(k), y(k-1))$. Hence, it can be expressed with the following rules for a fuzzy model, as shown in Figure 11.30:

1. **If** $u(k)$ is small **and** $y(k-1)$ is small **Then** $y(k)$ is small
2. **If** $u(k)$ is medium **and** $y(k-1)$ is small **Then** $y(k)$ is medium
3. **If** $u(k)$ is medium **and** $y(k-1)$ is medium **Then** $y(k)$ is high
4. **If** $u(k)$ is high **and** $y(k-1)$ is high **Then** $y(k)$ is high

It can also be represented by the fuzzy associative memory (FAM) table (Table 11.6).

11.12.3 Multiple Input Multiple Output Systems

Conventional linear mathematical models of electrical machines are quite accurate in the operating range for which they were developed, but their outputs degrade for slightly different operating conditions, because these models are unable to cope up with the non-linearity of the system. If a nonlinear model is developed for these systems, the model complexity increases. It is difficult to simulate a complex system model and the computation time is unbearably long. To deal with system nonlinearity, soft computing tools, like ANNs, fuzzy systems, and genetic algorithms, are used. The accuracy of an ANN model depends on the availability of sufficient and accurate historical (numerical) input–output

TABLE 11.6

FAM Table for First-Order System

$u(k)/y(k-1)$	Small	Medium	High
Small	Small		
Medium	Medium	High	
High			High

data for training the network. Also, there is no exact method available to find suitable ANN training method. The performance of a standard gradient descent training algorithm depends on initial weights, learning parameters, and the quality of data. Most of the time in real life situations, it is difficult to get sufficient and accurate data for training an ANN. In such situations, the operator experience may be helpful, which may be used to develop fuzzy logic-based models. Neither mathematical nor ANN models are transparent for illustration purposes, because in these models, information is buried in model structures (differential or difference equations, or ANN connections). Hence, it is not easy to illustrate these models to others. Fuzzy models are quite simple and easy to explain due to their simple structures and complete information in the form of rules.

Fuzzy logic is applied to a great extent in controlling processes, plants, and various complex and ill-defined systems, due to its inherent advantages like simplicity, ease in design, robustness, and adaptability. Also it is established that this approach works very well, especially when the systems are not clearly understood and transparent. In this section, this approach is used for the modeling and simulation of dc machines to predict their characteristics. The effects of different connectives (aggregation operators), like intersection, union, averaging, and compensatory operators, different implications, different compositional rules, different membership functions of fuzzy sets and their percentage overlapping, and different defuzzification methods have also been studied.

11.13 Takagi–Sugeno–Kang Fuzzy Models

The Mamdani fuzzy models described above have certain drawbacks as follows.

They are useful only when the input and the output both are qualitative (fuzzy), solely based on an expert's knowledge about the system. If the expert's knowledge about the system is faulty, the model developed is bad. The quantitative observation of the functioning of the system is not specifically used for the structure or parameters of the model. It does not contain an explicit form of objective knowledge about the system if such knowledge cannot be expressed and/or incorporated into the fuzzy set framework. This kind of knowledge is often available and is typically in the form of general conditions on the physical structure of the system. Sugeno and his coworkers proposed an alternative type of fuzzy reasoning, which does not permit the recognition of these types of conditions. These types of fuzzy models have a rule base that contains fuzzy input variables defined by some fuzzy sets, and the consequent part is functional type, for example

If *input$_1$* is *small* ... and *input$_n$* is *medium* then $y_i = f(input_i, w_i)$

The overall output is inferred from the weighted sum of the individual outputs generated by each rule.

The main advantage of the TSK model lies in the representation of qualitative and quantitative information in its rule base, which helps in describing complex technological processes. The above representation allows us to decompose a complex system into simpler linear or nonlinear subsystems. The overall model of a nonlinear system is a collection of logical models of subsystems.

In realistic situations, however, such a disjoint (crisp) decomposition is not possible, due to the inherent lack of natural boundaries in the system and also due to the fragmentary nature of available knowledge about the system. The TSK model allows us to replace the crisp decomposition by a fuzzy decomposition.

The first significant application of fuzzy logic in the modeling of complex systems, especially in the duplicating of an operator's experience in the area of control engineering (Mamdani, 1974, 75, 76; Tong, 1978), was a manifestation of the great power of this novel approach for dealing with complexities of the real world.

11.14 Adaptive Neuro-Fuzzy Inferencing Systems

An alternative method of developing fuzzy models inspired by classical systems theory and recent developments in ANNs is based on input–output data. In this method, model development consists of two major phases as follows.

1. Structure identification
   The identification of the structure of a fuzzy model includes the determination of input and output variables, the structure of the rules, number of rules in the fuzzy rule base, and the portioning of the input and output variables into fuzzy sets. It is one of the most difficult parts of model development and the most ill-defined process. It is more an art than a science, and not readily amenable to automated techniques. It is partially simplified if some expert's knowledge about the system is used with the observed data. Hence, it is useful for a system that is seen as a gray box rather than a black box.
2. Parameter identification
   Once a rough model is developed from the expert's knowledge, then input–output data are used to generate weights or probabilities associated with the importance of potential rules, which are related with the credibility of rules. The concept of learning the weights of the rules from data was devised by Tong (1978). Kasko (1997) developed a more general and computationally efficient method for modeling fuzzy systems.

Parameter identification is also closely related to the estimation of membership functions of fuzzy sets or fuzzy relations. The first works in this direction were based on Sanchez's fundamental results on the solutions of fuzzy relational equations (Sanchez, 1974). Real success in parameter identification of fuzzy models has been achieved after a paper was presented by Tagaki and Sugeno (1985) that introduced a new method of fuzzy reasoning. The successful coupling of fuzzy logic with neural networks has supplied a new and powerful tool for parameter identification of fuzzy models with the use of the back-propagation method.

The fuzzy logic model is developed on the basis of causal relationships between the variables. A causal relationship, as depicted by the causal-loop diagram of causal models, is often fuzzy in nature. For example, there is no simple calculation to find the answer of "how much load increase in an interconnected power system would result in a given change in power angle?" If one would like to answer this question, it is necessary to simulate a complex differential or a partial differential equation for the interconnected power system. This exercise is quite time consuming and cumbersome. It can be represented by a simple causal relationship, as shown in Figure 11.31, with appropriate fuzzy membership functions for linguistic values of load and power angles.

Modeling Using Fuzzy Systems

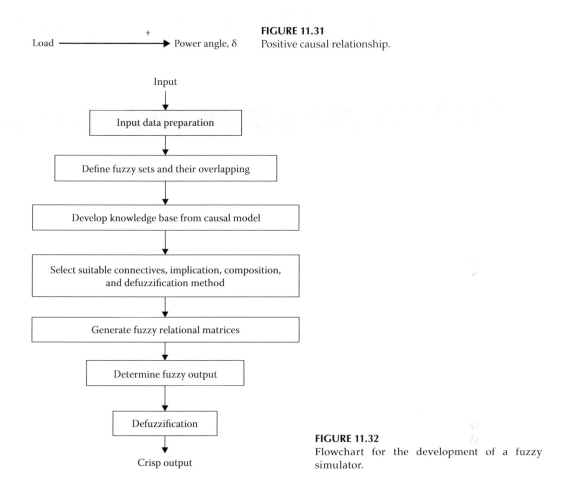

FIGURE 11.31
Positive causal relationship.

FIGURE 11.32
Flowchart for the development of a fuzzy simulator.

The system dynamics technique is used for modeling electrical machines (Chaturvedi, 1992, 1996). The knowledge gathered in a causal model is used to develop the fuzzy rule base. The flowchart for the development of a fuzzy simulator is shown in Figure 11.32.

Fuzzy logic model development needs the following vital decisions:

1. Selection of input and output variables
2. Normalization range of these variables
3. The number of fuzzy sets for each variable
4. Selection of membership functions for each fuzzy set
5. Determination of overlapping of fuzzy sets
6. Acquiring knowledge and defining appropriate number of rules
7. Selection of intersection operators
8. Selection of union operators
9. Selection of implication methods
10. Selection of compositional rules
11. Selection of defuzzification methods

1. Selection of variables, their normalization range, and the number of linguistic values

   The selection of variables depends on the problem situation and the type of analysis one would like to perform. For example, if one needs to conduct steady state analysis, the variables will be different from those for transient analysis. The reason behind this is that in transient analysis it is necessary to consider the dynamics of the system, while in steady state analysis we are only interested in the final output of the system, rather than how it comes to it. Hence, for transient analysis, more variables have to be handled. After identifying the variables for a particular system one can draw the casual-loop diagram for formulating the fuzzy rules.

   Once the variables are identified, it is also necessary to decide about the normalization range for input and output variables, and then specify the number of linguistic values for each variable. The general experience is that as the number of fuzzy sets increases, the computational time increases exponentially. Complex problems having numerous variables require enormous simulation time.

   A linguistic variable is fully characterized by a quintuple (V, T, X, G, M), as shown in Figure 11.33.

   V—name of the fuzzy or linguistic variable

   T—linguistic value/terms for linguistic variable, V

   X—universal set

   G—syntactic rule (grammar) for generating T

   M—semantic rule that assigns to each T

2. Selection of shapes of membership functions for each linguistic value

   A fuzzy membership function can be a triangular function, a trapezoidal function, an S-shaped function, a π-function, or an exponential function. For different fuzzy sets, one can use different fuzzy membership functions. The most popular membership function is the triangular function due to its simplicity and ease in calculations. In the present research work, modeling and simulation of the basic commutating machine have been performed using the triangular fuzzy membership function.

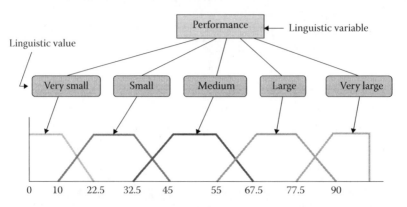

FIGURE 11.33
Quintuple for fuzzy (linguistic) variable.

Modeling Using Fuzzy Systems

3. Determination of overlapping of fuzzy sets
   As one fuzzy set reaches its end, it is not necessary that the next fuzzy set should continue from where the previous one ended. In other words, a particular value of the variable may belong to more than one fuzzy set. Thus, there is an overlap of two or more fuzzy sets. It is observed that overlapping of neighboring fuzzy sets affects the results to a great extent.

4. Selection of fuzzy intersection operators
   The intersection of two fuzzy sets, A and B, is specified in general by a binary operation on the unit, that is, a function of the form—$i: [0, 1] \times [0, 1] \rightarrow [0, 1]$.

   For each element, x, of the universal set, this function takes as its argument the pair consisting the element's membership grades in set A and in set B, and yields the membership grade of the element in the set constituting the intersection of A and B. Thus,

   $$(A \cap B)(x) = i[A(x), B(x)] \quad \forall x \in X$$

   Various fuzzy intersection operators are defined by different choices of parameter values, as given in Table 11.7.

5. Selection of fuzzy union operators
   Fuzzy unions are close parallels of fuzzy intersections. Like the fuzzy intersection, the general fuzzy union of two fuzzy sets, A and B, is specified by a function, $u: [0, 1] \times [0, 1] \rightarrow [0, 1]$.

   The argument of this function is the pair consisting the membership grade of some element, x, in fuzzy set A and the membership grade of that same element in fuzzy set B. This function returns the membership grade of the element in the set, $A \cup B$. Thus,

   $$(A \cup B)(x) = u[A(x), B(x)] \quad x \in X$$

TABLE 11.7

Fuzzy Intersection Operators (T-Norms)

Year	Name	Intersection Operator, i	Parameter
—	Schweizer and Sklar 2	$1 - [(1-a)^w + (1-b)^w - (1-a)^w(1-b)^w]^{1/w}$	$w > 0$
—	Schweizer and Sklar 3	$\exp(-(|\ln a|^w + |\ln b|^w)^{1/w})$	$w > 0$
—	Schweizer and Sklar 4	$ab/[a^w + b^w - a^w b^w]^{1/w}$	$w > 0$
1963	Schweizer and Sklar 1	$\{\max(0, a^{-w} + b^{-w} - 1)\}^{-1/w}$	$-\infty < w < \infty$
1978	Hamcher	$ab/[w + (1-w)(a+b-ab)]$	$w > 0$
1979	Frank	$\log_s[1 + \{(w^a - 1)(w^b - 1)/(w-1)\}]$	$w > 0, w \neq 1$
1980	Yager	$1 - \min\{1, [(1-a)^w + (1-b)^w]^{1/w}\}$	$w > 0$
1980	Dubois and Prade	$ab/\max(a, b, w)$	$w \notin [0, 1]$
1982	Dombi	$[1 + \{(1/a) - 1)^w + ((1/b) - 1)^w\}^{1/w}]^{-1}$	
1983	Weber	$\max(0, \{(a + b + wab - 1)/(1 + w)\})$	$w > -1$
1985	Yu	$\max[0, (1 + w)(a + b - 1) - wab]$	$w > -1$

TABLE 11.8

Fuzzy Unions (T-Conorms)

Year	Name	Union Operator, u	Parameter
—	Schweizer and Sklar 2	$[a^w + b^w - a^w b^w]^{1/w}$	$w > 0$
—	Schweizer and Sklar 3	$1 - \exp(-(|\ln(1-a)|^w + |\ln(1-b)|^w)^{1/w})$	$w > 0$
—	Schweizer and Sklar 4	$1 - [(1-a)(1-b)/[(1-a)^w + (1-b)^w - (1-a)^w(1-b)^w]^{1/w}$	$w > 0$
1963	Schweizer and Sklar 1	$1 - \{\max(0, (1-a)^w + (1-b)^w - 1)\}^{1/w}$	$w > 0$
1978	Hamcher	$[a + b + (w-2)ab]/[w + (1-w)ab]$	$w > 0$
1979	Frank	$1 - \log_s[1 + \{(w^a - 1)(w^b - 1)/(w - 1)\}]$	$w > 0, w \neq 1$
1980	Yager	$\min\{1, [a^w + b^w]^{1/w}\}$	$w > 0$
1980	Dubois and Prade	$1 - [(1-a)(1-b)/\max((1-a), (1-b), w)]$	$w \in [0, 1]$
1982	Dombi	$1/[1 + \{(1/a) - 1)^w + ((1/b) - 1)^w\}^{-1/w}]$	$w > 0$
1983	Weber	$\min(1, \{a + b + wab/(1-w)\}$	$w > -1$
1985	Yu	$\min[1, a + b + wab]$	$w > -1$

Various fuzzy unions are defined by different choices of parameters, as mentioned in Table 11.8. It is possible to define alternative compensatory operators (Mizumoto, 1983) by taking the convex combination of min (∩) and max (∩), as given below:

$$(x \cap y)^{(1-w)} \cdot (x \cup y)^w, \quad 0 \leq w \leq 1$$

Hence, in addition to the intersection (*t*-norm) and union (*t*-conorm) operators, the fuzzy simulator that has been developed also accommodates fuzzy compensatory operators, some of which can be obtained using *t*-norms, *t*-conorms, averaging operators, and compensatory operators. The 3D surfaces for compensatory operators for different parameters may be drawn with MATLAB programs.

6. Selection of implication methods

To select an appropriate fuzzy implication for approximate reasoning under each particular situation is a difficult problem. The logical connective implication (Ying 2002; Yaochu, 2003), that is, $P \rightarrow Q$ (P implies Q), in classical theory is $T(P \rightarrow Q) = T(P' \cup Q)$.

Various types of implication operators are summarized in Table 11.9.

7. Selection of compositional rules

Modus ponens deduction is used as a tool for making inferences in rule-based systems. A typical if-then rule is used to determine whether an antecedent (cause or action) infers a consequent (effect or reaction). Suppose that we have a rule of the form, IF A THEN B, where A is a set defined on universe X and B is a set defined on universe Y. This can be translated into a relation between sets A and B, that is, $R = (A \times B) \cup (A' \times Y)$. Now suppose that a new antecedent, say A_a, is known and we want to find its consequence, B_b:

$$B_b = A_a \circ R = A_a \circ ((A \times B) \cup (A' \times Y))$$

TABLE 11.9
List of Fuzzy Implications

Year	Name of Implication	Implication Function
1973	Zadeh	$\max[1-a, \min(a, b)]$
1969	Gaines–Rascher	1 if $a \leq b$
		0 if $a > b$
1976	Godel	1 if $a \leq b$
		0 if $a > b$
1969	Goguen	1 if $a \leq b$
		b/a if $a > b$
1938, 1949	Kleene–Dienes	$\max(1-a, b)$
1920	Lukasiewicz	$\min(1, 1-a+b)$
1987	Pseudo-Lukasiewicz–1	$\min[1, \{1-a+(1+w)b\}/(1+wa)]$ $w > -1$
1987	Pseudo-Lukasiewicz–2	$\min[1, (1-a^w+b^w)^{1/w}]$ $w > 0$
1935	Reichenbach	$1-a+ab$
1980	Willmott	$\min[\max(1-a), \max(a, 1-a),$ $\max(b, 1-b)]$
1986	Wu	1 if $a \leq b$
		$\min(1-a, b)$ if $a > b$
1980	Yager	1 if $a = b = 0$
		b^a others
1994	Klir and Yaun–1	$1-a+a^2b$
1994	Klir and Yaun–2	b if $a = 1$
		$1-a$ if $a \neq 1, b \neq 1$
		1 if $a \neq 1, b = 1$
	Mamdani	$\min(a, b)$
	Stochastic	$\min(1, 1-a+ab)$
	Correlation product	$a \cdot b$
1993	Vadiee	$\max(ab, 1-a)$

where the symbol, o, denotes the composition operation. Modus ponens deduction can also be used for the compound rule, IF A, THEN B, ELSE C, where this compound rule is equivalent to the relation:

$$R = (A \times B) \cup (A' \times C) \quad \text{or} \quad R = \max(\min(A, B), \min(A', C))$$

This is also called max–min compositional rule. There are various other compositional rules also, which are given in Table 11.10 (Chaturvedi, 2008).

8. Selection of defuzzification methods
   In the modeling and simulation of systems, the output of a fuzzy model needs to be a single scalar quantity, as opposed to a fuzzy set. The conversion of a fuzzy quantity to a precise quantity is called defuzzification, just as fuzzification is the conversion of a precise quantity to a fuzzy quantity. There are a number of methods in the literature, among them many that have been proposed by investigators in recent years, are popular for defuzzifying fuzzy outputs (Yager, 1993).
   The fuzzy logic system development software works as shown in Figure 11.34.

TABLE 11.10

Compositional Rule

1.	Max–min	$b = \max\{\min(a, r)\}$
2.	Min–max	$b = \min\{\max(a, r)\}$
3.	Min–min	$b = \min\{\min(a, r)\}$
4.	Max–max	$b = \max\{\max(a, r)\}$
5.	Godel	$r = 1$ if $a \leq r$
		$r = r$ otherwise
		$b = \max(1, r)$
6.	Max–product (max–dot)	$b = \max(a, r)$
7.	Max–average	$b = \max((a + r)/2)$
8.	Sum–product	$b = f(\Sigma(a, r))$

Note: a, b, and r are the membership values, $\mu_a(x)$, $\mu_b(y)$, and $\mu_r(x, y)$; f is a function.

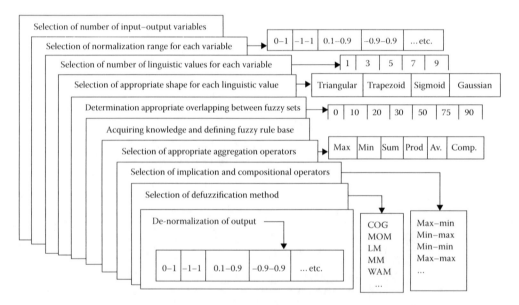

FIGURE 11.34
Different windows of a fuzzy system.

11.15 Steady State DC Machine Model

The schematic diagram of a dc machine is shown in Figure 11.35. It consists of three ports, that is, armature port, field port, and shaft port. The measurements are done at each port for the pair of complementary terminal variables with associated positive reference directions. In the motoring mode of operation of the dc machine at the field port, the field current is kept constant.

The steady state model of the dc machine contains fewer variables as compared to the transient model. In the case of a dc machine steady state model, suppose that we want

Modeling Using Fuzzy Systems

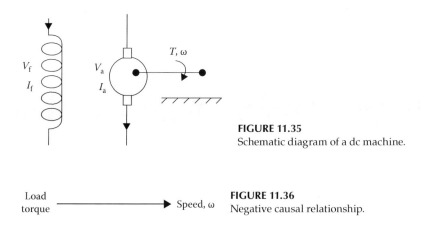

FIGURE 11.35
Schematic diagram of a dc machine.

FIGURE 11.36
Negative causal relationship.

to find the effect of load torque on motor speed. Then speed and load torque are the variables of interest. Hence, the causal relationship is identified between these variables. As load torque increases, machine speed reduces (see Figure 11.36), assuming that other variables are constant.

The causal relationships (links) are fuzzy in nature; hence, a fuzzy logic system could help in incorporating the beliefs and the perception of the modeler in a scientific way. It also provides a methodology for the qualitative analysis of system dynamics models. Since most of the concepts in natural language are fuzzy, a fuzzy set theoretic approach provides the best solution for such problems. It is also quite an effective way of dealing with complex situations. Hence, a model with an integrated fuzzy logic approach and the system dynamics technique is a natural choice for dealing with these types of circumstances. The system dynamics technique helps in identifying the relationship between variables and fuzzy logic helps in modeling these fuzzy-related variables.

Step 1 Identifying linguistic variables

The very first step in the modeling and simulation of this integrated approach is to identify variables for the given situation and then determine the causal links (relationships) between them keeping the remaining system variables unchanged.

For dc machine modeling under steady state conditions, the variables will be load torque, speed, and armature current.

Step 2 Defining ranges of linguistic variables

Once the variables are identified, the next step is to find the possible meaningful range of variation of these variables. This could be obtained by the experimental results or an operator's experience. The operator can easily tell about the variable range for a particular application and specific context. The typical range of variables is given in Table 11.10.

Step 3 Defining linguistic values for variables

Linguistic values for linguistic variables, that is, load torque (T_L), speed (ω), and armature current (I_a), have to be defined. Each of these variables are subdivided into an optimal number of linguistic values. In the case of a steady state model of a dc machine, five linguistic values are considered for each of these variables, such as very low (VL), low (L), medium (M), high (H), and very high (VH). The linear shape (membership function) of these linguistic values can be represented in the form of Table 11.11. Graphically, it can be shown in Figure 11.37.

TABLE 11.11

Range of Variables for a Steady State DC Machine Model

	Load Torque	Speed	Armature Current
Min value	0	150.88	3.0484
Max value	12	156.85	18.286

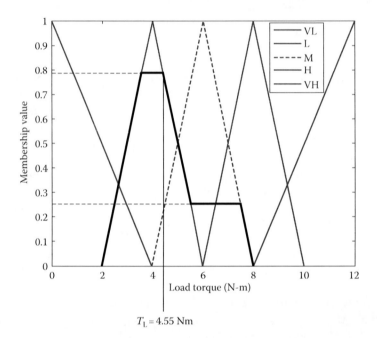

FIGURE 11.37
Fuzzification process.

$T_L = 4.55$ Nm

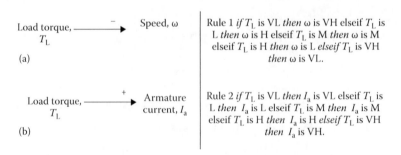

FIGURE 11.38
Causal links for a dc machine model and the corresponding fuzzy rules. (a) Negative causal link and (b) positive causal link.

Step 4 Defining rules

In the case of a dc machine, the causal link between variables may be drawn as shown in Figure 11.38. The arrow of the causal link represents the direction of influence and + or − signs show the effect. Depending on the signs, causal links can be categorized as positive or negative causal links.

Let us derive the fuzzy relational matrix for Rule 1 after max–min operations.

$$\text{Rule 1} = R(T_L, \omega)$$
$$= VL_{TL} \times VH_\omega + L_{TL} \times H_\omega + M_{TL} \times M_\omega + H_{TL} \times L_\omega + VH_{TL} \times VL_\omega$$

Rule 1

$$[1/0 + 0.75/1 + 0.5/2 + 0.25/3 + 0/4] \times [0/156.12 + 0.5/156.75 + 1/157.35$$
$$+ 0.5/158.04 + 0/158.85] + [0/2 + 0.50/3 + 1.0/4 + 0.50/5 + 0/6] \times [0/154.96$$
$$+ 0.5/155.56 + 1/156.12 + 0.5/156.75 + 0/157.35] + [0/4 + 0.50/5 + 1.0/6 + 0.50/7$$
$$+ 0/7] \times [0/153.71 + 0.5/154.33 + 1/154.96 + 0.5/155.56 + 0/156.12] + [0/6$$
$$+ 0.50/7 + 1.0/8 + 0.5/9 + 0/10] \times [0/152.22 + 0.5/153.00 + 1/153.71 + 0.5/154.33$$
$$+ 0/154.96] + [0/8 + 0.25/9 + 0.5/10 + 0.75/11 + 1/12]$$
$$\times [1/150.88 + 0.75/151.46 + 0.5/152.22 + 0.25/153.00 + 0/153.71];$$

Rule 1 =

	150.88	151.46	152.22	153.00	153.71	154.33	154.96	155.56	156.12	156.75	157.35	158.04	158.85
0										0.25	0.5	0.75	1.00
1										0.25	0.5	0.75	0.75
2										0.25	0.5	0.5	0.5
3								0.5000	0.5000	0.5000	0.2500	0.2500	0.2500
4								0.5	1	1			
5							0.5000	0.5	0.5	0.5			
6						0.5	1	0.5					
7				0.5	0.5	0.5	0.5	0.5					
8				0.5	1	0.5							
9	0.2500	0.2500	0.5000	0.5000	0.5000	0.5000							
10	0.5	0.5	0.25	0.25									
11	0.75	0.75	0.5	0.25									
12	1	0.75	0.5	0.25									

Similarly, other relational matrices can also be determined as given below:
Rule 2 $R(T_L, I_a)$

$$\begin{bmatrix}
 & 3.0484 & 4.3210 & 5.5905 & 6.8601 & 8.1271 & 9.3944 & 10.664 & 11.933 & 13.202 & 14.471 & 15.747 & 17.0151 & 18.286 \\
0 & 1 & 0.75 & 0.5 & 0.25 & & & & & & & & & \\
1 & 0.75 & 0.75 & 0.5 & 0.25 & & & & & & & & & \\
2 & 0.5 & 0.5 & 0.5 & 0.25 & & & & & & & & & \\
3 & 0.2500 & 0.2500 & 0.2500 & 0.5000 & 0.5000 & 0.5000 & & & & & & & \\
4 & & & & 0.5 & 0.5 & 0.5 & & & & & & & \\
5 & & & & 0.5 & 0.5 & 0.5000 & 0.5000 & 0.5 & & & & & \\
6 & & & & & 0.5 & 1 & 0.5 & & & & & & \\
7 & & & & & 0.5 & 0.5 & 0.5 & 0.5 & 0.5 & & & & \\
8 & & & & & & & 0.5 & 1 & 0.5 & & & & \\
9 & & & & & & 0.2500 & 0.2500 & 0.2500 & 0.5000 & 0.5000 & 0.5000 & & \\
10 & & & & & & & & & 0.25 & 0.25 & 0.5 & 0.5 & \\
11 & & & & & & & & & 0.25 & 0.5 & 0.75 & 0.75 & \\
12 & & & & & & & & & 0.25 & 0.5 & 0.75 & 1 & \\
\end{bmatrix}$$

Analysis

Let us take an example where the modeler is interested in studying the effect of load torque (say 4.55 N-m) on dc motor back EMF and armature current. First of all fuzzify the given crisp torque value into fuzzy linguistic values (fuzzification). Project $T_L = 4.55$ N-m to find the percentage of fuzzy linguistic terms.

Fuzzy load torque belongs to low and medium categories. It is 0.775 low and 0.275 medium torque, as shown in Figure 11.39.

$$\text{Fuzzy}(T_L) = [0/0 + 0/1 + 0/2 + 0.5/3 + 0.75/4 + 0.5/5 + 0.25/6 + 0.25/7 + 0/8$$
$$+ 0/9 + 0/10 + 0/11 + 0/12]$$

From these torque values, speed can be inferred from Rule 1.

$$\text{Fuzzy}(\omega) = \text{Fuzzy}(T_L) \circ \text{Rule 1}$$
$$= [0.25/153.00 + 0.25/153.71 + 0.5/154.33 + 0.5/154.96 + 0.5/155.56/ + 0.75/$$
$$156.12 + 0.75/156.75 + 0.25/157.35 + 0.25/158.04 + 0.25/158.85]$$

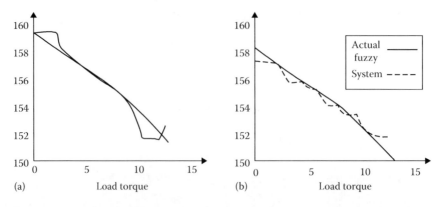

FIGURE 11.39
Study of different connectives with the Mamdani implication: (a) Yager connective ($w = 0.9$) and (b) max–min connective.

Modeling Using Fuzzy Systems

This fuzzified speed output is defuzzified using the weighted average method, as shown below:

$$\text{Crisp speed} = \frac{\sum_{i=1}^{n} W_i X_i}{\sum_{i=1}^{n} W_i}$$

$= (0.25 * 153.00 + 0.25 * 153.71 + 0.5 * 154.33 + 0.5 * 154.96$

$+ 0.5 * 155.56 + 0.75 * 156.12 + 0.75 * 156.75 + 0.25 * 157.35$

$+ 0.25 * 158.04 + 0.25 * 158.85) / (0.25 + 0.25 + 0.5 + 0.5 + 0.5 + 0.75$

$+ 0.75 + 0.25 + 0.25 + 0.25)$

$= 155.8388 \text{ rev/s}$

Similarly, armature current can also be determined for the given load torque using Rule 2:

$$\text{Rule 2 } R(T_L, I_a) \geq I_a = 8.98 \text{ A}$$

11.16 Transient Model of a DC Machine

A transient model provides information on the variation of different variables with respect to time in a system (transient behavior). Hence, the transient model needs to capture the dynamics of the system. To do that one needs to develop all possible causal links in the system and combine them to get a causal-loop diagram. The causal-loop diagram is the first step in modeling system dynamics using the system dynamics technique (Forrester, 1943). The causal-loop diagram for a dc machine is given in Chaturvedi (1993) and shown in Figure 11.40.

Field port variables are constant for a dc machine (basic commutating machines) in the electromechanical transducer mode of operation. Hence, the causal-loop diagram reduces to have only three causal loops after removing the causal loop of the field port (bottom portion of Figure 11.40). The reduced causal-loop diagram consists of two level variables, that is, armature current and angular speed; two rate variables (state variables), namely, rates of change of armature current and angular acceleration; exogenous variables, such as applied voltage and load torque; and machine parameters, like armature inductance, armature resistance, moment of inertia, and the damping coefficient of the machine. From this causal-loop diagram, fuzzy knowledge rules can be developed:

Rule 1 if I_a is VL then $\overset{o}{I}_a$ is VH elseif I_a is L then $\overset{o}{I}_a$ is H elseif I_a is M then $\overset{o}{I}_a$ is M elseif I_a is H then $\overset{o}{I}_a$ is L elseif I_a is VH then $\overset{o}{I}_a$ is VL

Rule 2 *if* ω *is* VL *then* $\overset{o}{\omega}$ *is* VH elseif ω is L *then* $\overset{o}{\omega}$ is H elseif ω is M *then* $\overset{o}{\omega}$ is M elseif ω is H *then* $\overset{o}{\omega}$ is L elseif ω is VH *then* $\overset{o}{\omega}$ is VL

Rule 3 *if* V_a is VL *then* $\overset{o}{I}_a$ is VL elseif **VA** is L then $\overset{o}{I}_a$ is L elseif V_a is M *then* $\overset{o}{I}_a$ is M elseif V_a is H then $\overset{o}{I}_a$ is H elseif V_a is VH *then* $\overset{o}{I}_a$ is VH

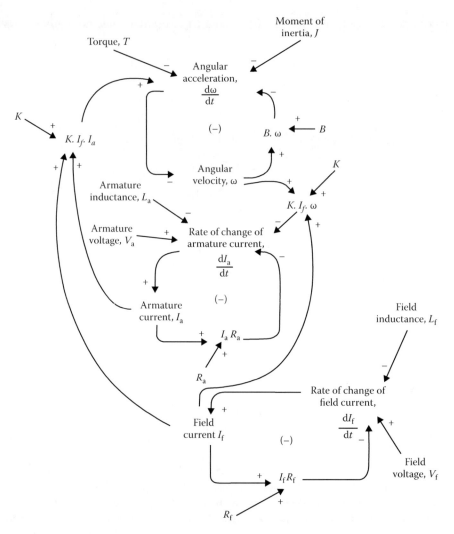

FIGURE 11.40
Causal-loop diagram for a dc machine with four negative causal loops.

Rule 4 *if* I_a *is* VL *then* $\acute{\omega}$ *is* VL *elseif* I_a *is* L *then* $\acute{\omega}$ *is* L *elseif* I_a *is* M *then* $\acute{\omega}$ *is* M *elseif* I_a *is* H *then* $\acute{\omega}$ *is* H *elseif* I_a *is* VH *then* $\acute{\omega}$ *is* VH

Rule 5 *if* T_L *is* VL *then* $\acute{\omega}$ *is* VH *elseif* T_L *is* L *then* $\acute{\omega}$ *is* H *elseif* T_L *is* M *then* $\acute{\omega}$ *is* M *elseif* T_L *is* H *then* $\acute{\omega}$ *is* L *elseif* T_L *is* VH *then* $\acute{\omega}$ *is* VL

Rule 6 *if* ω *is* VL *then* $\overset{o}{I}_a$ *is* VH *elseif* ω *is* L *then* $\overset{o}{I}_a$ *is* H *elseif* ω *is* M *then* $\overset{o}{I}_a$ *is* M *elseif* ω *is* H *then* $\overset{o}{I}_a$ *is* L *elseif* ω *is* VH *then* $\overset{o}{I}_a$ *is* VL

Rule 7 *if* $\overset{o}{I}_a$ *is* VL *then* **AC** *is* VL *elseif* $\overset{o}{I}_a$ *is* L *then* I_a *is* L *elseif* $\overset{o}{I}_a$ *is* M *then* I_a *is* M *elseif* $\overset{o}{I}_a$ *is* H *then* **AC** *is* H *elseif* $\overset{o}{I}_a$ *is* VH *then* I_a *is* VH

Rule 8 *if* $\acute{\omega}$ *is* VL *then* ω *is* VL *elseif* $\acute{\omega}$ *is* L *then* ω *is* L *elseif* $\acute{\omega}$ *is* M *then* ω *is* M *elseif* $\acute{\omega}$ *is* H *then* ω *is* H *elseif* $\acute{\omega}$ *is* VH *then* ω *is* VH

TABLE 11.12
Membership Function Values of Linguistic Variables

Linguistic Variables			Linguistic Values				
T_L	ω	I_a	Very Low	Low	Medium	High	Very High
0	150.88	3.0484	1	0	0	0	0
1	151.46	4.3210	0.75	0	0	0	0
2	152.22	5.5905	0.5	0	0	0	0
3	153.00	6.8601	0.25	0.5	0	0	0
4	153.71	8.1271	0	1.0	0	0	0
5	154.33	9.3944	0	0.5	0.5	0	0
6	154.96	10.664	0	0	1	0	0
7	155.56	11.933	0	0	0.5	0.5	0
8	156.12	13.202	0	0	0	1.0	0
9	156.75	14.471	0	0	0	0.5	0.25
10	157.35	15.747	0	0	0	0	0.5
11	158.04	17.0151	0	0	0	0	0.75
12	158.85	18.286	0	0	0	0	1.0

The above-developed fuzzy logic model for a dc machine has been simulated for the machine parameters given in Table 11.12.

1. Study of different connectives
   The fuzzy logic model is simulated for different t-norms and t-conorms, as given in the literature (Dubois and Prade, 1980, 85, 92; Klir, 1989, 95; Zimmerman, 1991), for example, Yager, Weber, Yu, Dubois and Prade, Hamcher, etc., with the Mamdani implication. The results are compared with actual results. All except Weber connectives are found to give satisfactory results in the range of load torque from 4 to 8 N-m. Dubois and Prade connectives perform better in the range from 1 to 11 N-m of load torque, but not in the complete range. It is considered expedient to try compensatory operators as connective. Hence, simulation has been performed for compensatory operators with the Gaines–Rascher implication and max–min compositions.

2. Study of different implication methods
   The dc machine fuzzy logic model is simulated for different implication methods such as Goguen, Kleene-Dienes, Godel, and Wu.

3. Different compositional rules
   The effects of different compositional rules on the fuzzy model simulation of a dc machine have been studied. The results obtained for various compositional rules are tabulated in Tables 11.13 through 11.16.

4. Different defuzzification methods
   A defuzzification method gives a crisp output from a fuzzy output. The effects of different defuzzification methods can be studied for a dc machine model with a compensatory operator as a connective, the Gaines–Rascher implication, and a max–min composition.

TABLE 11.13
Machine Parameters of a DC Machine

Armature resistance	R_a	= 0.40 ohm
Armature inductance	L_a	= 0.1 H
Damping coefficient	B	= 0.01 N-ms
Moment of inertia	J	= 0.0003 N-m s^2
Torque constant	K	= 13

Operating conditions

Applied voltage	V_a	= 125 V
Load torque	T_L	= 1 N-m

TABLE 11.14
Effects of Compositional Rules for Load Torque = 1 N-m

Composition	Speed (Actual Value = 158.04)	Armature Current (Actual Value = 4.321)
max(a, b) − min(a, b)	157.5641	10.67
max(a, b) − product(a, b)	156.8617	7.3117
max(a, b) − max(0, a + b − 1)	158.23083	3.9428
max(a + b − ab) − min(a, b)	157.0845	7.7276
min(1, a + b) − min(a, b)	157.4973	7.8753
min(1, a + b) − product(a, b)	158.339	5.985

TABLE 11.15
Effects of Compositional Rules for Load Torque = 4.55 N-m

Composition	Speed (Actual Value = 155.76)	Armature Current (Actual Value = 8.7647)
max(a, b) − min(a, b)	155.805	8.98
max(a, b) − product(a, b)	155.9298	8.7244
max(a, b) − max(0, a + b − 1)	155.13403	8.12858
max(a + b − ab) − min(a, b)	155.7086	10.2293
min(1, a + b) − min(a, b)	155.635	10.6725
min(1, a + b) − product(a, b)	155.7437	8.84

5. Effects of overlapping between membership functions
   The effects of different degrees of overlap between fuzzy sets have been studied and the simulation results are summarized in Table 11.17. It is found that for a fuzzy set, a slope of 0.5 and a membership variation of 0.1 give relatively better results.

Once the system model is developed, the system is controlled using a fuzzy controller. Fuzzy logic controllers are appropriate for systems with the following situations:

TABLE 11.16

Effects of Compositional Rules for Load Torque = 6.55 N-m

Composition	Speed (Actual Value = 154.46)	Armature Current (Actual Value = 11.304)
max(a, b) − min(a, b)	154.2701	10.67
max(a, b) − product(a, b)	154.6919	13.095
max(a, b) − max(0, $a + b − 1$)	154.7079	10.6642
max($a + b − ab$) − min(a, b)	154.3165	11.09594
min(1, $a + b$) − min(a, b)	154.5325	9.345
min(1, $a + b$) − product(a, b)	154.610	11.334

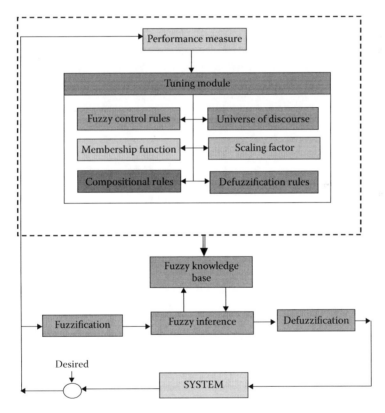

FIGURE 11.41
Block diagram for adaptive fuzzy control.

1. One or more continuous variables are present, which need to be controlled.
2. An exact mathematical model of the system does not exist. If it does exist then it is too complex to use for on-line controls.
3. Serious nonlinearities and delays are present in the system, which make the system model computationally expensive.
4. The system is time varying.

TABLE 11.17

Effects of Different Degrees of Overlap and Supports of Fuzzy Sets on Fuzzy Models

Slope of Fuzzy Set	Membership Variation	Speed	Armature Current
(a) Load torque = 1 N-m			
0.25	0–1	158.0175	4.3447
0.5	0–1	158.002	4.3452
0.375	0.25–1	157.8051	3.9294
0.25	0.5–1	157.653	3.7149
0.1875	0.25–1	157.4557	5.4967
0.125	0.5–1	157.3017	5.7984
(b) Load torque = 4.55 N-m			
0.25	0–1	155.48	9.4932
0.5	0–1	155.825	8.7613
0.375	0.25–1	155.707	9.015
0.25	0.5–1	155.6383	9.1841
0.1875	0.25–1	155.2684	9.93936
0.125	0.5–1	155.1192	10.2529
(c) Load torque = 6.55 N-m			
0.25	0–1	154.766	11.002
0.5	0–1	154.6325	11.2985
0.375	0.25–1	154.5292	11.5117
0.25	0.5–1	154.4275	11.7215
0.1875	0.25–1	154.8207	10.8849
0.125	0.5–1	154.8794	10.7594

5. The system requires an operator's intervention.
6. A written database of system information is not sufficiently available, but system expertise are available (i.e., a mental database is present). Hence, quantitative modeling is not possible, but a qualitative model could be developed.

In a fuzzy logic controller, the adaptation mechanism is to adjust the approximator causing it to match with some unknown nonlinear controller that will stabilize the plant and make a closed-loop system to achieve its performance objective, as shown in Figure 11.42.

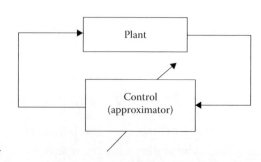

FIGURE 11.42
Direct adaptive control.

Modeling Using Fuzzy Systems 585

FIGURE 11.43
Cooling control for air conditioners.

Note that we call this scheme a "direct" adaptation scheme, since there is a direct adjustment of the parameters of the controller without identifying a model of the plant. This type of control is quite popular.

Neural networks and fuzzy systems can be used as "approximators" in the adaptive scheme. Neural networks are used for mapping nonlinear functions. Their parameters, for instance, weights and biases, are adjusted to map different shaped nonlinearities, as shown in Figure 11.43.

In recent years, the ANN (Guan, 1996; Liu, 2003) and the fuzzy set theoretic approaches (Gibbard, 1988; Hsu, 1990; Hassan, 1991; Ortmeyer, 1995; Wang, 1995; Gawish, 1999; Hosseinzadeh, 1999; Swaroop, 2005) have been proposed for many modeling and control applications. Both these techniques have their own advantages and disadvantages. The integration of these approaches can give improved results. Fuzzy sets can represent knowledge in an interpretable manner. Neural networks (Chaturvedi, 2004a,b,c) have the capability of interpolation over the entire range for which they have been trained, and also possess the capability of adaptability not possessed by fixed parameter devices designed and tuned for one operating condition. A generalized neuron-based PSS (Chaturvedi, 2004a,b,c, 2005) has been proposed to overcome the problems of the conventional ANN, and combining the advantages of self-optimizing adaptive control strategy and the quick response of the GN, make it more suitable for modeling complex systems. The main issue of ANN models is their training. To properly train an ANN, one needs sufficient and accurate data for training, efficient training algorithms, and an optimal structure of the ANN. It is really difficult to get appropriate and accurate training data for real-life problems. It is also difficult to select an optimal size of the ANN for a particular problem.

To overcome these problems, fuzzy logic models are proposed where no numerical training data is required and the operator's experience can be used. This makes fuzzy logic models very attractive for ill-defined systems or systems with uncertain parameters. With the help of fuzzy logic concepts, an expert's knowledge can be used directly to design a model. Fuzzy logic allows one to express the knowledge with subjective concepts, such as very big, too small, moderate, and slightly deviated, which are mapped to numeric ranges (Zadeh, 1973). Fuzzy logic applications for power system problems have been reported in the literature (Gibbard, 1988; Hsu, 1990; Hassan, 1991; Ortmeyer, 1995; Wang, 1995; Gawish, 1999; Hosseinzadeh, 1999; Swaroop, 2005). Due to its lower computation burden and its ability to accommodate uncertainties in the plant model, fuzzy logic appears to be suitable for implementation. Fuzzy models can be implemented through simple microcomputers with A/D and D/A converters (Hiyama, 1993; El-Metwally, 1996).

Example—Cooling Control of Air Conditioners

In the air-conditioning system shown in Figure 11.43, the temperature sensor output is given to the FKBS. The preprocessing of data from the temperature sensor is done for normalization. The preprocessed data is given to the fuzzification block for the conversion of crisp input values to fuzzy input values. A fuzzy membership function is defined for temperature and cooling, representing the meaning of the linguistic value of input and output variables of the system. The next step is to define the fuzzy rules, which form the relationship between temperature and cooling. In the problem, temperature is antecedent and cooling is consequent. After rule formation, individual

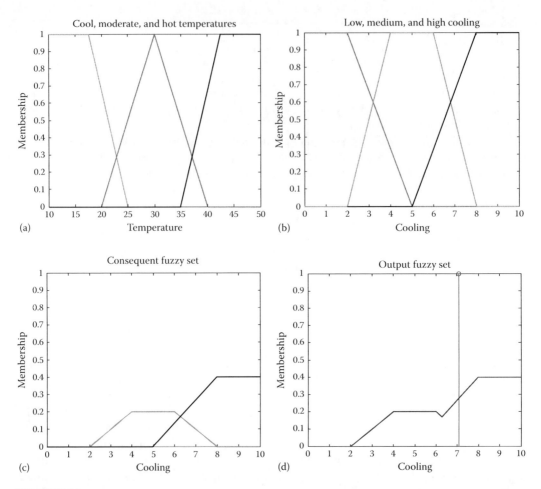

FIGURE 11.44
(a) Membership function of input "temperature." (b) Membership function of output "cooling." (c) Outputs of two rules that are fired. (d) Aggregated output of a fuzzy system with one input "temperature."

rule-based inference is done to compute the overall value of cooling on the individual contribution of each rule in the rule base. The last step applied is defuzzification, to convert the set of fuzzy inference cooling output into crisp cooling output as shown in Figure 11.44. The MATLAB code is written to perform each step in the fuzzy system clearly.

```
%Matlab Program for Cooling control system for Air Conditioners
x = [10:.01:50]; % Temperature
y = [0:.01:10]; % cooling
% Temperature membership function
cool_mf = trapmf(x,[10 10 17.5 25]);
moderate_mf = trimf(x,[20 30 40]);
hot_mf = trapmf(x,[35 42.5 50 50]);
antecedent_mf = [cool_mf;moderate_mf;hot_mf];
figure(1);
grid on;
plot(x,antecedent_mf)
title('Cool, Moderate and Hot Temperatures')
xlabel('Temperature')
ylabel('Membership')
```

Modeling Using Fuzzy Systems

```
% cooling
low_mf = trapmf(y,[0 0 2 5]);
medium_mf = trapmf(y,[2 4 6 8]);
high_mf = trapmf(y,[5 8 10 10]);
consequent_mf = [low_mf;medium_mf;high_mf];
figure(2);
grid on;
plot(y,consequent_mf)
title('Low, Medium and High high cooling')
xlabel('cooling')
ylabel('Membership')

temp=38
antecedent1 = cool_mf(find(x==temp));
antecedent2 = moderate_mf(find(x == temp));
antecedent3 = hot_mf(find(x == temp));
antecedent = [antecedent1;antecedent2;antecedent3]

consequent1 = low_mf.*antecedent1;
consequent2 = medium_mf.*antecedent2;
consequent3 = high_mf.*antecedent3;
figure(3);
plot(y,[consequent1;consequent2;consequent3])
axis([0 10 0 1.0])
title('Consequent Fuzzy Set')
xlabel('cooling')
ylabel('Membership')

Output_mf=max([consequent1;consequent2;consequent3]);
figure(4);
plot(y,Output_mf)
axis([0 10 0 1])
title('Output Fuzzy Set')
xlabel('cooling')
ylabel('Membership')

output=sum(Output_mf.*y)/sum(Output_mf);
output;
hold on;
stem(output,1);
```

The fuzzy system for the air conditioner has given a cooling output of 7.07 at 38°C using centroid defuzzification. If we are interested in the output for the complete range of temperature from 1°C to 50°C, then we have to implement a loop, as shown in the MATLAB code. In the given case, we have increased the range of the membership function for the temperature as per requirement as shown in Figure 11.45.

```
% Fuzzy model for cooling system
x = [0:1:60]; % Temperature
y = [0:1:10]; % cooling
% Temperature membership function
cool_mf = trapmf(x,[0 0 17.5 25]);
moderate_mf = trimf(x,[20 30 40]);
hot_mf = trapmf(x,[35 42.5 60 60]);
antecedent_mf = [cool_mf;moderate_mf;hot_mf];
figure(1);
grid on;
```

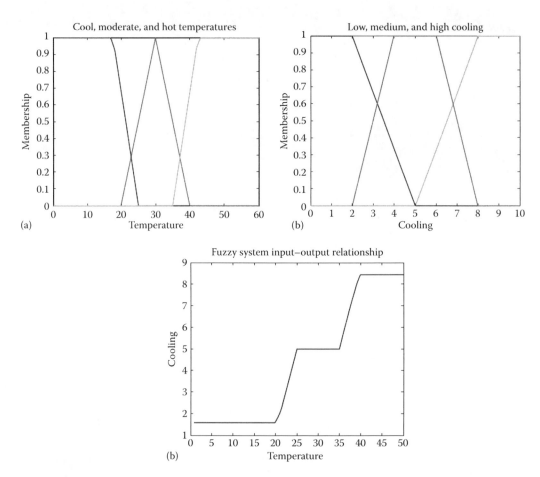

FIGURE 11.45
(a) Membership functions of input "temperature." (b) Membership functions of output "cooling." (c) Input–output relationship.

```
plot(x,antecedent_mf)
title('Cool, Moderate and Hot Temperatures')
xlabel('Temperature')
ylabel('Membership')

% cooling
low_mf = trapmf(y,[0 0 2 5]);
medium_mf = trapmf(y,[2 4 6 8]);
high_mf = trapmf(y,[5 8 10 10]);
consequent_mf = [low_mf;medium_mf;high_mf];
figure(2);
grid on;
plot(y,consequent_mf)
title('Low, Medium and High high cooling')
xlabel('cooling')
ylabel('Membership')

outputs=zeros(size([1:1:50]));
for temp=1:1:50
antecedent1 = cool_mf(find(x==temp));
```

```
    antecedent2 = moderate_mf(find(x == temp));
    antecedent3 = hot_mf(find(x == temp));
    antecedent = [antecedent1;antecedent2;antecedent3]

    consequent1 = low_mf.*antecedent1;
    consequent2 = medium_mf.*antecedent2;
    consequent3 = high_mf.*antecedent3;
    figure(3);
    plot(y,[consequent1;consequent2;consequent3])
    axis([0 10 0 1.0])
    title('Consequent Fuzzy Set')
    xlabel('cooling')
    ylabel('Membership')

    Output_mf=max([consequent1;consequent2;consequent3]);
    figure(4);
    plot(y,Output_mf)
    axis([0 10 0 1])
    title('Output Fuzzy Set')
    xlabel('cooling')
    ylabel('Membership')
    output=sum(Output_mf.*y)/sum(Output_mf);
    outputs(temp)=output;
    end
    plot([1:1:50],outputs)
    title('Fuzzy System Input Output Relationship')
    xlabel('Temperature')
    ylabel('cooling')
```

Considering the same system but with two input variables, which are temperature and humidity, the rule-based system changes (ref. Figures 11.46 and 11.47).

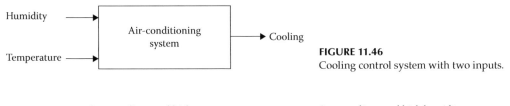

FIGURE 11.46
Cooling control system with two inputs.

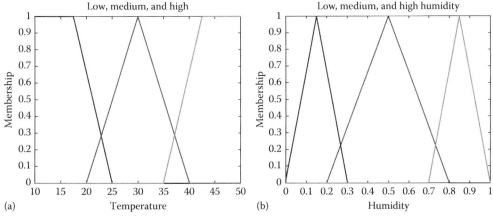

FIGURE 11.47
(a) Membership functions of input "temperature." (b) Membership functions of input "humidity."

(*continued*)

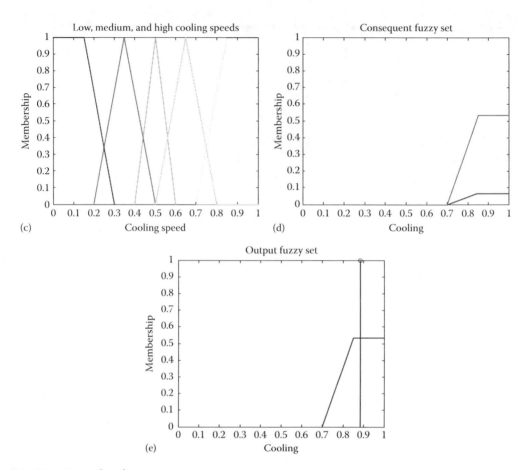

FIGURE 11.47 (continued)
(c) Membership functions of output. (d) Outputs of two rules that are fired. (e) Final output of fuzzy system.

```
%MATLAB Program for simulation Cooling system using Fuzzy
x = [10:.01:50]; % Temperature
y = [0:.001:1]; % humidity
z = [0:.001:1]; %cooling

% Membership functions for Temperature
templow_mf = trapmf(x,[10 10 17.5 25]);
tempmed_mf = trimf(x,[20 30 40]);
temphigh_mf = trapmf(x,[35 42.5 50 50]);
antecedent_mf = [templow_mf;tempmed_mf;temphigh_mf];
figure(1);
grid on;
plot(x,antecedent_mf)
title('low, medium and high')
xlabel('Temperature')
ylabel('Membership')

% Membership functions for humidity
humlow_mf = trimf(y,[0 .15 .3]);
hummed_mf = trimf(y,[.2 .5 .8]);
humhigh_mf = trimf(y,[.7 .85 1]);
```

Modeling Using Fuzzy Systems

```
antecedent_mf = [humlow_mf;hummed_mf;humhigh_mf];
figure(2);
grid on;
plot(y,antecedent_mf)
title('low, medium, and high humidity')
xlabel('humidity')
ylabel('Membership')

% Membership functions for cooling
coollow_mf = trapmf(z,[0 0 .15 .3]);
coolmediumlow_mf = trimf(z,[.2 .35 .5]);
coolmedium_mf = trimf(z,[.4 .5 .6]);
coolmediumhigh_mf= trimf(z,[.5 .65 .8]);
coolhigh_mf=trapmf(z,[.7 .85 1 1]);
consequent_mf =
[coollow_mf;coolmediumlow_mf;coolmedium_mf;coolmediumhigh_mf;coolhigh_
mf];
figure(3);
grid on;
plot(z,consequent_mf)
title('Low, Medium and High cooling Speeds')
xlabel('cooloing Speed')
ylabel('Membership')

temp=46
humid=.78
% Fuzzy Rule base
antecedent1 = min(templow_mf(find(x==temp)),humlow_mf(find(y==humid)));
antecedent2 = min(templow_mf(find(x==temp)),hummed_mf(find(y==humid)));
antecedent3 = min(templow_mf(find(x==temp)),humhigh_mf(find(y==humid)));

antecedent4 = min(tempmed_mf(find(x==temp)),humlow_mf(find(y==humid)));
antecedent5 = min(tempmed_mf(find(x==temp)),hummed_mf(find(y==humid)));
antecedent6 = min(tempmed_mf(find(x==temp)),humhigh_mf(find(y==humid)));

antecedent7 = min(temphigh_mf(find(x==temp)),humlow_mf(find(y==humid)));
antecedent8 = min(temphigh_mf(find(x==temp)),hummed_mf(find(y==humid)));
antecedent9 = min(temphigh_mf(find(x==temp)),humhigh_mf(find(y==humid)));

antecedent =
[antecedent1;antecedent2;antecedent3;antecedent4;antecedent5;antecedent6;
antecedent7;antecedent8;antecedent9]

consequent1 = coollow_mf.*antecedent1;
consequent2 = coollow_mf.*antecedent2;
consequent3 = coolmediumlow_mf.*antecedent3;

consequent4 = coolmediumlow_mf.*antecedent4;
consequent5 = coolmedium_mf.*antecedent5;
consequent6 = coolmediumhigh_mf.*antecedent6;

consequent7 = coolmediumhigh_mf.*antecedent7;
consequent8 = coolhigh_mf.*antecedent8;
consequent9 = coolhigh_mf.*antecedent9;

figure(4);
plot(z,[consequent1;consequent2;consequent3;consequent4;consequent5;cons
equent6;consequent7;consequent8;consequent9])
```

```
axis([0 1 0 1.0])
title('Consequent Fuzzy Set')
xlabel('Cooling')
ylabel('Membership')

Output_mf=max([consequent1;consequent2;consequent3;consequent4;consequen
t5;consequent6;consequent7;consequent8;consequent9]);
figure(5);
plot(z,Output_mf)
axis([0 1 0 1])
title('Output Fuzzy Set')
xlabel('Cooling')
ylabel('Membership')

output=sum(Output_mf.*y)/sum(Output_mf);
output;
hold on;
stem(output,1);
```

11.17 Fuzzy System Applications for Operations Research

Engineers and scientists are sometimes required to select, from among a set of feasible solutions, the solution that is best in some sense, such as the least expensive solution or the solution with the highest accuracy and the best performance. This process is referred to as optimal solution or simply optimization.

This section is intended to present the state of the art concerning the application of fuzzy sets to operations research, such as optimization. The conceptual framework for optimization in a fuzzy environment is reviewed and particularized to fuzzy linear programming. Optimization models in operations research assume that data are precisely known, that constraints delimit a crisp set of feasible decisions, and that criteria are well defined and easy to formalize. However, in the real world, such assumptions are not true. Hence, the simple linear programming method of optimization is not quite effective. Fuzzy linear programming is described in this section.

1. Problem formulation

    Let X be a set of alternatives that contains the solution of a given multi-criteria optimization problem. Bellman and Zadeh pointed out that in a fuzzy environment, goals and constraints formally have the same nature and can be represented by fuzzy sets on X. Let C_i be the fuzzy domain delimited by the ith constraint ($i = 1, ..., m$) and G_j the fuzzy domain associated with the jth goal ($j = 1, ..., n$). G_j is, for instance, the optimizing set of an objective function, g_j. When goals and constraints have the same importance, Bellman and Zadeh (1970) called the fuzzy set (D) on X, as the *fuzzy decision*:

    $$D = \left(\bigcap_{i=1,...m} C_i \right) \cap \left(\bigcap_{j=1,...n} G_j \right) \tag{11.37}$$

    that is,

    $$\forall_x \in X, \quad \mu_D(x) = \min\left[\min_{i=1,m} \mu_{C_i}(x), \min_{j=1,n} \mu_{G_j}(x) \right] \tag{11.38}$$

Modeling Using Fuzzy Systems

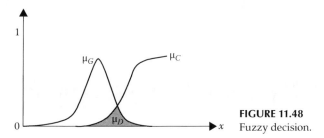

FIGURE 11.48
Fuzzy decision.

A fuzzy decision is shown in Figure 11.48 that corresponds to a constraint "x should be substantially greater than x_0" and an objective function, g, whose optimizing set is G.

The final decision, x_f, can be chosen in the set

$$M_f = \{(x_f, \mu_D(x_f) \geq \mu_D(x), \forall x \in X)\}$$

M_f is called the *maximal decision set*.

When criteria and constraints have unequal importance, membership functions can be weighted by x-dependent coefficients, α_i and β_j, such that

$$\forall x \in X, \quad \sum_{i=1}^{m} \alpha_i(x) + \sum_{j=1}^{n} \beta_j(x) = 1 \tag{11.39}$$

and according to Bellman and Zadeh (1970)

$$\mu_D(x) = \sum_{i=1}^{m} \alpha_i(x)\mu_{Ci}(x) + \sum_{j=1}^{n} \beta_j(x)\mu_{Gj}(x) \tag{11.40}$$

Note that D satisfies the property:

$$\left(\bigcap_{i=1,m} C_i\right) \cap \left(\bigcap_{j=1,n} G_j\right) \subseteq D \subseteq \left(\bigcup_{j=1,m} C_i\right) \cup \left(\bigcup_{j=1,n} G_j\right)$$

However, other aggregation patterns for the μ_{Ci} and the μ_{Gj} may be worth considering. When criteria and constraints refer to different sets, X and Y, respectively, and there is some causal link between X and Y, a fuzzy decision can still be constructed. X is, for instance, a set of causes constrained by C_i, $i = 1, \ldots, m$, and Y a set of effects on which is defined a set of fuzzy goals, G_j, $j = 1, \ldots, n$. Let R be a fuzzy relation on $X \times Y$. The fuzzy decision, D, can be defined on X by aggregation of the fuzzy domains, C_i, and the fuzzy goals, G_j o R^{-1}, induced from G_j.

The definition of an optimal decision by maximizing μ_D (in the sense of Equation 11.37) is not always satisfactory, especially when $\mu_D(x_f)$ is very low. It indicates that goals and constraints are more or less contradictory, and thus x_f cannot be a good solution. For such a situation, Asai et al. (1975) have proposed the following approach: Choose an alternative that better satisfies the constraints and substitutes an attainable short-range goal for the unattainable original one. More specifically, we must find a pair, (x_c, x_G), where x_c is a short-range optimal decision and X_G a short-range estimated goal, and (x_c, x_G) maximizes:

$$\mu_D(x, x') = \min(\mu_c(x), \mu_G(x'), \mu_R(x, x')) \tag{11.41}$$

C and G are, respectively, the fuzzy constrained domain and the fuzzy goal (we take $m = n = 1$ for simplicity), and R expresses a fuzzy tolerance on the discrepancy between the immediate optimal decision, x_c, and the fuzzy goal, G; x_G is the most reasonable objective because it is a trade-off between a feasible decision and G. Asai et al. (1975) discussed the choice of R and found that a likeness relation was most suitable with respect to some natural intuitive assumptions. The authors generalized their approach to the N-period case where N short-range decisions must be chosen together with N short-range goals and μ_c and μ_G may be time dependent (see Asai et al., 1975). Some definitions pertaining to time dependency in fuzzy set theory in the scope of planning may be found in Lientz (1972).

2. Fuzzy linear programming

    a. *Soft constraints*

    We start with the following problem:

    Minimize $Z = gx$

    Subjected to $Ax \leq b, x \geq 0$

where

g is a vector of coefficients of the objective function
b is a vector of constraints
A is the matrix of coefficients of constraints

The fuzzy version of this problem is (Zimmermann, 1976, 1978)

$$gx \lesssim Z_0, \quad A_x \lesssim b, \quad x \geq 0 \tag{11.42}$$

The symbol $\lesssim$ denotes a relaxed version of $<$ and assumes the existence of a vector, μ, of membership functions, μ_i, $i = 0, \ldots, m$, defined as follows: Let a_{ij} and b_i be the coefficients of A and b, respectively; then, for $l = 0, \ldots, m$:

$$\mu_i\left(\sum_{j=1}^{n} a_{ij} x_j\right) = \begin{cases} 1 & \text{for } \sum_{j=1}^{n} a_{ij} x_j \leq b_i \\ 1 - \frac{1}{d} \sum_{j=1}^{n}(a_{ij} x_j - b_i) & \text{for } b_i \leq \sum_{j=1}^{n} a_{ij} x_j \leq b_i + d_i \\ 0 & \text{for } \sum_{j=1}^{n} a_{ij} x_j > b_i + d_i \end{cases}$$

with $b_0 = Z_0$ and $a_{oj} = g_j$ (ref. Figure 11.49). d_i is a subjectively chosen constant expressing a limit of the admissible violation of the constraint, i. Z_0 is a constant to be determined.

Modeling Using Fuzzy Systems

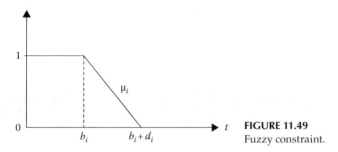

FIGURE 11.49
Fuzzy constraint.

The fuzzy decision of Equation 11.42 is D, such that

$$\mu_D(x) = \min_i \mu_i \left(\sum_{i=1}^{m} \alpha_{ij} x_j \right)$$

The maximization of μ_D is equivalent to the linear program:
Maximize $x_n + 1$

$$\text{Subjected to } x_{n+1} \leq \mu_i \left(\sum_{i=1}^{n} \alpha_{ij} x_j \right) \quad \text{where } i = 0, \ldots, m \quad (11.43)$$

The constant, $Z_0 + d_0$, is determined by solving the above equation (Equation 11.43) without the constraint, $I = 0$; let $\underline{x}$ be its solution. We state $Z_0 + d_0 = g\underline{x}$ and Z_0 is defined as the optimal value of the objective function in which b is replaced by $b_i + d_i$, $\forall_i$. The constraints, $Ax > b$ or $Ax = b$, can be softened in a similar way (Sommer and Pollatschek, 1976; Sularia, 1977).

3. Fuzzy constraints with fuzzy coefficients
   What happens to a linear constraint, $A_i x = b_i$, when the coefficients, a_{ij} and b_i, become fuzzy numbers $\tilde{A}_i x$ can be calculated by means of extended addition. The symbol = can be understood in two different ways:

   a. As a strict equality between A_{ix} and b_i (equality of the membership functions). This equality can be weakened into an inclusion, $A_{ix} \subseteq b_i$, which also reduces to an equality in the nonfuzzy case; the fuzziness of b_i is interpreted as a maximum tolerance for the fuzziness of $A_i x$.

   b. As an approximate equality between $A_i x$ and b.

Both points of view will be successively investigated. We assume here that the variables are positive ($x > 0$), and a_{ij} and b_i are fuzzy numbers.

   a. Tolerance constraints
      Consider the system of linear fuzzy constraints:

$$A_i x \subseteq b_i, \quad i = 1, m$$

Since the coefficients are fuzzy numbers, we can write symbolically:

$$A_i = (A_i \, \underline{A}_i \, \overline{A}_i)$$

where A_i, $\underline{A}_i$, and $\overline{A}_i$ are vectors of mean values and left and right spreads. Since T_i are positive, the system is equivalent to

$$A_{ix} = b_i, \quad A_{ix} \leq b_i, \quad A_{ix} < b_i, \quad i = 1,\ldots, m, \quad x \geq 0$$

which is an ordinary linear system of equalities and inequalities. According to the value of m and the number, n, of variables involved, it may or may not have solutions. Owing to this result, the "robust programming problem" (Negoita, 1976) is

Maximize gx

Subjected to $\tilde{A}_i x \subseteq b_i$, $i = 1, \ldots, m$; $x > 0$

which can be turned into a classical linear programming. This approach seems more tractable than that of Negoita (1976).

b. Approximate equality constraints

Let a and b be two fuzzy numbers such that a is greater than b, denoted $a > b$, as soon as $a - b > a + b$. a is said to be approximately equal to b if and only if neither $a \geq b$ nor $b \geq a$ holds.

A system of approximate equalities in the above sense can be considered, i.e.:

$$\tilde{A}_i x \approx b_i, \quad i = 1,\ldots, m$$

where A_i is a vector of fuzzy numbers, b_i, which are R-L fuzzy numbers in which "$\approx$" denotes approximate equality. This fuzzy system is equivalent to the nonfuzzy one:

$$b_i - A_i x \leq b_i + A_i x \quad \text{when } O \leq b_i - A_i x$$

$$A_i x - b_i \leq b_i + A_i x \quad \text{when } O \leq A_i x - b_i$$

The above approach assumes the existence of an equality threshold.

An alternative approach can be, as in a, to define the constraint domain associated with an approximate equality, $A_i x \approx b_i$, by the membership function, m_i, such that $\mu_i(x) = \text{hgt}(A_i x \cap b_i)$. More specifically:

$$\mu_i(x) = \begin{cases} R \, \dfrac{b_i - A_i x}{b_i + A_i x} & \text{if } b_i - A_i x \geq 0 \\ L \, \dfrac{A_i x - b_i}{b_i + A_i x} & \text{if } A_i x - b_i \geq 0 \end{cases}$$

The problem of finding x_j, maximizing min $i = 1, \ldots, m$ $\mu_i(x)$, the optimal decision with respect to the m fuzzy constraints, can be thus reduced to a nonlinear program.

This approach can be extended to fuzzy linear objective functions and to linear approximate inequality constraints.

Example—Students' Performance Evaluation System Using ANFIS

In this example, we deal with the analysis of students' performance based on students' attendance, and their internal and external scores. This information is used to train the ANN and the mathematical operation is performed by fuzzy logic. It is implemented for the real-time evaluation of students and has found the trends of students' performances quite close to the actual trend.

	Inputs		Outputs
Attendance	Internal	External	Performance
8.24	7.96	3.32	6.1
8.84	8.44	3.06	5.95
7.6	7.64	3.16	5.57
9.36	8.8	4.18	6.93
9.24	7.9	6.2	8.15
8.45	7.54	5.3	7.9
8.65	9.4	4.2	6.9
8.05	7.88	3.1	5.56
7.6	7.6	4.1	6.42
7.5	8.1	5.2	7.36
8.84	9.24	4.6	7.11
8.13	7.12	2.58	5.39
8.16	7.68	2.98	5.8
9.4	9.52	4.6	7.2
9.14	9.16	4.1	7.14
9.2	8.4	4.7	7.25
8.5	8.9	5.2	7.34
8.46	8.12	2.9	5.95
9.55	9.64	4.7	7.63
8.42	8.69	6.1	8.1
8.33	8.1	5.7	7.6
8.1	8.2	5.3	7.7
7.6	7.3	4.2	6.7
8.95	8	2.82	5.62
8.7	8.6	4.3	7.62

To start ANFIS on the MATLAB, we type *anfisedit* on the command prompt of MATLAB. The ANFIS window appears on screen, as shown in Figure 11.50.

To load training data in the ANFIS environment, select the type *training* and press the *load data* option. The following window appears, as shown in Figure 11.51.

The loaded data are displayed in the ANFIS window as small circles, as shown in Figure 11.52. These data are used for training the system.

The next step is to generate the fuzzy system, as shown in Figure 11.53.

After defining the rules for the membership function, we set the parameters for training, using a neural network. Therefore, a neural network is used for generalization while a fuzzy one is used for specialization. The training error is shown in Figure 11.54.

The original training data and the ANFIS output are checked and shown in Figure 11.55.

We can predict any evaluation outputs for different inputs using fuzzy rules, as shown in Figure 11.56.

The ANFIS surfaces to show the effects of different pairs of inputs on students' performances are shown in Figures 11.57 through 11.59.

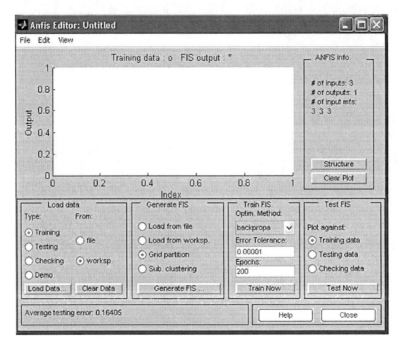

FIGURE 11.50
ANFIS window.

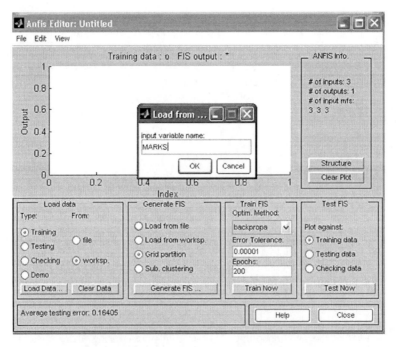

FIGURE 11.51
ANFIS editor with load data.

Modeling Using Fuzzy Systems

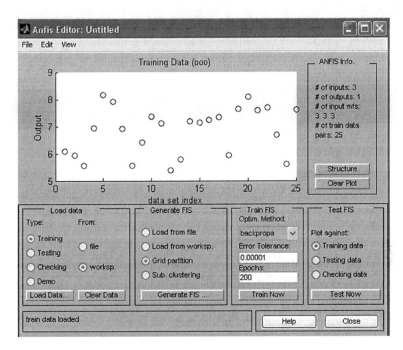

FIGURE 11.52
Training data.

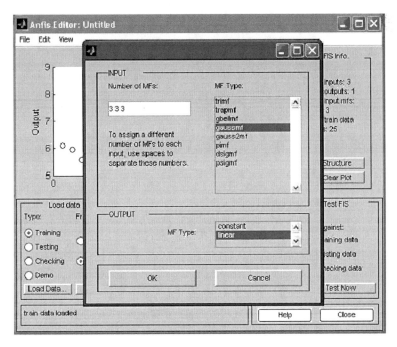

FIGURE 11.53
ANFIS editor with ANN structure window.

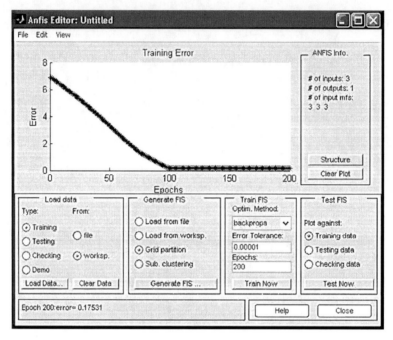

FIGURE 11.54
ANFIS training performance.

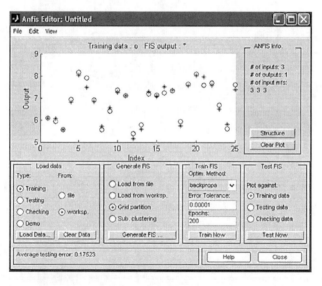

FIGURE 11.55
Training and checking data plot.

Modeling Using Fuzzy Systems

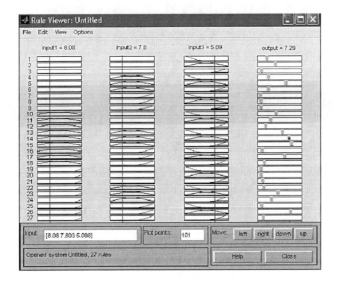

FIGURE 11.56
Rule viewer of ANFIS.

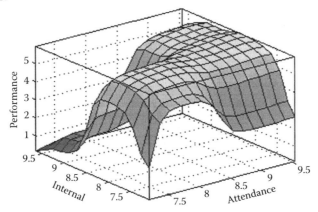

FIGURE 11.57
3D surface showing effects of attendance and internal evaluation of performances.

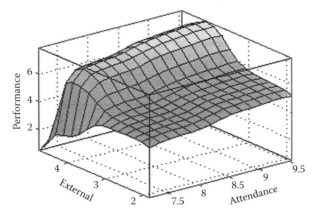

FIGURE 11.58
3D surface showing effects of attendance and external evaluation of performances.

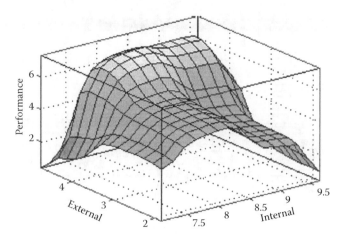

FIGURE 11.59
3D surface showing effects of internal and external evaluations of performances.

11.18 Review Questions

1. Consider a simple first-order plant model given by

$$\dot{x}(t) = Ax(t) + bu(t)$$

   where values of A and B are dependent on system parameters, and $u(t)$ is the input vector.
   Develop a fuzzy logic controller for this simple system and compare its performance with a conventional PI controller.

2. Explain qualitatively the influence of a membership function, crossover points, and the method of defuzzification, on fuzzy controllers.

3. Discuss the roles of normalization and scaling in generating a set of rules for the knowledge base.

4. List out various steps involved in the development of a fuzzy logic model for a given system.

5. Distinguish between self-organizing and fixed rules-based fuzzy logic approaches for modeling.

6. Explain the following rule:

If	PE = NB	and	CPE = not (NB or NM)	and	SE = ANY
and	CSE = ANY	then	HC = PB		
elseif	PF = NB or NM	and	CPE = NS	and	SE = ANY
and	CSE = ANY	then	HC = PM		

   (a) Represent the rule using max and min operators.
   (b) Represent the rule in a narrative style.

11.19 Bibliography and Historical Notes

Prof. Lotfi A. Zadeh is the father of fuzzy logic. He first coined the word "fuzzy" by publishing a seminal paper on fuzzy sets in 1965. Research progress in fuzzy logic is nicely summarized in a collection edited by Gupta, Saridis, and Gaines. E. Mamdani, which gives a survey of fuzzy logic control and discusses several important issues in design, development, and analysis of the fuzzy logic controller. Bart Kasko introduced the geometric interpretation of fuzzy sets and investigated the related issues in the late 1980s.

Fuzzy logic offers the option to try to model the nonlinearity of the functioning of the human brain when several pieces of evidence are combined to make an inference. There are several publications available on fuzzy aggregation operators, of which a few notable ones are Yager (1977, 1981, 1988, 1991, 1996), Trillas (1983), Czogala (1984), Dubois and Prade (1985, 1992), Combettes (1993), Cubillo (1995), Bandemer (1995), Klir and Yuan (1995), Bloch (1996), Radko (2000), Kaymak and Sousa (2003), Mendis (2006), and Ai-Ping (2006). The different compositional operators are described in Zeng (2005) and Wang (2006).

Fuzzy implication rules and the GMP were first introduced by Zadeh in 1975. This area soon became one of the most active research topics for fuzzy logic researchers. Gains (1976, 1983, 1985) described a wide range of issues regarding fuzzy logic reasoning, including fuzzy implications.

A book edited by T. Terano, M. Sugeno, and Y. Tuskamoto (1984) contains a chapter that describes several good applications of fuzzy expert systems. Negoita (1985) also wrote a book on fuzzy expert systems.

Pin (1993) implemented fuzzy behavioral approaches to mobile robot navigation, using special-purpose fuzzy inference chips.

The literature dealing with the use of neural networks for learning membership functions and inference rules is rapidly growing; the following are a few relevant references: Takagi and Hayashi (1991), Wang and Mendel (1992), Jang (1993), and Wang (1994). The overview of the role of fuzzy sets in approximate reasoning was prepared by Dubois and Prade (1991), and Nakanishi et al. (1993).

Fuzzy computer hardware is also one of the most growing areas. There is long list of publications in computer hardware (Gupta and Yamakawa, 1988; Diamond et al.). An important contribution to fuzzy logic hardware is a design and implementation of a fuzzy memory to store one-digit fuzzy information by Hirota and Ozawa (1989).

In the area of communication, the fuzzy set theoretic approach is used as fuzzy adaptive filters for nonlinear channel equalization (Wang and Mandel, 1993) and signal detection (Saade and Schwarzlander, 1994).

It is also used for the analysis of fuzzy cognitive maps (Styblinski and Meyer, 1991), and in robotics (Palm, 1992; Nedugadi, 1993; Chung, 1994), vision, identification, reliability and risk analysis, and medical diagnosis.

12

Discrete-Event Modeling and Simulation

12.1 Introduction

There are many systems which respond to environmental changes in different ways such as the aircraft and factory systems. The movement of the aircraft occurs smoothly, whereas the changes in the factory occur discontinuously. The ordering of raw materials or the completion of a product, for example, occurs at specific points in time.

Systems such as the aircraft, in which the changes are predominantly smooth, are called continuous systems. Systems like the factory, in which changes are predominantly discontinuous, are called discrete systems. Few systems are wholly continuous or discrete. The aircraft, for example, may make discrete adjustments to its trim as altitude changes, while, in the factory examples, machining proceeds continuously, even though the start and finish of a job are discrete changes. However, in most systems one type of change predominates so that systems can usually be classified as being discrete.

In addition, in the factory system, if the number of parts is sufficiently large, there may be no point in treating the number as a discrete variable. Instead, the number of parts might be represented by a continuous variable with the machining activity controlling the rate at which parts flow from one state to another. This is, in fact, the approach of a modeling technique called system dynamics.

There are also systems that are intrinsical but information about them is only available at discrete points in time. These are called sampled-data system. The study of such systems includes, the problem of determining the effects of the discrete sampling, especially when the intention is to control the system on the basis of information gathered by the sampling.

This ambiguity in how a system might be represented illustrates an important point. The description of a system, rather than the nature of the system itself determines what type of model will be used. A distinction needs to be made because the general programming methods are used to simulate continuous models and discrete models differently. However, no specific rules can be given as to how a particular system is to be represented. The purpose of the model, coupled with the general principle that a model should not be more complicated than it is needed, will determine the level of detail, and the accuracy with which a model needs to be developed. Weighing these factors and drawing on the experience of knowledgeable people will decide the type of model needed.

The rapid evolution of computing, communication, and sensor technologies has brought about the proliferation of "new" dynamic systems, mostly technological and often highly complex. Examples are all around us: air traffic control systems; automated manufacturing systems; computer and communication networks; embedded and networked systems; and software systems. The "activity" in these systems is governed by operational rules designed by humans; their dynamics are therefore characterized by asynchronous occurrences of discrete events. These features lend themselves to the term discrete event system (DES) for this class of dynamic systems. So what is meant by a DES?

A DES is defined as a dynamic system which evolves in accordance with the occurrence of physical events. This kind of system requires control and coordination to ensure the orderly flow of events. Many DES models have already been reported in literature, for example, Markov chains and Markov jump processes, Petri nets, queuing networks, finitely recursive processes, models based on minmax algebra, discrete-event simulation, and generalized semi-Markov processes (Cao and Ho, 1990).

Discrete-event models can further be distinguished along at least two dimensions from traditional dynamic system models—how they treat passage of time (stepped vs event-driven) and how they treat coordination of component elements (synchronous vs asynchronous). Some event-based approaches enable more realistic representation of loosely coordinated semiautonomous processes, while classical models such as differential equations and cellular automata tend to impose strict global coordination on such components. Event-based simulation is inherently efficient since it concentrates processing attention on events significant changes in states that are relatively rare in space and time, rather than continually processing every component at every time step.

Discrete-event modeling is a mathematical procedure that is created to describe a dynamic process. Then the model is simulated so that it predicts possible situations that can be used to evaluate and improve system performance. Discrete-event modeling and simulation is used to create predictions of the system states during time intervals, which can be modified to examine what and if situations. For example, a common use is to evaluate a waiting line, called a queue. The question that is often asked is how long on average would a customer have to wait in a line to get to the service and if the wait time is excessively high how it can be reduced. Adding additional servers is one solution but how many servers should be used is another common question. Modeling and simulation can help answer these questions without actually creating a physical situation that has to be measured, which could be an expensive process.

In stochastic discrete-event simulation paradigm, the dynamics of a complex system are captured by discrete (temporal) state variables, where an event is a combined process of a large number of state transitions between a set of state variables accomplished within the event execution time. The key idea is to segregate the complete state space into a disjoint set of independent events that can be executed simultaneously without any interaction. The application of discrete event-based system modeling techniques in large-scale computer and communication networks has demonstrated the accuracy of this approach for higher order system dynamics within the limits of input data, state partitioning algorithms, uncertainty of information propagation, and highly mobile entities.

12.2 Some Important Definitions

As per the consideration of the different views, a system in general can be described as a set of objects that depend on each other or interact with each other in some way to fulfill some task. While doing this, the system might be influenced by the system's environment. So it is very essential to set a distinction between the system and its environment and hence it is worthwhile to define some important terms.

Entity: An entity is an object of interest in the system, whereby the selection of the object of interest depends on the purpose and level of abstraction of the study.

Attribute: Attributes are used to describe the properties of the system's entities.

An entity may have attributes that pertain to that entity alone. Thus, attributes should be considered as local values. In the example, an attribute of the entity could be the time of arrival. Attributes of interest in one investigation may not be of interest in another investigation. Thus, if red parts and blue parts are being manufactured, the color could be an attribute. However, if the time in the system for all parts is of concern, the attribute of color may not be of importance.

State: The state of the system is a set of variables that is capable of characterizing the system at any time.

Event: An event is an instantaneous incidence that might result in a state change. Events can be endogenous (generated by the system itself) or exogenous (induced by the system's environment).

Systems can be classified as continuous-state or as discrete-state systems (Banks et al., 2001). In continuous-state systems, the state variables change continuously over time. By contrast, the state of discrete-state systems only changes at arbitrary but instantaneous points in time.

Resources: A resource is an entity that provides service to dynamic entities. The resource can serve one or more than one dynamic entity at the same time, that is, operates as a parallel server. A dynamic entity can request one or more units of a resource. If denied, the requesting entity joins a queue or takes some other action (i.e., diverted to another resource, ejected from the system). (Other terms for queues include files, chains, buffers, and waiting lines.) If permitted to capture the resource, the entity remains for a time, then releases the resource.

There are many possible states of resource. Minimally, these states are idle and busy. But other possibilities exist including failed, blocked, or starved states.

List processing: Entities are managed by allocating them to resources that provide service, by attaching them to event notices thereby suspending their activity into the future, or by placing them into an ordered list. Lists are used to represent queues.

Lists are often processed according to FIFO (first-in-first-out), but there are many other possibilities. For example, the list could be processed by LIFO (last-in-first-out), according to the value of an attribute, or randomly, to mention a few. An example where the value of an attribute may be important is in SPT (shortest processing time) scheduling. In this case, the processing time may be stored as an attribute of each entity. The entities are ordered according to the value of that attribute with the lowest value at the head or front of the queue.

Activities and delays: An activity is duration of time whose duration is known prior to commencement of the activity. Thus, when the duration begins, its end can be scheduled. The duration can be a constant, a random value from a statistical distribution, the result of an equation, input from a file, or computed based on the event state. For example, a service time may be a constant 10 min for each entity; it may be a random value from an exponential distribution with a mean of 10 min; it could be 0.9 times a constant value from clock time 0 to clock time 4 h, and 1.1 times the standard value after clock time 4 h; or it could be 10 min when the preceding queue contains at most four entities and 8 min when there are five or more in the preceding queue.

A delay is an indefinite duration that is caused by some combination of system conditions. When an entity joins a queue for a resource, the time that it will remain in the queue may be unknown initially since that time may depend on other events that may occur. An example of another event would be the arrival of a rush order that preempts the

TABLE 12.1

Entities, Attributes, and Activities for Different Systems

Systems	Entities	Attributes	Activities
Market	Customer	Shopping items	Checking out
Telephone system	Messages	Length	Transmission
Bank	Customer	Balance status	Depositing or withdrawing money
Traffic	Cars	Speed and distance	Driving
Quality control system	Products	Quality	Checking
Factory system	Products	Orders remain	Arrival of orders

resource. When the preemption occurs, the entity using the resource relinquishes its control instantaneously. Another example is a failure necessitating repair of the resource.

Discrete-event simulations contain activities that cause time to advance. Most discrete-event simulations also contain delays as entities wait. The beginning and ending of an activity or delay is an event.

Table 12.1 shows the entities, attributes, and activities for different systems.

Deterministic vs. stochastic activates: If the outcome of an activity can be described completely in terms of its input, the activity is called deterministic. Often the activity is a part of the system environment because the exact outcome at any point of time is not known. For example, consider a manufacturing industry. The production of items depends on a number of factors like random power failure, worker strikes, machine failure, etc. If the activity is truly stochastic, there is no known explanation for its randomness. Sometimes, it requires too much details to describe an activity fully, the activity is said to be stochastic.

Exogenous vs. endogenous activities: Endogenous activities are those which are occurring within the system and exogenous activities describe the activities in the environment that affect the system. If a system has no exogenous activities, it is called a closed system, while an open system is one that has exogenous activities. Consider an example of a factory, which receives the orders. The arrival of orders may be considered to be outside to the influence of the factory and therefore part of environment. Hence it may be considered as an exogenous activity.

Discrete-event simulation model: Discrete-event simulation model can be defined as one in which the state variables change only at those discrete points in time at which events occur. Events occur as a consequence of activity times and delays. Entities may compete for system resources, possibly joining queues while waiting for an available resource. Activity and delay times may "hold" entities for durations of time.

A discrete-event simulation model is conducted over time ("run") by a mechanism that moves simulated time forward. The system state is updated at each event along with capturing and freeing of resources that may occur at that time.

An orthogonal classification of systems is based on the distinction whether the progress of the dynamic system state is pushed forward by time or by the occurrence of events (Cassandras and Lafortune, 1999). While in *time-driven* systems all state changes are synchronized by the system clock, in *event-driven* systems events occur asynchronously and possibly concurrently/simultaneously.

Thus, a DES is a discrete state, event-driven (not time-driven) system, that is, the state changes of the system depend completely on the appearance of discrete events over time.

Modeling techniques: Various approaches to the modeling of (discrete-event) dynamic systems exist (Ramadge and Wonham, 1989). These approaches can be divided into logical models and timed models. The latter are also known as performance models and can further be split into the classes of deterministic (or nonstochastic) and stochastic models.

Logical models and nonstochastic timed models are primarily used in control theory (Thistle, 1994). These models mainly consider the logical aspects of the system, for example, the order of events.

Evaluation techniques: Various evaluation techniques exist for the modeling techniques mentioned in Bolch et al. (2006). These evaluation techniques include analytical/numerical methods (e.g., Markovian analysis, algorithms for product-form queuing networks) as well as (discrete-event) simulation.

Especially when dealing with very complex systems, the following drawbacks can be observed:

1. With Markovian analysis, it is necessary to create the whole state space (e.g., in the form of generator matrices of continuous-time Markov chains). Due to the finite memory of computer systems, the calculation of results heavily suffers from state space explosion. Furthermore, Markovian analysis can only be done for models that fulfill the Markovian property (i.e., state sojourn times have to be memoryless, i.e., have to be exponentially or geometrically distributed).

2. Algorithms for product-form queuing networks require queuing network models in product form, which is a hard constraint. Furthermore, queuing networks lack means for handling synchronization and concurrency of system components.

3. To get narrow confidence intervals, it is necessary to simulate complex systems with long simulation runs and/or with a large number of simulation runs, especially in the presence of rare events. Thus, although simulation is a versatile tool because it can handle stochastic processes that fall into the class of generalized semi-Markov processes, it is a very time-consuming method for deriving results.

Of course, workarounds have been presented to these and further problems (Trivedi et al., 1994; Wuechner et al., 2005; Bolch et al., 2006). These include largeness avoidance, state space truncation, lumping, decomposition, fluid models, stiffness avoidance, stiffness tolerance, phase approximations, and simulation speedup techniques.

12.3 Queuing System

An important application of discrete system simulation is in studying the dynamics of waiting line queues as shown in Figure 12.1. It is often observed in many real-life situations, for example buying tickets, stopping at red traffic light, standing in a queue to pay money at a supermarket cash counter, etc. Therefore, many systems contain queue as subsystems. There are many reasons to study a queuing system as mentioned below:

1. To determine the optimal usage of service facility
2. Waiting time for the customers which is a cause for their dissatisfaction
3. Number of persons in the queue at a particular time to decide the waiting space required
4. Number of customers served

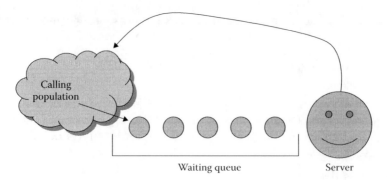

FIGURE 12.1
Queuing system.

There are two important attributes which determine the properties of a waiting queue, namely, arrival rate and service rate. The arrival rate of customers is the average number of customers who join the queue per unit time and the service rate is the average number of customers served in per unit time by the server. Customers normally refers to people, machine, trucks, patients, pallets, airplanes, cases, orders, dirty clothes, etc., and the server refers to receptionist, repair person, mechanics, clerk at counter, doctor, automatic packer, CPU of computer, washing machine, etc.

- *Calling population* is the population where the customers are coming, may be finite or infinite (based on their arrival rate is defined).
- *System capacity* is the maximum number of customers in the queue. Sometimes it is limited, sometimes unlimited.
- *Arrival process* characterizes by inter-arrival times of successive customers. Inter-arrival time may be scheduled time or random time (probability distribution). Scheduled arrival like patients to physician after appointment, scheduled airline flight arrival to airport. Random arrivals like arrival of people to restaurant, banks, PCO, arrival of demand, orders etc.

Queue behavior

- This refers to customer action in queue.
  Leave when line is too long (Balk)
  Leave after sometimes when see that line is too slow (Renege)
  Move from one line to another (Jockey)

Measure of performance

- Average time spent in system
- Server utilization
- Waiting time/customer
- Number of customers served

There are two kinds of queuing system simulation:

1. Discrete time-oriented simulation
2. Discrete event-oriented simulation

The discrete time-oriented simulation refers to the simulation for equal slices of time and the discrete event-oriented simulation examines only major events.

12.4 Discrete-Event System Simulation

In this simulation, each event occurs at an instant in time and marks a change of state in the system. The methods for analysis of the simulation models are numerical methods rather than by analytical methods, that is, analyzing a model is opposed to the analytical approach, where the method of analyzing the system is purely theoretical. Analytical methods employ the deductive reasoning of mathematics to "solve" the model. For example, some inventory models employ differential calculus to determine the minimum-cost policy. Numerical methods employ computational procedures to "solve" mathematical models, that is, execution of a computer program that gives information about the system being investigated. In these simulation models which employ numerical methods, models are "run" rather than solved. An artificial history of the system is generated based on the assumptions, and observations are collected to be analyzed and to estimate the true system performance measures. A common exercise is to build discrete-event simulations model for a customer arriving at a bank to be served by a teller.

A common exercise in learning how to build discrete-event simulations is to model a queue, such as day-to-day operation of a customer arriving at a bank to be served by a teller. In this example, the system entities are customer-queue and tellers. The system events are customer-arrival and customer-departure. (The event of teller-begins-service can be part of the logic of the arrival and departure events.) The system states, which are changed by these events, are number-of-customers-in-the-queue (an integer from 0 to n) and teller-status (busy or idle). The random variables that need to be characterized to model this system stochastically are customer-inter-arrival-time and teller-service-time.

12.5 Components of Discrete-Event System Simulation

1. Clock
   The simulation should keep track of the current simulation time, in whatever units the measurement is being done for the system being modeled. In discrete-event simulations, as opposed to real time simulations, time "hops" between events as they are instantaneous—the clock skips to the next event start time as the simulation proceeds.

2. Events list
   The simulation maintains the events list. There are pending simulated events which need to be simulated themselves. The events are modeled as sequences of events giving the starting time and ending time of discrete event. The pending list is removed in a chronological order regardless of the way in which the statistics are entered. In the case of the bank, the event customer arrival at instant of time t would, if the customer queue was empty and teller was idle, include the creation of the subsequent event customer departure to occur at time $t + s$, where s is a number generated from the service-time distribution.

3. Random-number generators
   The simulation needs to generate random variables of variety of distributions depending on the system model. This is carried out by one or more pseudorandom number generators. The pseudorandom numbers as opposed to true random numbers benefits the simulation when there is a need to rerun with exactly the same behavior. The event-based random number generates random numbers from a specific distribution parameter and initial seed. It generates a random number in an event-based manner each time an entity arrives for simulation. The seed is reset to the value of the initial seed parameter each time a simulation starts making the random behavior repeatable.

4. Statistics
   The simulation particularly keeps track of the system's statistics, which quantify the aspects of interest. For instance, in the case of the bank, it is of interest to track the mean service times.

5. Ending condition
   Theoretically a discrete-event simulation could keep running forever. So, the simulation designer has to decide when the simulation will end. Typical choices are "at time instant t" or "after processing n number of events" or, more generally, "when statistical measure X reaches the value x."

The main loop of a discrete-event simulation begins with the "**Start**" command in which it initializes simulation, state variables, clock, and schedule an initial event. The "**Do loop**" or "**While loop**" execution is the simulation when the case is FALSE and it includes setting of clock to next event time, executing next event and removing from the events list and finally updating statistics as shown in Figure 12.2. When the "**End**" command is executed, it will print the results.

Example 12.1: Simulate a Cash Withdrawal at an ATM Counter

Consider the ATM counter for discrete system simulation. The following components may be taken for this system.

1. System states
   $Q(t)$—Number of customers in waiting line
   $S(t)$—Server status at time t
2. Entities—The customer and server explicitly modeled
3. Events
   A—Arrival
   D—Departure
   E—Stopping event

Discrete-Event Modeling and Simulation

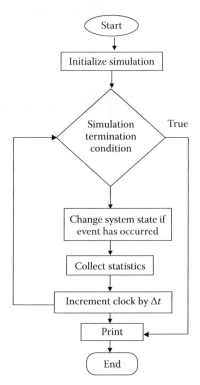

FIGURE 12.2
Flow chart of discrete-event simulation.

4. Event Notice
    (A, t)—An arrival event to occur at future time t
    (D, t)—A customer departure at future time t
    (E, T_{sim})—The simulation stop event at future time T_{sim}
5. Activities
    Inter-arrival time, service time has to be defined
6. Delay—Customer time spent in the system

The life cycle of a customer at an ATM counter is shown in Figure 12.3

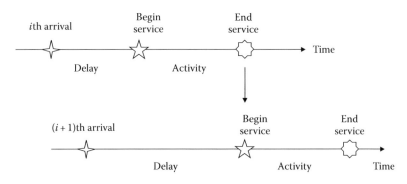

FIGURE 12.3
Life cycle of a customer at an ATM counter.

The Pseudo-code for arrival event is written below:

```
% Pseudo code for arrival event
      Arrival event occur at Clock =t
      If S(t) ==1 then
         Q=Q+1;
      Else
         S(t) = 1
         Generate service time Ts
         Schedule new departure at t+Ts
      End
      Generate Inter-arrival time Ta
      Schedule next arrival at t+Ta
      Collect statistics
             Time advance control routine
Return
```

Similarly departure event Pseudo-code may also be written as follows:

```
% Pseudo code for departure event
      departure event occur at Clock =t
      If Q(t) >0 then
         Q=Q-1;
```

TABLE 12.2

Inter-Arrival Time and Their Service Time of Customers

Inter-arrival time	0	5	3	1	1
Service time	3	2	1	2	1

TABLE 12.3

Simulation Results for an ATM Counter

\multicolumn{3}{c}{States}	\multicolumn{2}{c}{Cumulative Statistics}	
Clock	$Q(t)$	$S(t)$	B	MQ	Description
0	0	1	0	0	First arrival at Ta = 0
					Service time = 3
3	0	0	3	0	First departure
5	0	1	3	0	Second arrival Ta = 5
					Service time Ts = 2
7	0	0	5	0	Second departure
8	0	1	5	0	Third arrival Ta = 8
					Service time Ts = 1
9	0	1	6	0	Third departure
9	0	1	6		Fourth arrival Ta = 9
					Service time Ts = 2
10	1	1	8	1	Fifth arrival Ta = 10
					Service time Ts = 1
11	0	1	8	1	Fourth departure
12	0	0	9	1	Fifth departure

```
          Generate service time Ts
          Schedule new departure at t+Ts
      Else
          S(t) = 0
      End
      Collect statistics
      Time advance control routine
Return
```

Let us consider the inter-arrival time of customers and their service time as shown in Table 12.2. For these input data, the discrete system simulation results of ATM machine are shown in Table 12.3.

12.6 Input Data Modeling

Input data collection and analysis require major time and resource commitment in discrete-event simulation. In the simulation of a queuing system, typical input data are the distribution of time between arrivals and service time. In real-life applications, determining appropriate distributions for input data is a major task from the standpoint of time and resource requirements. Wrong or faulty input data lead to wrong results and mislead the modeler. The procedure used for input data modeling is as follows:

1. Data collection—Collect the data for real-life system. Unfortunately, sometimes it is not possible to collect sufficient data due to system limitations or many other reasons. In such situations expert knowledge and opinion is used to develop some fuzzy systems to generate useful and good data.
2. Identify the distribution of data—When data are collected or obtained, develop a frequency distribution or histogram of data.
3. Select the most appropriate distribution family and their parameters.
4. Check the goodness of fit.

12.7 Family of Distributions for Input Data

The exponential, normal, and Poisson distributions are frequently used. There are literally hundreds of distributions that may be created. Some of them are

1. Binomial distribution
2. Negative binomial distribution
3. Poisson distribution
4. Normal distribution
5. Lognormal distribution
6. Exponential distribution
7. Gamma distribution

8. Beta distribution
9. Erlang distribution
10. Weibull distribution
11. Discrete or uniform distribution
12. Triangular distribution
13. Empirical distribution

Never ignore physical characteristics of the system/process while selecting distribution function such as the system or process is bounded or unbounded; discrete or continuous valued, etc.

The last step in the process is the testing of the distribution hypothesis. The Kolomogrov–Smirnov (K–S) and chi-square goodness of fit tests can be applied to many distribution assumptions. When a distributional assumption is rejected, another distribution is tried. When all else fails, the empirical distribution may be used in the model.

12.8 Random Number Generation

Random numbers are useful for simulating discrete system simulation. Random sequences with a variety of statistical properties can be generated using different techniques.

12.8.1 Uniform Distribution

The simplest type of random sequence is a sequence of numbers as interval [a b] where each number in the interval is likely to occur. A sequence of this form is said to be uniformly distributed. The probability density function $p(x)$ for a uniform distribution is shown in Figure 12.4.

The area under a section of the probability density function is the probability that a random number will fall in the segment. The most widely used numerical method for generating a sequence of uniform distributed random integer is the linear congruential method based on the formula given in Equation 12.1.

$$u_{k+1} = (\alpha u_k + b) \% \gamma \qquad (12.1)$$

where % denotes the modulo operator. That is, $a\%b$ is the remainder after dividing a by b. The integers α, β, and γ are called multiplier, increment, and modulus, respectively.

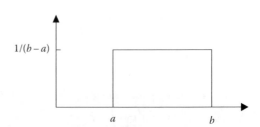

FIGURE 12.4
Probability density of a uniform distribution.

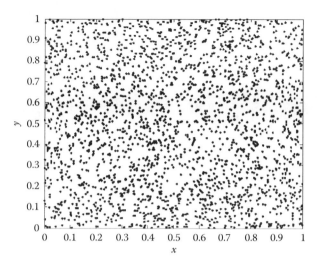

FIGURE 12.5
Uniform distribution in two-dimensional space.

```
Pseudo code for uniform rand number generator
Pick n≫1 and seed the random number generator, RAND
For k=1 to n
      { Set zk=RAND }
For k=1 to m do
{ p=1+nRAND/(umax+1)
Xk=a+(b-a)zp/umax
Zp=RAND
}
```

In MATLAB® there is a built-in function for uniform random number generation

```
≫rand(1);         % single value
≫rand(1,50);      % one dimension array
≫rand(10);        % square matrix
```

Then a uniform distribution is used to generate a pair of random numbers x and y using MATLAB function *rand(2500,2)* and plotted on x–y plane, as shown in Figure 12.5.

12.8.2 Gaussian Distribution of Random Number Generation

Although uniform distribution of random number is useful in many applications, there are other instances where it is more appropriate to generate random numbers, which are not restricted to an interval of finite length. Nonuniform distribution of random numbers can be constructed from a uniform distribution over the interval [0, 1].

The most widely used nonuniform distribution of random numbers is the Gaussian distribution, also called the normal distribution. The probability distribution in two dimensional space is shown in Figure 12.6.

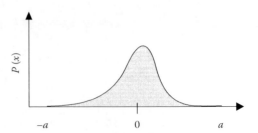

FIGURE 12.6
Probability density of Gaussian distribution.

$$p(x) = \frac{1}{\sigma\sqrt{2\pi}} \exp\left[\frac{-(x-\mu)^2}{2\sigma^2}\right] \quad (12.2)$$

where
   μ is the mean of distribution of random number
   σ is the standard deviation of distribution of random number

The mean is the point about which the bell-shaped curve is centered, and standard deviation is the measure of spread of random numbers about the mean. It is possible to transform a uniform distribution into a Gaussian distribution using the following pair of random numbers with the Box–Muller method.

$$x_1 = \cos(2\pi u_2)\sqrt{-2\ln u_1} \quad (12.3)$$

$$x_1 = \sin(2\pi u_2)\sqrt{-2\ln u_1} \quad (12.4)$$

In MATLAB, *randn* is a built-in function for generating normal random number generation. Then a uniform distribution is used to generate a pair of random numbers x and y using MATLAB function *randn*(2500,2) and plotted on x–y plane, as shown in Figure 12.7.

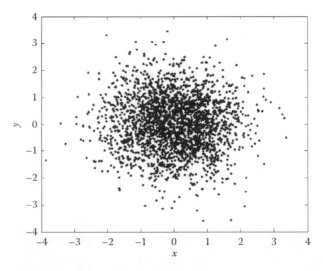

FIGURE 12.7
Gaussian distribution function in two-dimensional space.

12.9 Chi-Square Test

Suppose that N_i is the number of events observed in the ith bin, and that n_i is the number expected according to some known distribution. Note that the N_is are integer, while the n_is may not be. Then the chi-square test is

$$x^2 = \sum_i \frac{(N_i - n_i)^2}{n_i} \qquad (12.5)$$

A large value of x^2 indicates that the hypothesis is null.

12.10 Kolomogrov–Smirnov Test

The K–S test is applicable to unbinned distributions that are functions of single independent variable, that is, to data sets where each data point can be associated with single number. The list of data points can be easily converted to an unbiased estimator $S_N(x)$ of cumulative distribution function of probability distribution form which it was drawn:

If N events are located at values x_i, where $i = 1, 2, \ldots, N$, then $S_N(x)$ is the function giving the fraction of data points to the left of a given value x. This function is the obvious constant between consecutive (i.e., sorted in ascending order) x_i's, and jumps by the same constant $1/N$ at each x_i.

The K–S test is a simple measure of the maximum value of the absolute and difference between two cumulative distribution functions. Thus, for comparing one data set's $S_N(x)$ to a known cumulative distribution function $P(x)$, the K–S statistic is

$$D = \max_{-\infty < c < \infty} |S_N(x) - P(x)| \qquad (12.6)$$

12.11 Review Questions

1. Explain discrete system simulation.
2. Classify discrete system simulation.
3. Explain the following terms:
    a. Clock
    b. Event
    c. Entity
    d. Attribute
4. What do you mean by input data modeling in discrete-event models.
5. Write a MATLAB program for arrival and departure activities of a discrete system simulation.
6. Write a MATLAB program for drawing the histogram for gamma distribution.

Appendix A

A.1 What Is MATLAB®?

MATLAB is a high-performance language for technical computing. It integrates computation, visualization, and programming in an easy-to-use environment where problems and solutions are expressed in a familiar mathematical notation. Typical uses include

- Math and computation
- Algorithm development
- Modeling, simulation, and prototyping
- Data analysis, exploration, and visualization
- Scientific and engineering graphics
- Application development, including graphical user interface building

MATLAB is an interactive system whose basic data element is an array that does not require dimensioning. This allows you to solve many technical computing problems, especially those with matrix and vector formulations, in a fraction of the time it would take to write a program in a scalar noninteractive language such as C or Fortran. The name MATLAB stands for *matrix laboratory*. In industry, MATLAB is the tool of choice for high-productivity research, development, and analysis. MATLAB features a family of application-specific solutions called *toolboxes*. Very important to most users of MATLAB, toolboxes allow *learning* and *applying* specialized technology. Toolboxes are comprehensive collections of MATLAB functions (M-files) that extend the MATLAB environment to solve particular classes of problems. Areas in which toolboxes are available include signal processing, control systems, neural networks, fuzzy logic, wavelets, simulation, and many others.

A.2 Learning MATLAB

Learning MATLAB is just like learning how to drive a car. You can learn all the rules but to become a good driver you have to get out on the road and drive. The easiest and best way to learn MATLAB is to use MATLAB.

A.3 The MATLAB System

The MATLAB system consists of five main parts:

A.3.1 Development Environment

This is the set of tools and facilities that help to you use MATLAB functions and files. Many of these tools are graphical user interfaces. It includes the MATLAB desktop and Command Window, a command history, and browsers for viewing help, the workspace, files, and the search path.

A.3.2 The MATLAB Mathematical Function Library

This is a vast collection of computational algorithms ranging from elementary functions like sum, sine, cosine, and complex arithmetic, to more sophisticated functions like matrix inverse, matrix eigenvalues, Bessel functions, and fast Fourier transforms, etc.

A.3.3 The MATLAB Language

This is a high-level matrix/array language with control flow statements, functions, data structures, input/output, and object-oriented programming features. It allows both "programming in the small" to rapidly create quick and dirty throwaway programs, and "programming in the large" to create complete large and complex application programs.

A.3.4 Handle Graphics

This is the MATLAB graphics system. It includes high-level commands for two-dimensional (2-D) and three-dimensional (3-D) data visualization, image processing, animation, and presentation graphics. It also includes low-level commands that allow to fully customize the appearance of graphics as well as to build complete graphical user interfaces on your MATLAB applications.

Appendix A 623

A.3.5 The MATLAB Application Program Interface (API)

This is a library that allows us to write C and Fortran programs that interact with MATLAB. It include facilities for calling routines from MATLAB (dynamic linking), calling MATLAB as a computational engine, and for reading and writing MAT-files.

A.4 Starting and Quitting MATLAB

On a Microsoft Windows platform, to start MATLAB, double-click the MATLAB shortcut icon on your Windows desktop. On a UNIX platform, to start MATLAB, type MATLAB at the operating system prompt.

To end your MATLAB session, select **Exit MATLAB** from the **File** menu in the desktop, or type quit in the Command Window. To execute specified functions each time MATLAB quits, such as saving the workspace, we can create and run a finish .m script.

A.5 MATLAB Desktop

When MATLAB starts, the MATLAB desktop appears, containing tools (graphical user interfaces) for managing files, variables, and applications associated with MATLAB. The first time MATLAB starts, the desktop appears as shown in the following illustration, although your Launch Pad may contain different entries. You can change the way your desktop looks by opening, closing, moving, and resizing the tools in it. You can also move tools outside of the desktop or return them back inside the desktop (docking). All the desktop tools provide common features such as context menus and keyboard shortcuts. You can specify certain characteristics for the desktop tools by selecting **Preferences** from the **File** menu. For example, you can specify the font characteristics for the Command Window text. For more information, click the **Help** button in the **Preferences** dialog box.

A.6 Desktop Tools

This section provides an introduction to MATLAB's desktop tools. You can also use MATLAB functions to perform most of the features found in the desktop tools.

A.6.1 Command Window

Use the **Command Window** to enter variables and run functions and M-files as shown in Figure A1.

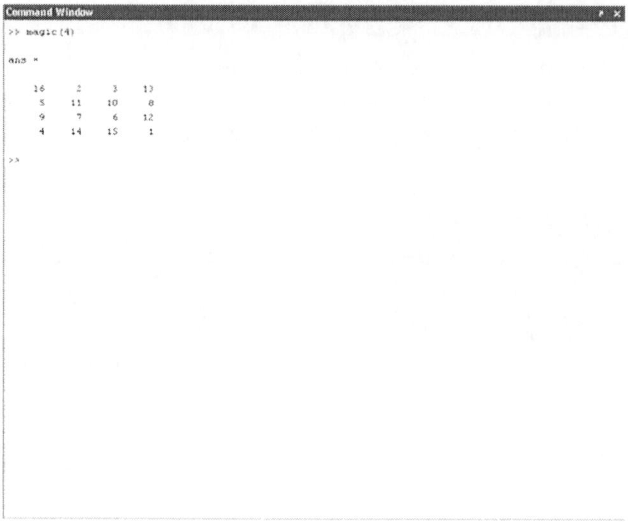

FIGURE A1
Command Window.

A.6.2 Command History

Lines you enter in the Command Window are logged in the **Command History** window. In the Command History, you can view previously used functions, and copy and execute selected lines.

A.6.2.1 Running External Programs

External programs can run from the MATLAB Command Window. The exclamation point character! is a shell escape and indicates that the rest of the input line is a command to the operating system. This is useful for invoking utilities or running other programs without quitting MATLAB. On Linux, for example, !emacs magik.m invokes an editor called emacs for a file named magik.m. When you quit the external program, the operating system returns control to MATLAB.

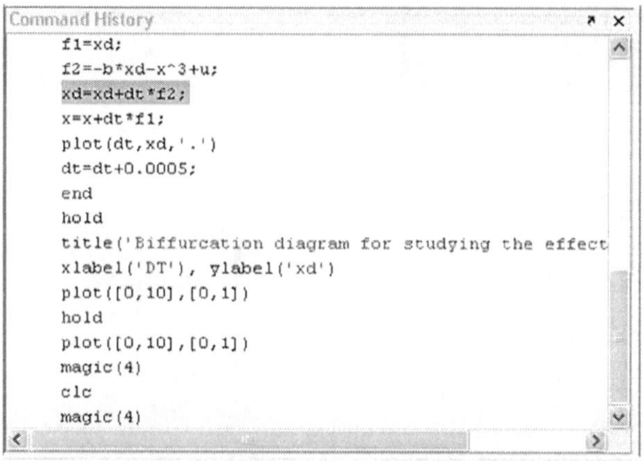

FIGURE A2
Command History Window.

Appendix A

A.6.2.2 Launch Pad

MATLAB's **Launch Pad** provides easy access to tools, demos, and documentation as shown in Figure A3.

A.6.2.3 Help Browser

Use the Help browser to search and view documentation for all MathWorks products. The Help browser is a Web browser integrated into the MATLAB desktop that displays HTML documents. To open the Help browser, click the help button in the toolbar, or type help browser in the Command Window.

The Help browser consists of two panes, the Help Navigator, which you use to find information, and the display pane, where you view the information as shown in Figure A4.

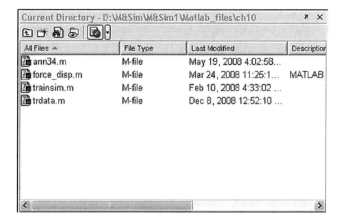

FIGURE A3
Launch Pad.

FIGURE A4
Help Browser.

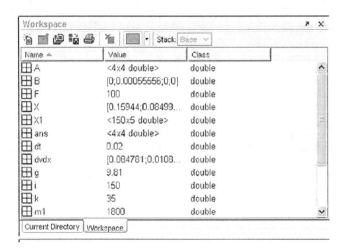

FIGURE A5
Workspace Window.

A.6.2.4 Current Directory Browser

MATLAB file operations use the current directory and the search path as reference points. Any file you want to run must either be in the current directory or on the search path. A quick way to view or change the current directory is by using the **Current Directory** field in the desktop toolbar.

To search for, view, open, and make changes to MATLAB-related directories and files, use the MATLAB Current Directory browser. Alternatively, you can use the functions dir, cd, and delete.

A.6.2.5 Workspace Browser

The MATLAB workspace consists of the set of variables (named arrays) built up during a MATLAB session and stored in memory. We may add variables to the workspace by using functions, running M-files, and loading saved workspaces. To view the workspace and information about each variable, use the Workspace browser as shown in Figure A5 or use the functions who and whos. To delete variables from the workspace, select the variable and select **Delete** from the **Edit** menu. Alternatively, use the clear function. The workspace is not maintained after you end the MATLAB session. To save the workspace to a file that can be read during a later MATLAB session, select **Save Workspace As** from the **File** menu, or use the save function. This saves the workspace to a binary file called a MAT-file, which has a .mat extension. There are options for saving to different formats. To read in a MAT-file, select **Import Data** from the **File** menu or use the load function.

A.6.2.6 Array Editor

Double-click on a variable in the Workspace browser to see it in the Array Editor as shown in Figure A6. Use the Array Editor to view and edit a visual representation of 1-D or 2-D numeric arrays, strings, and cell arrays of strings that are in the workspace, change values of array elements, change the display format.

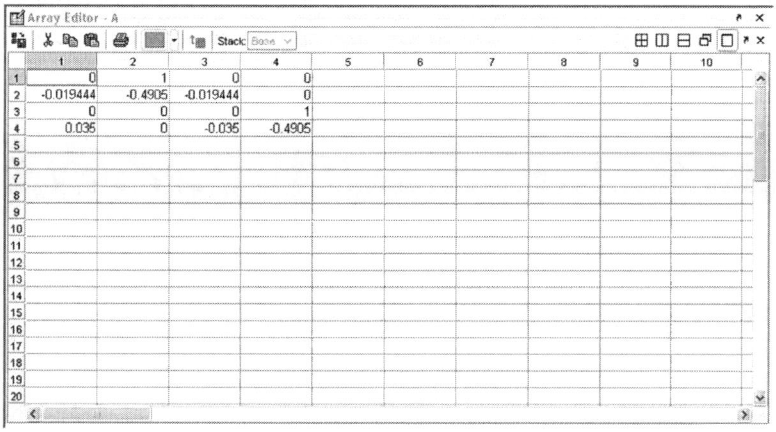

Appendix A   627

FIGURE A6
Array Editor.

A.6.2.7 Editor/Debugger

Use the Editor/Debugger to create and debug M-files, which are programs you write to run MATLAB functions. The Editor/Debugger provides a graphical user interface for basic text editing, as well as for M-file debugging. Any text editor can be used to create M-files, such as Emacs, and can use preferences (accessible from the desktop **File** menu) to specify that editor as the default. If another editor is used, still the MATLAB Editor/Debugger can be used for debugging. To view contents of an M-file, one can display it in the Command Window by using the type function.

A.6.2.8 Other Development Environment Features

Additional development environment features are

- Importing and Exporting Data: Techniques for bringing data created by other applications into the MATLAB workspace, including the Import Wizard, and packaging MATLAB workspace variables for use by other applications.
- Improving M-File Performance: The Profiler is a tool that measures where an M-file is spending its time. Use it to help you make speed improvements.
- Interfacing with Source Control Systems: Access your source control system from within MATLAB, Simulink, and State flow.
- Using Notebook: Access MATLAB's numeric computation and visualization software from within a word processing environment (Microsoft Word).

A.7 Entering Matrices

The best way to get started with MATLAB is to learn how to handle matrices. Start MATLAB and follow along with each example. The matrices can be entered into MATLAB in several different ways:

- Enter an explicit list of elements
- Load matrices from external data files
- Generate matrices using built-in functions
- Create matrices with your own functions in M-files

Start by entering matrix as a list of its elements. You have only to follow a few basic conventions:

- Separate the elements of a row with blanks or commas
- Use a semicolon (;) to indicate the end of each row
- Surround the entire list of elements with square brackets, [ ]. To enter a matrix, simply type in the Command Window

```
A = [16 3 2 13; 5 10 11 8; 9 6 7 12; 4 15 14]
```

MATLAB displays the matrix you just entered.

```
A =
    16     3     2    13
     5    10    11     8
     9     6     7    12
     4    15    14     1
```

This exactly matches the numbers in the engraving. Once the matrix is entered, it is automatically remembered in the MATLAB workspace. You can refer to it simply as *A*. Now that you have *A* in the workspace, take a look at what makes it so interesting. Why is it magic?

sum, transpose, and diag

There are some special properties magic square matrix such as ways of summing its elements. If you take the sum along any row or column, or along either of the two main

Appendix A

diagonals, you will always get the same number. Let's verify that using MATLAB. The first statement to try is sum(A) MATLAB replies with ans = 34 34 34 34.

When you do not specify an output variable, MATLAB uses the variable ans, short for *answer*, to store the results of a calculation. You have computed a row vector containing the sums of the columns of A. Sure enough, each of the columns has the same sum, the *magic* sum, 34. How about the row sums? MATLAB has a preference for working with the columns of a matrix, so the easiest way to get the row sums is to transpose the matrix, compute the column sums of the transpose, and then transpose the result. The transpose operation is denoted by an apostrophe or single quote, '. It flips a matrix about its main diagonal and it turns a row vector into a column vector. So A' produces

```
ans =
      16   5   9   4
       3  10   6  15
       2  11   7  14
      13   8  12   1
```

And

sum(A')' produces a column vector containing the row sums

```
ans =
      34
      34
      34
      34
```

The sum of the elements on the main diagonal is easily obtained with the help of the diag function, which picks off that diagonal.

diag(A) produces

```
ans =
      16
      10
       7
       1
```

and **sum(diag(A))** produces

```
    ans = 34
```

The other diagonal, the so-called *antidiagonal*, is not so important mathematically, so MATLAB does not have a readymade function for it. But a function originally intended for use in graphics, fliplr, flips a matrix from left to right.

```
    sum(diag(fliplr(A)))
    ans = 34
```

A.8 Subscripts

The element in row i and column j of A is denoted by A(i,j). For example, A(4,2) is the number in the fourth row and second column. For our magic square, A(4,2) is 15. So it is possible to compute the sum of the elements in the fourth column of A by typing

$$A(1,4) + A(2,4) + A(3,4) + A(4,4)$$

This produces ans = 34

But this is not the most elegant way of summing a single column. It is also possible to refer to the elements of a matrix with a single subscript, A(k). This is the usual way of referencing row and column vectors. But it can also apply to a fully 2-D matrix, in which case the array is regarded as one long column vector formed from the columns of the original matrix. So, for our magic square, A(8) is another way of referring to the value 15 stored in A(4,2). If you try to use the value of an element outside of the matrix, it is an error.

$$t = A(4,5)$$

Index exceeds matrix dimensions.

On the other hand, if you store a value in an element outside of the matrix, the size increases to accommodate the newcomer.

```
X = A;
X(4,5) = 17
3-7
X =
     16   3   2  13   0
      5  10  11   8   0
      9   6   7  12   0
      4  15  14   1  17
```

A.9 The Colon Operator

The colon (:) is one of MATLAB's most important operators. It occurs in several different forms. The expression 1:10 is a row vector containing the integers from 1 to 10.

1 2 3 4 5 6 7 8 9 10

To obtain non-unit spacing, specify an increment. For example, 100:–7:50 is

100 93 86 79 72 65 58 51

and 0:pi/4:pi is 0 0.7854 1.5708 2.3562 3.1416

Subscript expressions involving colons refer to portions of a matrix. A(1:k,j) is the first k elements of the jth column of A. So sum(A(1:4,4)) computes the sum of the fourth column. But there is a better way. The colon by itself refers to *all* the elements in a row or column of

Appendix A

a matrix and the keyword end refers to the *last* row or column. So sum(A(:,end)) computes the sum of the elements in the last column of *A*.

ans = 34

Why is the magic sum for a 4-by-4 square equal to 34? If the integers from 1 to 16 are sorted into four groups with equal sums, that sum must be sum(1:16)/4 which, of course, is

ans = 34

A.10 The Magic Function

MATLAB actually has a built-in function that creates magic squares of almost any size. Not surprisingly, this function is named magic.

```
B = magic(4)
B =
    16   2   3  13
     5  11  10   8
     9   7   6  12
     4  14  15   1
```

This matrix is almost the same as the one in the engraving and has all the same "magic" properties; the only difference is that the two middle columns are exchanged. To make this *B* into *A*, swap the two middle columns.

```
A = B(:,[1 3 2 4])
```

This says "for each of the rows of matrix *B*, reorder the elements in the order 1, 3, 2, 4." It produces

```
A =
    16   3   2  13
     5  10  11   8
     9   6   7  12
     4  15  14   1
```

Why would he go to the trouble of rearranging the columns when he could have used MATLAB's ordering? No doubt he wanted to include the date of the engraving, 1514, at the bottom of his magic square.

A.11 Expressions

Like most other programming languages, MATLAB provides mathematical *expressions*, but unlike most programming languages, these expressions involve entire matrices. The building blocks of expressions are

- Variables
- Numbers
- Operators
- Functions

A.11.1 Variables

MATLAB does not require any of type declarations or dimensional statements. When MATLAB encounters a new variable name, it automatically creates the variable and allocates the appropriate amount of storage. If the variable already exists, MATLAB changes its contents and, if necessary, allocates new storage. For example, num_students = 25 creates a 1-by-1 matrix named num_students and stores the value 25 in its single element. Variable names consist of a letter, followed by any number of letters, digits, or underscores. MATLAB uses only the first 31 characters of a variable name. MATLAB is case sensitive; it distinguishes between uppercase and lowercase letters. A and a are *not* the same variable. To view the matrix assigned to any variable, simply enter the variable name.

A.11.2 Numbers

MATLAB uses conventional decimal notation, with an optional decimal point and leading plus or minus sign, for numbers. *Scientific notation* uses the letter e to specify a power-of-ten scale factor. *Imaginary numbers* use either i or j as a suffix. Some examples of legal numbers are

3 -99 0.0001
9.6397238 1.60210e-20 6.02252e23
1i -3.14159j 3e5i

All numbers are stored internally using the *long* format specified by the IEEE floating-point standard. Floating-point numbers have a finite *precision* of roughly 16 significant decimal digits and a finite *range* of roughly 10^{-308} to 10^{+308}.

A.11.3 Operators

Expressions use familiar arithmetic operators and precedence rules.

A.11.4 Functions

MATLAB provides a large number of standard elementary mathematical functions, including abs, sqrt, exp, and sin. Taking the square root or logarithm of a negative number is not an error; the appropriate complex result is produced automatically. MATLAB also provides many more advanced mathematical functions, including Bessel and gamma functions. Most of these functions accept complex arguments. For a list of the elementary mathematical functions like +, −, *, /, \, ^ etc., type

```
help elfun
```

For a list of more advanced mathematical and matrix functions, type

```
helps specfun
```

Appendix A

Some of the functions, like sqrt and sin, are built-in. They are part of the MATLAB core so they are very efficient, but the computational details are not readily accessible. Other functions, like gamma and sinh, are implemented in M-files. You can see the code and even modify it if you want. Several special functions provide values of useful constants. Infinity is generated by dividing a nonzero value by zero, or by evaluating well-defined mathematical expressions that *overflow*, that is, exceed realmax. Not a number is generated by trying to evaluate expressions like 0/0 or Inf-Inf that do not have well-defined mathematical values. The function names are not reserved. It is possible to overwrite any of them with a new variable, such as eps = 1.e–6 and then use that value in subsequent calculations. The original function can be restored with clear eps.

```
pi      3.14159265…
i       Imaginary unit, √−1
j       Same as i
eps     Floating-point relative precision, 2^-52
realmin Smallest floating-point number, 2^-1022
realmax Largest floating-point number, (2-ε) 2^1023
Inf     Infinity
NaN     Not-a-number
```

A.11.4.1 Generating Matrices

MATLAB provides four functions that generate basic matrices. Here are some examples.

```
Z = zeros(2,4)
Z =
    0   0   0   0
    0   0   0   0
F = 5*ones(3,3)
F =
    5   5   5
    5   5   5
    5   5   5
N = fix(10*rand(1,10))
N =
    4  9  4  4  8  5  2  6  8  0
A = randn(4,4)
A =
    1.0668    0.2944   -0.6918   -1.4410
    0.0593   -1.3362    0.8580    0.5711
   -0.0956    0.7143    1.2540   -0.3999
   -0.8323    1.6236   -1.5937    0.6900
```

A.12 The Load Command

The load command reads binary files containing matrices generated by earlier MATLAB sessions, or reads text files containing numeric data. The text file should be organized as a rectangular table of numbers, separated by blanks, with one row per line, and an equal number of elements in each row. Store the file under the name **filename.dat**. Then the command **load filename.dat** reads the file and creates a variable, *x*.

M-files

You can create your own matrices using M-files, which are text files containing MATLAB code. Use the MATLAB Editor or another text editor to create a file containing the same statements you would type at the MATLAB command line. Save the file under a name that ends in .m. Store the file under the name **filename.m**. Then the statement **filename** reads the file and creates a variable, contained by the file.

Concatenation

Concatenation is the process of joining small matrices to make bigger ones. The pair of square brackets, [ ], is the concatenation operator. For an example, start with the 4-by-4 magic square, A, and form B = [A A+32; A+48 A+16].

Deleting rows and columns

Any rows and columns can be deleted from a matrix using just a pair of square brackets. Start with

```
X = A;
```

Then, to delete the second column of X, use

```
X(:,2) = []
```

If you delete a single element from a matrix, the result isn't a matrix anymore. So, expressions like X(1,2) = [ ] result in an error.

Adding a matrix to its transpose produces a *symmetric* matrix.
A + A'
The determinant of this particular matrix happens to be zero, indicating that the matrix is *singular*.

```
d = det(A)
X = inv(A)           Inverse of Matrix A.
RCOND = 1.175530e-017.
e = eig(A)  → Eigenvalues of Matrix A.
poly(A) → the coefficients in the characteristic polynomial

mu = mean(D) → Average value of each column of matrix D
sigma = std(D) → Standard deviation of each column of matrix D
B = A - 8.5  → A scalar is subtracted from a matrix A by subtracting it from each element
```

Controlling command window input and output

So far, you have been using the MATLAB command line, typing commands and expressions, and seeing the results printed in the Command Window. This section describes how to

- Control the appearance of the output values
- Suppress output from MATLAB commands
- Enter long commands at the command line
- Edit the command line

Appendix A

A.13 The Format Command

The format command controls the numeric format of the values displayed by MATLAB. The command affects only how numbers are displayed, not how MATLAB computes or saves them.

```
        x = [4/3 1.2345e-6]
format short
        1.3333 0.0000
format short e
        1.3333e+000 1.2345e-006
format short g
        1.3333 1.2345e-006
format long
        1.33333333333333 0.00000123450000
format long e
        1.333333333333333e+000 1.234500000000000e-006
format long g
        1.33333333333333 1.2345e-006
format bank
        1.33 0.00
format rat
        4/3 1/810045
format hex
        3ff5555555555555 3eb4b6231abfd271
```

If you want more control over the output format, use the sprintf and fprintf functions.

A.14 Suppressing Output

If you simply type a statement and press **Return** or **Enter**, MATLAB automatically displays the results on screen. However, if you end the line with a semicolon, MATLAB performs the computation but does not display any output. This is particularly useful when you generate large matrices. For example,

```
A = magic(100);
```

A.15 Entering Long Command Lines

If a statement does not fit on one line, use three periods, ..., followed by **Return** or **Enter** to indicate that the statement continues on the next line. For example,

```
s = 1 - 1/2 + 1/3 - 1/4 + 1/5 - 1/6 + 1/7 ...
        - 1/8 + 1/9 - 1/10 + 1/11 - 1/12;
```

A.16 Basic Plotting

MATLAB has extensive facilities for displaying vectors and matrices as graphs, as well as annotating and printing these graphs. This section describes a few of the most important graphics functions and provides examples of some typical applications.

A.16.1 Creating a Plot

The plot function has different forms, depending on the input arguments. If y is a vector, plot(y) produces a piecewise linear graph of the elements of y versus the index of the elements of y. If you specify two vectors as arguments, plot(x,y) produces a graph of y versus x. For example, these statements use the colon operator to create a vector of x values ranging from zero to 2π, compute the sine of these values, and plot the result.

```
x = 0:pi/100:2*pi;
y = sin(x);
plot(x,y)
```

Now label the axes and add a title. The characters \pi create the symbol π.

```
xlabel('x = 0:2\pi')
ylabel('Sine of x')
title('Plot of the Sine Function','FontSize',12)
```

A.16.2 Multiple Data Sets in One Graph

Multiple x–y pair arguments create multiple graphs with a single call to plot. MATLAB automatically cycles through a predefined (but user settable) list of colors to allow discrimination between each set of data. For example, these statements plot three related functions of x, each curve in a separate distinguishing color.

```
y2 = sin(x-.25);
y3 = sin(x-.5);
plot(x,y,x,y2,x,y3)
```

The legend command provides an easy way to identify the individual plots.

```
legend('sin(x)','sin(x-.25)','sin(x-.5)')
```

A.16.3 Plotting Lines and Markers

If you specify a marker type but not a line style, MATLAB draws only the marker. For example, plot(x,y,'ks') plots black squares at each data point, but does not connect the markers with a line. The statement plot(x,y,'r:+') plots a red dotted line and places plus sign markers at each data point. You may want to use fewer data points to plot the markers than you use to plot the lines. This example plots the data twice using a different number of points for the dotted line and marker plots.

```
x1 = 0:pi/100:2*pi;
x2 = 0:pi/10:2*pi;
plot(x1,sin(x1),'r:',x2,sin(x2),'r+')
```

Imaginary and complex data

When the arguments to plot are complex, the imaginary part is ignored *except* when plot is given a single complex argument. For this special case, the command is a shortcut for a plot of the real part versus the imaginary part. Therefore, plot(Z) where **Z** is a complex vector or matrix, is equivalent to plot(real(Z),imag(Z)).

For example

```
t = 0:pi/10:2*pi;
plot(exp(i*t),'-o')
axis equal
```

A.16.4 Adding Plots to an Existing Graph

The hold command enables to add plots to an existing graph. When you type hold on, MATLAB does not replace the existing graph when you issue another plotting command; it adds the new data to the current graph, rescaling the axes if necessary. For example, these statements first create a contour plot of the peaks function, then superimpose a pseudocolor plot of the same function.

```
[x,y,z] = peaks;
contour(x,y,z,20,'k')
hold on
pcolor(x,y,z)
shading interp
hold off
```

The hold on command causes the pcolor plot to be combined with the contour plot in one figure.

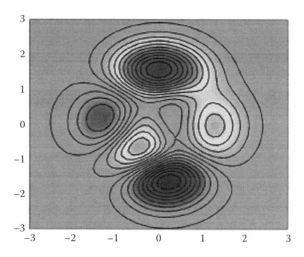

A.16.5 Multiple Plots in One Figure

The subplot command enables you to display multiple plots in the same window or print them on the same piece of paper. Typing subplot(m,n,p) partitions the figure window into an *m*-by-*n* matrix of small subplots and selects the *p*th subplot for the current plot.

The plots are numbered first along the top row of the figure window, then the second row, and so on. For example, these statements plot data in four different subregions of the figure window.

```
t = 0:pi/10:2*pi;
[X,Y,Z] = cylinder(4*cos(t));
subplot(2,2,1); mesh(X)
subplot(2,2,2); mesh(Y)
subplot(2,2,3); mesh(Z)
subplot(2,2,4); mesh(X,Y,Z)
```

A.16.6 Setting Grid Lines

The grid command toggles grid lines on and off. The statement **grid on** turns the grid lines on and **grid off** turns them back off again.

A.16.7 Axis Labels and Titles

The xlabel, ylabel, and zlabel commands add x-, y-, and z-axis labels. The title command adds a title at the top of the figure and the text function inserts text anywhere in the figure. A subset of TeX notation produces Greek letters. You can also set these options interactively.

```
t = -pi:pi/100:pi;
y = sin(t);
plot(t,y)

axis([-pi pi -1 1])
xlabel('-\pi \leq {\itt} \leq \pi')
ylabel('sin(t)')
title('Graph of the sine function')
text(1,-1/3,'{\itNote the odd symmetry.}')
```

A.16.8 Saving a Figure

To save a figure, select **Save** from the **File** menu. To save it using a graphics format, such as TIFF, for use with other applications, select **Export** from the **File** menu. You can also save from the command line—use the saveas command, including any options to save the figure in a different format.

A.16.9 Mesh and Surface Plots

MATLAB defines a surface by the z-coordinates of points above a grid in the x–y plane, using straight lines to connect adjacent points. The **mesh** and **surf** plotting functions display surfaces in three dimensions. **mesh** produces wireframe surfaces that color only the lines connecting the defining points. **surf** displays both the connecting lines and the faces of the surface in color.

```
[X,Y] = meshgrid(-8:.5:8);
R = sqrt(X.^2 + Y.^2) + eps;
```

Appendix A 639

```
Z = sin(R)./R;
mesh(X,Y,Z,'EdgeColor','black')
```

By default, MATLAB colors the mesh using the current colormap. However, this example uses a single-colored mesh by specifying the EdgeColor surface property. See the surface reference page for a list of all surface properties. You can create a transparent mesh by disabling hidden line removal. hidden off.

Example—Colored Surface Plots

A surface plot is similar to a mesh plot except the rectangular faces of the surface are colored. The color of the faces is determined by the values of Z and the colormap (a colormap is an ordered list of colors). These statements graph the *sinc* function as a surface plot, select a colormap, and add a color bar to show the mapping of data to color.

```
surf(X,Y,Z)
colormap hsv
colorbar
```

See the colormap reference page for information on colormaps.

Surface plots with lighting

Lighting is the technique of illuminating an object with a directional light source. In certain cases, this technique can make subtle differences in surface shapes that are easier to see. Lighting can also be used to add realism to 3-D graphs. This example uses the same surface as the previous examples, but colors it red and removes the mesh lines. A light object is then added to the left of the "camera" (that is the location in space from where you are viewing the surface). After adding the light and setting the lighting method to phong, use the view command to change the view point so you are looking at the surface from a different point in space (an azimuth of –15 and an elevation of 65°). Finally, zoom in on the surface using the toolbar zoom mode.

```
surf(X,Y,Z,'FaceColor','red','EdgeColor','none');
camlight left; lighting phong
view(-15,65)
```

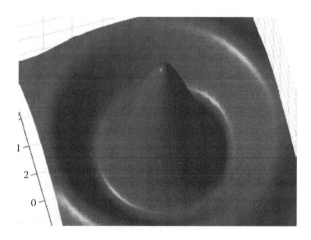

A.17 Images

2-D arrays can be displayed as *images*, where the array elements determine brightness or color of the images. For example, the statements load durer whos

Name	Size	Bytes	Class
X	648 × 509	2,638,656	double array
caption	2 × 28	112	char array
map	128 × 3	3,072	double array

load the file durer.mat, adding three variables to the workspace. The matrix X is a 648-by-509 matrix and the map is a 128-by-3 matrix that is the colormap for this image.

A.18 Handle Graphics

When we use a plotting command, MATLAB creates the graph using various graphics objects, such as lines, text, etc. All graphics objects have properties that control the appearance and behavior of the object. MATLAB enables to query the value of each property and set the value of most properties. Whenever MATLAB creates a graphics object, it assigns an identifier (called a handle) to the object. Handle Graphics is useful if you want to

- Modify the appearance of graphs
- Create custom plotting commands by writing M-files that create and manipulate objects directly

A.18.1 Setting Properties from Plotting Commands

Plotting commands that create lines or surfaces enable you to specify property name/property value pairs as arguments. For example, the command

```
plot(x,y,'LineWidth',1.5)
```

plots the data in the variables x and y using lines having a LineWidth property set to 1.5 points (one point = 1/72 in.). You can set any line object property this way.

A.18.2 Different Types of Graphs

MATLAB supports a variety of graph types that enable you to present information effectively. The type of graph you select depends, to a large extent, on the nature of data. The following list can help you select the appropriate graph:

- Bar and area graphs are useful to view results over time, comparing results, and displaying individual contribution to a total amount.
- Pie charts show individual contribution to a total amount.
- Histograms show the distribution of data values.

Appendix A 641

- Stem and stairstep plots display discrete data.
- Compass, feather, and quiver plots display direction and velocity vectors.
- Contour plots show equivalued regions in data.
- Interactive plotting enables you to select data points to plot with the pointer.
- Animations add an addition data dimension by sequencing plots.

A.18.2.1 Bar and Area Graphs

Bar and area graphs display vector or matrix data. These types of graphs are useful for viewing results over a period of time, comparing results from different datasets, and showing how individual elements contribute to an aggregate amount. Bar graphs are suitable for displaying discrete data, whereas area graphs are more suitable for displaying continuous data.

Function	Description
bar	Displays columns of m-by-n matrix as m groups of n vertical bars
barh	Displays columns of m-by-n matrix as m groups of n horizontal bars
bar3	Displays columns of m-by-n matrix as m groups of n vertical 3-D bars
bar3h	Displays columns of m-by-n matrix as m groups of n horizontal 3-D bars
area	Displays vector data as stacked area plots graphs

Types of bar graphs

MATLAB has four specialized functions that display bar graphs. These functions display 2- and 3-D bar graphs, and vertical and horizontal bars.

Function Description	MATLAB Command
Two-dimensional	bar
Two-dimensional horizontal	barh
Three-dimensional vertical	bar3
Three-dimensional horizontal	bar3h

Grouped bar graph

By default, a bar graph represents each element in a matrix as one bar. Bars in a 2-D bar graph, created by the bar function, are distributed along the x-axis with each element in a column drawn at a different location. All elements in a row are clustered around the same location on the x-axis.

Pie charts

Pie charts display the percentage that each element in a vector or matrix contributes to the sum of all elements. pie and pie3 create 2-D and 3-D pie charts.

Example—Pie Chart

Here is an example using the pie function to visualize the contribution that three products make to total sales. Given a matrix X where each column of X contains yearly sale figures for a specific product over a 5 year period.

```
X = [19.3 22.1 51.6; 34.2 70.3 82.4; 61.4 82.9 90.8; 50.5 54.9 59.1;
     29.4 36.3 47.0];
```

Sum each row in X to calculate total sales for each product over the 5 year period.

```
x = sum(X);
```

You can offset the slice of the pie that makes the greatest contribution using the explode input argument. This argument is a vector of zero and nonzero values. Nonzero values offset the respective slice from the chart.

First, create a vector containing zeros.

```
explode = zeros(size(x));
```

Then find the slice that contributes the most and set the corresponding explode element to 1.

```
[c,offset] = max(x);
explode(offset) = 1;
```

The explode vector contains the elements [0 0 1]. To create the exploded pie chart, use the statement

```
h = pie(x,explode); colormap summer
```

Removing a piece from a pie chart

When the sum of the elements in the first input argument is equal to or greater than 1, pie and pie3 normalize the values. So, given a vector of elements x, each slice has an area of $xi/sum(xi)$, where xi is an element of x. The normalized value specifies the fractional part of each pie slice.

When the sum of the elements in the first input argument is less than 1, pie and pie3 do not normalize the elements of vector **x**. They draw a partial pie.

For example

```
x = [.19 .22 .41];
pie(x)
```

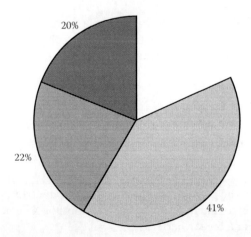

Appendix A

Histograms

MATLAB's histogram functions show the distribution of data values. The functions that create histograms are hist and rose.

Function	Description
hist	Displays data in a Cartesian coordinate system
rose	Displays data in a polar coordinate system

The histogram functions count the number of elements within a range and display each range as a rectangular bin. The height (or length when using rose) of the bins represents the number of values that fall within each range.

Histograms in Cartesian coordinate systems

The hist function shows the distribution of the elements in Y as a histogram with equally spaced bins between the minimum and maximum values in Y. If Y is a vector and is the only argument, hist creates up to 10 bins. For example,

```
yn = randn(10000,1);
hist(yn)
```

generates 10,000 random numbers and creates a histogram with 10 bins distributed along the x-axis between the minimum and maximum values of yn.

A typical 3-D graph

This table illustrates typical steps involved in producing 3-D scenes containing either data graphs or models of 3-D objects. Example applications include pseudocolor surfaces illustrating the values of functions over specific regions and objects drawn with polygons and colored with light sources to produce realism.

Step	Typical Code
1. Prepare your data	Z = peaks(20);
2. Select window and position	figure(1) plot region within window subplot(2,1,2)
3. Call 3-D graphing function	h = surf(Z);
4. Set colormap and shading	colormap hot algorithm shading interp set(h,'EdgeColor','k')

5. Add lighting
   light('Position',[-2,2,20]) lighting phong material([0.4,0.6,0.5,30])
      set(h,'FaceColor',[0.7 0.7 0],...
      'BackFaceLighting','lit')

6. Set viewpoint
   view([30,25]) set(gca,'CameraViewAngleMode','Manual')

7. Set axis limits and tick marks
   axis([5 15 5 15 -8 8]) set(gca'ZTickLabel','Negative||Positive')

8. Set aspect ratio `set(gca,'PlotBoxAspectRatio',[2.5 2.5 1])`
9. Annotate the graph with axis labels, legend, and text
   ```
   xlabel('X Axis')
   ylabel('Y Axis')
   zlabel('Function Value')
   title('Peaks')
   ```
10. Print graph `set(gcf,'PaperPositionMode','auto')`
    `print dps2`

Line plots of 3-D data

The 3-D analog of the plot function is plot3. If **x**, **y**, and **z** are three vectors of the same length, plot3(x,y,z) generates a line in 3-D through the points whose coordinates are the elements of **x**, **y**, and **z** and then produces a 2-D projection of that line on the screen. For example, these statements produce a helix.

```
t = 0:pi/50:10*pi;
plot3(sin(t),cos(t),t)
axis square; grid on
```

A.19 Animations

MATLAB provides two ways of generating moving, animated graphics:

- Continually erase and then redraw the objects on the screen, making incremental changes with each redraw.
- Save a number of different pictures and then play them back as a movie.

A.20 Creating Movies

If you increase the number of points in the Brownian motion example to something like $n = 300$ and $s = 0.02$, the motion is no longer very fluid; it takes too much time to draw each time step. It becomes more effective to save a predetermined number of frames as bitmaps and to play them back as a *movie*. First, decide on the number of frames, say

```
nframes = 50;
```

Next, set up the first plot as before, except using the default EraseMode (normal).

```
x = rand(n,1)-0.5;
y = rand(n,1)-0.5;
h = plot(x,y,'.');
set(h,'MarkerSize',18);
axis([-1 1 -1 1])
axis square
grid off
```

Generate the movie and use getframe to capture each frame.

```
for k = 1:nframes
x = x + s*randn(n,1);
y = y + s*randn(n,1);
set(h,'XData',x,'YData',y)
M(k) = getframe;
end
```

Finally, play the movie 30 times.

```
movie(M,30)
```

A.21 Flow Control

MATLAB has several flow control constructs:

- If statements
- Switch statements
- For loops
- While loops
- Continue statements
- Break statements

A.21.1 If

The if statement evaluates a logical expression and executes a group of statements when the expression is *true*. The optional elseif and else keywords provide for the execution of alternate groups of statements. An end keyword, which matches the if terminates the last group of statements. The groups of statements are delineated by the four keywords—no braces or brackets are involved. MATLAB's algorithm for generating a magic square of order n involves three different cases: when n is odd, when n is even but not divisible by 4, or when n is divisible by 4. This is described by

```
if rem(n,2) ~= 0
      M = odd_magic(n)
elseif rem(n,4) ~= 0
      M = single_even_magic(n)
else
      M = double_even_magic(n)
end
```

In this example, the three cases are mutually exclusive, but if they were not, the first *true* condition would be executed. It is important to understand how relational operators and if statements work with matrices. To check for equality between two variables, we might use

if A == B, ...

This is legal MATLAB code, and does what you expect when *A* and *B* are scalars. But when *A* and *B* are matrices, A == B does not test *if* they are equal, it tests *where* they are equal; the result is another matrix of 0s and 1s showing element-by-element equality. In fact, if *A* and *B* are not the same size

then A == B is an error.

The proper way to check for equality between two variables is to use the isequal function:

if isequal(A,B), ...

A.21.2 Switch and Case

The switch statement executes groups of statements based on the value of a variable or expression. The keywords case and otherwise delineate the groups. Only the first matching case is executed. There must always be an end to match the switch.

The logic of the magic squares algorithm can also be described by

```
switch (rem(n,4)==0) + (rem(n,2)==0)
   case 0
          M = odd_magic(n)
   case 1
          M = single_even_magic(n)
   case 2
          M = double_even_magic(n)
   otherwise
          error('This is impossible')
end
```

Note: Unlike the C language switch statement, MATLAB's switch does not fall through. If the first case statement is *true*, the other case statements do not execute. So, break statements are not required.

A.21.2.1 For

The for loop repeats a group of statements a fixed, predetermined number of times. A matching end delineates the statements.

```
for n = 3:32
   r(n) = rank(magic(n));
end
r
```

The semicolon terminating the inner statement suppresses repeated printing, and the "r" after the loop displays the final result. It is a good idea to indent the loops for readability, especially when they are nested.

```
for i = 1:m
   for j = 1:n
          H(i,j) = 1/(i+j);
   end
end
```

A.21.2.2 While

The while loop repeats a group of statements an indefinite number of times under control of a logical condition. A matching end delineates the statements. Here is a complete program, illustrating while, if, else, and end, that uses interval bisection to find a zero of a polynomial.

```
a = 0; fa = -Inf;
b = 3; fb = Inf;
while b-a > eps*b
   x = (a+b)/2;
   fx = x^3-2*x-5;
   if sign(fx) == sign(fa)
         a = x; fa = fx;
   else
         b = x; fb = fx;
   end
end
x
```

The result is a root of the polynomial $x^3 - 2x - 5$, namely

```
x =
    2.09455148154233
```

The cautions involving matrix comparisons that are discussed in the section on the if statement also apply to the while statement.

A.21.2.3 Continue

The continue statement passes control to the next iteration of the 'for' or 'while' loop in which it appears, skipping any remaining statements in the body of the loop. In nested loops, continue passes control to the next iteration of the for or while loop enclosing it. The example below shows a continue loop that counts the lines of code in the file, magic.m, skipping all blank lines and comments. A continue statement is used to advance to the next line in magic.m without incrementing the count whenever a blank line or comment line is encountered.

```
fid = fopen('magic.m','r');
count = 0;
while ~feof(fid)
   line = fgetl(fid);
   if isempty(line) | strncmp(line,'%',1)
         continue
   end
   count = count + 1;
end
disp(sprintf('%d lines',count));
```

A.21.2.4 Break

The break statement is used to exit early from a for or while loop. In nested loops, break exits from the innermost loop only. Here is an improvement on the example from the previous section. Why is this use of break a good idea?

```
        a = 0; fa = -Inf;
        b = 3; fb = Inf;
        while b-a > eps*b
                x = (a+b)/2;
                fx = x^3-2*x-5;
                if fx == 0
                        break
                elseif sign(fx) == sign(fa)
                        a = x; fa = fx;
                else
                        b = x; fb = fx;
                end
        end
        x
```

A.22 Other Data Structures

This section introduces you to some other data structures in MATLAB including

- Multidimensional arrays
- Cell arrays
- Characters and text
- Structures

A.22.1 Multidimensional Arrays

Multidimensional arrays in MATLAB are arrays with more than two subscripts. They can be created by calling zeros, ones, rand, or randn with more than two arguments. For example,

```
R = randn(3,4,5);
```

creates a 3-by-4-by-5 array with a total of $3 \times 4 \times 5 = 60$ normally distributed random elements. A 3-D array might represent 3-D physical data, say the temperature in a room, sampled on a rectangular grid. Or, it might represent a sequence of matrices, $A(k)$, or samples of a time-dependent matrix, $A(t)$. In these latter cases, the (i, j)th element of the kth matrix, or the $t$kth matrix, is denoted by A(i,j,k). MATLAB's and Dürer's versions of the magic square of order 4 differ by an interchange of two columns. Many different magic squares can be generated by interchanging columns. The statement

```
p = perms(1:4);
```

generates the $4! = 24$ permutations of 1:4. The kth permutation is the row vector, p(k,:). Then

```
        A = magic(4);
```

Appendix A

```
M = zeros(4,4,24);
for k = 1:24
M(:,:,k) = A(:,p(k,:));
end
```

stores the sequence of 24 magic squares in a 3-D array, M. The size of M is size(M)

```
ans = 4 4 24
```

It turns out that the third matrix in the sequence is Dürer's.

```
M(:,:,3)
ans =
       16    3    2   13
        5   10   11    8
        9    6    7   12
        4   15   14    1
```

The statement

```
sum(M,d)
```

computes sums by varying the d_{th} subscript. So

```
sum(M,1)
```

is a 1-by-4-by-24 array containing 24 copies of the row vector
34 34 34 34
and

```
sum(M,2)
```

is a 4-by-1-by-24 array containing 24 copies of the column vector
34
34
34
34
Finally

```
S = sum(M,3)
```

adds the 24 matrices in the sequence. The result has size 4-by-4-by-1, so it looks like a 4-by-4 array.

```
S =
      204  204  204  204
      204  204  204  204
      204  204  204  204
      204  204  204  204
```

A.22.2 Cell Arrays

Cell arrays in MATLAB are multidimensional arrays whose elements are copies of other arrays. A cell array of empty matrices can be created with the cell function. But, more often, cell arrays are created by enclosing a miscellaneous collection of things in curly braces, {}. The curly braces are also used with subscripts to access the contents of various cells. For example,

```
C = {A sum(A) prod(prod(A))}
```

produces a 1-by-3 cell array. The three cells contain the magic square, the row vector of column sums, and the product of all its elements. When C is displayed, you see

```
C =
[4x4 double]   [1x4 double]   [20922789888000]
```

This is because the first two cells are too large to print in this limited space, but the third cell contains only a single number, 16!, so there is room to print it. Here are two important points to remember. First, to retrieve the contents of one of the cells, use subscripts in curly braces. For example, C{1} retrieves the magic square and C{3} is 16!. Second, cell arrays contain *copies* of other arrays, not *pointers* to those arrays. If you subsequently change A, nothing happens to C. 3-D arrays can be used to store a sequence of matrices of the *same* size. Cell arrays can be used to store a sequence of matrices of *different* sizes. For example,

```
M = cell(8,1);
for n = 1:8
M{n} = magic(n);
end
M
```
produces a sequence of magic squares of different order.
```
M =
    [ 1]
    [ 2x2 double]
    [ 3x3 double]
    [ 4x4 double]
    [ 5x5 double]
    [ 6x6 double]
    [ 7x7 double]
    [ 8x8 double]
```

A.22.3 Characters and Text

Enter text into MATLAB using single quotes. For example,

```
s = 'Hello'
```

The result is not the same kind of numeric matrix or array we have been dealing with up to now. It is a 1-by-5 character array.

Internally, the characters are stored as numbers, but not in floating-point format. The statement

Appendix A

```
a = double(s)
```

converts the character array to a numeric matrix containing floating-point representations of the ASCII codes for each character. The result is

```
a =
    72  101  108  108  111
```

The statement

```
s = char(a)
```

reverses the conversion. Converting numbers to characters makes it possible to investigate the various fonts available on computer. The printable characters in the basic ASCII character set are represented by the integers 32:127. (The integers less than 32 represent nonprintable control characters.) These integers are arranged in an appropriate 6-by-16 array with

```
F = reshape(32:127,16,6)';
```

The printable characters in the extended ASCII character set are represented by F+128. When these integers are interpreted as characters, the result depends on the font currently being used. Type the statements

```
char(F)
char(F+128)
```

and then vary the font being used for the MATLAB Command Window. Select **Preferences** from the **File** menu. Here is one example of the kind of output you might obtain.

Concatenation with square brackets joins text variables together into larger strings. The statement

```
h = [s, ' world']
joins the strings horizontally and produces
h =
    Hello world
The statement
v = [s; 'world']
```

joins the strings vertically and produces

```
v =
    Hello
    world
```

Note that a blank has to be inserted before the "w" in h and that both words in "v" need to have the same length. The resulting arrays are both character arrays:

h is 1-by-11 and v is 2-by-5.

To manipulate a body of text containing lines of different lengths, you have two choices—a padded character array or a cell array of strings. The char function accepts any number of lines, adds blanks to each line to make them all the same length, and forms a character array with each line in a separate row. For example,

```
S = char('A','rolling','stone','gathers','momentum.')
```

produces a 5-by-9 character array.

```
S =
    A
    rolling
    stone
    gathers
    momentum.
```

There are enough blanks in each of the first four rows of S to make all the rows the same length. Alternatively, you can store the text in a cell array. For example,

```
C = {'A';'rolling';'stone';'gathers';'momentum.'}
```

is a 5-by-1 cell array.

```
C =
    'A'
    'rolling'
    'stone'
    'gathers'
    'momentum.'
```

You can convert a padded character array to a cell array of strings with

```
C = cellstr(S)
```

and reverse the process with

```
S = char(C)
```

Structures

Structures are multidimensional MATLAB arrays with elements accessed by textual *field designators*. For example,

```
S.name = 'Ed Plum';
S.score = 83;
S.grade = 'B+'
```

creates a scalar structure with three fields.

```
S =
name: 'Ed Plum'
score: 83
grade: 'B+'
```

Like everything else in MATLAB, structures are arrays, so you can insert additional elements. In this case, each element of the array is a structure with several fields. The fields can be added one at a time

Appendix A 653

```
S(2).name = 'Toni Miller';
S(2).score = 91;
S(2).grade = 'A-';
```

or an entire element can be added with a single statement.

```
S(3) = struct('name','Jerry Garcia',…
'score',70,'grade','C')
```

Now the structure is large enough that only a summary is printed.

```
S =
1x3 struct array with fields:
name
score
grade
```

There are several ways to reassemble the various fields into other MATLAB arrays. They are all based on the notation of a *comma separated list*. If you type

```
S.score
```

it is the same as typing

```
S(1).score, S(2).score, S(3).score
```

This is a comma separated list. Without any other punctuation, it is not very useful. It assigns the three scores, one at a time, to the default variable ans and dutifully prints out the result of each assignment. But when you enclose the expression in square brackets,

```
    [S.score]
```

it is the same as

```
    [S(1).score, S(2).score, S(3).score]
```

which produces a numeric row vector containing all of the scores.

```
ans =
     83  91  70
```

Similarly, typing

```
S.name
```

just assigns the names, one at time, to ans. But enclosing the expression in curly braces,

```
{S.name}
```

creates a 1-by-3 cell array containing the three names.

```
ans =
    'Ed Plum'  'Toni Miller'  'Jerry Garcia'
    And
    char(S.name)
```

calls the char function with three arguments to create a character array from the name fields

```
ans =
    Ed Plum
    Toni Miller
    Jerry Garcia
```

A.23 Scripts and Functions

MATLAB is a powerful programming language as well as an interactive computational environment. Files that contain code in the MATLAB language are called M-files. You can create M-files using a text editor, then use them as you would any other MATLAB function or command. There are two kinds of M-files:

- Scripts, which do not accept input arguments or return output arguments. They operate on data in the workspace.
- Functions, which can accept input arguments and return output arguments. Internal variables are local to the function. If you are a new MATLAB programmer, just create the M-files that you want to try out in the current directory. As you develop more of your own M-files, you will want to organize them into other directories and personal toolboxes that you can add to MATLAB's search path. If you duplicate function names, MATLAB executes the one that occurs first in the search path. To view the contents of an M-file, for example, myfunction.m, use type myfunction.

A.23.1 Scripts

When you invoke a script, MATLAB simply executes the commands found in the file. Scripts can operate on existing data in the workspace, or they can create new data on which to operate. Although Scripts do not return output arguments, any variables that they create remain in the workspace, to be used in subsequent computations. In addition, Scripts can produce graphical output using functions like plot. For example, create a file called magicrank.m that contains these MATLAB commands.

```
% Investigate the rank of magic squares
r = zeros(1,32);
for n = 3:32
r(n) = rank(magic(n));
end
r
bar(r)
Typing the statement
magicrank
```

Appendix A 655

This file causes MATLAB to execute the commands, compute the rank of the first 30 magic squares, and plot a bar graph of the result. After execution of the file is complete, the variables n and r remain in the workspace.

A.23.2 Functions

Functions are M-files that can accept input arguments and return output arguments. The name of the M-file and of the function should be the same. Functions operate on variables within their own workspace, separate from the workspace you access at the MATLAB command prompt. A good example is provided by rank. The M-file rank.m is available in the directory toolbox/matlab/matfun. The file rank is:

```
function r = rank(A,tol)
          s = svd(A);
          if nargin==1
                  tol = max(size(A)') * max(s) * eps;
          end
          r = sum(s > tol);
```

The first line of a function M-file starts with the keyword function. It gives the function name and order of arguments. In this case, there are up to two input arguments and one output argument. The next several lines, up to the first blank or executable line, are comment lines that provide the help text. These lines are printed when you type help rank. The first line of the help text is the H1 line, which MATLAB displays when you use the lookfor command or request help on a directory. The rest of the file is the executable MATLAB code defining the function. The variable s introduced in the body of the function, as well as the variables on the first line, r, A, and tol, are all *local* to the function; they are separate from any variables in the MATLAB workspace. This example illustrates one aspect of MATLAB functions that is not ordinarily found in other programming languages—a variable number of arguments. The rank function can be used in several different ways.

```
rank(A)
     r = rank(A)
     r = rank(A,1.e-6)
```

Many M-files work this way. If no output argument is supplied, the result is stored in ans. If the second input argument is not supplied, the function computes a default value. Within the body of the function, two quantities named nargin and nargout are available which tell you the number of input and output arguments involved in each particular use of the function. The rank function uses nargin, but does not need to use nargout.

A.23.2.1 Global Variables

If you want more than one function to share a single copy of a variable, simply declare the variable as global in all the functions. Do the same thing at the command line if you want the base workspace to access the variable. The global declaration must occur before the variable is actually used in a function. Although it is not required, using capital letters for the names of global variables helps distinguish them from other variables. For example, create an M-file called falling.m.

```
function h = falling(t)
global GRAVITY
h = 1/2*GRAVITY*t.^2;
Then interactively enter the statements
global GRAVITY
GRAVITY = 32;
y = falling((0:.1:5)');
```

The two global statements make the value assigned to GRAVITY at the command prompt available inside the function. You can then modify GRAVITY interactively and obtain new solutions without editing any files.

A.23.2.2 Passing String Arguments to Functions

You can write MATLAB functions that accept string arguments without the parentheses and quotes. That is, MATLAB interprets

```
foo a b c as
foo('a','b','c')
```

However, when using the unquoted form, MATLAB cannot return output arguments. For example, legend apples oranges creates a legend on a plot using the strings apples and oranges as labels. If you want the legend command to return its output arguments, then you must use the quoted form.

```
[legh,objh] = legend('apples','oranges');
```

In addition, you cannot use the unquoted form if any of the arguments are not strings.

A.23.2.3 Constructing String Arguments in Code

The quoted form enables you to construct string arguments within the code.

The following example processes multiple data files, August1.dat, August2.dat, and so on. It uses the function int2str, which converts an integer to a character, to build the filename.

```
for d = 1:31
s = ['August' int2str(d) '.dat'];
load(s)
% Code to process the contents of the d-th file
end
```

A.23.2.4 A Cautionary Note

While the unquoted syntax is convenient, it can be used incorrectly without causing MATLAB to generate an error. For example, given a matrix A

```
A =
     0  -6   -1
     6   2  -16
    -5  20  -10
```

The eig command returns the eigenvalues of A.

```
eig(A)
ans =
        -3.0710
        -2.4645+17.6008i
        -2.4645-17.6008i
```

The following statement is not allowed because A is not a string; however, MATLAB does not generate an error.

```
eig A
ans =
        65
```

MATLAB actually takes the eigenvalues of ASCII numeric equivalent of the letter A (which is the number 65).

A.23.2.5 The Eval Function

The eval function works with text variables to implement a powerful text macro facility. The expression or statement eval(s) uses the MATLAB interpreter to evaluate the expression or to execute the statement contained in the text string s. The example of the previous section could also be done with the following code, although this would be somewhat less efficient because it involves the full interpreter, not just a function call.

```
for d = 1:31
s = ['load August' int2str(d) '.dat'];
eval(s)
% Process the contents of the d-th file
End
```

A.23.2.6 Vectorization

To obtain the most speed out of MATLAB, it is important to vectorize the algorithms in M-files. Where other programming languages might use for or DO loops, MATLAB can use vector or matrix operations. A simple example involves creating a table of logarithms.

```
x = .01;
for k = 1:1001
  y(k) = log10(x);
  x = x + .01;
end
```

A vectorized version of the same code is

```
x = .01:.01:10;
y = log10(x);
```

For more complicated code, vectorization options are not always so obvious. When speed is important, however, you should always look for ways to vectorize your algorithms.

A.23.2.7 Preallocation

If vectorization is not possible for a piece of code, then for loops can be made faster by preallocating any vectors or arrays in which output results are stored. For example, this code uses the function zeros to preallocate the vector created in the for loop. This makes the for loop execute significantly faster.

```
r = zeros(32,1);
for n = 1:32
        r(n) = rank(magic(n));
end
```

Without the preallocation in the previous example, the MATLAB interpreter enlarges the **r** vector by one element each time through the loop. Vector preallocation eliminates this step and results in faster execution.

A.23.2.8 Function Handles

You can create a handle to any MATLAB function and then use that handle as a means of referencing the function. A function handle is typically passed in an argument list to other functions, which can then execute, or evaluate, the function using the handle. Construct a function handle in MATLAB using the *at* sign, @, before the function name. The following example creates a function handle for the sin function and assigns it to the variable fhandle.

```
fhandle = @sin;
```

Evaluate a function handle using the MATLAB feval function. The function plot_fhandle, shown below, receives a function handle and data, and then performs an evaluation of the function handle on that data using feval.

```
function x = plot_fhandle(fhandle, data)
plot(data, feval(fhandle, data))
```

When you call plot_fhandle with a handle to the *sin* function and the argument shown below, the resulting evaluation produces a sine wave plot.

```
plot_fhandle(@sin, -pi:0.01:pi)
```

A.23.2.9 Function Functions

A class of functions, called "function functions," works with nonlinear functions of a scalar variable. That is, one function works on another function. The function functions include

- Zero finding
- Optimization
- Quadrature
- Ordinary differential equations

Appendix A

MATLAB represents the nonlinear function by a function M-file. For example, here is a simplified version of the function humps from the MATLAB/demos directory.

```
function y = humps(x)
y = 1./( (x-.3).^2 + .01) + 1./( (x-.9).^2 + .04) - 6;
```

Evaluate this function at a set of points in the interval $0 \leq x \leq 1$ with x = 0:.002:1;

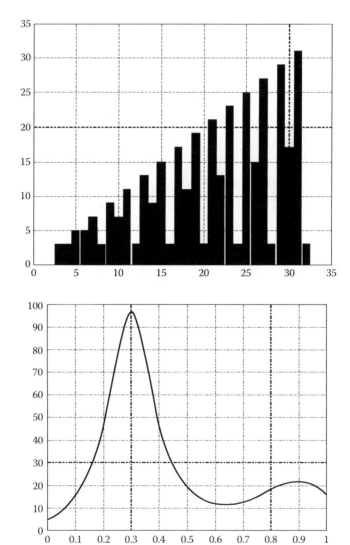

```
y = humps(x);
```

Then plot the function with

```
plot(x,y)
```

The graph shows that the function has a local minimum near $x = 0.6$. The function fminsearch finds the *minimizer*, the value of x where the function takes on this minimum. The first argument to fminsearch is a function handle to the function being minimized and the second argument is a rough guess at the location of the minimum.

```
p = fminsearch(@humps,.5)
p =
    0.6370
```

To evaluate the function at the minimizer

```
humps(p)
ans =
    11.2528
```

Numerical analysts use the terms *quadrature* and *integration* to distinguish between numerical approximation of definite integrals and numerical integration of ordinary differential equations. MATLAB's quadrature routines are quad and quadl. The statement

```
Q = quadl(@humps,0,1)
```

computes the area under the curve in the graph and produces

```
Q =
    29.8583
```

Finally, the graph shows that the function is never zero on this interval. So, if you search for a zero with

```
z = fzero(@humps,.5)
```

you will find one outside of the interval

```
z =
    -0.1316
```

Appendix B: Simulink

B.1 Introduction

Simulink® is a software package used for modeling, analyzing, and simulating a wide variety of dynamic systems. Simulink provides a graphical interface for constructing the models. It has a library of standard components, which makes block diagram representation easier and quicker. Simulink is a ready-access learning tool for simulating operational problems found in the real world because simulation algorithms and parameters can be changed in the middle of simulation with intuitive results. It is particularly useful for studying the effect of nonlinearities on the behavior of the system.

B.2 Features of Simulink

1. A comprehensive library for creating linear, nonlinear, discrete, or multi-input/output systems.
2. Mask facility for creating custom blocks.
3. Unlimited hierarchical model structure.
4. Scalar and vector connections.
5. Interactive simulations with live display.
6. One can easily perform what-if analyses by changing model parameters.
7. Simulink block library can be extended with special purpose block sets.
8. Custom blocks and block libraries can be created by using your own icons and user interfaces from MATLAB®, FORTRAN, or C code.

B.3 Simulation Parameters and Solvers

First select the simulation parameters from the simulation dialog box, as shown in Figure B1.

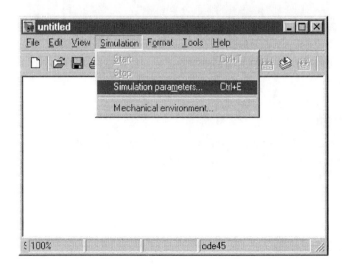

FIGURE B1
Simulink editor.

Simulink parameter dialog box is displayed, which uses four pages to manage simulation parameters as shown in Figure B2.

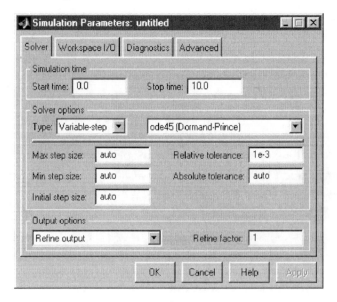

FIGURE B2
Simulink parameter selection window.

1. *Solver page*: The solver page allows to
   - Set the start and stop times.
   - Choose the solver and specify solver parameters—one can select between variable step and fixed step solvers.
   - Output options—It helps in controlling the simulation output. These are refine output, produce additional output, and produce specified output only.

Appendix B

2. *Workspace input/output page*: It allows
   - Loading input from the workspace
   - Saving the output to the workspace
3. *Diagnostics page*: It allows to select the level of warning messages displayed during a simulation.
4. *Advanced page*: It is used for advanced settings such as optimization and model verification blocks control.

B.4 Construction of Block Diagram

On clicking the Simulink icon, the Simulink block library containing seven icons and five pull down and five pull down menus head appears. The model of the system, the input of the system, and the output of the system must be specified. For better understanding, consider the following examples.

Example 1

Model the equation that converts Celsius temperature to Fahrenheit. Obtain a display of Fahrenheit–Celsius temperature graph over a range of 0°C–100°C.
Tf = 9/5 * Tc + 32
The blocks needed to build the model as shown in Figure B3 are

- Ramp block to input the temperature signal
- A constant block for 32
- A gain block to multiply the input signal by 9/5
- A sum block to add the quantities
- A scope block to display the output. The construction of Simulink model is shown in Figures B4 and B5

When click Simulink icon or write Simulink on command prompt in MATLAB Simulink browser opens as shown in Figures B6 and B7. The example of continous system simulation model and its response is shown in Figures B8 and B9 respectively.

Simulink: Model analysis and construction functions.

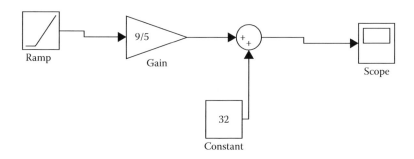

FIGURE B3
Fahrenheit-Celsius temperature converter model.

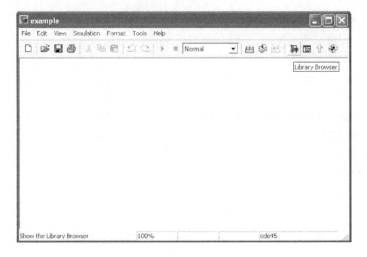

FIGURE B4
Simulink model.

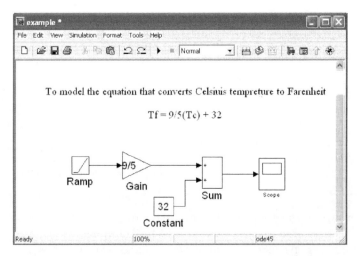

FIGURE B5
Simulink model for Example 1.

Simulation

sim—Simulate a Simulink model.
sldebug—Debug a Simulink model.
simset—Define options to SIM options structure.
simget—Get SIM options structure.

Linearization and trimming

linmod—Extract linear model from continuous-time system.
linmod2—Extract linear model, advanced method.
dlinmod—Extract linear model from discrete-time system.
trim—Find steady state operating point.

Appendix B

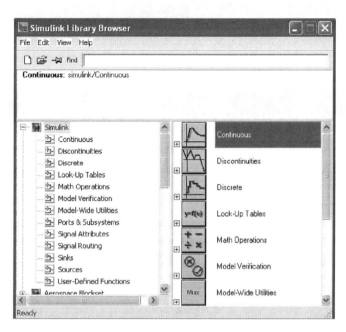

FIGURE B6
Simulink library browser.

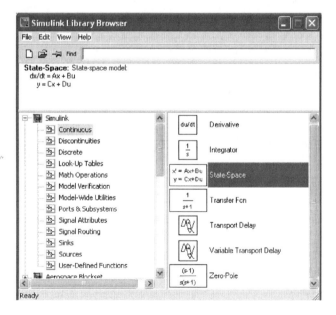

FIGURE B7
Simulink library browser of continuous system blocks.

Model construction

close_system—Close model or block.
new_system—Create new empty model window.
open_system—Open existing model or block.

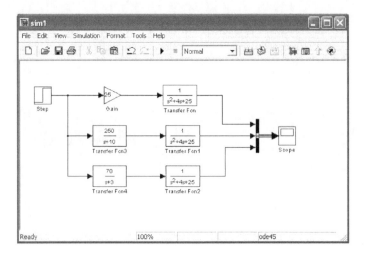

FIGURE B8
Simulink model system simulation.

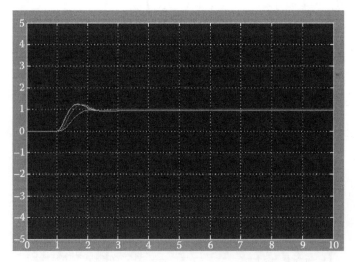

FIGURE B9
Simulation results.

load_system—Load existing model without making model visible.
save_system—Save an open model.
add_block—Add new block.
add_line—Add new line.
delete_block—Remove block.
delete_line—Remove line.
find_system—Search a model.
hilite_system—Hilite objects within a model.
replace_block—Replace existing blocks with a new block.
set_param—Set parameter values for model or block.
get_param—Get simulation parameter values from model.
add_param—Add a user-defined string parameter to a model.

Appendix B   667

delete_param—Delete a user-defined parameter from a model.
bdclose—Close a Simulink window.
bdroot—Root level model name.
gcb—Get the name of the current block.
gcbh—Get the handle of the current block.
gcs—Get the name of the current system.
getfullname—Get the full path name of a block.
slupdate—Update older 1.x models to 3.x.
addterms—Add terminators to unconnected ports.
boolean—Convert numeric array to boolean.
slhelp—Simulink user's guide or block help.

Masking
hasmask—Check for mask.
Hasmaskdlg—Check for mask dialog.
hasmaskicon—Check for mask icon.
iconedit—Design block icons using ginput function.
maskpopups—Return and change masked block's popup menu items.
movemask—Restructure masked built-in blocks as masked subsystems.

Library
libinfo—Get library information for a system.

Diagnostics
sllastdiagnostic—Last diagnostic array.
sllasterror—Last error array.
sllastwarning—Last warning array.
sldiagnostics—Get block count and compile stats for a model.

Hardcopy and printing
frameedit—Edit print frames for annotated model printouts.
print—Print graph or Simulink system; or save graph to M-file.
printopt—Printer defaults.
orient—Set paper orientation.

B.5 Review Questions

1. Describe the main features of the MATLAB simulation method.
2. Discuss the basics of the MATLAB simulation method with reference to any six of the following main features:
   a. Entering simple matrix
   b. Statements and variables
   c. WHO and permanent variables
   d. Numbers and arithmetic expressions
   e. Complex numbers and matrices

      f. Output format
      g. The help facility
      h. Quitting and saving the workspace
      i. Matrix operations and functions
      j. Relational and logical operations
      k. Control flow construct
      l. M-file and functions
      m. Other features–graphics, running external programs

3. Explain the following command lines written in MATLAB:
    a. `A = [1 4 7 ; 2 5 8; 3 6 9]`
    b. `A = [1 2 ; 3 4] + i*[5 7; 6 8]`
    c. `format long x = [2/3 1.3215e-6]`
    d. `save temp x y z`
    e. `load`
    f. `x = [1 2]; y = [4 5];z = x./y`
    g. `[v,d] = daig(A)`
    h. `A(1:5,:)`
    i. `n = 1 while prod(1:n) < 1.0e20, n = n+1; end`
    j. `x.j. if y = [0 0.4 0.8 1.5 2.0 0.7 0.1] plot(y)`

4. Write the name of any five toolboxes available in MATLAB.

5. If x = [1 4 7 4; 3 9 6 5; 2 5 8 7; 2 9 5 6], then what will be the output of following commands:

    `A = x(:, 2:3)`

    `B = x(1,:) + x(3,: )`

6. Write the syntax in MATLAB for finding the following mathematical expression:
$F = 4x^2 + 2y^2 + 2\sin\sqrt{x^2 + y^3}$

7. Find the output for the following MATLAB program:
```
X = [1 2 3];
Y = [2; 1; 5];
Z = [2 1 0];
A = X + Z
C = X .* Z
D=X./Y'
E=X'.Y
```

8. Write the syntax for the following operations:
    a. Single command for eigen value and eigen vector
    b. Convolution for two polynomial *a* and *b*
    c. Clear all variables and functions from workspace
    d. Clear command window, command history is lost
    e. Command for displaying text on command window

Appendix B

    f. Generate random matrix of size 3 × 4
    g. Generate identity matrix of order 4 × 4
    h. For finding the exponential of x^2

9. What will be the answer of 4\3 in MATLAB when format short? How is it different from 4/3?

10. What will be the output on the screen for the following MATLAB program:

```
for I=1:3
for J=1:3
A(I,J)=(I^2+J);
end
end
A1=A*eye(3)
```

Appendix C: Glossary

C.1 Modeling and Simulation

Adaptive control: A control methodology in which control parameters are continuously and automatically adjusted in response to be measured/estimated process variables to achieve near-optimum system performance.

Adaptive model: A model whose parameters or properties are adjusted online (continuously during execution) to satisfy an objective.

Admittance: The reciprocal of the impedance of an electric circuit.

Analog-to-digital converter (ADC): A device that converts analog input voltage signals into digital form.

Attribute: Attributes are used to describe the properties of the system's entities.

Bit error rate (BER): The probability of a single transmitted bit being incorrectly determined upon reception.

Bounded-input bounded-output (BIBO): A signal that has a certain value at a certain instant in time, and this value does not equal infinity at any given instant of time. A bounded output is the signal resulting from applying the bounded-input signal to a stable system.

CASE: Computer-aided software engineering. A general term for tools that automate various phases of the software engineering life cycle.

Chaos theory: Where the response and development of systems are studied under changes of their initial conditions.

Chaos: Erratic and unpredictable dynamic behavior of a deterministic system that never repeats itself. Necessary conditions for a system to exhibit such behavior are that it be nonlinear and have at least three independent dynamic variables.

Chaotic behavior: A highly nonlinear state in which the observed behavior is very dependent on the precise conditions that initiated the behavior. The behavior can be repeated (i.e., it is not random), but a seemly insignificant change, such as voltage, current, noise, temperature, and rise times will result in dramatically different results, leading to unpredictability. The behavior may be chaotic under all conditions, or it may be well behaved (linear to moderately nonlinear) until some parametric threshold is exceeded, at which time the chaotic behavior is observed. In a mildly chaotic system, noticeable deviations resulting from small changes in the initial conditions may not appear for several cycles or for relatively long periods. In a highly chaotic system, the deviations are immediately apparent.

Characteristics equation: The relation formed by equating to zero the denominator of a transfer function.

Closed-loop system: Any system having two separate paths inside it. The first path conducts the signal flow from the input of that system to the output of that same system (forward path). The second path conducts the signal flow from the output to the input of the system (feedback path), thus establishing a feedback loop for the system.

Common-mode rejection ratio (CMRR): A measure of quality of an amplifier with differential inputs and defined as the ratio between the common-mode gain and the differential gain.

Complexity: The intricate pattern of interwoven parts and knowledge required.

Computer prototype: To refine idea or decision on the computer before implementing it in the real world. In the early 1990s, Boeing used computer simulations to rapid-prototype its 777 aircraft without building a physical prototype in order to save time and money. The consequence was that the new product will be built faster, better, and cheaper than the previously developed one.

Computer simulation: A set of computer programs that allows one to model the important aspects of the behavior of the specific system under study. Simulation can aid the design process by, for example, following one to determine appropriate system design parameters or aid the analysis process by, for example, allowing one to estimate the end-to-end performance of the system under study.

Computer simulation: A set of computer programs that allows one to model the important aspects of the behavior of the specific system under study. Simulation can aid the design process by, for example, allowing one to determine appropriate system design parameters or aid the analysis process by, for example, allowing one to estimate the end-to-end performance of the system under study.

Controllability: A property that in the linear system case depends upon the A, B matrix pair which ensures the existence of some control input that will drive any arbitrary initial state to zero in finite time.

Co-tree: The complement of a tree in a network.

Cut set: A minimal subsystem, the removal of which cuts the original system into two connected subsystems.

Damped oscillation: It is an oscillation in which the amplitude decreases with time.

Damping: A characteristic built into systems that prevents rapid or excessive corrections that may lead to instability or oscillatory conditions.

Damping coefficient: For a simple mechanical viscous damper (dashpot), the force F may be related to the velocity v by $F = -cv$, where c is the viscous damping coefficient, given in units of newton-seconds per meter.

Data: Any information, represented in binary that a computer receives, processes, or outputs.

Data model: An integrated set of tools to describe the data and its structure, data relationships, and data constraints.

Database computer: A special hardware and software configuration aimed primarily at handling large databases and answering complex queries.

Database management system (DBMS): A software system that allows for the definition, construction, and manipulation of a database.

Database: A shared pool of interrelated data.

Decomposition: An operation performed on a complex system whose purpose is to separate its constituent parts or subsystems in order to simplify the analysis or design procedures. Decomposition is performed to make their model traceable, for example, by dividing them into smaller parts.

Decomposition: An operation performed on a complex system whose purpose is to separate its constituent parts or subsystems in order to simplify the analysis or design procedures. For optimization algorithms, decomposition is reached by resolving the objective function or constraints into smaller parts, for example, by partitioning the matrix of constrains in linear programs followed by the solution of a number of low dimensional linear programs and coordination by Lagrange multipliers.

Delay: The time required for an information, signal, or object to propagate along a defined path.

Detectability: A linear system is said to be detectable if its unstable part is observable.

Appendix C 673

Digital computer: A collection of digital devices including an arithmetic logic unit (ALU), read-only memory (ROM), random-access memory (RAM), and control and interface hardware.

Direct memory access (DMA): The process of sending data from an external device into the computer memory with no involvement of the computer's central processing unit.

Discrete time approximation: An approximation used to obtain the time response of a system based on division of time into small increments.

Distributed database: A collection of multiple, logically interrelated databases distributed over a computer network.

Disturbance signal: An unwanted input signal that affects the system's output signal.

Dominant poles (roots): The poles of a system (roots of the characteristics equation) that cause the dominant transient response of the system.

Eigenvalue: The multiplicative scalar associated with an eigen function or an eigenvector. It is the root of the characteristic equation $|\lambda_i I - A| = 0$.

Eigenvector: For a linear system A, any vector x whose direction is unchanged when operated upon by A.

Entity: An entity is an object of interest in the system, whereby the selection of the object of interests depends on the purpose and level of abstraction of the study.

Error: The occurrence of an incorrect value in some unit of information within a system.

Essential model: A software engineering model which describes the behavior of a proposed software system independent of implementation aspects.

Event: An event is an instantaneous incidence that might result in a state change. Events can be endogenous (generated by the system itself) or exogenous (induced by the system's environment).

Expert systems: A computer program that emulates a human expert in a well-bounded domain of knowledge.

Failure rate: The failure rate, λ, is the (predicted or measured) number of failures per unit time for a specified part or system operating in a given environment. It is usually assumed to be constant during the working life of a component or system.

Failure: A deviation in the expected performance of a system.

Fault avoidance: A technique that attempts to prevent the occurrence of faults.

Fault tolerance: The ability to continue the correct performance of functions in the presence of faults.

Fault: A physical defect, imperfection, or flaw that occurs in hardware or software.

Feedback control: The regulation of a response variable of a system in a desired manner using measurements of that variable in the generation of the strategy of manipulation of the controlling variables.

Gaussian noise: A noise process that has a Gaussian distribution for the measured value at any time instant.

Hedges: Linguistic terms that intensify, dilute, concentrate, or complement a fuzzy set.

Homogeneous solution: A system of linear constant-coefficient differential equations has a complete solution that consists of the sum of a particular solution and a homogeneous solution. The homogeneous solution satisfies the original differential equation with the input set to zero. Analogous definitions exist for difference equations.

Implementation model: A software engineering model which describes the technical aspects of a proposed system within a particular implementation environment.

Impulse response: The response of a system when the input is an impulse function.

Information model: A software engineering model which describes an application domain as a collection of objects and relationships between those objects.

Maintainability, $M(t)$: The probability that an inoperable system will be restored to an operational state within the time t.

Mapping: A transformation, particularly between abstract spaces.

Mathematical model: Description of the behavior of a system using mathematical equations.

Mean time to failure: This figure is used to give an expected working lifetime for a given part, in a given environment.

If the failure rate λ is constant, then MTTF = $\frac{1}{\lambda}$.

Mean time to repair: The MTTR figure gives a prediction for the amount of time taken to repair a given part or system.

Module structure chart: A component of the implementation model; it describes the architecture of a single computer program.

Negative feedback: The output signal is fed back so that it is subtracted from the input signal.

Nonlinear model: A model that includes nonlinear differential equations.

Nonlinear response: The characteristic of certain physical systems that some output property changes in a manner more complex than linear in response to some applied input.

Nonlinear Schrödinger equation: The fundamental equation describing the propagation of short optical pulses through a nonlinear medium, so called because of a formal resemblance to the Schrödinger equation of quantum mechanics.

Nonlinear system: A system that does not obey the principle of superposition and homogeneity.

Normal tree: A tree that contains all the independent across sources, the maximum number of storage type, the minimum number of delay type, and none of the independent through sources.

Object collaboration model: A component of the essential model; it describes how objects exchange messages in order to perform the work specified for a proposed system.

Object: An "entity" or "thing" within the application domain of a proposed software system.

Objective function: When optimizing a structure toward a certain result, that is, during optimization routines, the objective function is a measure of the performance which should be maximized or minimized (to be extremized).

Observability: The property of a system that ensures the ability to determine the initial state vector by observing system outputs for a finite time interval. For linear systems, an algebraic criterion that involves system and output matrices can be used to test this property. A linear system is said to be observable if its state vector can be reconstructed from finite-length observations of its input and output.

Observer (or estimator): A linear system whose state output approximates the state vector of a different system, rejecting noise and disturbances in the process.

Off-line testing: Testing process carried out while the tested circuit is not in use.

Online testing: Concurrent testing to detect errors while circuit is in operation.

Open circuit impedance: The impedance into an N-port device when the remaining ports are terminated in open circuits.

Open loop system: A system in which the output does not have any influence on the input given to that system.
Parameter estimation: The procedure of estimation of model parameters based on the model's response to certain test inputs.
Performability, $P(L, t)$: The probability that a system is performing at or above some level of performance, L, at the instant of time t.
Physical prototype: They could be mock-ups of future products carved out of foam to give designers and users an idea of how the finished product might function and an intuition for how it might feel.
Positive feedback: The output signal is fed back so that it adds to the input signal.
Positive-(semi)definite: A positive-(semi)definite matrix is a symmetric matrix A such that for any nonzero vector x, the quadratic form $x^T A x$ is positive (nonnegative).
Reliability, $R(t)$: The conditional probability that a system has functioned correctly throughout an interval of time, $[t_0, t]$, given that the system was performing correctly at time t_0.
Reliability: Reliability $r(t)$ is the probability that a component or system will function without failure over a specified time period under stated conditions.
Resources: A resource is an entity that provides service to dynamic entities.
Risk: Uncertainty embodied in the system produce unintended consequences.
Robust control: Control of a dynamical system so that the desired performance is maintained despite the presence of uncertainties and modeling inaccuracies.
Robustness of model: A mathematical model is said to be robust if small changes in the parameter lead to small changes in the behavior of model. The decision is made by using sensitivity analysis for the models.
Root-mean-squared (RMS) error: The square root of the mean squared error.
Safety, $S(t)$: The probability that a system will either perform its functions correctly or will discontinue its functions in a well-defined, safe manner.
Sensitivity: A property of a system indicating the combined effect of component tolerances on overall system behavior, the effect of parameter variations on signal perturbations, and the effect of model uncertainties on system performance and stability. For example, in radio technology, sensitivity is the minimum input signal required by the receiver to produce a discernible output. The sensitivity of a control system could be measured by a variety of sensitivity functions in time, frequency, or performance domains. A sensitivity analysis of the system may be used in the synthesis stage to minimize the sensitivity and thus aim for insensitive or robust design.
Signal-to-noise ratio (SNR): The ratio between the signal power and the noise power at a point in the signal traveling path.
Simulation: It is imitation of operation of a real-world process or system over time. This is done by iterative process of developing model and conducting experiments on it.
Singular value: Singular values are nonnegative real numbers that measure the magnification effect of an operator in the different basis directions of the operator's space.
Stabilizable: A system is stabilizable if all its unstable modes are controllable.
State equation: The equation that describes the relationship between the derivation of the state variables as a function of the state variables, inputs, and parameter are called state equation.
State transition matrix: Matrix operation which determines the transition of any initial state $x(0)$ at $t = 0$ to a state $x(t)$ at time t is known as state transition matrix.

State variables: A set of variables that completely summarize the system's status in the following sense. If all states x_i are known at time t_0, then the values of all states and outputs can be determined uniquely for any time $t_1 > t_0$, provided the inputs are known from t_0 onward. State variables are components in the state vector. State space is a vector space containing the state vectors.

State: The state of the system is a set of variables that is capable of characterizing the system at any time.

Steady state error: The difference between the desired reference signal and the actual signal in steady state, that is, when time approaches infinity.

Steady state response: That part of the response which remains as time approaches infinity.

String theory: Where the physicists are trying to develop a "theory of everything."

Synthesis: The process by which new physical configurations are created. The combining of separate elements or devices to form a coherent whole.

Systems engineering: An approach to the overall life cycle evolution of a product or system. Generally, the systems engineering process comprises a number of phases. There are three essential phases in any systems engineering life cycle: formulation of requirements and specifications, design and development of the system or product, and deployment of the system. Each of these three basic phases may be further expanded into a larger number. For example, deployment generally comprises operational test and evaluation, maintenance over an extended operational life of the system, and modification and retrofit (or replacement) to meet new and evolving user needs.

Time domain: The mathematical domain that incorporates the time response and the description of a system in terms of time.

Time varying systems: A system for which one or more parameters may vary with time.

Transfer function: The ratio of system output to system input in frequency domain (s-domain).

Transient response: That part of the response which vanishes as time approaches infinity.

Transient response: The response of a system as a function of time.

Transition matrix $\Phi(t)$: The matrix exponential function that describes the unforced response of the system.

Validation: It is a process in which model behavior is compared with the actual system and error is reduced to an acceptable level. It is also called calibration. This tells us whether the model is faithful or not?

Verification: Verification is concerned with "Is the model implemented correctly?"

Zero-input response: That part of the response due to the initial condition only.

Zero-state response: That part of the response due to the input only.

C.2 Artificial Neural Network

Activation/threshold function: It is a function which controls the neuron output.

Adaptive: A system that can be modified during operation to meet specified criteria.

Artificial neural network: A set of nodes called neurons and a set of connections between the neurons that form a network is called aritifical neural network.

Appendix C

Artificial neuron: An elementary analog of a biological neuron with weighted inputs, an internal threshold, and a single output. When the activation of the neuron equals or exceeds the threshold, the output takes the value C_1, which is an analog of the firing of a biological neuron. When the activation is less than the threshold, the output takes on the value 0 (in the binary case) or –1 (in the bipolar case) representing the quiescent state of a biological neuron. It mimics the behavior of biological neuron with the help of an electronic circuit.

Axon: Output channel.

Back propagation: ANN models where error at output layer is propagated back to modify the weights.

Bias: It is the connection strength for a fixed input.

Biological neuron: The tiny processing cell in the human brain.

Cell body: Accumulator (with threshold function).

Dendrites: Input receptors in neuron.

Epoch/iteration: A cycle of processing in a neural network, which contains forward calculation for determining neural output as well as backward calculation to update the weights.

Global minima: There is no other value of x in the domain of the function f, where the value of the function is smallest.

Gradient descent system: A system that attempts to reach its stable state by moving consistently down the steepest portion of its energy surface.

Hamming distance: The number of digit positions in which the corresponding digits of two binary words of the same length differ. The minimum distance of a code is the smallest Hamming distance between any pair of code words. For example, if the sequences are 1010110 and 1001010, then the Hamming distance is 3.

Hidden layer: A layer of neurons in a multilayer perceptron network that is intermediate between the output layer and the input layer.

Hidden layer: An array of neurons positioned between the input and output layers.

Input layer: An array of neurons to which an external input or signal is presented. It is intended to perform intellectual operations in a manner not unlike that of the neurons in the human brain. In particular, artificial neural networks have been designed and used for performing pattern recognition operations.

Local minima: During learning of ANN the network could not reach to its absolute minima, which is called local minima.

Noise: A distortion of an input.

Output layer: An array of neurons to which output from the network is taken.

Perceptron: A single-layer neural network that that can solve only the linearly separable problems.

Simulated annealing: The process of introducing and then reducing the amount of random noise introduced into the weights and inputs of a neural network. The process is analogous to methods used in solidification, where the system starts with a high temperature to avoid local energy minima, and then gradually the temperature is lowered according to a particular algorithm.

Supervised learning: A learning process where the output for the given input is known and used for modifying the weights. In this learning, examples are used.

Training: The process of changing weights or rather refining weights is called learning/training.

Unsupervised learning: Learning in the absence of external information on outputs.

Weight: It is connection strength between different neurons situated at different layers.
Working memory: A component or a place of computer system where the intermediate results of intelligent system are temporarily stored.

C.3 Fuzzy Systems

Adaptive fuzzy system: A fuzzy system that does not require rules from a human expert; it generates and tunes its own rules. A neuro-fuzzy system or fuzzy neural system are adaptive fuzzy systems.
Antecedent: The clause that implies the other clause in a conditional statement, that is, the *if* part in the *if-then* rule.
Approximate reasoning: An inference procedure used to derive conclusions from a set of fuzzy if-then rules and some conditions (facts). The most used approximate reasoning methods are based on the generalized modus ponens.
Artificial intelligence: The discipline devoted to produce systems that perform tasks which would require "intelligence" if performed by a human being.
Automatic knowledge acquisition: A branch of machine learning devoted to explicating the principles of the induction of rule bases.
Backtracking: The process of backing up through a series of inferences in the face of unacceptable results.
Backward chaining: An inference mechanism which works from a goal and attempts to satisfy a set of initial conditions. Also referred to as goal-directed chaining.
Cognition: An intelligent process by which knowledge is gained about perceptions or ideas.
Consequent: The resultant clause in a conditional statement, that is, the *then* part in the *if-then* rule.
Convex fuzzy set: Fuzzy sets whose α cuts are crisp sets for all $\alpha \in [0,1]$.
Crisp set: The classical or crisp set is the collection of items and the items can be member or nonmember of that set.
Database system: A system which marries the properties of database systems with the properties of expert systems.
Defuzzification: The process of transforming a fuzzy set into a crisp set or a real-valued number.
Degree of membership: An expression of confidence or certainty that an element belongs to a fuzzy set. It varies from 0 (no membership) to 1 (complete membership).
Domain expert: The person who provides the expertise on which a knowledge base is modified.
Domain: A bounded area of knowledge. A pool of values used to define columns of a relation.
Equivalence relation: Relation that is reflexive, symmetric, and transitive.
Expert system: A computer system that achieves high levels of performance in areas that for human beings require large amounts of expertise.
Expert system: A computer program that emulates a human expert in a well-bounded domain of knowledge.
Expertise: A set of capabilities underlying skilled performance in some task area.

Appendix C

Fact: A relationship between objects.
Forward chaining: An inference mechanism that works from a set of initial conditions to a goal. Also referred to as a data-directed chaining. Making inferences by matching the condition sides of the IF-THEN rules to the facts at hand. It works from facts to conclusions; it also called antecedent mode, event-driven mode, or data-driven mode of inference.
Fuzzification: Changing crisp number or set to a fuzzy number or set.
Fuzzifier: A fuzzy system that transforms a crisp (nonfuzzy) input value in a fuzzy set. The most used fuzzifier is the *singleton fuzzifier*, which interprets a crisp point as a fuzzy singleton. It is normally used in fuzzy control systems.
Fuzziness: The degree or extent of imprecision that is naturally associated with a property, process, or concept.
Fuzzy aggregation operator: It is an operator which aggregated two or more fuzzy sets. Boundary conditions for fuzzy aggregation operator are $h[0, 0, 0, \ldots 0] = 0$ and $h[1, 1, 1\ldots, 1] = 1$. $h[0,1]n \to [0,1]$, where $n \geq 2$ and h is a continuous monotonically increasing aggregation function.
Fuzzy identification: A process of determining a fuzzy system or a fuzzy model. A typical example is identification of fuzzy dynamic models consisting of determination of the number of fuzzy space partitions, determination of membership functions, and determination of parameters of local dynamic models.
Fuzzy input–output model: Input–output models involving fuzzy logic concepts. A typical example is a fuzzy dynamic model consisting of a number of local linear transfer functions connected by a set of nonlinear membership functions.
Fuzzy logic: A logic that deals with the variables, which is not restricted to binary states (0 or 1 only), but has a degree of truth between 0 and 1.
Fuzzy modeling: Combination of available mathematical description of the system dynamics with its linguistic description in terms of IF-THEN rules. In the early stages of fuzzy logic control, fuzzy modeling meant just a linguistic description in terms of IF-THEN rules of the dynamics of the plant and the control objective. Typical examples of fuzzy models in control application include Mamdani model, Takagi–Sugeno–Kang model, and fuzzy dynamic model.
Fuzzy modifier: An added description of a fuzzy set that leads to an operation that changes the shape (mainly the width and position) of a membership function.
Fuzzy parameter estimation: A method that uses fuzzy interpolation and fuzzy extrapolation to estimate fuzzy grades in a fuzzy search domain based on a few cluster center grade pairs. An application of this method is to estimate mining deposits.
Fuzzy relation: A relation which has degree of membership between 0 (not relation) and 1 (fully related). It is a subset of Cartesian product of several crisp sets.
Fuzzy sets: A set that allows its elements to have degrees of membership. An ultra fuzzy set has its membership function itself as a fuzzy set.
Fuzzy systems: A system whose variable (fuzzy) values are linguistic terms.
Heuristic: A rule of thumb. A mechanism with no guarantee of success.
Inference engine: A device or component carrying out the operation of fuzzy inference, that is, combining fuzzy IF-THEN rules in a fuzzy rule base into a mapping from a fuzzy set in the input universe of discourse to a fuzzy set in the output universe of discourse. It is a part of knowledge base system that makes inferences from the knowledge base.

Inference: The process of generating conclusions from conditions or new facts from known facts.

Information: It is the interpreted data. Information is data with attributed meaning in context.

Knowledge base system: A system containing knowledge which can perform tasks that require intelligence if done by human beings.

Knowledge base: An artificial intelligence database that is made up not merely of files of uniform content, but of facts, inferences, and procedures corresponding to the type of information needed for problem solution.

Knowledge engineering: A person, analogous to the system analyst in traditional computing, who builds a knowledge base system. This is the process of developing an expert system.

Knowledge representation: The process of mapping the knowledge of some domain into a computational medium.

Knowledge source: Any source for knowledge—documents, manuals, tape recording, etc.

Knowledge: It is derived from information by integrating information with existing knowledge. The same data may be interpreted differently by different people depending on their existing knowledge.

Linguistic variables: Common language expression used to describe a condition or a situation such as "hot," "cold," etc. It can be expressed using fuzzy set defined by the designer.

Membership function: The mapping that associates each element in a set with its degree of membership. It can be expressed as discrete values or as continuous functions. The commonly used membership functions are triangular, sigmoid, Gaussian, and trapezoidal.

Natural language processing: Processing of natural language (English, for example) by a computer to facilitate human communication with the computer—or for other purposes, such as language translation.

Parallel processing: Simultaneous processing, as opposed to the sequential processing in a conventional (Von-Neumann) type of computer architecture.

Production rules: An IF-THEN rule having a set of conditions and a set of consequent conclusions.

Rule: A mechanism for generating new facts.

Singleton: A set that has one member only.

Syllogism: A deductive argument in logic whose conclusion is supported by two premises.

C.4 Genetic Algorithms

Allele: One of a pair or series of alternative genes that occur at a given locus in chromosome: one constraining form of genes (bit value or feature value).

Carriers: A heterozygous individual with both recessive and dominant alleles in allelic pair.

Chromosome: Microscopically observable thread-like structures that are the main carriers of hereditary information (coded string).

Crossover: A genetic process which results in gene exchange by combining the different chromosomes selected from previous generation (parents).

Deoxyribonucleic acid (DNA): A chemical known as genetic material from which the genes are composed.
Diploid: An organism or cell having a set of two genomes.
Dominance: Applied one member of an allelic pair that has the ability to manifest itself at the exclusion of the expression of the other alleles.
Fitness: Survivability of a living being in a particular environment. It is the objective function value.
Gametes: They are reproductive cells.
Gene: A hereditary determiner specifying a biological function; a unit of inheritance located in a fixed place on chromosome. It has feature, character, or detector.
Genetic algorithm: An optimization technique that searches for parameter values by mimicking natural selection and the laws of genetics. A genetic algorithm takes a set of solutions to a problem and measures the "goodness" of those solutions. It then discards the "bad" solutions and keeps the "good" solutions. Next, one or more genetic operators, such as mutation and crossover, are applied to the set of solutions. The "goodness" metric is applied again and the algorithm iterates until all solutions meet a certain criteria or a specific number of iterations have been completed.
Genome: A complete set of chromosomes inherited as a unit from one parent. It is the complete string of all variables.
Genotype: Actual gene constitution for a trait (string structure).
Heterozygous: An organism carrying unlike alleles in an allelic pair.
Homozygous: An organism carrying same alleles in an allelic pair.
Lethals: An allele that has an influence on viability of an organism in such a way that the organism is unable to live, known as lethal gene. (String which disappears under specified conditions.)
Mutation: Sudden change in genetic material or gene in chromosome. It is just flipping a bit within a string.
Phenotype: Visible expression of traits. In GA, it is the set of parameters or a decoded string.

Bibliography

Al-Baiyat, S.A. and Bettayeb, M. (1993), A new model reduction scheme for k-power bilinear systems. *Proceedings of the IEEE Conference on Decision and Control*, San Antonio, TX.

Al-Baiyat, S.A., Bettayeb, M., and Al-Saggaf, U.M. (1994), New model reduction scheme for bilinear systems. *International Journal of Systems Science*, 25:1631–1642.

Alexander, B. (2000), The definition of system. *Kybernetes*, 29(4):444–451.

Alligood, K.T., Sauer, T.D., and Yorke, J.A. (1997), *Chaos: An Introduction to Dynamical Systems*. New York: Springer-Verlag.

Amizadeh, F. and Jamshidi, M. (1994), *Soft Computing, Fuzzy Logic, Neural Networks, and Distributed Artificial Intelligence*, vol. 4. Englewood Cliffs, NJ: Prentice Hall.

Antsaklis, P.J. and Passino, K.M. (Eds.) (1993), *An Introduction to Intelligent and Autonomous Control*. Norwell, MA: Kluwer Academic Publishers.

Aoki, M. (1968), Control of large dynamic system by aggregation, *IEEE Transactions on Automatic Control*, AC-13:246–256.

Arbib, M.A. (Ed.) (2003), *Handbook of Brain Theory and Neural Networks*, 2nd edn. Cambridge, MA: MIT Press.

Aris, R. (1999), *Mathematical Modeling: A Chemical Engineer's Perspective*. San Diego, CA: Academic Press.

Armstrong, D. (1981), What is consciousness? In *The Nature of Mind*. Ithaca, NY: Cornell University Press.

Asai, K., Tanaka, H., and Okuda, T. (1975), Decision-making and its goal in a fuzzy environment. In Zadeh, L.A., Fu, K.S., Tanaka, K., and Shimura, M. (Eds.), *International Fuzzy Sets and Their Applications to Cognitive and Decision Processes*. New York: Academic Press, pp. 227–256.

Auger, P.M. (1991), Introduction to the special issue on hierarchy theory and its applications, *International Journal of General Systems*, 18(3):189–190.

Badii, R. and Politi, A. (1997), *Complexity: Hierarchical Structures and Scaling in Physics*. Cambridge, U.K.: Cambridge University Press.

Bahm, A. (1981), Five types of systems philosophy, *International Journal of General Systems*, 6(4):233–238.

Bailey, K.D. (1994), *Sociology and the New Systems Theory: Toward a Theoretical Synthesis*. New York: State of New York Press.

Baker, G.L. (1996), *Chaos, Scattering and Statistical Mechanics*. Cambridge, U.K.: Cambridge University Press.

Bandler, J.W., Markettos, N.D., and Sinha, N.K. (1973), Optimal system modeling using recent gradient method, *International Journal of Systems Science*, 4:257–262.

Banerjee, S. (1999), National AIDS Control Programme, *National Journal on Social Welfare*, 3:15–20.

Banks, J., Carson II, J.S., Nelson, B.L., and Nicol, D.M. (2001), *Discrete-Event System Simulation*, 3rd edn. Englewood Cliffs, NJ: Prentice Hall.

Beer, S. (1985), *Diagnosing the System for Organization*. New York: John Wiley & Sons.

Beltrami, E. (1987), *Mathematics for Dynamic Modeling*. New York: Academic Press.

Bender, E.A. (2000), *An Introduction to Mathematical Modeling*, Mineola, NY: Dover.

Bernhard, J.A. and Marios, C.A. (2000), System dynamics modelling in supply chain management: Research review, *Proceedings of the 32nd Conference on Winter Simulation*. Orlando, FL.

Berr, J.S., Burghes, D.N., Huntley, I.D., and Moscardini, A.O. (1987), *Mathematical Modeling Course*. New York: John Wiley & Sons.

Bertalanffy, L. von. (1969a), *General System Theory*. New York: George Braziller, pp. 39–40.

Bezdek, J.C. (1993), Fuzzy models—What are they and why?, *IEEE Transactions on Fuzzy Systems*, 1(1):1–6, February.

Bhat, N., Minderman, P., McAvoy, T., and Wang, N. (1990), Modeling chemical process systems via neural network computation, *IEEE Control Systems Magazine*, 10(3):24.

Bird, R.J. (2003), *Chaos and Life: Complexity and Order in Evolution and Thought*. New York: Columbia University Press, 352 pp.

Bolch, G., Greiner, S., de Meer, H., and Trivedi, K. (2006), *Queuing Networks and Markov Chains*, 2nd edn. New York: John Wiley & Sons.

Bowden, K. (1991), On general physical systems theories, *International Journal of General Systems*, 18(1):61–79.

Buckley, W.F. (1967), *Sociology and Modern Systems Theory*. Englewood Cliffs, NJ: Prentice Hall.

Burkhardt, H. (1981), *The Real World and Mathematics*. Glasgow, Scotland: Blackie.

Cannon, R.H., Jr. (1967), *Dynamics of Physical Systems*. New York: McGraw-Hill.

Cao, X.-R. and Ho, Y.-C. (1990), Models of discrete event dynamic systems, *IEEE Control Systems Magazine*, 10(4):69–76, June.

Carpenter, G.A. and Grossberg, S. (1996), Learning, categorization, rule formation, and prediction by fuzzy neural networks. In Chen, C.H. (Ed.), *Fuzzy Logic and Neural Network Handbook*. New York: McGraw-Hill, Inc., pp. 1.3–1.45.

Cassandras, C.G. and Lafortune, S. (1999), *Introduction to Discrete Event Systems*. Dodrecht, the Netherlands: Kluwer Academic Publishers, September.

Chalmers, D. (1996), *The Conscious Mind*. Oxford, U.K.: Oxford University Press.

Chaturvedi, D.K. (1992), Modelling and simulation of power systems: An alternative approach, Doctoral thesis. Dayalbagh Educational Institute (Deemed University), Dayalbagh, Agra, India.
Chaturvedi, D.K. (1997), Modelling and simulation of electrical power systems: An alternative approach, Doctoral thesis. Dayalbagh Educational Institute (Deemed University), Dayalbagh, Agra, India.
Chaturvedi, D.K. (2005), Dynamic model of HIV/AIDS population of Agra region, *International Journal of Environmental Research and Public Health, USA*, 2(3):420–429.
Chaturvedi, D.K. (2008), *Soft Computing Techniques and Its Applications to Electrical Engineering*. Germany: Springer-Verlag.
Chaturvedi, D.K. and Gupta, B.R. (1995), Simulation of temperature variation in parachute inflation, *Journal of the Institution of Engineers (India)*, AS, 76:29–31, September.
Chaturvedi, D.K. and Malik, O.P. (2005), A generalized neuron based adaptive power system stabilizer for multimachine environment, *IEEE Transactions on Power Systems*, 20(1):358–366.
Chaturvedi, D.K. and Neeraj, V. (1995), *Modelling and Simulation of Production Distribution System, National Systems Conference*. Coimbatore, India: PSG College of Technology, pp. 620–623, December 14–16.
Chaturvedi, D.K. and Satsangi, P.S. (1992), System dynamics modelling and simulation of basic commutating electric machines: An alternative approach, *Journal of the Institution of Engineers (India)*, EL, 73:6–10, April.
Chaturvedi, D.K. and Satsangi, P.S. (1993), Innovative approach for predicting the performance characteristics of synchronous generators, *Journal of the Institution of Engineers (India)*, El, 74:109–113, November.
Chaturvedi, D.K. and Sharma, R.K. (1996), Modeling and simulation of force generated in stanchion system of aircraft arrester barrier system, *International Journal of Modeling, Measurements, and Control, France*, B, 64(2):33–51.
Chaturvedi, D.K. and Sharma, R.K. (1996), Modelling and simulation of force generated in stanchion system of aircraft arrester barrier system, *AMSE Journal on Modelling, Measurements and Control, France*, B, 64(2):33–51.
Chaturvedi, D.K., Satsangi, P.S., and Kalra, P.K. (1997), Short term load forecasting using generalized neural network (GNN) approach, *Journal of the Institute of Engineers (India)*, 78:83–87, August.
Chaturvedi, D.K., Satsangi, P.S., and Kalra, P.K. (1998), A fuzzy simulation model of basic commutating electrical machines, *International Journal of Engineering Intelligent Systems*, 6(4):225–236, December.
Chaturvedi, D.K., Satsangi, P.S., and Kalra, P.K. (1999a), Applications of generalized neural network to load frequency control problem, *Journal of the Institution of Engineers (India)*, EL, 80:41–47, August.
Chaturvedi, D.K., Satsangi, P.S., and Kalra, P.K. (1999b), Development of fuzzy simulator for DC machine modelling, *Journal of the Institution of Engineers (India)*, EL, 80:53–58, August.
Chaturvedi, D.K., Satsangi, P.S., and Kalra, P.K. (1999c), Flexible neural network models for electrical machine, *Journal of the Institution of Engineers (India)*, EL, 80:13–16, May.
Chaturvedi, D.K., Satsangi, P.S., and Kalra, P.K. (1999d), Load frequency control: A generalized neural network approach, *Electric Power and Energy Systems, Elsevier Science*, 21:405–415.
Chaturvedi, D.K., Satsangi, P.S., and Kalra, P.K. (1999e), New neuron model for simulating rotating electrical machines and load forecasting problems, *International Journal on Electric Power System Research, Elsevier Science, Ireland*, 52:123–131.
Chaturvedi, D.K., Mohan, M., and Saxena, A.K. (2001a), Development of HIV infected population model using system dynamics technique, *International Journal of Modeling, Measurements, and Control, France*, 22(2):1–14.
Chaturvedi, D.K., Singh, P., Mohan, M., Gaur, S.K., and Mishra, D.S. (2001b), Development of HIV model and its simulation, *Journal of Health Management*, 3(1):65–84.
Chaturvedi, D.K., Satsangi, P.S., and Kalra, P.K. (2001c), Fuzzified neural network approach for load forecasting problems, *International Journal on Engineering Intelligent Systems, CRL Publishing, U.K.*, 9(1):3–9, March.
Chaturvedi, D.K., Malik, O.P., and Kalra, P.K. (2004a), Experimental studies of generalized neuron based power system stabilizer, *IEEE Transactions on Power Systems*, 19(3):1445–1453, August.

Chaturvedi, D.K., Malik, O.P., and Kalra, P.K. (2004b), A generalized neuron based adaptive power system stabilizer, *IEE Proceedings of Generation, Transmission & Distribution*, 15(2):213–219, March 2004.

Chaturvedi, D.K., Malik, O.P., and Kalra, P.K. (2004c), A generalized neuron based PSS in a multi-machine power system, *IEEE Transactions on Energy Conversions, USA*, 19(3):625–632, September.

Chaturvedi, D.K., Malik, O.P., and Kalra, P.K. (2004d), Application of generalized neuron based power system stabilizer in a multi-machine power system (research note), *International Journal of Engineering Transactions B, Iran*, 17(2):131–140, July 2004.

Checkland, P.B. (1997), *Systems Thinking, Systems Practice*. Chichester, U.K.: John Wiley & Sons, Ltd.

Checkland, P.B. (1999), *Soft Systems Methodology in Action*. Chichester, U.K.: John Wiley & Sons Ltd.

Checkland, P.B. (2006), *Learning for Action: A Short Definitive Account of Soft Systems Methodology, and Its Use Practitioners, Teachers and Students*. Chichester, U.K.: Wiley (with John Poulter).

Checkland, P.B. and Winter, M.C. (2000), *The Relevance of Soft Systems Thinking, Human Resource Development International*, 3(3):411–417.

Chen, C.F. (1974), Model reduction of multivariable control system by means of matrix continued fraction, *International Journal of Control*, 29(2):225–238.

Chen, C.F. and Shieh, L.S. (1969), Continued fraction inversion by Routh algorithm, *IEEE Transactions on Circuit Theory*, CT-16(2):197–202.

Chen, C.F. and Shieh, L.S. (1972), A novel approach to linear model simplification, *International Journal of Control*, 22(2):231–238.

Cochin, I. (1980), *Analysis and Design of Dynamic Systems*. New York: Harper & Row.

Contwell, A. (1991), *AIDS: The Mystery and the Solution*. Los Angeles, CA: Aries Rising Press.

Cox, E. (1998), *The Fuzzy Systems Handbook—A Practitioner's Guide to Building, Using, and Maintaining, Fuzzy Systems*, 2nd edn. Boston, MA: AP Professional.

Croall, I.F. and Mason, J.P. (Eds.) (1991), *Industrial Applications of Neural Networks*. New York: Springer-Verlag.

Cumberbatch, E. and Fitt, A. (2001), *Mathematical Modeling: Case Studies from Industry*. Cambridge, U.K.: Cambridge Press.

D'Agostino, R.B., Shuman, L.J., and Wolfe, H. (1984), *Mathematical Modeling: Applications in Emergency Health Services*. New York: Haworth Press.

D'Souza, A. (1988), *Design of Control Systems*. Englewood Cliffs, NJ: Prentice Hall.

Davison, E.J. (1968), A new method for simplifying large linear dynamic system, *IEEE Transactions on Automatic Control*, AC-13:214–215, April.

Dayal, B. and Satsangi, P.S. (2008), Managing complexity in systems through neutral processes of nominal group technique and interpretive structural modelling: A novel application from literary field, *National Systems Conference (NSC 2008)*, I.I.T. Roorkee, India.

Dayal, B. and Srivastava, M. (2008), Variety management in modelling a system: A case study of fictional system representation by the Hindi Movie "Sarkar Raj," *National Systems Conference (NSC 2008)*, I.I.T. Roorkee, India.

Dayhoff, J. (1991), *Neural Network Architectures: An Introduction*. New York: Van Nostrand Reinhold.

Dennett, D.C. (1978), *Brainstorms*. Cambridge, MA: MIT Press.

Dennett, D.C. (1984), *Elbow Room: The Varieties of Free Will Worth Having*. Cambridge, MA: MIT Press.

Dennett, D.C. (1991), *Consciousness Explained*. Boston, MA: Little, Brown and Company.

Dennett, D.C. (2003), *Freedom Evolves*. New York: Viking.

Devaney, R.L. (2003), *An Introduction to Chaotic Dynamical Systems*, 2nd edn. Boulder, CO: Westview Press.

Douglas, K.L. and Euel, W.E. (Ed.) (1997), *Chaos Theory in the Social Sciences: Foundations and Applications*. Ann Arbor, MI: University of Michigan Press.

Dubois, D. and Prade, H. (1980), *Fuzzy Sets and Systems: Theory and Applications*. New York: Academic Press.

Dubois, D. and Prade, H. (1991), Fuzzy sets in approximate reasoning, Part I and II: Inference with possibility distributions, *Fuzzy Sets and Systems*, 40(1):143–202.
Duda, R., Hart, P., and Stork, D. (2001), *Pattern Classification*, 2nd edn. New York: John Wiley & Sons.
Dudai, Y. (1989), *The Neurobiology of Memory*, Oxford, U.K.: Oxford University Press.
Dullerud, G.E. and Paganini, F.A. (2000), *Course in Robust Control Theory: A Convex Approach*. Berlin: Springer.
Eyechoff, P. (1974), *System Identification: Parameter and State Estimation*. London, England: Wiley.
Feldman, D.A. and Johnson, T.M. (Eds.) (1986), *The Social Dimensions of AIDS: Methods and Theory*. New York: Praeger Publisher.
Feldmann, P. and Freund, R.W. (1995), Efficient linear circuit analysis by Pade approximation via the Lanczos process, *IEEE Transactions CAD*, 14:639–649.
Fielding, C. and Flux, P.K. (2003), *Non Linearities in Flight Control Systems*. Warton, U.K.: Royal Aeronautical Society.
Fishman, G.S. (1978), *Principles of Discrete Event Simulation*. New York: John Wiley.
Flood, R.L. (1999), *Rethinking the Fifth Discipline: Learning within the Unknowable*. London, U.K.: Routledge.
Forrester, J.W. (1968), *Principles of System*. Cambridge, MA: Wright Allen Press, Inc.
Forrester, J.W. (1968), *Industrial Dynamics*, Cambridge, MA: MIT Press, 1968.
Fujimoto, R.M., (2003), Parallel simulation: Distributed simulation systems, *Workshop on Parallel and Distributed Simulation*, 97.
Franklin, G.F. and Powell, J.D. (1986), *Feedback Control of Dynamic Systems*. Reading, MA: Addison-Wesley.
Genesio, R. and Milaness, M. (1976), A note on the derivation can use of reduced order models, *IEEE Transactions on Automatic Control*, AC-21(1):118–122, February.
George, J.K. (1969), *Approach to General Systems Theory*. New York: Van Nostrand Reinhold Co.
Gershenfeld, N. (1998), *The Nature of Mathematical Modeling*. Cambridge, U.K.: Cambridge University Press.
Glavaski, S., Marsden, J.E., and Murray, R.M. (1998), Model reduction, centering, and the Karhunen–Loeve expansion. *Proceedings of the IEEE Conference on Decision and Control*, Tampa, FL, 37: 2071–2076.
Glover, K. (1984), All optimal Hankel-norm approximations of linear multivariable systems and their L-infinity error bounds, *International Journal of Control*, 39:1115–1193.
Gollub, J.P. and Baker, G.L. (1996), *Chaotic Dynamics*. Cambridge, U.K.: Cambridge University Press.
Gong, V. (Ed.) (1985), *Understanding AIDS: A Comprehensive Guide*. New Brunswick, NJ: Rutgers University Press.
Goodman, M.R. (1983), *Study Notes on System Dynamics*. London, England: The MIT Press.
Gray, W.S. and Mesko, J. (1998), Energy functions and algebraic gramians for bilinear systems. *Proceedings of the 4th IFAC Nonlinear Control Systems Design Symposium*, Enschede, the Netherlands.
Gruca, A. and Bertrand, P. (1978), Approximation of high order systems by low order models with delays, *International Journal of Control*, 28(6):953–965.
Gupta, S. (Ed.) (1986), *AIDS: Associated Syndromes*. New York: Plenum Press.
Gutman, P.O., Mannerfelt, C.F., and Monlander, P. (1982), Contribution to the model reduction problem, *IEEE Transactions on Automatic Control*, AC-27(2):454–455.
Gutzwiller, M. (1990), *Chaos in Classical and Quantum Mechanics*. New York: Springer-Verlag.
Hadamard, J. (1898), Les surfaces à courbures opposées et leurs lignes géodesiques, *Journal de Mathématiques Pures et Appliquées*, 4:27–73.
Hanselman, D. and Littlefield, B. (2000), *Mastering Matlab 6: Comprehensive Tutorial and Reference*. Englewood Cliffs, NJ: Prentice Hall.
Haykin, S. (1998), *Neural Networks*. New Delhi, India: Pearson Education.
Hickin, J.D. and Sinha, N.K. (1980), A new method for reducing multivariable systems, *IEEE Transactions on Automatic Control*, AC-25:1121–1127.

Hipel, K.W. and Fraser, N.M. (1980), Metagame analysis of the garrison conflict, *Water Resources Research*, 16:629–637.
Hoover, W.G. (1999), *Time Reversibility, Computer Simulation, and Chaos*. Singapore: World Scientific.
Hristu-Varsakelis, D. and Kyrtsou, C. (2008), Evidence for nonlinear asymmetric causality in US inflation, metal and stock returns, *Discrete Dynamics in Nature and Society*, Article ID 138547, 7, doi:10.1155/2008/138547.
Hritonenko. N. et al. (2003), *Applied Mathematical Modelling of Engineering Problems*. Dordrecht, the Netherlands: Kluwer Academic Publisher.
Hsu, C.S. and Hou, D. (1991), Reducing unstable linear control systems via real Schur transformation, *Electronics Letters*, 27(11):984–986.
Humpert, B. (1994), Improving backpropagation with a new error function, *Neural Network*, 7(8):1191–1192.
Ives, R.L. (1958), Neon oscillator rings, *Electronics*, 31:108–115, October.
Jamshidi, M. (1983), *Large Scale System Modelling and Control*, New York: North Holland.
Jang, J.S. and Sun, C. (1995), Neuro-fuzzy modeling and control, *Proceedings of IEEE* 83(3):378–406.
Jang, J.-S.R., Sun, C.-T., and Mizutani, E. (1996), *Neuro-Fuzzy and Soft Computing*. New York: Prentice Hall.
Jin, Y. (2000), Fuzzy modeling of high-dimensional systems: Complexity reduction and interpretability improvement. *IEEE Transactions on Fuzzy Systems*, 8(2):212–221.
Jones, W.P. and Hoskins, J. (1987), Backpropagation: A generalized delta rule, *Byte*, 12:155–163, October.
Jose, M.B. et al. (Eds.) (2003), *Advances in Soft Computing: Engineering Design and Manufacturing*. Germany: Springer Verlag, August 22, 2003.
Kailath, T. (1980), *Linear Systems*. Englewood Cliffs, NJ: Prentice Hall.
Kaku, M. (2008), *Physics of Impossible*. New York: Doubleday.
Kalra, P.K. and Batra, J.L. (1990), Management of flexible manufacturing—An expert systems approach. In Lal, G.K. and Singh, L.P. (Eds.), *Challenges of Change*. New Delhi: Tata McGraw-Hill Publishing Company, 56–76.
Kalra, P.K. and Srivastav, S.C. (1990), Application of expert system for electric energy management system: A feasibility study. In Lal, G.K. and Singh, L.P. (Eds.), *Challenges of Change*. New Delhi: Tata McGraw-Hill Publishing Company, 76–84.
Kalra, P.K., Srivastava, S.C., and Chaturvedi, D.K. (1992), Possible applications of neural nets to power system operation and control, *International Journal of Power System Research*, 25:83–90.
Kerns, K.J. and Yang, A.T. (1998), Stable and efficient reduction of large, multi-port rc network by pole analysis via congruence transformations, *IEEE Transactions on Computer-Aided Design of Integrated Circuits and Systems*, 16(7):734–744, July.
Kiel, L.D. and Elliott, E.W. (1997), *Chaos Theory in the Social Sciences*. Cambridge, MA: Perseus Publishing.
Klir, G.J. (1969), An approach to general systems theory, New York: Van Nostrand Reinhold Co.
Klir, G.J. and Folger, T.A. (2000), *Fuzzy Sets, Uncertainty and Information*. New Delhi: Prentice Hall of India.
Koenig, H.E. and Blackwell, W.A. (1961), *Electromechanical System Theory*. New York: McGraw-Hill Book Company.
Koenig, H.E., Tokad, Y., and Kesavan, H.K. (1966), *Analysis of Discrete Physical Systems*. New York: McGraw-Hill Book Company.
Kosko, B. (1997), *Fuzzy Engineering*. Upper Saddle River, NJ: Prentice Hall.
Kumar, S. and Satsangi, P.S. (1992), System dynamics simulation of Hopfield neural networks, *International Journal of Systems Science*, 23(9):1517–1525.
Kumar, A. and Singh, V. (1978), An improved algorithm for continued fraction inversion, *IEEE Transactions on Automatic Control*, AC-23(5):938–940.
Kuo, B.C. (1990), *Automatic Control Systems*. New Delhi, India: Prentice Hall of India.
Kyrtsou, C. (2008), Re-examining the sources of heteroskedasticity: The paradigm of noisy chaotic models, *Physica A*, 387:6785–6789.

Lal, M. and Mitra, R. (1974), Simplification of large system dynamic using a moment evaluation algorithm, *IEEE Transactions on Automatic Control*, 19(5):602–603, October.
Lathi, B.P. (1965), *Signals, Systems, and Communication*. New York: Wiley.
Lathi, B.P. (1987), *Signals and Systems*. Carmichael, CA: Berkeley–Cambridge Press.
Leigh, J.R. (1985), *Applied Digital Control*. Englewood Cliffs, NJ: Prentice Hall.
Leitch, R.R. (1989), A review of the approaches to the qualitative modelling of complex systems. In Linkens, D.A. (Ed.), *Trends in Information Technology*. London, U.K.: Peter Peregrinus.
Lentz, B.P. (1972), On time-dependent fuzzy sets. *Information Science*, 4:367–376.
Lin, C.C. and Segel, L.A. (1988), *Mathematics Applied to Deterministic Problems in the Natural Sciences*. Philadelphia, PA: Society for Industrial and Applied Mathematics (SIAM).
Lipmann, R.P. (1989), Pattern classification using neural networks, *IEEE Communications Magazine*, 27(11):47–54, November.
Ljung, L. (1999), *System Identification*. Englewood Cliffs, NJ: Prentice Hall.
Lorenz, E.N. (1963), Deterministic non-periodic flow, *Journal of the Atmospheric Sciences*, 20:130–141.
Lorenz, E. (1996), *The Essence of Chaos*. Seattle, WA: University of Washington Press.
Luenberger, D.G. (1979), *Introduction to Dynamic Systems*. New York: Wiley.
Maki, D.P. (1973), *Applications, With Emphasis on the Social, Life, and Management Sciences*, Prentice Hall.
Maki, D.P. and Thomson, M. (1973), *Mathematical Models and Applications: With Emphasis on the Social Life and Management Sciences*. Englewood Cliffs, NJ: Prentice Hall.
Mamdani, E.H. (1974), Application of fuzzy algorithms for control of simple dynamic plant, *Proceedings of IEEE*, 121(12):1585–1588.
Mamdani, E.H. (1976), Advances in the linguistic synthesis of fuzzy controller, *International Journal of Man–Machine Studies*, 8(6):669–678.
Mamdani, E.H. and Assilian, S. (1975), An experiment in linguistic synthesis with a fuzzy logic controller, *International Journal of Man–Machine Studies*, 7(1):1–13.
Mohan, M., Chaturvedi, D.K., and Kalra, P.K. (2003a), Development of new neuron structure for short term load forecasting, *International Journal of Modeling and Simulation, AMSE Periodicals*, 46(5):31–52.
Mohan, M., Chaturvedi, D.K., Satsangi, P.S., and Kalra, P.K. (2003b), Neuro-fuzzy approach for development of new neuron model, *International Journal of Soft Computing—A Fusion of Foundations, Methodologies and Applications*, Springer-Verlag, 8(1):19–27, October.
Mark, M.M. (2007), *Mathematical Modeling*. Amsterdam, the Netherlands: Elsevier.
Marshel, S.A. (1966), An approximation method for reducing the order of a large system, *Control Engineering*, 10:642–648.
McLean, D. (1999), Aircraft flight control system, Department of Aeronautics and Astronautics, *The Aeronautical Journal*, 103:159–165.
Michalewicz, Z. (1992), *Genetic Algorithms + Data Structure = Evolution Programs*. Berlin, Germany: Springer-Verlag.
Minsky, M.L. and Papert, S.A. (1969), *Perceptron*. Cambridge, MA: MIT Press.
Mishra, R.N. and Wilson, D.A. (1980), A new algorithm for optimal reduction of multivariable systems, *International Journal of Control*, 31(3):443–466.
Moon, F. (1990), *Chaotic and Fractal Dynamics*. New York: Springer-Verlag.
Moore, B.C. (1981), Principal component analysis in linear systems: Controllability, observability, and model reduction. *IEEE Transactions on Automatic Control*, 26(1):17–32.
Murray, S.K. (1987), *Mathematical Modelling: Classroom Notes in Applied Mathematics*. Philadelphia, PA: Society for Industrial and Applied Mathematics (SIAM).
Nagrath, I.J. and Gopal, M. (2001), *Control System Engineering*. New Delhi, India: New Age International (P) Limited.
Nakamura, S. (1993), Applied numerical methods in C, Englewood Cliffs, Prentice Hall, NJ.
Negoita, C.V. and Sularia, M. (1976), On fuzzy programming and tolerances in planning, *Economic Computation and Economic Cybernatics Studies and Research*, 1:3–14.

Newman, A.J. and Krishnaprasad, P.S. (1998a), Nonlinear model reduction for RTCVD. *Proceedings of the 32nd Conference on Information Sciences and Systems*, Princeton, NJ.

Newman, A.J. and Krishnaprasad, P.S. (1998b), Computation for nonlinear balancing. *Proceedings of the IEEE Conference on Decision and Control*, Tampa, FL, pp. 4103–4104.

Odabasioglu, A., Celik, M., and Pileggi, L.T. (1998), PRIMA: Passive reduced-order interconnect macromodeling algorithm, *IEEE Transactions on CAD*, 17(8):645–654.

Ogata, K. (1990), *Modern Control Engineering*, 2nd edn. Englewood Cliffs, NJ: Prentice Hall.

Ott, E. (2002), *Chaos in Dynamical Systems*. New York: Cambridge University Press.

Padulo, L. and Arbib, M.A. (1974), *System Theory*. Philadelphia, PA: Saunders.

Panos, D. (1988), *AIDS and the Third World*. London, U.K.: Panos Institute.

Parathasarthy, R. and Harpreet, S. (1975), On continued fraction inversion by Routh's algorithm, *IEEE Transactions on Automatic Control*, AC-20:278–279.

Pavel, L. and Fairman, F.W. (1997), Controller reduction for nonlinear plants—An L_2 approach, *International Journal of Robust and Nonlinear Control*, 7:475–505.

Pedrycz, W. (1984), Identification in fuzzy systems, *IEEE Transactions on Systems, Man, and Cybernetics*, 14:361–366.

Peierls, R. (1980), Model-making in physics, *Contemporary Physics*, 21(1):3–17.

Peng, L. and Pileggi, L.T. (2003), NORM: Compact model order reduction of weakly nonlinear systems, *IEEE/ACM Design Automation Conference*, Anaheim, CA, pp. 427–477.

Phillips, J.R., Daniel, L., and Silveira, L.M. (2003), Guaranteed passive balanced transformation for model order reduction, *IEEE Transactions on Computer-Aided Design of Integrated Circuits and Systems*, 22(8):1027–1041.

Poincaré, J.H. (1890), Sur le problème des trois corps et les équations de la dynamique. Divergence des séries de M. Lindstedt, *Acta Mathematica*, 13:1–270.

Pouvreau, D. and Drack, M. (2007), On history of Ludwig von Beranlanffy's "General Systemology" and on its relationship to cybernetics, *International Journal of General Systems*, 36(3):281–337.

Prasad, R., Pal, J., and Pant, A.K. (1995), Multivariable system approximation using polynomial derivates, *Journal of the Institution of Engineers (India), Electrical Engineering Division*, 76:186–188.

Ramadge, P.J.G. and Wonham, W.M. (1989), The control of discrete event systems, *Proceedings of the IEEE*, 77(1):81–98.

Renee, S. (Ed.) (1988), *Blaming Others: Prejudice, Race and World-Wide AIDS*. London, U.K.: Panos Institute.

Roe, P.H. (2009), Creation, the universe, and systems, *International Journal on Literature and Theory, Literary Paritantra (Systems)*, 1(1 and 2):25–28.

Ross, T.J. (1995), *Fuzzy Logic with Engineering Applications*. New York: McGraw-Hill, Inc.

Rowley, C.W. and Marsden, J.E. (2000), Reconstruction equations and the Karhunen–Loeve expansion for systems with symmetry, *Physica D*, 142:1–19.

Roychowdhury, J. (1999), Reduced-order modeling of time-varying systems, *IEEE Transactions on Circuits and Systems II: Analog and Digital Signal Processing*, 46(10):1273–1288.

Rozsa, P. and Sinha, N.K. (1974), Efficient algorithm for irreducible realization of a rational matrix, *International Journal of Control*, 21:273–284.

Sage, A.P. (1977), *Methodology for Large Scale System*. New York: McGraw-Hill Book Company.

Sage, A.P. (Ed.) (1990), *Concise Handbook on Information Systems Challenges of Change*. New Delhi, India: Tata McGraw-Hill Publishing Company, pp. 171–181.

Saini, A., Chaturvedi, D.K., and Saxena, A.K. (2006), Optimal load flow: A GA-fuzzy system approach, *International Journal of Emerging Electrical Power System Research*, 5(2):1–21.

Salthe, S.N. (1991), Two forms of hierarchy theory in western discourses, *International Journal of General Systems*, 18(3):251–264.

Sanchez, E. (1976), Resolution of composite fuzzy relation equations, *Information and Control*, 30:38–48.

Sandefur, J.T. (1993), *Discrete Dynamical Modeling*, McGraw-Hill.

Satsangi, P.S. (1985), Management by qualitative systems analysis, *Prabandh*, 1:11–24.

Satsangi, P.S. (2006), Generalizing physical systems through applied systems research from "real physical" systems through "conceptual" socio-economic-environmental systems to "esoteric" creational systems, *International Journal of General Systems*, 35(2):127–167.

Satsangi, P.S. (2008), Generalized physical system theory for applies systems research, *National Systems Conference (NSC 2008)*, I.I.T. Roorkee, India.

Satsangi, P.S. (2009), Linear graph theoretic system paradigm—A learning system modeling methodology, *International Journal on Literature and Theory, Literary Paritantra (Systems)*, 1(1 and 2):1–16.

Satsangi, P.S. and Ellis, J.B. (1971), General systems from network systems: A philosophy of modelling, *International Journal of Systems Science*, 2(1):1–16.

Scherpen, J.M.A. (1993), Balancing for nonlinear systems, *Systems and Control Letters*, 21(2):143–153.

Scherpen, J.M.A. (1996), H-infinity balancing for nonlinear systems, *International Journal of Robust and Nonlinear Control*, 6:645–668.

Scherpen, J.M.A. and van der Schaft, A.J. (1994), Normalized co-prime factorizations and balancing for unstable nonlinear-systems, *International Journal of Control*, 60(6):1193–1222.

Schilling, R.J. and Harris, S.L. (2000), *Applied Numerical Methods for Engineers Using Matlab and C*. Pacific Grove, CA: Brooks/Cole.

Schmidt, J.W. and Taylor, R.E. (1970), *Simulation and Analysis of Industrial Systems*. Homewood, IL: Richard D. Irwin.

Shamash, Y. (1974), Order reduction of linear systems by Pade approximation methods, *IEEE Transactions on Automatic Control*, AC-19:615–616.

Shieh, L.S. and Goldman, M.J. (1974), Continued fraction expansion and inversion of the Cauer third form, *IEEE Transactions Circuits and Systems*, CAS 21:341–345.

Singh, M.G. and Titlied, A. (1979), *Handbook of Large Scale Systems Engineering Applications*. New York: North Holland.

Sinha, N.K. and Bereznai, G.T. (1977), Optimal approximation of high order systems by low order models, *International Journal of Control*, 13(3):88–90, February.

Skyttner, L. (2006), *General Systems Theory: Problems, Perspective, Practice*. Singapore: World Scientific Publishing Company, ISBN:9-812-56467-5.

Smith, P. (1998), *Explaining Chaos*. Cambridge, U.K.: Cambridge University Press.

Sommer, G. and Pollatschek, M.A. (1978), A fuzzy programming approach to an air pollution regulation problem, in R. Trap et al. (eds.): *Progress of Cybernetics and Systems Research*, Vol. III, pp. 303–313, Hemisphere Publ. Corp., Washington, D.C.

Sprott, J.C. (2003), *Chaos and Time-Series Analysis*. Oxford, U.K.: Oxford University Press.

Srinivasa Rao, S.R. (2003), Mathematical modeling of AIDS epidemic in India, *Current Science*, 84(9):1192–1197.

Stachowicz, M.S. and Kochanska, M.E. (1986), Analysis of the application of fuzzy relations in modeling, *Proceedings of the North American Fuzzy Information Society '86*, New Orleans, LA.

Stachowicz, M.S. and Kochanska, M.E. (1987), Fuzzy modeling of the process, *Proceedings of Second International Fuzzy Systems Association Congress*, Tokyo, Japan.

Sterman, J.D. (2000), *Business Dynamics: Systems Thinking and Modeling for a Complex World*. New York: McGraw-Hill.

Stewart, I. (1990), *Does God Play Dice?: The Mathematics of Chaos*. Cambridge, MA: Blackwell Publishers.

Stewart, R. (2004), *Simulation—The Practice of Model Development and Use*. Chichester, U.K.: Wiley.

Strogatz, S.H. (1994), *Nonlinear Dynamics and Chaos*. Reading, MA: Addison Wesley.

Strogatz, S. (2000), *Nonlinear Dynamics and Chaos*. Cambridge, MA: Perseus Publishing.

Sugeno, M. and Yasukawa, T. (1993), A fuzzy logic based approach to qualitative modelling, *IEEE Transactions on Fuzzy Systems*, 1:7–31.

Sularia, M. (1977), On fuzzy programming in planning, *Kyhernetr*, 6:229–230.

Tagaki, T. and Sugeno, M. (1985), Fuzzy identification of systems and its applications to modeling and control, *IEEE Transactions on Systems, Man, and Cybernetics*, 15:116–132.

Tan, S.X.-D. and He, L. (2007), *Advanced Model Order Reduction Techniques in VLSI Design*. New York: Cambridge University Press.
Tél, T. and Gruiz, M. (2006), *Chaotic Dynamics: An Introduction Based on Classical Mechanics*. Cambridge, U.K.: Cambridge University Press.
Thayer, R.H. (1990), *Software System Engineering*, E80304-6/90/0000/00TI. IEEE CS Press, Los Alamitos, CA, pp. 77–116.
Thistle, J.G. (1994), Logical aspects of control of discrete event systems: A survey of tools and techniques. In Cohen, G. and Quadrat, J.-P. (Eds.), *International Conference on Analysis and Optimization of Systems—Discrete Event Systems*, ser. *Lecture Notes in Control and Information Sciences*, 199. Berlin: Springer-Verlag, pp. 3–15, June.
Thomas, G. (1994), *AIDS in India*. Jaipur, India: Rawat Publications.
Thompson, C.W.N. (1991), Systems analysis, engineering and management program of Northwestern University, *International Journal of General Systems*, 19(1):25 and 30.
Tong, R.M. (1978), Synthesis of fuzzy models for industrial processes—Some recent results, *International Journal of General Systems*, 4:143–163.
Towill, D.R. (1963), Low order modelling techniques, tools or toys, *Conference on Computer Aided Control System Design*, University of Cambridge, Cambridge, U.K., April.
Trivedi, K.S., Ciardo, G., Malhotra, M., and Garg, S. (1994), Dependability and performability analysis using stochastic Petri nets. In Cohen, G., and Quadrat, J.-P. (Eds.), *Proceedings of 11th International Conference on Analysis and Optimization of Systems—Discrete Event Systems*, ser. *Lecture Notes in Control and Information Sciences*, 199. Berlin: Springer-Verlag, pp. 144–157, June.
Tufillaro, N.B., Abbott, T., and Reilly, J. (1992), *An Experimental Approach to Nonlinear Dynamics and Chaos*. New York: Addison-Wesley.
Uvarova, L.A. (2001), *Mathematical Modeling—Problems, Methods, Applications*, 1st edn., Springer.
Uvarova, L.A. and Latyshev, A.V. (2001), *Mathematical Modeling: Problems, Methods, Applications*. New York: Kluwer Academic/Plenum Publishers.
van der Pol, B. and van der Mark, J. (1927), Frequency demultiplication, *Nature*, 120:363–364.
Vester, F. (2007), *The Art of Interconnected Thinking. Ideas and Tools for Tackling with Complexity*, MCB-Verlag Munchen.
Warfield, J.N. (1974), Toward interpretation of complex structural models, *IEEE Transaction on S.M.C.*, 4(5):405–417.
Warfield, J.N. (1990), *A Science of Generic Design: Managing Complexity through Systems Design*, vol. 1. Salinas, CA: Intersystem Publications.
Warfiled, J.N. (2006), *An Introduction to Systems Science*. Singapore: World Scientific Company.
Weinberg, G.M. (2001—revised), *An Introduction to General Systems Thinking*. New York: Dorset House, ISBN 0-932-63349-8.
(1993, 1994), *AIDS Manual*, 1st and 2nd edn. WHO Geneva, Switzerland: World Health Organization.
Wills, M.J. (1999), *Proportional-Integral-Derivative (PID) Controls*. Newcastle upon Tyne, U.K.: Department of Chemical Process Engineering, University of Newcastle.
Wilson, B. (1984), *System Concepts, Methodologies and Applications*. Chichester, U.K.: John Wiley & Sons.
Wilson, B. (1990), *Systems: Concepts, Methodologies and Applications*, 2nd edn. Chichester, U.K.: Wiley, ISBN 0-471-92716-3.
Wilson, B. (2001), *Soft Systems Methodology: Conceptual Model Building and Its Contribution*. Chichester, U.K.: Wiley, ISBN 0-471-89489-3.
Wilson, D.A. and Mishra, R.N. (1979), Optimal reduction of multivariable systems, *International Journal of Control*, 29(2):267–278.
Winter, M.C. and Checkland, P.B. (2003), Soft systems: A fresh perspective for project management, *Civil Engineering*, 156(4):187–192.
Wu, F., Yang, X.H., Packard, A., and Becker, G. (1996), Induced L-norm control for LPV systems with bounded parameter variation rates, *International Journal of Robust and Nonlinear Control*, 6(9–10):983–998.

Wuechner, P., de Meer, H., Barner, J., and Bolch, G. (2005), MOSEL-2—A compact but versatile model description language and its evaluation environment. In Wolfinger, B. and Heidtmann, K. (Eds.), *Proceedings of MMBnet'05*, University of Hamburg, Hamburg, pp. 51–59, September.

Yager, R.R. and Filev, D.P. (2002), *Essentials of Fuzzy Modelling and Control*. New York: John Wiley & Sons, Inc.

Yang, X.-S. (2008), *Mathematical Modelling for Earth Sciences*. Edinburgh, Scotland: Dudedin Academic.

Zadeh, L.A. (1965), Fuzzy sets, *Information and Control*, 8:338–353.

Zadeh, L.A. (1969), *Systems Theory*. New Delhi, India: Tata McGraw-Hill.

Zadeh, L.A. (1975a), The concept of a linguistic variable and its applications to approximate reasoning—I, *Information Sciences*, 8:199–249.

Zadeh, L.A. (1975b), The concept of a linguistic variable and its applications to approximate reasoning—II, *Information Sciences*, 8:301–357.

Zadeh, L.A. (1975c), The concept of a linguistic variable and its applications to approximate reasoning—III, *Information Sciences*, 9:43–80.

Zadeh, L.A. (1983), The role of fuzzy logic in the management of uncertain in expert system, *Fuzzy Sets and Systems*, 118:199–227.

Zadeh, L. and Desoer, C. (1963), *Linear System Theory: The State Space Approach*, W. Linvill, L.A. Zadeh, and G. Dantzia, Eds., McGraw-Hill Series in System Science, New York: McGraw-Hill.

Zaslavsky, G.M. (2005), *Hamiltonian Chaos and Fractional Dynamics*. Oxford, U.K.: Oxford University Press.

Zimmerman, H.J. (1987), *Fuzzy Sets, Decision Making and Expert Systems*. Boston, MA: Kluwer Academic.

Zimmermann, H. (1991), *Fuzzy Set Theory and Its Applications*. Boston, MA: Kluwer Academic Publishers.

Zimmerman, H.J. (1992), *Fuzzy Mathematical Programming*. New York: John Wiley & Sons.

Zurada, J. (1992), *Introduction to Artificial Neural Systems*. St. Paul, MN: West Publishing Co.

Index

A

Abstract model, 35–36
Adams–Bashforth method, 178–180
Adams–Bashforth predictor multistep method, 412–413
Adams–Moulton corrector method, 413
Adaptive neuro-fuzzy inferencing systems
　causal-loop diagram, 568
　compositional rules, 572–573
　defuzzification methods, 573–574
　development models, 568
　flowchart, 568–569
　implication methods, 572, 573
　intersection operators, 571
　linguistic variable, 570
　logic system development, 573–574
　max–min rule, 573, 574
　overlapping, 571
　parameter identification, 568
　T-Conorms, 571–572
　T-Norms, 571
　union operators, 571–572
Aggregation method, 228–232
AIDS/HIV population modeling
　awareness level, 365, 366
　causal links, 365, 366
　causal loop, 366, 367
　dynamo model, 368, 369
　flow diagram, 368
　time profile, 369, 370
　variable identification, 365
　variable list, 366, 367
Aircraft arrester barrier system
　aircraft energy absorbing system, 385, 386
　dynamic model
　　causal loop, 390
　　net area, 389, 390
　experimental data *vs.* simulated data, 391, 393
　flow diagram, 390, 391
　MATLAB program, 393–394
　subsystems, 385–386
　system dynamic technique
　　stages of, 387
　　stanchion systems, forces, 387, 389
　tension time profile, 391–393
Analogical reasoning, 28

Analysis of systems, 18
ANFIS
　ANN structure window, 597, 599
　3D surface showing effects, 600–601
　load data, 597–598
　original training data, 600
　real-time evaluation, 596–597
　rule viewer, 600
　training data, 597–598
　training performance, 597, 599
　window, 597
ANN, *see* Artificial neural network
Application program interface (API), 623
Applied systems engineering, 29
Aristotelian's view, 19
Artificial neural network (ANN)
　artificial neuron, 504–505
　biological neuron, 503–504
　testing phase
　　back propagation, 509
　　building model, 508–509
　　model, 506–508
　　neural network application, 510–512
　　physical system's modeling, 512–515
　　rainfall prediction, 523–525
　　teacher evaluation system, 517–523
　　training algorithm, 509–510
　　weighing machine model, 516–517, 518, 519, 520
　training phase
　　error minimization process, 506
　　neuron characteristics, 505
　　pattern and preprocessing, 506
　　stopping criteria, 506
　　topology, 505
Atomism, 25

B

Backlash effects
　constant input, 443, 447
　pilots control column, 442–443
　ramp input, 443, 449
　sinusoidal input, 443, 449
　step input, 443, 448
　waveform generator input, 443, 450

693

Balanced realization-based reduction method
    controllability and observability grammians, 236–237
    models, 237–238
    properties of, 238
Balanced truncation, 238–244
Bertalanffy's contribution, 18–19
Box–Muller method, 618
Business system environment, 4
Butterworth filter, 428–429

C

Calay Hamilton method, 172
Cartesian product, 541–542
Cascaded systems, 7
Cauer's first form, 255–259
Cause–effect relationships
    negative causal links
        ambient temperature, 336, 337
        heat transfer, 335, 336
    positive causal links
        negative causal loop, 337, 338
        positive causal loop, 337–338
        S-shaped growth, 338–339
Cell arrays, 650
CFE, *see* Continued fraction expansion
Chaotic system
    definition, 479–480
    first-order continuous-time system
        autonomous time invariant system, 484
        function trajectory *vs.* x, 484–485
        linearization, 485–486
    historical prospective
        analysis and prediction, 484
        cryptography, 482
        dynamics and chaos, 481
        Henri Poincaré, 481
        linear and nonlinear systems, 483
        orbits and periodicity, 482
        stabilization and control, 483
        synthesis, 484
        vs. systematic knowledge handling, 481–482
    meaning, 478
    scientific meaning, 478–479
Chemical reactor, 425–426
Chi-square test, 619
Classical reduction method (CRM), 234
Classification of system
    complexity, 6
    interactions, 6–7
    nature and type of components, 7
    time frame, 5–6
    uncertainty
        linear *vs.* nonlinear systems, 8–9
        static *vs.* dynamic systems, 8
Colon (:) operator, 630–631
Command History
    Array Editor, 626–627
    current directory browser, 626
    editor debugger, 627
    environment features, 627
    launch pad, 625
    running external programs, 624
    workspace Window, 626
Command Window, 623–624
Commutating machine
    armature current *vs.* time, 375
    basic machine model, 371
    causal loop, 372
    dynamo equation, 373
    flow diagram, 372, 373
    MATLAB program, 373–374
    operation modes, 371–372
    rate of change of armature current (RAC), 375, 376
    speed *vs.* time, 375
Conceptual systems, 6
    component postulate, 121
    four-terminal component, 119, 120
    fundamental axiom, 121
    system postulate
        circuit, 124–127
        cutset, 122–124
        system graph, 121–122
        vertex, 127
    three-terminal component, 119, 120
    two-terminal component, 119, 121
Connectionist system, 28–29
Continued fraction expansion (CFE), 252–259
Continuous system
    definition, 5
    simulation
        demand of, 424–425
        MATLAB program, 423–424
        model development, 423
        net volume, 424
        solution, 423
        water reservoir system, 422–423
Continuous-time and discrete-time systems, 13–14
Continuous *vs.* discrete models, 37–38
Controllability, state space model, 180–181
Coupled systems, 7

Index

CRM, *see* Classical reduction method
Cushioned package system
    cushioned package response, 287, 288
    f–v analogy, 285
    MATLAB program, 286

D

D'Alembert's principle, 277–278
Data structures
    cell arrays, 650
    characters and text
        ASCII codes, 651
        cell array, 652
        char function, 654
        comma separated list, 653
        floating-point format, 650–651
        padded character array, 651–652
        scalar structures, 652–653
        single quotes, 650
        square brackets, 651
    multidimensional arrays, 648–649
Dead-zone nonlinearities
    comparison, 440–441
    constant input, 441, 443
    ramp input, 441, 443
    Simulink model, 441
    sinusoidal input, 441–442
    step input, 441–442
Decision function
    cause–effect relationships
        negative causal links, 335–337
        positive causal links, 337–338
    computing sequence
        level equations, 341
        rate equations, 342
        straight line approximation, 340
    dynamo equations, 339–340
    interconnected network, 338–340
Defuzzification methods, 554–555, 573–574
Degenerative system model development
    electrical system
        components, 145
        state space model, 148–151
        system graph and f-tree, 144–146
    mechanical system, 160–162
    model development, 146–148
    multiterminal components, 157–159
    nondegenerative systems, 151–157
    time varying and nonlinear components, 162–166
Descriptive model, 36

Deterministic systems, 7
    vs. stochastic activates, 608
    vs. stochastic models/systems, 15–16, 37
Diagonalization, 170–171
Differentiation method
    approximation criterion, 265–266
    MATLAB program, 267
    results comparison, 264
    Simulink model, 267, 268
    system responses, 267, 268, 270–273
Discrete-event modeling and simulation
    chi-square test, 619
    components
        clock, 611
        events list and random-number generators, 612
        flow chart, 613
        life cycle, ATM counter, 613–614
        Pseudo-code, 614–615
        statistics and ending condition, 612
    definitions
        activities and delays, 607–608
        deterministic *vs.* stochastic activates, 608
        discrete-event simulation (DES) model, 608
        entity and attribute, 607
        evaluation techniques, 609
        exogenous *vs.* endogenous activities, 608
        list processing, 607
        modeling techniques, 609
        state, resources and event, 607
    distributions, 615–616
    input data modeling, 615
    Kolomogrov–Smirnov test, 619
    queuing system, 609, 610–611
    random number generation
        Gaussian distribution, 617–618
        uniform distribution, 616–617
    sampled-data system, 605
Discrete-event simulation (DES) model, 608
Discrete models/systems, 5, 37–38
Discrete-time systems, 13–14
Discretization error, 418–419
Distributed parameter systems, 13
Dominant eigenvalue method
    characteristic roots' locations, 225
    limitations of, 228
    vs. nondominant eigenvalues, 226
    poles locations effects, 225
    system stability, 224
Dynamic system modeling, 8, 37, 565–566

Dynamo equations
　AIDS/HIV population modeling
　　awareness level, 365, 366
　　causal links, 365, 366
　　causal loop, 366, 367
　　dynamo model, 368, 369
　　flow diagram, 368
　　time profile, 369, 370
　　variable identification, 365
　　variables list, 366, 367
　commutating machine
　　armature current *vs.* time, 375
　　basic machine model, 371
　　causal loop, 372
　　dynamo equation, 373
　　flow diagram, 372, 373
　　MATLAB program, 373–374
　　operation modes, 371–372
　　rate of change of armature current (RAC), 375, 376
　　speed *vs.* time, 375
　heroin addiction modeling
　　addiction rate *vs.* time characteristic, 346, 347
　　causal loop, 346
　　flow diagram, 346
　infected population modeling
　　causal loop, 354
　　flow diagram, 354, 355
　　MATLAB program, 355
　　simulation model, 356
　inventory control system
　　causal loop, 348, 349
　　flow diagram, 348, 350
　　inventory responses, 350
　　order rate responses, 350, 351
　market advertising interaction model
　　aims of, 356
　　causal loop, 357
　　dynamo equations, 358–359
　　flow diagram, 357, 358
　　postevaluation criterion, 356–357
　population problem modeling
　　flow diagram, 347, 348
　　net birth rate *vs.* time, 347, 349
　　positive causal loop, 347, 348
　production distribution system
　　causal loop, 360, 361
　　departments of, 359, 360
　　dynamo equations, 362–363
　　flow diagram, 360, 362
　　production rate, 363, 364
　　rate variables *vs.* time, 363, 364
　rat population
　　casual loop, 352
　　death rate, 353
　　dynamo equation, 352
　　flow diagram, 352
　　rat birth rate (RBR), 352, 353

E

Electrical systems
　components, 112
　degenerative system model
　　components, 145
　　state space model, 148–151
　　system graph and f-tree, 144–146
　mathematical models
　　components, 78–79
　　expression derivation, 83–84
　　Kirchhoff's current and voltage laws, 79
　　state model, 80–82
Electromechanical systems, 84–87
Emerqentism, 25
Empirical balanced truncation, 246
Empirical model, 248–249
Esoteric systems, 6
Euler's method, 175–177
Exogenous *vs.* endogenous activities, 608
Expert system, 28

F

Fahrenheit-Celsius temperature
　construction, 663, 665
　converter model, 663
　diagnostics, 665
　hardcopy and printing, 665
　library browser, 663, 666
　linearization and trimming, 664
　masking and library, 664
　simulation, 663–664, 667
FCS, *see* Flight control system
Feedback model, 37
First-order continuous-time system
　autonomous time invariant system, 484
　function trajectory *vs.* x, 484–485
　linearization, 485–486
Flight control system (FCS)
　backlash effects
　　constant input, 443, 447
　　pilots control column, 442–443
　　ramp input, 443, 449
　　sinusoidal input, 443, 449

Index 697

step input, 443, 448
waveform generator input, 443, 450
basic control surfaces, 435, 436
components, 437
cumulative effects, 443, 450, 451
dead-zone nonlinearities
comparison, 440–441
constant input, 441, 443
ramp input, 441, 443
Simulink model, 441
sinusoidal input, 441–442
step input, 441–442
fuzzy controller
basic structure, 461
components, 462–469
modeling, 438–439
PID controller
cascade system, 454–456
P, I, D, PD, PI, PID and fuzzy controllers, 456–461, 470–472
root locus method, 445–447, 449–454
principle, 435–436
saturation nonlinearity effects
constant input, 442, 447
ramp input, 442, 445
servo amplifier, 441–442
Simulink model, 442, 444
sinusoidal input, 442, 446
step input, 442, 445
waveform generator input, 442, 446
Simulink model, 440
tuned fuzzy system, 469, 472
fuzzy rules table, 473, 475
input-1 membership function, 469, 473
input-2 membership function, 469, 474
output membership function, 469, 474
rule format, 469, 475
surface view, 473, 476
Flow control statement
break statement, 647–648
continue statement, 647
if statement, 645–646
for loop, 646
switch and case, 646
while loop, 647
Flow-rate variables, 334–335
Fluid systems, 44–46, 87–92
Force–current (*f–i*) analogous drawing rule
analogous quantities, 279
cushioned package system
cushioned package response, 287, 288
f–v analogy, 285
MATLAB program, 285

equations of motion
d'Alembert's principle, 280
f–i analogy, 280–281
f–v analogy, 280
lever-type coupling, 283–284
log chippers
combined constraint equation, 292–293
f–v and *f–i* analogy, 295, 296
Gauss–Jorden elimination, 292
incidence matrix, 292
log chipping system, 290
lumped model, 290, 291
mechanical network, 295, 296
reduced incidence matrix, 292
Simulink model, 294, 295
space equation, 294
terminal equation, 293–294
topological constraints, 291–292
variable identification, 290–291
mechanical coupling devices
friction wheel, 281
f–v and *f–v* analogy, 281–283
translatory motion, 282
speedometer cable
f–i and *f–v* analogy, 297, 298
lumped model, 297
mechanical network, 297, 298
truck–trailer system
f–i analogy, 288, 289
f–v analogy, 288, 290
systematic and equivalent mechanical system, 287, 289
Force–voltage (*f–v*) analogous drawing rule, 278–279
Frequency domain simplification techniques
balanced realization-based reduction method
controllability and observability grammians, 236–237
models, 237–238
properties of, 238
balanced truncation
Lyapunov equations, 239
properties of, 241
theorems, 241–243
unstable systems reduction, 243–244
classical reduction method (CRM), 234
continued fraction expansion (CFE), 235
continued fraction inversion, 255–258
Routh array, 253
differentiation method
approximation criterion, 265–266
MATLAB program, 267

results comparison, 264
  Simulink model, 267, 268
  system responses, 267, 268, 270–273
 frequency-weighted balanced model
    reduction
  empirical balanced truncation, 246
  empirical model, 248–249
  linearized model, 247–248
  mechanical system, 246–247
  steady state matching, 246
 moment-matching method, 235
 Pade approximation method, 234
 routh stability criterion
  denominator stability array, 260
  model response comparison, 262
  numerator stability array, 260
 stability criterion method (SCM), 234
 stability preservation method (SPM), 234
 time moment matching
  responses comparison, 252
  Routh array, 250
Frequency-weighted balanced models
 empirical balanced truncation, 246
 empirical model, 248–249
 linearized model, 247–248
 mechanical system, 246–247
 steady state matching, 246
Friction, 434
Fundamental axiom, 121
Fuzzy controller
 basic structure, 461
 components, 462–463
  application results, 466, 469
  3D characteristics, 463
  defuzzification, 464–465, 466, 468–469
  P, I, D, PD, PI and fuzzy controllers, 469–472
  rule base, 464
  rule editor and viewer, 464, 467
  rule format, 464–465, 467
  surface view, 466, 468
Fuzzy knowledge-based system (FKBS), 556
Fuzzy systems, 8
 adaptive neuro-fuzzy inferencing systems
  causal-loop diagram, 568
  compositional rules, 572–573
  defuzzification methods, 573–574
  development models, 568
  flowchart, 568–569
  implication methods, 572, 573
  intersection operators, 571
  linguistic variable, 570
  logic system development, 573–574
  max–min rule, 573, 574
  overlapping, 571
  parameter identification, 568
  T-Conorms, 571–572
  T-Norms, 571
  union operators, 571–572
 applications, 558–559
  multiple input multiple output systems, 566–567
  multiple input single output (MISO) systems, 564–566
  single input single output (SISO) systems, 559–564
 approximate reasoning
  classical implication, 547
  compositional rule, 548–549
  conjunction rule, 546
  disjunction rule, 546
  entailment rule, 546
  fuzzy relation schemes, 549–550
  generalized modus ponens (GMS) tautology, 547–548
  inference rule, 545
  Larsen's product operator, 553–554
  Mamdani's minimum operator, 550–551
  membership function, 546
  negation rule, 547
  projection rule, 546
  simple crisp inference, 545–546
  steps, 550
  Tsukamoto inference, 551–553
  Zadeh theory, 545–546
 Cartesian product, 541–542
 characteristics, 540
 defuzzification methods, 554–555
 features, 531–532
 fuzzy relation
  composition, 543–545
  inverse relation, 543
  matrices, 543
  R, binary relation, 542
 fuzzy sets
  characteristic (membership) function, 531
  degree of truth, 528–530
  list of elements, 530
  rule method, 531
 operations
  bounded sum, 536–537
  complement, 532–533
  concentration, 533–534
  dilation, 534–535
  intensification, 535–536
  intersection, 532
  linguistic hedges, 538–539

Index

strong α-cut, 537–538
union, 532
operations research
  ANFIS, 596–602
  coefficients, 595–596
  linear programming, 594–595
  optimization models, 592
  problem formulation, 592–593
properties, 541
rule-based systems, 556–558
steady state DC machine model
  analysis, 578–579
  causal relationships, 575
  definition rule, 576–578
  defuzzification process, 576
  identifying linguistic variables, 575
  Mamdani implication, 579
  range definition, 575
  schematic diagram, 574–575
  values, 575–576
Takagi–Sugeno–Kang models, 567–568
transient models
  air-conditioning system, 585–591
  ANN models, 585
  causal-loop diagram, 579–580
  closed-loop system, 584–585
  compositional rules, 581–582
  connectives, 581
  defuzzification methods, 581
  direct adaptive control, 584
  implication methods, 581
  knowledge rules, 579–580
  linguistic variables, 581
  overlapping effects, 582–584

G

Gaussian distribution, 617–618
Generalized modus ponens (GMS) tautology, 547–548
Graph theory
  cycle, 305–306
  interaction graph, 302, 303
  net and loop, 305
  nondirected interaction graph, 302
  parallel lines, 306
  properties of relations, 306–307
  self-interaction matrix, 301, 302
  simple antenna and servomechanism, 302, 303

H

Hankel matrix method, 233–234
Hankel–Norm model order reduction, 234

Hard and soft systems, 16–17
Heat generated modeling, parachute
  causal loop, 383, 384
  dynamo equations, 384–385
  flow diagram, 383, 384
  utility decelerator packing press facility, 383
Henri Poincaré, 481
Heroin addiction modeling
  addiction rate *vs.* time characteristic, 346, 347
  causal loop, 346
  flow diagram, 346
Hierarchically nested set of systems, 2–3
Holism, 25
Homogeneity, 10
Hooke's law, 8–9
Hybrid system, 5–6
Hydraulic system
  components, 112, 134
  f-circuit equation, 136–137
  f-cutset equation, 134–136
  mathematical model
    capacitance, 88
    dynamic response, 44–46
    inertance, 88–89
    liquid level system, 89–92
    resistance, 87
  open-circuit and short-circuit parameters measurement, 105–107
  short circuit parameters, 108–109
  system graph and formulation tree, 134
  system identification, 133
  terminal graph, 105–106
Hysteresis, 435

I

Independent systems, 7
Infected population modeling
  causal loop, 354
  flow diagram, 354, 355
  MATLAB program, 355
  simulation model, 356
Intent structure system
  edge adjacency matrix, 312
  edge digraph, 313
  reachability matrix, 311
Interconnected components of system, 2
Interpretive structural modeling (ISM)
  contextual relation SSIM
    edge adjacency graph, 317, 318
    edge adjacency matrix, 317
    interpretive structural model, 317, 318
    reachability matrix, 316

graph theory
    cycle, 305–306
    directed interaction graph, 302, 303
    loop and net, 305
    nondirected interaction graph, 302
    parallel lines, 306
    properties of relations, 306–307
    self-interaction matrix, 301, 302
    simple antenna and servomechanism, 302, 303
intent structure system
    edge adjacency matrix, 312
    edge digraph, 313
    reachability matrix, 311
lower triangularization method, 317, 319
model exchange isomorphism (MEI), 307–308
paternalistic family system
    adjacency matrix, 321, 322
    edge adjacency matrix, 321, 322
    interpretive structural from, 321, 323
    lower triangular matrix, 321
    SSIM form, 320, 321
structured self-interaction matrix (SSIM), 310–311
system directed graphs, 313–314
system subordinate matrix, 308–310
system undirected graphs, 314, 315
Inventory control system
    causal loop, 348, 349
    flow diagram, 348, 350
    inventory responses, 350
    order rate responses, 350, 351
Iterated function system (IFS), 479

K

Kolomogrov–Smirnov test, 619

L

Laplace method, 171–172
Large and complex applied system engineering
    analogical reasoning, 28
    artificial intelligence, 28
    attributes, 26
    connectionist representations, 28–29
    management control system, 27
    metagame theory, 27
    process control system, 27
    qualitative models, 28
    software system engineering, 29
    subsystems, 26–27
    systems philosophy, 25
Large-scale system, 2, 26
Larsen's product operator, 553–554
Lever-type coupling, 283–284
Liapunov stability, 184–186
Linear differential equation, 10–11
Linear graph theoretic approach
    current and voltage equations, 115
    definitions
        branch and chord/link, 117
        circuit/loop, 117
        coforest, 118
        complement of subgraph, 117
        connected graph and cotree, 117
        cutest, 118
        degree of vertex, 117
        edge and edge sequence, 116
        edge train, 117
        forest, 118
        fundamental circuit (f-circuit), 118
        fundamental cutset (f-cutset), 118–119
        incidence set, 116, 119
        isomorphism, 116
        Lagrangian tree, 118
        linear graph and multiplicity, 116
        nullity, 118
        path, 117
        rank, 118
        separate part, 117
        subgraph, 116
        system graph, 118
        terminal graph, 117
        tree, 117
        vertex, 116
    force/torque equation, 115
Linear perfect couplers, 110–111
Linear systems
    homogeneity, 10
    mathematical viewpoint
        linear differential equation, 10–11
        nonlinear differential equations, 11
    vs. nonlinear systems, 8–9
    superposition theorem, 9–10
Linear systems analogous
    advantages of, 277
    d'Alembert's principle, 277–278
    dual networks, 277, 278
    force–current $(f$–$i)$ analogous drawing rule
        analogous quantities, 279–281
        cushioned package system, 284–288
        equations of motion, 279–281
        lever-type coupling, 283–284

Index

log chippers, 290–296
mechanical coupling devices, 281–283
speedometer cable, 296–298
truck–trailer system, 287–290
force–voltage *(f–v)* analogous drawing rule, 278–279
Linear time invariant systems, 186
Linear *vs.* nonlinear system, 434
Load command
    concatenation, 634
    filename.dat, 633
    input and output, 634
    M-files, 634
    rows and columns, 634
Log chippers
    combined constraint equation, 292–293
    f–v and *f–i* analogy, 295, 296
    Gauss–Jorden elimination, 292
    incidence matrix, 292
    log chipping system, 290
    lumped model, 290, 291
    mechanical network, 295, 296
    reduced incidence matrix, 292
    Simulink model, 294, 295
    space equation, 294
    terminal equation, 293–294
    topological constraints, 291–292
    variable identification, 290–291
Lower triangularization method, 317, 319
Lumped *vs.* distributed parameter systems, 13

M

Machine computation, 170
Mamdani's minimum operator, 550–551
Managerial and socioeconomic system, 327–328
Market advertising interaction model
    aims of, 356
    causal loop, 357
    dynamo equations, 358–359
    flow diagram, 357, 358
    postevaluation criterion, 356–357
Markovian analysis, 609
Mathematical modeling
    electrical systems, 78–84
    electromechanical systems, 84–87
    fluid systems, 44–46, 87–92
    mechanical systems
        rotational, 64–77
        translational motion, 46–64
    output equation, 44
    state-variables, equation, 43–44
    thermal systems, 92–99

Mathematical *vs.* descriptive model, 36
MATLAB
    air-conditioning system, 586–591
    aircraft arrester barrier system, 393–394
    airplane braking using parachute, 98–99
    animations, 644
    application program interface (API), 623
    basic plotting
        add plots to an existing graph, 637
        axis labels and titles, 638
        colored surface plots, 639
        creation, 636
        grid lines, 638
        light source, 639
        lines and markers, 636–637
        mesh and surface plots, 638–639
        multiple data sets in one graph, 636
        multiple plots in one figure, 637–638
        save, 638
    Butterworth filter, 428
    chemical reaction, 426
    colon (:) operator, 630–631
    commutating machine, 373–374
    continuous system simulation, 423–424
    creating movies, 644–645
    cushioned package system, 286
    data structures
        cell arrays, 650
        characters and text, 650–654
        multidimensional arrays, 648–649
    desktop, 623
    development environment, 622
    differentiation method, 267
    dynamic response, 46
    eighth-order system, 263
    entering matrices
        elements, 628
        steps, 627–628
        sum, transpose and diag, 628–629
        workspace, 628
    expressions
        functions, 632–633
        numbers, 632
        operators, 632
        variables, 631–632
    flow control statement
        break statement, 647–648
        continue statement, 647
        if statement, 645–646
        for loop, 646
        switch and case, 646
        while loop, 647
    format command, 635

functions
  cautionary note, 656–657
  construct string arguments, 656
  eval function, 657
  function functions, 658–660
  function handles, 658
  global variables, 655–656
  M-file rank.m, 655
  passing string arguments, 656
  preallocation, 658
  vectorization, 657
fuzzy toolbox, 560
graphics system, 622
handle graphics
  bar and area graphs, 641
  histogram functions, 643
  line plots, 644
  pie chart, 641–642
  properties, 640
  types, 640–641
  typical 3-D graph, 643–644
images, 640
infected population modeling, 355
language, 622
learning, 621
load command
  concatenation, 634
  filename.dat, 633
  input and output, 634
  M-files, 634
  rows and columns, 634
long command lines, 635
magic function, 630–631
mathematical function library, 622
meaning, 621
mechanical coupler, 61
mechanical system
  with three masses, two springs, and dashpot, 63–64
  translational system, 50
  with two masses and two springs, 54–55
pendulum problem, 420–421
Piston–Crank mechanism, 97
randn, 618–619
rocket dynamics, 427
root locus method, 447
rotational mechanical system, 66–67
scripts, 654–655
simple mass–dashpot and spring system, 476–477
Simpson's rule, 408
simulation, 494, 495, 496
starting and quitting, 623

state model
  electrical system, 80–81
  mechanical system, 51–53
state space model, 207
subscripts, 630
suppressing output, 635
system stability, 486
thermal system, 94–95
tools
  Command History, 624–627
  Command Window, 623–624
train systems, 58–59
water reservoir system, 423–424
weighing machine model, 517
Mechanical coupling devices
  friction wheel, 281
  f–v and f–v analogy, 281–283
  translatory motion, 282
Mechanical systems
  degenerative system model development, 160–162
  rotational models
    capacitor size determination, 73–75
    car velocity, 71–72
    3-D projectile trajectory, 72–73
    equation derivation, 65–67
    flight of model rocket, 75–77
    gear-train systems, 68–70
    mathematical model, 67–68
    torsional spring, 64–65
  translational models
    dashpot, 48
    mass, 48–49
    mathematical model, 49–50
    springs, 47–48
    state model, 50–53
    three masses, two springs, and dashpot, 62–64
    train systems, 56–61
    two masses and two springs, 53–55
Metagame theory, 27
Modeling
  aircraft models, 38–39
  characteristics, 38
  classification
    continuous *vs.* discrete models, 37–38
    deterministic *vs.* stochastic models, 37
    mathematical *vs.* descriptive model, 36
    open *vs.* feedback model, 37
    physical *vs.* abstract model, 35–36
    pictorial representation, 36
    static *vs.* dynamic model, 37
    steady state *vs.* transient model, 37

Index   703

component postulate, 40
component terminal equation, 45
conservation laws, 45
definition, 32–33, 38–39
evaluation, 41
fundamental axiom, 40
inputs, 38–39
mathematical modeling
   electrical systems, 78–84
   electromechanical systems, 84–87
   fluid systems, 44–46, 87–92
   hydraulic system, 44–46
   mechanical systems, 46–77
   output equation, 44
   state, state variables and state equation, 43–44
   thermal systems, 92–99
methods, complex systems, 34–35
necessity, 33–34
n-terminal component graph, 40
process, 39
two-terminal components
   accumulator type, 42–43
   delay type elements, 42
   different systems comparison, 41–42
   dissipater type, 41–42
   sources/drivers, 43
variables, 40
Model order reduction
   applications of, 273
   large-scale systems, 219–220
   methods of
      frequency domain, 234–273
      time domain simplification techniques, 223–234
   vs. model simplification, 220–221
   need of, 221
   principle of, 221–222
Moment-matching method, 235
Multidimensional arrays, 648–649
Multi-input multi-output (MIMO) system, 141
Multiple input multiple output systems, 566–567
Multiple input single output (MISO) systems, 564–566
Multi-terminal components, 113–114

N

Neural network applications
   data mining, 510–511
   financial risk, 511–512
   HR management, 512

   industrial applications, 511
   medical, science and marketing, 512
   operational analysis and energy, 512
Nondegenerative systems, 151–157
Nonlinear differential equations, 11
Nonlinear system
   flight control system (FCS)
      backlash effects, 442–443, 447, 448, 449, 450
      basic control surfaces, 435, 436
      components, 437
      cumulative effects, 443, 450, 451
      dead-zone nonlinearities, 440–441, 442, 443
      fuzzy controller, 461–469
      modeling, 438–439
      PID controller, 445–461
      principle, 435–436
      saturation nonlinearity effects, 441–442, 444, 445, 446, 447
      Simulink model, 440
      tuned fuzzy system, 469–473, 474–476
   linear *vs.* nonlinear system, 8–9, 434
   system dynamics techniques, 328
   simple mass–dashpot and spring system, 475–478
   types, 434–435
Numerical methods
   Adams–Bashforth predictor multistep method, 412–413
   Adams–Moulton corrector method, 413
   characteristics, 413
   comparison, 413–414
   discretization error, 418–419
   multistep function, 405–406
   one-step Euler's method, 410
   rectangle rule, 406
   round off error, 415–418
   Runge–Kutta fourth-order method, 411–412
   Runge–Kutta methods, 410–411
   Simpson's rule, 407–410
   single-step methods, 405
   step size *vs.* error, 418
   trapezoid and tangent formulae, 406–407
   truncation error, 414–415

O

Observability, state space model, 181–182
Ohm's law, 8–9
One-step Euler's method, 410
Open *vs.* feedback model, 37
Operations research

ANFIS, 596–602
coefficients, 595–596
linear programming, 594–595
optimization models, 592
problem formulation, 592–593
Optimal order reduction, 233
Organicism, 25

P

Pade approximation method, 234
Parachute deceleration device
canopies, 376
canopy stress distribution, 378
functional phases, 376
inflation
circular parachutes, 378
stages of, 377
trajectory
causal loop, 378, 379
dynamo equations, 379–383
flow diagram, 379, 380
Philosophy, systems
Aristotle's investigations, 19
Bertalanffy's contribution, 18–19
method of science, 20
Newtonian mechanics, 19–20
problems of science, 20–21
teleology, 19
types, 25
Physical systems
definition, 6
modeling
automobile modeling, 512–513
comparison, 513–514
electrical systems, 78–84
electromechanical systems, 84–87
error, 513, 515
fluid systems, 87–92
hydraulic system, 44–46
Laplace transform, 513
linear perfect coupler, 128
mechanical systems, 46–77
output equation, 44
pseudo power or quasi power, 129–131
speed profile, 513
state-variables, equation, 43–44
steps, 127–128
thermal systems, 92–99
theory, 103
Physical vs. abstract model, 35–36
PID controller
cascade system
arrangement and comparation, 455, 456
comments, 455
oscillations (T_{er}), 455
Routh array, 454
stable operation, 455
P, I, D, PD, PI, PID, and fuzzy controllers
arrangement and results, 460–461
integral controller, 457–458
linear and nonlinear systems, 456–457
proportional controller, 457, 458
proportional-derivative controller, 459–460
proportional-integral controller, 457, 459
root locus method
characteristic equation, 446, 449–450
comments, 454
compensated system, 449
compensator, 449, 452
MATLAB, 447
proportional controller, 446
specifications, 445
stability check, 453–454
steady state behavior, 451–452
Pitch control system
Backlash effects
constant input, 443, 447
pilots control column, 442–443
ramp input, 443, 449
sinusoidal input, 443, 449
step input, 443, 448
waveform generator input, 443, 450
components
altitude controller, 437
block diagram, 437
rate gyro, 437
servo unit, 437
vertical gyro, 437
cumulative effects, 443, 450, 451
dead-zone nonlinearities
comparison, 440–441
constant input, 441, 443
ramp input, 441, 443
Simulink model, 441
sinusoidal input, 441–442
step input, 441–442
modeling, 438–439
PID controller
cascade system, 454–456
P, I, D, PD, PI, PID, and fuzzy controllers, 456–461, 470–472
root locus method, 445–447, 449–454
saturation nonlinearity effects
constant input, 442, 447
ramp input, 442, 445
servo amplifier, 441–442

Index 705

  Simulink model, 442, 444
  sinusoidal input, 442, 446
  step input, 442, 445
  waveform generator input, 442, 446
 servomotor, 438
 Simulink model, 440–441
Population problem modeling
 flow diagram, 347, 348
 net birth rate *vs.* time, 347, 349
 positive causal loop, 347, 348
Production distribution system
 causal loop, 360, 361
 departments of, 359, 360
 dynamo equations, 362–363
 flow diagram, 360, 362
 production rate, 363, 364
 rate variables *vs.* time, 363, 364
Proportional-integral-derivative (PID), 435

Q

Qualitative models, 28
Queuing system, 609, 610–611

R

Rainfall prediction, neural network, 523–525
Rat population
 casual loop, 352
 death rate, 353
 dynamo equation, 352
 flow diagram, 352
 rat birth rate (RBR), 352, 353
Real-world chaotic systems, 480
Reductionism, 20–22
Refutation, 20
Repeatability, 20
Rocket dynamics, 427–428
Rotary mechanical system, 112
Round off error, 415–418
Routh stability criterion
 denominator stability array, 260
 model response comparison, 262
 numerator stability array, 260
Runge–Kutta fourth-order method, 411–412
Runge–Kutta methods, 410–411

S

Saturation nonlinearity effects
 constant input, 442, 447
 ramp input, 442, 445
 servo amplifier, 441–442
 Simulink model, 442, 444

  sinusoidal input, 442, 446
  step input, 442, 445
  type, 435
  waveform generator input, 442, 446
SCM, *see* Stability criterion method
Sensitivity, state space model, 182–184
Simple mass–dashpot and spring system
 damping equations, 475
 MATLAB program, 476–477
 results, 477–478
 system equations, 476
Simpson's rule, 407–410
Simulation
 advantages, 402
 analysis and decision making, 404
 applications, 404–405
 Butterworth filter, 428–429
 chemical reactor, 425–426
 continuous system simulation
  demand, water, 424–425
  MATLAB program, 423–424
  model development, 423
  net volume, 424
  solution, 423
  water reservoir system, 422–423
 effectiveness, 404
 efficiency, 403
 infected population modeling, 356
 numerical methods
  Adams–Bashforth predictor multistep method, 412–413
  Adams–Moulton corrector method, 413
  characteristics, 413
  comparison, 413–414
  discretization error, 418–419
  multistep function, 405–406
  one-step Euler's method, 410
  rectangle rule, 406
  round off error, 415–418
  Runge–Kutta fourth-order method, 411–412
  Runge–Kutta methods, 410–411
  Simpson's rule, 407–410
  single-step methods, 405
  step size *vs.* error, 418
  trapezoid and tangent formulae, 406–407
  truncation error, 414–415
 pendulum problem
  case study, 421
  force equation, 419
  MATLAB functions, 420–421
  results, 421–422
  Simulink model, 421–422

risk reduction, 404
rocket dynamics, 427–428
uses, 403
water pollution problem, 429–430
Simulink
   dead-zone nonlinearities, 441
   differentiation method, 267, 268
   Fahrenheit-Celsius temperature
      construction, 663, 665
      converter model, 663
      diagnostics, 665
      hardcopy and printing, 665
      library browser, 663, 666
      linearization and trimming, 664
      masking and library, 664
      simulation, 663–664, 667
   features, 661
   flight control system (FCS), 440
   log chipper modeling, 294, 295
   pendulum problem, 421–422
   pitch control system, 440–441
   simulation parameters and solvers, 661–663
Single input single output (SISO) systems
   algorithm, 561
   develop fuzzy rules, 564
   hysteresis curve, 559
   input variables, 563
   Mamdani fuzzy model, 560, 562
   MATLAB fuzzy toolbox, 560
   membership functions, 560–561, 562
   output variables, 564
Single port and multiport systems
   elements, 109
   linear perfect couplers, 110–111
   mathematical representation, 109–110
   multiterminal components, 113–114
   two-terminal components, 112–113
Soft operations research (OR), 17
Soft systems, 16–17
Soft systems methodology (SSM), 16–17
Software system engineering, 29
Spreading activation process, 28
Stability criterion method (SCM), 234
Stability preservation method (SPM), 234
Stanchion system modeling, *see* Aircraft arrester barrier system
State equations solutions, 173–175; *see also* State transition matrix
State of system, 1
State space model
   automobile motion, 142–144

using computer program
   input data preparation, 187
   state equations formulation, 187–192
conceptual system
   component postulate, 121
   four-terminal component, 119, 120
   fundamental axiom, 121
   system postulate, 121–127
   three-terminal component, 119, 120
   two-terminal component, 119, 121
controllability, 180–181
degenerative system
   electrical system, 144–146, 148–151
   mechanical system, 160–162
   model development, 146–148
   multiterminal components, 157–159
   nondegenerative systems, 151–157
   time varying and nonlinear components, 162–166
element information, 187
hydraulic system
   components, 134
   f-circuit equation, 136–137
   f-cutset equation, 134–136
   system graph and formulation tree, 134
   system identification, 133
incidence matrix, priority level, 187
Liapunov stability, 184–186
linear time invariant systems, 186
multi-input multi-output (MIMO) system, 141
observability, 181–182
parameters computation, 105–109
physical systems
   linear perfect coupler, 128
   pseudo power or quasi power, 129–131
   steps, 127–128
   theory, 103
sensitivity, 182–184
single port and multiport systems
   elements, 109
   linear perfect couplers, 110–111
   mathematical representation, 109–110
   multiterminal components, 113–114
   two-terminal components, 112–113
standard elements, 187, 193
state equations solutions, 173–175 (*see also* State transition matrix)
subassembly, mechanical system, 137–138
system analysis techniques
   free body diagram method, 115
   Lagrangian technique, 115
   linear graph theoretic approach, 115–119
   system structure, 114

Index

system components and interconnections, 103–105
theorems, 138–141
topological restrictions
  accumulator or storage type elements, 132–133
  across driver, 133
  delay type elements, 132
  dissipater type elements, 132
  gyrator, 131
  open circuit element ("A" type), 132
  perfect coupler, 131
  short circuit element ("A" type), 132
  through driver, 133
State transition matrix
  Adams–Bashforth method, 178–180
  classical method of solution, 166–167
  coefficient matrix, 173
  computation
    Calay Hamilton method, 172
    diagonalization, 170–171
    Laplace method, 171–172
    machine computation, 170
  definition, 167
  Euler's method, 175–177
  nonhomogenous state equation, 167–168
  properties, 168–169
  time response, 168–169
Static *vs.* dynamic models/systems, 8, 37
Steady state DC machine model, fuzzy systems
  analysis, 578–579
  causal relationships, 575
  definition rule, 576–578
  defuzzification process, 576
  identifying linguistic variables, 575
  Mamdani implication, 579
  range definition, 575
  schematic diagram, 574–575
  values, 575–576
Steady state matching, 246
Steady state *vs.* transient model, 37
Step size *vs.* error, 418
Stochastic systems/models, 7, 37, 15–16
Structuralism, 25
Structured self-interaction matrix (SSIM), 310–311
Subspace projection method, 232
Superposition theorem, 9–10
Synthesis of systems, 18
System analysis techniques
  free body diagram method, 115
  Lagrangian technique, 115
  linear graph theoretic approach, 115–119
  system structure, 114
Systematic knowledge handling *vs.* Chaotic system, 481–482
System components and interconnections, 103–105
System dynamic (SD) model structure
  characteristics, 333–334
  decision function
    cause–effect relationships, 335–338
    computing sequence, 340–342
    dynamo equations, 339–340
    interconnected network, 338–339
  flow-rate variables
    definition of, 334
    four concepts, 335
  level variables, 334
System dynamics techniques
  advantages of, 332–333
  aircraft arrester barrier system
    aircraft energy absorbing system, 385, 386
    dynamic model, 389–390
    experimental data *vs.* simulated data, 391, 393
    flow diagram, 390, 391
    MATLAB program, 393–394
    subsystems, 385–386
    system dynamic technique, 387–389
    tension time profile, 391–393
  dynamo equations
    AIDS/HIV population modeling, 363–371
    commutating machine, 371–376
    heroin addiction problem modeling, 346–347
    infected population modeling, 354–356
    inventory control problem, 348–351
    market advertising interaction model, 356–360
    population problem modeling, 346–349
    production distribution system, 359–364
    rat population, 351–354
  flow diagrams symbols
    auxiliary variables, 345
    information takeoff, 345
    levels, 344
    parameters, 345, 346
    source and sinks, 344, 345
  heat generated modeling, parachute
    causal loop, 383, 384
    dynamo equations, 384–385
    flow diagram, 383, 384
    utility decelerator packing press facility, 383

managerial and socioeconomic system, 327–328
parachute deceleration device
  canopy stress distribution, 378
  functional phases, 376
  parachute canopies, 376
  parachute inflation, 377–378
  parachute trajectory, 378–383
sources of information
  mental database, 329
  numerical database, 330
  written/spoken database, 330
strength of
  decision-making processes, 332
  information feedback control theory, 331–332
  system dynamics (SD) methodology, 331
system analysis, 332
system dynamic (SD) model structure
  characteristics, 333–334
  decision function, 335–342
  flow-rate variables, 334–335
  level variables, 334
traditional management, 328
types of equations
  auxiliary equations, 343–344
  level equation, 342
  rate equation, 343
Systems
  analysis, 18
  boundary, 3
  classification
    complexity, 6
    interactions, 6–7
    nature and type of components, 7
    time frame, 5–6
    uncertainty, 7–9
  components and their interactions, 3–4
  continuous-time and discrete-time systems, 13–14
  definition, 1
  deterministic *vs.* stochastic systems, 15–16
  environment, 4–5
  examples, 2
  hard and soft systems, 16–17
  hierarchically nested set, 2–3
  interconnected components, 2
  large and complex applied system engineering
    analogical reasoning, 28
    artificial intelligence, 28
    attributes, 26
    connectionist representations, 28–29
    management control system, 27
    metagame theory, 27
    process control system, 27
    qualitative models, 28
    software system engineering, 29
    subsystems, 26–27
    systems philosophy, 25
  linear systems
    homogeneity, 10
    mathematical viewpoint, 10–11
    superposition theorem, 9–10
  lumped *vs.* distributed parameter systems, 13
  modeling (*see* Modeling)
  philosophy, 18–21
  synthesis, 18
  systems thinking
    Bertalanffy's articulates, 21
    definitions, 22–23
    examples, 22–24
    magic *vs.* science, 21
  time-varying *vs.* time-invariant systems, 12
System state model (SYSMO), 187, 193

T

Takagi–Sugeno–Kang models, 567–568
T-Conorms, 571–572
Teacher evaluation system
  ANN model, 519, 521–522
  block diagram, 518, 521
  comparison, 519, 523
  current education system, 517, 520
  training error, 519, 523
Thermal systems
  components, 112
  mathematical models
    airplane deceleration with brake parachute, 97–99
    capacitance, 92–93
    Piston-Crank mechanism, 96–97
    resistance, 92
    supplied heat *vs.* room temperature, 93–95
Three-terminal components, 119, 120
Time domain simplification techniques
  aggregation method
    cases, 230
    limitations of, 232
  dominant eigenvalue method
    characteristic roots' locations, 225
    limitations of, 228
    vs. nondominant eigenvalues, 226

Index

poles locations effects, 225
system stability, 224
Hankel matrix method, 233–234
Hankel–Norm model order reduction, 234
optimal order reduction, 233
subspace projection method, 232
Time moment matching
responses comparison, 252
Routh array, 250
Time varying and nonlinear components, 162–166
Time-varying *vs.* time-invariant systems, 12
T-Norms, 571
Transient models
air-conditioning system, 585–591
ANN models, 585
causal-loop diagram, 579–580
closed-loop system, 584–585
compositional rules, 581–582
connectives, 581
defuzzification methods, 581
direct adaptive control, 584
implication methods, 581
knowledge rules, 579–580
linguistic variables, 581
overlapping effects, 582–584
vs. steady state, 37
Translational mechanical system models
dashpot, 48
mass, 48–49
spring, mass and dashpot
mathematical model, 49–50
state model, 50–53
springs, 47–48
three masses, springs and dashpot, 62–64
train systems
free body diagram and Newton's law, 56–57
mechanical coupler, 59–61
simulation results, 58
state-variable and output equations, 57–59
two masses and two springs, 53–55
Translatory motion mechanical system, 112
Truck–trailer system
f–i analogy, 288, 289
f–v analogy, 288, 290
systematic and equivalent mechanical system, 287, 289
Truncation error, 414–415
Tsukamoto inference, 551–553
Tuned fuzzy system, 469, 472
fuzzy rules table, 473, 475
input-1 membership function, 469, 473
input-2 membership function, 469, 474
output membership function, 469, 474
rule format, 469, 475
surface view, 473, 476
Two-terminal components, 112

V

Vandermonde's matrix, 170–171

W

Water pollution problem, 429–430
Water reservoir system, *see* Continuous system simulation
Weighing machine models
data fitting, 517, 519
displacement *vs.* force, 517, 518
error reduction, 517, 518
Laplace transform, 516
prediction, 517, 520
random input force, 516–517

Z

Zadeh theory, 545–554